Statistics for Chemical Engineers

Build a firm foundation for studying statistical modeling, data science, and machine learning with this practical introduction to statistics, written with chemical engineers in mind.

Key Features

- Introduces a data–model–decision approach to applying statistical methods to real-world chemical engineering challenges.
- Establishes links between statistics, probability, linear algebra, calculus, and optimization, covering classical and modern topics of statistics such as uncertainty quantification, risk modeling, and decision-making under uncertainty.
- Over 100 worked examples using MATLAB and Python demonstrate how to apply theory to practice, with over 70 end-of-chapter problems to reinforce student learning.
- Introduces key topics using a modular structure, which supports learning at a range of paces and levels.

Requiring only a basic understanding of calculus and linear algebra, this textbook is the ideal introduction for undergraduate students in chemical engineering and a valuable preparatory text for advanced courses in data science and machine learning with chemical engineering applications.

Victor M. Zavala is the Baldovin-DaPra Professor of Chemical and Biological Engineering at the University of Wisconsin–Madison and a senior computational mathematician at Argonne National Laboratory. He is the recipient of the Harvey Spangler Award for Innovative Teaching and Learning Practices from the College of Engineering at the University of Wisconsin–Madison and of the Presidential Early Career Award for Scientists and Engineers (PECASE).

Cambridge Series in Chemical Engineering

Books in the Series

Baldea and Daoutidis, *Dynamics and Nonlinear Control of Integrated Process Systems*
Chamberlin, *Radioactive Aerosols*
Chau, *Process Control: A First Course with Matlab*
Cussler, *Diffusion: Mass Transfer in Fluid Systems, Third Edition*
Cussler and Moggridge, *Chemical Product Design, Second Edition*
De Pablo and Schieber, *Molecular Engineering Thermodynamics*
Deen, *Introduction to Chemical Engineering Fluid Mechanics*
Denn, *Chemical Engineering: An Introduction*
Denn, *Polymer Melt Processing: Foundations in Fluid Mechanics and Heat Transfer*
Dorfman and Daoutidis, *Numerical Methods with Chemical Engineering Applications*
Duncan and Reimer, *Chemical Engineering Design and Analysis: An Introduction 2E*
Fan, *Chemical Looping Partial Oxidation Gasification, Reforming, and Chemical Syntheses*
Fan and Zhu, *Principles of Gas-Solid Flows*
Fox, *Computational Models for Turbulent Reacting Flows*
Franses, *Thermodynamics with Chemical Engineering Applications*
Grossmann, *Advanced Optimization for Process Systems Engineering*
Leal, *Advanced Transport Phenomena: Fluid Mechanics and Convective Transport Processes*
Lim and Shin, *Fed-Batch Cultures: Principles and Applications of Semi-Batch Bioreactors*
Litster, *Design and Processing of Particulate Products*
Maravelias, *Chemical Production Scheduling*
Marchisio and Fox, *Computational Models for Polydisperse Particulate and Multiphase Systems*
Mewis and Wagner, *Colloidal Suspension Rheology*
Morbidelli, Gavriilidis, and Varma, *Catalyst Design: Optimal Distribution of Catalyst in Pellets, Reactors, and Membranes*
Nicoud, *Chromatographic Processes*
Noble and Terry, *Principles of Chemical Separations with Environmental Applications*
Orbey and Sandler, *Modeling Vapor-Liquid Equilibria: Cubic Equations of State and their Mixing Rules*
Petyluk, *Distillation Theory and its Applications to Optimal Design of Separation Units*
Pfister, Nicoud, and Morbidelli, *Continuous Biopharmaceutical Processes:* Chromatography, Bioconjugation, and Protein Stability
Ramkrishna and Song, *Cybernetic Modeling for Bioreaction Engineering*
Rao and Nott, *An Introduction to Granular Flow*
Russell, Robinson, and Wagner, *Mass and Heat Transfer: Analysis of Mass Contactors and Heat Exchangers*
Schobert, *Chemistry of Fossil Fuels and Biofuels*
Shell, *Thermodynamics and Statistical Mechanics*
Sirkar, *Separation of Molecules, Macromolecules and Particles: Principles, Phenomena and Processes*
Slattery, *Advanced Transport Phenomena*
Subramaniam, *Green Catalysis and Reaction Engineering: An Integrated Approach with Industrial Case Studies*
Varma, Morbidelli, and Wu, *Parametric Sensitivity in Chemical Systems*
Vassiliadis et al., *Optimization for Chemical and Biochemical Engineering*
Weatherley, *Intensification of Liquid–Liquid Processes*
Wolf, Bielser, and Morbidelli, *Perfusion Cell Culture Processes for Biopharmaceuticals*
Zavala, *Statistics for Chemical Engineers*
Zhu, Fan, and Yu, *Dynamics of Multiphase Flows*

Statistics for Chemical Engineers

From Data to Models to Decisions

Victor M. Zavala
University of Wisconsin–Madison

CAMBRIDGE
UNIVERSITY PRESS

CAMBRIDGE
UNIVERSITY PRESS

Shaftesbury Road, Cambridge CB2 8EA, United Kingdom

One Liberty Plaza, 20th Floor, New York, NY 10006, USA

477 Williamstown Road, Port Melbourne, VIC 3207, Australia

314–321, 3rd Floor, Plot 3, Splendor Forum, Jasola District Centre, New Delhi – 110025, India

103 Penang Road, #05–06/07, Visioncrest Commercial, Singapore 238467

Cambridge University Press is part of Cambridge University Press & Assessment, a department of the University of Cambridge.

We share the University's mission to contribute to society through the pursuit of education, learning and research at the highest international levels of excellence.

www.cambridge.org
Information on this title: www.cambridge.org/highereducation/isbn/9781009541893

DOI: 10.1017/9781009541916

When citing this work, please include a reference to the DOI 10.1017/9781009541916

First published 2026

A catalogue record for this publication is available from the British Library

A Cataloging-in-Publication data record for this book is available from the Library of Congress

ISBN 978-1-009-54189-3 Hardback

Additional resources for this publication at www.cambridge.org/zavala

Cover image: iStock / Getty Images aleksey_rezin

To Katie, Owen, and Frankie

To Irma, Karla, and Victor

To all my students, friends, and collaborators

Contents

Preface

Background and Motivation

This book aims to *change the way we think about statistics* and *teach statistics* in chemical engineering. Statistics is one of the pillars of modern science and engineering and of emerging topics such as data science and machine learning; despite this, its scope and relevance has remained misunderstood and underappreciated in chemical engineering education (and in engineering education at large). Specifically, statistics is often taught by placing emphasis on *data analysis*. I would like to convince the reader that statistics is much more than that; particularly, statistics is a *mathematical modeling* paradigm that complements physical modeling paradigms used in chemical engineering (e.g., thermodynamics, transport phenomena, conservation, and reaction kinetics). Statistics, in particular, can help model random phenomena that might not be predictable from physics alone (or from deterministic physical laws), quantify the uncertainty of predictions obtained with physical models, discover physical models from data, and create models directly from data (in the absence of physical knowledge).

The *fusion* of statistical and physical modeling paradigms provides a powerful framework for analyzing data, extracting knowledge from data, and making informed predictions and decisions. In addition, I would like to convince the reader that statistics is a *foundational* topic in that it *fundamentally transforms the way* we perceive the world and it touches every aspect of chemical engineering. More broadly, I would like to convince the reader that statistics provides a *conceptual framework* that can help chemical engineers do what they do best: abstract, understand, design, and control *complex systems*.

The desire to write this book came from my personal experience in learning statistics in college and in identifying the significant gaps in my understanding of statistics throughout my professional career. Similar feelings are often shared with me by professionals working in industry and academia. Like many chemical engineers, I took a course in statistics in college that covered classical topics such as random variables, descriptive statistics, regression, design of experiments, and basic probability. This course, while I found it to be interesting, felt disconnected from the rest of the chemical engineering curriculum. Specifically, with the exception of linear regression and design of experiments, I did not encounter other major uses of statistics in the curriculum. This left me with a perception that statistics was an intellectual "curiosity." Throughout my professional career, I have been exposed to a broad range of applications in which knowledge of statistics has proven to be essential: uncertainty quantification, quality control, risk assessment, modeling of random phenomena, process monitoring, forecasting, machine learning, computer vision, and decision-making under uncertainty. These are applications that are pervasive in industry and academia. It is also important to recognize that the field of statistics continues to evolve and many new concepts/tools have become available; for example, the fields of uncertainty quantification, Bayesian analysis, statistical learning, and decision-making under uncertainty have experienced significant growth in recent years.

I believe that *there is a need to modernize the scope and approach to teach statistics.* This should be done carefully by finding a suitable blend of classical and new topics, finding points of connection with the rest of the chemical engineering curriculum, and thinking about what are the unique skills and interests of chemical engineers. As such, a pedagogical question that motivates this book is:

How can we better teach statistics to chemical engineers?

Statistics is typically taught by mathematics/statistics departments to a broad range of engineers; as such, it might be difficult for instructors to standardize the teaching material and make explicit connections to applications that are of interest to chemical engineers. Specifically, it is important to understand that chemical engineers are trained to use *physical knowledge* (e.g., thermodynamics, reaction kinetics, and transport phenomena) to analyze systems and make decisions (e.g., design an experiment or a chemical process). As such, when teaching statistics, it is important to emphasize when and how these tools can complement (or substitute) physics. Moreover, it is important to remember that a *key skill of chemical engineers* is the ability to develop mathematical abstractions (models) to analyze complex systems, and by *complex*, I mean systems that involve heterogeneous phenomena (e.g., reactions, flows, heating/cooling, and separation) and that might involve different scales (e.g., molecular level, unit level, process level, and enterprise level). As such, when teaching statistics, it is important to emphasize how this can provide tools to facilitate the modeling/understanding of complex systems. It is also important to remember that, when chemical engineers analyze data, they are ultimately interested in understanding phenomena and extracting underlying principles; in other words, engineers aim to attribute physical meaning/origin of behavior that helps them make decisions (e.g., design a new material or microbe to conduct a function). As such, when teaching statistics, it is important not to discard the physical and decision-making context. Finally, with the advent of machine learning and data science, it is important to remember that statistics (together with calculus and linear algebra) provides key mathematical fundamentals that are the *building blocks* of these tools. As such, when teaching statistics, it is important to explain how foundations can be used to develop or select tools and to analyze the outcomes of such tools.

Statistics is typically taught early in the chemical engineering curriculum, together with other mathematics courses such as calculus, linear algebra, and numerical methods. Teaching statistics early in the curriculum is important, as it provides the foundations and tools that are needed in other courses. However, if taught too early (e.g., sophomore year), instructors will be limited in the content that they can cover, as students do not yet have the proper foundations to cover advanced topics. For example, if students have not taken linear algebra or multivariate calculus, it will be difficult for them to understand principal component analysis and parameter estimation. Moreover, if statistics is taught too late (e.g., senior year), it will be difficult for students to appreciate the connections between statistics and the rest of the curriculum. I thus think that a suitable timing to teach statistics is in the junior year, or alternatively, one could spread the course into smaller modules that are taught in different years. Teaching statistics in the junior year has several *strategic benefits*; for instance, given that statistics makes extensive use of linear algebra, numerical methods, and calculus, it can provide a venue to reinforce these concepts. For instance, data science and machine learning can help students better understand and appreciate the relevance of all the mathematical concepts learned so far in the curriculum. Moreover, by teaching statistics in the junior year, students will be in a better position to use statistical tools in core courses (e.g., parameter estimation in reactor engineering, data analysis in laboratories, uncertainty quantification in process design, and data-driven modeling in process control). Moreover, by teaching statistics at the junior level, students can begin to question the origin and quality of data and models (e.g., empirical vs. first-principles); this can help reinforce the understanding of physical principles and can help them appreciate aspects of complex systems that are less/more difficult to predict.

Design and Organization

The *design of this book* follows a *data–models–decisions* pipeline. The intent of this design is to emphasize that statistics is a modeling paradigm that maps data to models and models to decisions; this design also aims to "connect the dots" between different branches of statistics and engineering. The focus on the pipeline is also important in reminding the reader that understanding the application context matters. For instance, the data and type of model used for process design can be quite different from the data and type of model used for experimental design. Similarly, the nature of the decision and the data available influence the type of model used. The book design is also intended for the reader to understand the close interplay between statistical and physical modeling; specifically, we emphasize how statistics provides tools to model aspects of a system that cannot be fully predicted from physics. Moreover, we emphasize how physical systems might naturally exhibit random behavior due to uncontrollable aspects (e.g., defects of materials, noise, fluctuations, and vibrations).

The book design is also intended to help the reader appreciate how statistics provides an *important foundation* for a broad range of modern tools of data science and machine learning (e.g., neural networks and logistic regression). More broadly, the book emphasizes how statistics provides a way to think about the world. For instance, we discuss how statistical thinking is fundamentally different from deterministic thinking (which ignores uncertainty). Making this distinction is extremely important, as chemical engineering courses tend to follow a *deterministic mindset* (e.g., there is no uncertainty/variability in the data and the model used for analysis is perfect). In this context, the book discusses how ambiguity can arise when one faces uncertainty, as key variables of interest (e.g., cost and carbon emissions) are no longer numbers (they are distributions) and thus cannot be compared so easily. Moreover, we discuss how statistics provides tools that can help make decisions that *mitigate/control uncertainty*.

The design of the book is also intended to *reinforce mathematical fundamentals (linear algebra, optimization, and calculus)* that are found in a broad range of applications; I believe that statistics and data science provide an excellent framework for doing this, as one can relate mathematical concepts (e.g., eigenvalues) to statistical concepts (e.g., information content). Statistics also provides a suitable framework because it connects different mathematical concepts that are often treated separately (e.g., estimation connects linear algebra and optimization).

The book content is structured in three major modules/parts: **Part I:** Basic Concepts (Chapters 1–3), **Part II:** Intermediate Concepts (Chapters 4–5), and **Part III:** Advanced Concepts (Chapters 6–7). The chapter contents can be summarized as follows:

- **Chapter 1** provides a *"big picture" introduction* to statistics by following the *data–model–decisions pipeline*. This aims to explain how statistics provides tools to use data to model uncertainty and variability, understand how uncertainty propagates through systems, and make decisions that help mitigate and control the effect of uncertainty. This chapter also introduces fundamental concepts of random variables; this discussion emphasizes how a random variable is a model (a theoretical model) that can be used to explain actual data of a system that we observe. Moreover, this discussion introduces key elements of random variable models, such as probability density functions, domains, parameters, and summarizing statistics (e.g., mean, variance, quantiles, and probability of loss). This chapter also discusses *structural models*, which are models that capture systematic relationships between variables.
- **Chapter 2** discusses how to *characterize uncertainty using univariate (single-variable) random variable models*. Here, we highlight the properties that different models have, what type of behavior they aim to model, and when it is appropriate to use them. Moreover, we emphasize connections between models and behavior that is typically observed in physical systems (e.g., due to aspects that cannot be fully controlled or predicted). This chapter also highlights connections between random variable models (e.g., some models are generalizations or special cases of other models). This discussion aims to highlight how the selection of statistical models is directly analogous to the selection of physical models (e.g., equations

of state or reaction kinetic models) in that one needs to understand the assumptions that each model makes and under what circumstances they can be (or not) used.

- **Chapter 3** discusses how to *model uncertainty using multivariate random variable models*. Here, we highlight concepts of correlation and covariance and of joint and conditional probabilities that help explain how random variables are interrelated. This discussion is intended for the reader to understand how connections between random variables can be used to gain understanding of a complex system, develop models, and make predictions. This chapter will also show how to compute probabilities of complex events and how events can be used to model decision-making logic. This chapter also highlights that there are few multivariate models available (e.g., multivariate Gaussian) but that one can still conduct useful analysis based on data alone.
- **Chapter 4** discusses how to *estimate the parameters for random variable models* from data using a variety of techniques. This chapter also shows how to compute estimates for theoretical quantities of random variables (e.g., expected value, variance, and probabilities) from available data and how to quantify the accuracy of such estimates. This discussion will lead us to important computational tools such as Monte Carlo approximations and important asymptotic properties (e.g., the law of large numbers, the central limit theorem, and the extreme value theorem) that explain how estimates behave as we accumulate data.
- **Chapter 5** discusses how to *estimate parameters for structural models* from data. Here, we will highlight how structural models exploit interconnections between variables to make predictions and how estimation techniques aim to separate the structural form of the model (the "trend") from inherent random effects present in the data (e.g., sensor noise). This discussion will also show how one can use random variable models to quantify the uncertainty of structural model predictions. This chapter discusses a powerful, optimization-based estimation technique known as maximum likelihood estimation in depth and applies it to linear and nonlinear models. This chapter also introduces the notion of information content and highlights how information is inherently linked to data availability and quality and how it can be quantified using optimization and linear algebra techniques. We also discuss the importance of incorporating prior knowledge in the estimation procedure (in the form of constraints and regularization terms) in order to obtain better parameter estimates and avoid spurious behavior (e.g., overfitting or predictions that do not obey physics). The chapter closes with a discussion of Bayesian estimation, which provides an alternative paradigm to maximum likelihood estimation that can help capture a broader set of settings (e.g., reconcile data to physics).
- **Chapter 6** discusses *advanced tools of statistical data analysis and learning*. Here, the goal is to understand how to apply the fundamentals of statistics from previous chapters to analyze complex datasets and to build sophisticated predictive models. This chapter explains how to use different types of techniques (eigenvalue decompositions, convolutions, and Fourier transforms) to

extract information from different types of data (e.g., images and time series). In addition, the chapter discusses advanced machine learning models (logistic regression, kernel models, and neural networks) and shows how to use statistics to help explain their design and behavior.

- **Chapter 7** discusses how to *make decisions in the face of uncertainty*. Here, we discuss how to model different attitudes toward risk, how to compare decisions, and how to obtain optimal decisions. Specifically, we will see that making decisions under uncertainty is challenging because variables of interest (e.g., cost) are no longer numbers but are random variables; as such, selecting a suitable decision requires the comparison of the probability density functions or of summarizing statistics. We will also introduce a technique known as *stochastic optimization* that aims to make decisions that shape the distribution of variables of interest in desirable ways. An important goal of this discussion will be to convince the reader that *deterministic* decision-making paradigms can lead to decisions that are vulnerable/non-robust when facing different types of circumstances. Finally, the book closes with a discussion on *Bayesian optimization*, which is a powerful technique that combines diverse topics covered in the book. Specifically, we will see that this technique is a *closed-loop learning* paradigm that aims to strategically collect data to learn (to develop a predictive model) and to make decisions. This learning paradigm mimics how humans and living systems naturally leverage data (e.g., collected from our sensory systems) to improve predictions and decisions.

The book is intended to provide a framework to develop a *statistical modeling* course that complements physical modeling courses covered in chemical engineering education. Along these lines, the book aims to place *statistical modeling at the core of chemical engineering*. Specifically, it is essential for engineers to model random phenomena and understand the origins of such phenomena, properly quantify uncertainty and risk of models/predictions when making critical decisions, and understand what data is most useful (or not useful) when making a decision. This is becoming increasingly relevant as societal problems become increasingly complex and higher volumes of data become available (e.g., better sensors and instrumentation are becoming available). It is also important to highlight that this book differs from other engineering statistics books in that it places emphasis on modeling and chemical engineering applications; moreover, the book covers classical and modern topics of statistics and places emphasis on mathematical fundamentals.

Audience

The book is primarily intended for undergraduate chemical engineers (ideally at the junior level) who have a basic background in calculus, linear algebra, and programming. The content of the book can be covered in a single semester (by adjusting the level of depth/difficulty in each topic), but the book has been designed in a modular form so that the content can be spread throughout the curriculum: sophomore year

(Part I), junior year (Part II), and senior year (Part III). The content should also be useful for undergraduate statistics courses in other engineering fields (e.g., mechanical, electrical, industrial, and biomedical engineering). Some advanced content can also be used in graduate courses (primarily for students who seek to reinforce their knowledge of statistics and mathematical fundamentals) and for developing short courses for industrial practitioners.

Approach

In discussing the different topics, we will make special emphasis on the *key role that data plays* in helping us select appropriate models and make decisions. In this context, we emphasize how to determine if sufficient and good-quality data is available to develop models and make decisions; moreover, we will discuss how to best select data to conduct different types of functions (e.g., estimation vs. optimization). We also place emphasis on Monte Carlo simulations, which provide a general framework to use data to quantify uncertainty and make decisions. In discussing the different topics, we will also make emphasis on the deep connections that exist between statistics and other mathematical fields such as calculus (e.g., optimization, integration, and differential equations) and linear algebra (e.g., vectors, matrices, and eigenvalues). This will highlight how statistics, data science, and machine learning provide an excellent venue to reinforce and understand mathematical concepts. For instance, we discuss how second derivatives allow us to measure information content and how eigenvalues can help reveal redundancies in data. Reinforcing concepts of optimization and linear algebra is particularly important, as such concepts are broadly applicable in science and engineering.

We provide examples to connect all the topics discussed with a broad range of subjects of interest in chemical engineering. These examples are accompanied by their *software implementation*, which can help the reader understand how to implement the concepts in a practical setting and can help reinforce knowledge of computer programming and numerical computations. It is important to recognize that statistics, when seen from a perspective of modeling, can become rather mathematical in nature; as such, having a blend of theory and computational views is important. The book has a mathematical feel; this was both intentional and necessary; one of the reasons for this is to ensure that the reader understands the mathematics behind the software tools available. This mathematical understanding is necessary to interpret results, understand the applicability of specific tools, and facilitate troubleshooting.

A broad range of exercise problems are provided to reinforce the concepts and help show the reader how to apply statistical tools to solve realistic problems; some of these problems have been derived from real data and settings (in some instances based on research by our research group). The selection of examples and exercises is biased by my own experience, but I am hoping that the principles taught are broadly applicable.

People often ask me if textbooks (such as this one) should continue to be used in education, now that we have access to myriad online resources. In my opinion, a textbook conveys not only knowledge but also a thought process that some students (and instructors) can find beneficial. Moreover, a textbook aims to present a coherent vision of a subject (something difficult to do by simple compilation of resources); in other words, a textbook aims to "connect the dots" between topics/subjects. The book is designed with this in mind; it aims to present an end-to-end and coherent vision on how statistical principles can be used for a wide variety of applications.

Statistics is a broad area of mathematics, and it is inevitable that some topics needed to be left out. For instance, this book does not cover topics of stochastic processes (e.g., Markov chains and time series), statistical mechanics, and information theory. Moreover, the book only covers basic topics of machine learning and data science. However, I am hoping that the book provides the necessary foundations to explore these more advanced topics; again, the intention of this book is to connect the dots between different topics (as opposed to going deep into each topic).

Online Resources

The book incorporates documented code for all examples. A solution manual and code for exercises are also available for instructors. The availability of this code is intended for students to quickly see how to apply abstract concepts and develop "computational thinking" needed for actual implementation.

Acknowledgments

There are many people that helped me put together this book. First and foremost, I want to thank all my undergraduate and graduate students at the University of Wisconsin–Madison; the design of this book was largely inspired by the many things that I have learned while teaching different courses in chemical engineering.

I want to thank all my current and former PhD trainees and postdocs (particularly David Cole, Leo Gonzalez, Ugo Ikegwu, Lisa Je, Bruce Jiang, Kibaek Kim, Jiaze Ma, Ashley McCullough, Amy Qin, Alex Smith, Jaron Thompson, Elvis Umana, Bo-Xun Wang, and Weiqi Zhang) for all their help in proofreading earlier versions of the book and for their help in developing exercise problems and example code. Many of the concepts and problems discussed in this book are inspired by the research conducted at the Scalable Systems Laboratory (a.k.a. ZavaLab) at the University of Wisconsin–Madison.

The design of this book has been inspired by the work of Mihai Anitescu, Larry Biegler, Richard Braatz, Ignacio Grossmann, Joe Qin, Tunde Ogunnaike, and Jim Rawlings (and so many others). I have learned so much from you about statistics, optimization, data science, and linear algebra, and I hope that this is well captured in the contents of this book.

I want to thank my colleagues in the Department of Chemical and Biological Engineering at the University of Wisconsin–Madison and the Mathematics and Computer Science Division at Argonne National Laboratory for creating collaborative environments that have helped me appreciate the many applications of statistics and data science and for creating an environment that welcomes and fosters textbook writing. I dedicate this book to the legendary textbook writers of Wisconsin and Argonne, who have been a great source of inspiration: Bird, Box, Hill, Hougen, Lightfoot, Mangasarian, Moré, Rawlings, Stewart, and Wright. Special thanks go to Mike Graham, Dan Klingenberg, and Thatcher Root, who encouraged and supported me in developing a statistics course and book targeted at chemical engineers in Wisconsin. I also would like to acknowledge financial support from the Hougen Fellowship, which was instrumental in allocating time for writing a significant portion of this book.

I would like to thank my family (particularly Katie Zavala) for supporting me throughout the development of this book; this has been one of the most challenging projects that I have pursued in my career. I have spent many hours (often during weekends) writing and polishing this book.

Funding Support

I acknowledge financial support from the Department of Chemical and Biological Engineering at the University of Wisconsin–Madison under the Hougen program; this program provided funding for a research leave under which significant portions of this book were developed and written. I acknowledge partial financial support from the US National Science Foundation (NSF) under grants CBET-1748516 and IIS-1837812. The design of this book was motivated by research conducted under an NSF CAREER grant (CBET-1748516), and a significant portion of the technical content is inspired and based on research conducted under an NSF BIGDATA grant (IIS-1837812). I acknowledge partial support from the US Department of Energy, Office of Science, Advanced Scientific Computing Research, under contract number DE-AC02-06CH11357. This funding contributed toward my sabbatical year at Argonne National Laboratory in which parts of this book were developed and written.

Sources and Permissions

The following are data/content sources and associated permissions that I would like to explicitly acknowledge.

- The equilibrium Gibbs reactor model that is used to motivate diverse examples and exercises in the book is an adaptation of the equilibrium model presented by Seider, Seader, Lewin, and Widagdo [1, chapter 7].
- The problem and data of Exercise 2.9 (sieve analysis) were based on the content reported in [2].
- The problem of Exercise 2.7 (stress failure) uses data reported in [3]. The problem is also inspired by the tutorial notes reported in [4].
- The problem of Exercise 2.11 (residence time distribution) is an adaptation of the example problem reported by Professor Scott Fogler in his textbook [5].
- Exercise 2.1 was inspired by the experiment reported in [6].
- The flow cytometry examples and exercises in Chapters 3 and 6 are based on research reported in [7]. The data obtained from this research is used with permission from the authors. I thank Bruce Jiang for his help in putting together these examples/exercises; I also thank Nicholas Abbott for the collaboration that led to this paper.
- The thermostat examples provided in Chapter 3 are adaptations and extensions of an example provided in the textbook of Professor Tunde Ogunnaike [8].

- The electricity market Exercises 3.2, 3.5, and 3.9 use public data reported by the Electric Reliability Council of Texas (ERCOT) [9]. I thank Jiaze Ma for his help in putting together these exercises.
- The brain analysis problem of Exercise 3.7 is based on data reported in [10]. I thank Alexander Smith for his help in identifying this dataset.
- The problem of Exercise 4.8 is based on the content and data reported in [11].
- The Hougen–Watson estimation examples of Chapters 5 and 6 use data that is reprinted (adapted) with permission from [12]. Copyright 1960, American Chemical Society.
- Exercise 5.1 (lake quality) uses public data reported by the Wisconsin Department of Natural Resources [13]. I thank David Cole for his help in putting together this example.
- Exercise 5.5 (predicting solubility) uses data that is reprinted (adapted) with permission from [14]. Copyright 1999, American Chemical Society. I thank Montgomery Laky and Daniel Laky for their help in putting together this example.
- Exercises 5.6 and 5.7 (predicting CMC) use data that is reprinted (adapted) with permission from [15]. Copyright 2021, American Chemical Society. I thank Amy Qin for her help in putting together these exercises.
- Exercise 5.8 uses public data reported by the US National Oceanic and Atmospheric Association [16].
- Exercise 5.11 and Example 7.11 (experimental design) use data reported in [17].
- Exercise 5.12 is based on research reported in [18]. The data obtained from this research is used with permission from the authors. I thank Raka Dastidar and George Huber for their help in putting together this exercise.
- Examples 6.6, 6.19, and 6.29 (liquid crystal droplets) use data that is reprinted (adapted) with permission from [19]. Copyright 2023, American Chemical Society. I thank Amy Qin for her help in putting together these examples; I also thank Claribel Acevedo-Velez, David Lynn, Reid Van Lehn, and Fengrui Wang for the collaboration that led to this paper.
- Exercises 6.1, 6.2, and 6.3 (plastic characterization) use data that is reprinted (adapted) with permission from [20]. Copyright 2022, American Chemical Society. I thank Bruce Jiang for his help in putting together these examples; I also thank Fei Long and Ezra Bar-Ziv for the collaboration that led to this paper.
- Exercise 6.5 (asset placement) uses public data reported by the California Independent System Operator (CAISO) [21].
- Exercise 6.8 (sulfide removal) uses data that is reprinted with permission from [22]. Copyright 2013, Springer Nature. I thank Ryan Ettie for identifying this dataset as part of his final project for my statistics course at the University of Wisconsin–Madison.
- Exercise 6.10 (optimizing footwear materials) uses data that is published in [23]. I thank Amy Qin for her help in putting together this exercise; I also thank Allen Roman and Tim Osswald for the collaboration that led to this paper.

- Exercise 6.11 (applying convolutional filters) uses data that is reprinted (adapted) with permission from [24]. Copyright 2023, American Chemical Society. I thank Bruce Jiang for his help in putting together this exercise; I also thank Nicholas Abbott, Manos Mavrikakis, Alexander Smith, and Reid Van Lehn for the collaboration that led to this paper.
- Exercise 7.10 is inspired by the model reported in [25]. I thank Jiaze Ma for his help in putting together this exercise.
- Exercises 7.4 and 7.5 are based on the model reported in [26]. I thank Javier Tovar-Facio for his help in developing this model.

Part I

Basic Concepts

1 Introduction to Statistics

1.1 What Is Statistics?

As chemical engineers, we are trained to use fundamental laws of physics and chemistry (e.g., conservation of mass, energy, and momentum) to make decisions (e.g., design and control of a material, a chemical process, a microorganism, or a solar panel). The discovery of such laws is often the result of extensive experimentation and collection of observations (*data*). Fundamental laws embed *knowledge* and are expressed as mathematical *models* (e.g., Fourier's law, Fick's law, and the laws of thermodynamics and conservation of mass/energy); knowledge captures relationships/connections between the variables of a system.

Models derived from fundamental laws are often referred to as *mechanistic models* and are expressed in the form of equations and abstractions of different forms (e.g., algebraic equations, differential equations, and network models). Such models provide a *concise* summary of previous observations and are often refined, as more experimental observations are made; as such, our understanding of phenomena and associated fundamental laws is evaluated and refined over time. In fact, some fundamental laws have been discarded altogether, as new observations that contradict them have been obtained, and new laws have been developed to replace them.

Mechanistic models allow engineers to make *predictions* under situations that might lie outside of previous experiments and observations. For example, we know that energy and mass are always conserved (we do not need to conduct experiments to verify this). The ability to predict behavior using models allows us to make *decisions*; specifically, if we understand relationships between variables, then we can manipulate them to obtain a desired outcome. Mechanistic models are thus an essential component that enables engineering practice, as they help understand, design, and control systems.

Unfortunately, our understanding of real-life systems has always been limited by diverse factors such as accessibility to instruments to make observations/measurements or due to the high complexity of systems under study. As such, *models are limited in their predictive power*. Despite these limitations, humans have been extremely successful at practicing engineering; in fact, in some cases, humans have designed and controlled systems without fundamental laws and predictive models (i.e., by pure trial-and-error experimentation). An important question that arises in this context is then:

> How can humans make predictions and decisions without the use of fundamental laws?

For instance, a toddler learns how to build stable block structures without any training and understanding of physics. This means that our brains have embedded capabilities to extract knowledge (learn) directly from observations (data). In other words, the brain learns by building predictive models, and data needed for learning is collected via our senses (e.g., vision and touch). Knowledge that is directly acquired from observations is known as *empirical*. Empirical models (nowadays called *data-driven* models) also encode knowledge; these models capture connections between variables and allow us to make predictions. Empirical models are easier to obtain than mechanistic models but tend to apply to situations that are similar to previous experiments and observations; in other words, empirical models are often not as *generalizable* as mechanistic models.

The inherent limitations of mechanistic and empirical models imply that our predictions always contain a certain degree of *uncertainty*. Uncertainty refers to any *unknown* aspects of a system that cannot be predicted. For example, in assembling block structures, a toddler might not be thinking about the impact of wind. Moreover, and this is important, engineered systems are affected by *random phenomena* that cannot be fully predicted or controlled. For example, the properties of a gas mixture are affected by random fluctuations of molecules, the time to failure of a material cannot be fully predicted because this is the result of microscopic defects that cannot be observed/controlled, and the exact time at which a corn kernel will "pop" cannot be predicted because this is affected by microscopic variations of kernel composition.

Quantifying uncertainty is essential in making decisions; in fact, we (as humans) do this all the time by using "our gut," by relying on experience and expert systems (e.g., market and weather forecasts), or by evaluating possible ("what if?") scenarios. For example, even if we cannot perfectly predict the failure time of a material, experience can give us a rough estimate of when a failure is expected to occur. Failing to account for uncertainty can lead to decisions that are vulnerable to unexpected/extreme events (decisions that are "risky"); for instance, it is believed that the Maya empire collapsed due in part to their inability to predict and prepare for extreme drought events (despite the fact that they had highly sophisticated calendars to keep track of seasonal weather patterns).

Engineering practice requires a *combination* of knowledge (mechanistic and empirical) and characterization of the unknown (uncertainty), as sketched in Figure 1.1. Known aspects of a system are captured by *predictive models* (e.g., from physics or experience), while unknown aspects of a system are captured by *uncertainty*

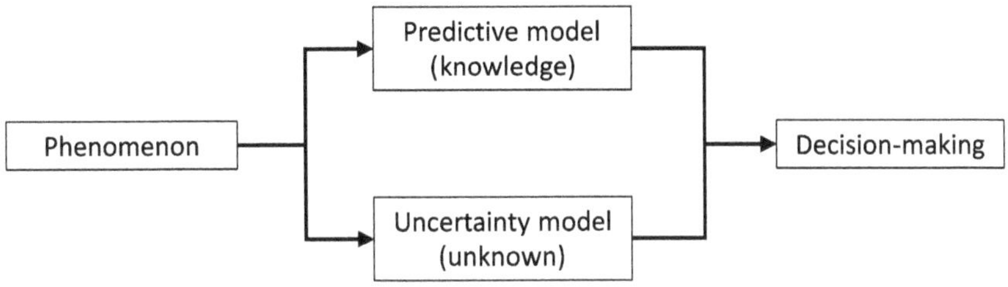

Figure 1.1 Real-life phenomena involve aspects that are known (predictable) and unknown (unpredictable), and we need to account for both aspects when making decisions.

models. Simply put, uncertainty models do not aim to predict outcomes/events of a system, but, rather, they aim to predict how "likely" it is that different events might be observed and they aim to predict the *range of possible events* that might be observed. For instance, an uncertainty model can help us determine the *most likely* failure time of a material and can help us determine how much failure times of a material vary (what is the range of failure times). Similarly, an uncertainty model can help us determine the likelihood of extreme events; this information can be used to develop strategies to *anticipate and mitigate* such events.

Engineers who use available knowledge (mechanistic or empirical) often require less experimentation (trial and error) when making a decision; in other words, their decisions are better *informed*. Conversely, engineers who lack knowledge might require more experimentation to make a decision. For instance, a toddler could save a lot of time building stable block structures if they knew that equilibrium is achieved by balancing forces.

Obtaining predictive models from data/observations (extracting knowledge from data) is not easy; for instance, developing a detailed mechanistic model can be prohibitive (from a technological or cost/time stand-point), and thus, an engineer might need to resort to simpler (less expensive) empirical models to make a decision. As expected, we would tend to trust the decision that is made with the better model. In other words, *gaining information comes at a cost*, and *reducing uncertainty comes at a cost*. Think about, for instance, the amount of resources (e.g., scientists, sensors, data, and computers) that have been used in developing and refining weather forecasting models or simulation models that are used to design chemical processes or the power grid. Engineers thus must balance model quality/uncertainty with data collection and model development cost when making decisions; moreover, engineers need to develop strategies to protect the system against uncertainty.

Statistics is a branch of mathematics that provides tools to:

- collect data, analyze data, and extract knowledge from data (in the form of predictive models),
- characterize and quantify the unknown aspects of a system (aspects that are not predictable),

- use characterizations of the unknown aspects of our system (uncertainty) to make decisions.

From an *engineering perspective*, statistics provides a framework that can help move *from data to knowledge (models) to decisions*. From a *scientific and philosophical perspective*, statistics provides a framework that can help us understand how humans and living systems naturally gather and process data to gain knowledge and make decisions.

Example 1.1 Noisy Flow Sensors

Consider a simple example to illustrate some of the concepts discussed so far; specifically, consider the mixing system shown in Figure 1.2. Here, we have a couple of input streams with flows F_1 and F_2 that are mixed to generate a single output stream with flow F_3. Assume that the streams are all in the liquid phase and have the same concentration (e.g., pure water) and temperature; also assume that mixing is instantaneous.

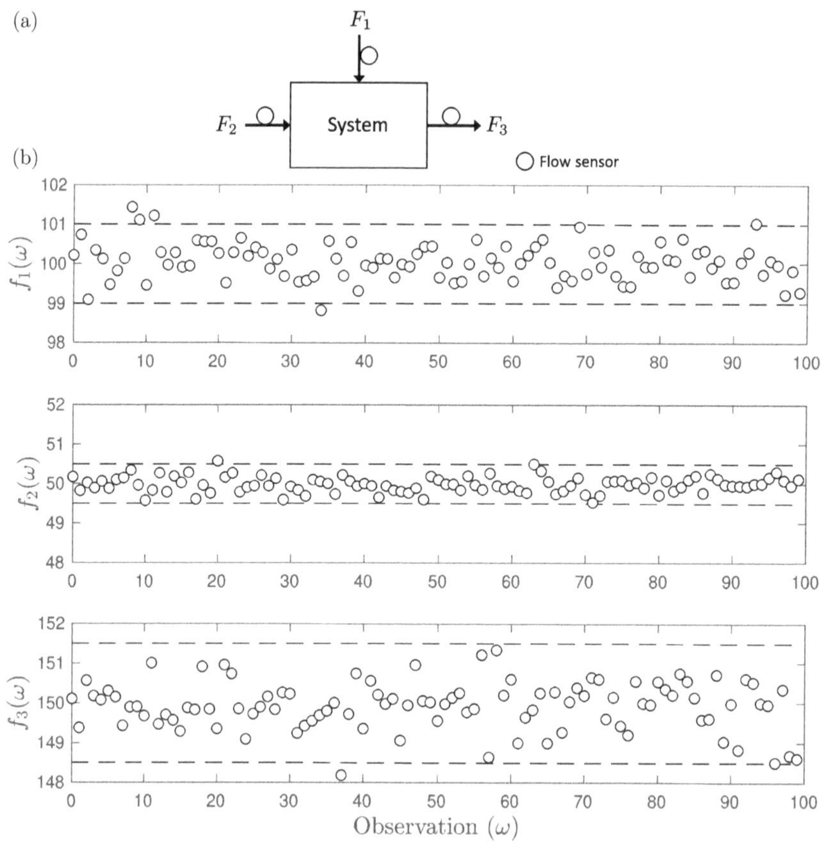

Figure 1.2 (a) Schematic representation of a mixing system with flow sensors. (b) Data for input and output flows of the mixer (100 observations). Horizontal lines indicate the ±1% margin of error for sensors.

The system is equipped with three *sensors* that measure the flow of all streams at any given time. We denote the values reported by the flow sensors at a given time instance ω as $f_1(\omega), f_2(\omega)$, and $f_3(\omega)$. Note that we make an explicit distinction between the input flows F_1, F_2, and F_3 and their observed values $f_1(\omega)$, $f_2(\omega)$, and $f_3(\omega)$. The reason for this will become apparent later in this chapter.

At a given time ω, the flow sensors report the following data:

$$f_1(\omega) = 101.5 \text{ gpm}, \quad f_2(\omega) = 50.5 \text{ gpm}, \quad f_3(\omega) = 151 \text{ gpm}.$$

Note that there is a mismatch between the total input flow $f_1(\omega) + f_2(\omega)$ and the output flow $f_3(\omega)$; the mismatch error is $\epsilon(\omega) = f_3(\omega) - f_2(\omega) - f_1(\omega) = -1$ gpm. This raises suspicion because we know that $f_1(\omega) + f_2(\omega) = f_3(\omega)$ must hold (from basic mass conservation laws). Mass conservation is the *known* aspect of this system (what is predictable), but there appears to be a *hidden* aspect of this system that is giving rise to a deviation from our predictions.

The fact that mass conservation does not appear to hold raises an important aspect of real-life applications: Data is collected using devices, instruments, or techniques that have *limited precision* or that are subject to *random* errors. For instance, the sensor manufacturer tells us that sensors have a margin of error of $\pm 1\%$. One might then ask: What is the origin of sensor errors? Is this error persistent, or does it follow a trend? In addition, one may ask: What does a margin of error $\pm 1\%$ actually mean? This seems to suggest that, for instance, the actual value of F_1 is in the range $[99, 101]$ gpm, but, does this happen all the time? Is it possible that a new measured value is outside this range?

To address such questions, we first note that sensors are physical devices that are subject to random effects (e.g., fluctuations that occur at molecular scales, vibrations of equipment, lack of reproducibility in data collection, and interference in communication channels). Sensor noise is the *unknown aspect* of our system; this is something that we cannot predict from mass conservation laws, but we might be able to characterize some of its behavior. In doing so, it seems prudent to collect more data (as opposed to using a single observation). Having more data will allow us to determine if errors are persistent or if there are trends in such errors. In Figure 1.2, we present 100 observations collected for the sensors. Here, we highlight the $\pm 1\%$ margin of error for the sensors. From the data, we see that the sensor measurements for F_1, F_2, and F_3 "gravitate" around the center values of $100, 50$, and 150 gpm. Interestingly, these center values satisfy conservation (even if the individual observations do not). This can also be visualized in Figure 1.3, where we see that the mismatch error gravitates around zero (where the mass balance closes). This reveals that the mass balance is closing (*on average*), but there is inherent noise in the sensor readings that is blurring this fact. This is why, in industrial practice, one often "filters out" sensor noise when verifying that a system is working properly. This is done using statistical estimation techniques that we will discuss later in this book; specifically, we will see how these techniques can help reconcile data to a physical model (e.g., mass conservation). These types of techniques are essential for closing mass and energy balances of complex industrial processes (which can contain hundreds of noisy sensors).

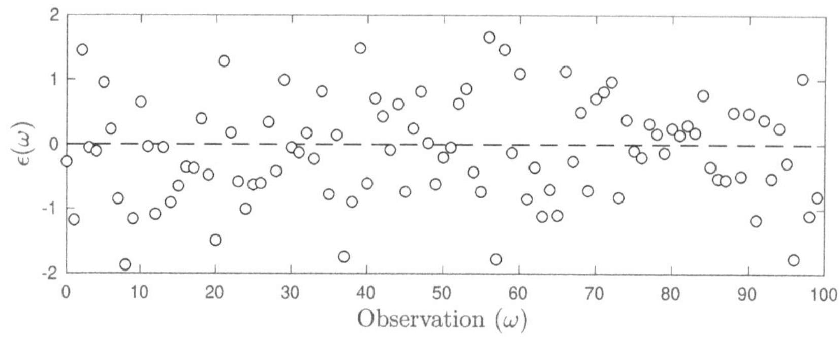

Figure 1.3 Conservation error between flows $\epsilon(\omega) = f_3(\omega) - f_2(\omega) - f_1(\omega)$.

In Figure 1.2, we note that some observations lie outside the margin of error provided by the manufacturer (this is particularly evident for the sensor of F_1). This highlights that the error margin provided by the manufacturer is not "absolute," in the sense that the observations will always lie within them. Upon further investigation, we realize that the margin of error reported by the manufacturer indicates that this is for a probability of 99%. This indicates that, for approximately 99 out of 100 observations, we will observe that the sensor measurements lie inside the margin of error. In Figure 1.2, however, we see that 4 observations out of 100 lie outside the margin of error for the sensor of F_1. This is why we say "approximately" because this estimate depends on the number of observations available. Later in the book, we will see that, as we collect more data, the percentage of observations that lie inside the margin will converge exactly to 99%. We will also see how one can derive margins of error (or precision levels) using the concept of "confidence intervals."

This example also indicates that we need to be careful in how we define *precision* for variables that exhibit random behavior. In addition, the example highlights how having *sufficient data* is critical to properly reveal trends and to reach conclusions. Moreover, this example highlights how a combination of engineering knowledge (conservation laws) and experimental data (observations) is needed to understand the system.

Script to reproduce results: `ch1_mixer_example` □

Example 1.2 Equilibrium (Gibbs) Reactor

We now consider a more sophisticated example to further highlight known and unknown aspects of a real system. We will be using this example throughout the book to illustrate a variety of concepts.

Consider an *experimental laboratory reactor* under which the following gas-phase reaction takes place (see Figure 1.4):

$$CO + 2H_2 \longleftrightarrow CH_3OH.$$

We define the set of species involved in the reaction as:

$$\mathcal{K} = \{CO, H_2, CH_3OH\}.$$

The reactor input molar flows for the species are defined as $\mu_k,\ k \in \mathcal{K}$. The reaction stoichiometric coefficients are given by $\gamma_{CO} = -1$, $\gamma_{H_2} = -2$, and $\gamma_{CH_3OH} = 1$. The reactor is run over a sufficiently long period of time so that the reaction reaches equilibrium. The output molar flows for the species are given by μ_k^{out}, $k \in \mathcal{K}$.

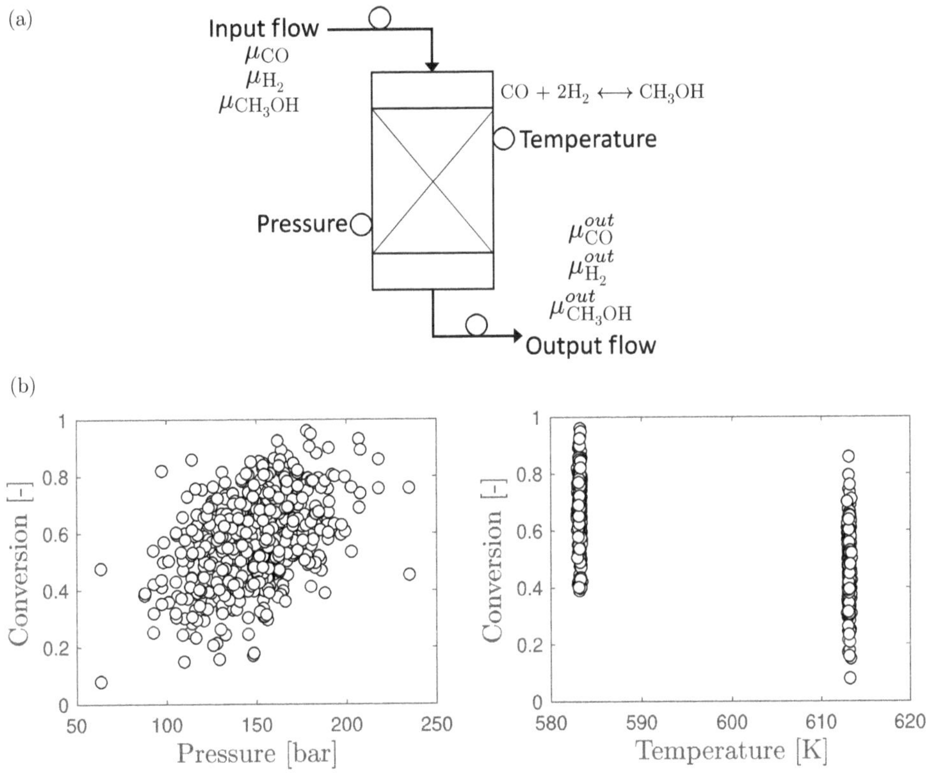

Figure 1.4 (a) Schematic of equilibrium reactor. (b) Effect of pressure and temperature on conversion.

The reactor operating pressure (denoted as P) and temperature (denoted as T) can be manipulated to control the reactor output (i.e., to increase the conversion of reactants CO and H_2 into the product CH_3OH). The control system of the reactor allows pressure to be varied over the continuous range [50, 250] bar and allows temperature to be operated in a couple of modes: low (583 K) or high (613 K).

Conversion is defined as the fraction of input flow CO that is consumed in the reaction; accordingly, this is calculated as:

$$\xi = 1 - \frac{\mu_{CO}^{out}}{\mu_{CO}}.$$

We are interested in running experiments to determine how temperature and pressure affect conversion (how these variables are related); we expect to use such knowledge to identify conditions that maximize the output flow of our product of interest, CH_3OH.

Assume for now that we only have data for input flows μ^k, temperature T, and pressure P (because we can control these), but we do not have data for the output flows μ^k_{out}; in other words, we have significant uncertainty about how the system conversion will respond. To resolve this uncertainty, we run experiments at different combinations of T and P to collect data for the outputs and figure out from this what works best.

The data resulting from our experiments is shown in Figure 1.4. Here, we can see that the conversion seems to increase with higher pressures and lower temperatures. We note, however, that these trends are not as clear as we might have hoped for. This indicates that sensor noise or some other hidden (unknown) factors might be impacting the behavior of the system (e.g., mixing effects or difficulty in reproducing the experiment perfectly every time).

After running the experiments, we now have some basic knowledge of how temperature and pressure affect the reactor conversion. We now might ask the following questions: Is this knowledge general? Can we expect the same pattern to hold at conditions that we have not tested experimentally? Addressing such questions purely based on experiments can require significant time, money (e.g., reactants and energy), and labor. Fortunately, plenty is already known about the behavior of equilibrium reactors from fundamental laws, and we can use such laws to help us save experimental time.

From conservation laws, we know that moles of reactant and product are conserved and thus:

$$\begin{aligned} \mu_{\mathrm{CO}}^{out} &= \mu_{\mathrm{CO}} + \gamma_{\mathrm{CO}} \cdot \xi \cdot \mu_{\mathrm{CO}} \\ \mu_{\mathrm{H_2}}^{out} &= \mu_{\mathrm{H_2}} + \gamma_{\mathrm{H_2}} \cdot \xi \cdot \mu_{\mathrm{CO}} \\ \mu_{\mathrm{CH_3OH}}^{out} &= \mu_{\mathrm{CH_3OH}} + \gamma_{\mathrm{CH_3OH}} \cdot \xi \cdot \mu_{\mathrm{CO}}. \end{aligned} \tag{1.1}$$

We can write these algebraic equations in the following compact form:

$$\mu_k^{out} = \mu_k + \gamma_k \cdot \xi \cdot \mu_{\mathrm{CO}}, \; k \in K. \tag{1.2}$$

In other words, the output molar flows are a function of the input molar flows and of the molar flows consumed (for reactants) or generated (for the product). Here, CO is used as the limiting reactant. These relationships tell us that if we know the input flows (which we have control over) and we know the conversion ξ, then we can predict the output flows of the reactor (without sensing them). This illustrates how fundamental laws embed knowledge and can help us reduce the amount of data needed. This knowledge also gives us some basic information on what we should expect to see from our experiments and thus narrows down the uncertainty surrounding the system.

Unfortunately, we do not have a sensor to measure conversion, but we can anticipate that this is somehow affected by temperature and pressure. Indeed, laws from thermodynamics (many of these based on the work of J. W. Gibbs in the 19th century) capture such a dependence. Under ideal gas conditions, we know that the following condition holds at equilibrium:

$$K_{eq} = \frac{(\mu_{CH_3OH} + \xi \cdot \mu_{CO})}{(\mu_{CO} - \xi \cdot \mu_{CO})(\mu_{H_2} - 2 \cdot \xi \cdot \mu_{CO})^2} \left(\frac{\mu_{out}}{P}\right)^2, \tag{1.3}$$

where:

$$\mu_{out} = \sum_{k \in \mathcal{K}} \mu_k^{out}, \tag{1.4}$$

is the total output flow. Now think about the implications of this new source of knowledge; if we knew the equilibrium constant K_{eq}, then we could solve Eq. (1.3) to obtain the conversion ξ, and with this, we could predict the reactor output flows. Moreover, the equation tells us that increasing pressure increases the ratio of products to reactants. Specifically, for fixed K_{eq}, increasing pressure requires increasing the numerator $\mu_{CH_3OH} + \xi \cdot \mu_{CO}$, which implies increasing ξ. We thus have that increasing pressure P has the effect of increasing ξ. Note that this is a general relationship and not just an observation from an experiment; in other words, Eq. (1.3) encodes knowledge about the system.

The only missing variable is the equilibrium constant K_{eq}; this quantity can also be obtained from fundamental laws of thermodynamics. Specifically, the Van't Hoff equation indicates that the equilibrium constant depends on temperature as:

$$\frac{\partial \log K_{eq}}{\partial T} = -\frac{\Delta H_0}{R \cdot T}, \tag{1.5}$$

where R is the universal gas constant and ΔH_0 is the enthalpy of reaction at standard conditions. These quantities can be obtained from experiments; fortunately, scientists have already performed such experiments, and these quantities are reported in databases and are therefore known.

The Van't Hoff equation tells us how the reaction equilibrium is affected by temperature. For the reaction of interest, we have that $\Delta H_0 < 0$ (the reaction is exothermic) and thus decreasing temperature (increasing $1/T$) increases K_{eq}. Increasing the value of K_{eq} in Eq. (1.3) has the effect of increasing the numerator $\mu_{CH_3OH} + \xi \cdot \mu_{CO}$, which implies increasing conversion ξ. We thus have that decreasing temperature increases conversion; again, this relationship is general.

We thus conclude that the available laws from thermodynamics and conservation provide *significant knowledge* about the system and will allow us to predict the outputs of the reactor given T, P, and the input flows μ_k. Specifically, these laws give us a *mechanistic model* that can be used to predict conversion and output flows as a function of T, P, and μ_k without the need to run experiments. This illustrates how fundamental laws embed substantial amounts of knowledge and can help save experimental time/cost. From the experimental data available, however, it is clear that our physical model (conservation and thermodynamic laws) cannot explain some of the variability present. Later in this book, we will learn techniques that allow us to separate the aspects of a system that are known and unknown (uncertain) using estimation techniques.

Advanced readers might have noticed that, in this example, equilibrium laws used only hold under special conditions; for example, activities for species are given by partial pressures only under ideal conditions (e.g., low pressures and high temperatures). As a result, we have uncertainty about what can occur at high pressures and low temperatures. This reveals that fundamental laws, while often provide a good picture of the system, are also often incomplete, and their predictions might still embed some degree of uncertainty. We could add a more general thermodynamic model to account for nonideal behavior and with this seek to mitigate our uncertainty, but other factors will still remain unknown (e.g., what if the reactor is not perfectly mixed? what if the sensors are not accurate at high pressures and temperatures?). This means that additional experiments might still be needed to validate our predictive model. Later in this book, we will discuss techniques that help us identify the best set of new experiments to conduct (based on the quality of the model and data that we have).

Script to reproduce results: `ch1_gibbs_data` □

The previous examples reveal a general limitation that is inherent in engineering practice: No matter how complex and sophisticated our predictive models and sensing devices are, we will *always* face a certain degree of uncertainty. As a result, some degree of experimentation and data might be needed to validate and refine models. Despite this limitation, we note that having some *prior knowledge* of fundamental laws can drastically reduce the number and types of experiments needed and can help mitigate uncertainty. Specifically, having no fundamental knowledge might force us to perform a large number of experiments and to measure many different variables to capture general behavior, while having fundamental knowledge can help focus experiments on aspects that cannot be explained by the model.

1.2 Random Variables

In statistics, we use *random variables (RVs)* to model unknown (uncertain) aspects of systems. By definition, an RV does not have a known value and exhibits variability. This *statistical* view of the world is a significant departure from the more traditional *deterministic* viewpoint commonly used in engineering practice (and, more broadly, in real life). Specifically, under a deterministic view of the world, we assume that variables have known and unique values (they do not exhibit variability).

Importantly, it is essential to recognize that a deterministic view of the world is an *assumption* that is often made to simplify analysis and is not always appropriate. Specifically, under a deterministic setting, we ignore the presence of uncertainty and associated variability (we assume full certainty). For example, think about the temperature reported by your home thermostat; this value is reported in a deterministic manner (as a single value) but ask yourself: Can we fully trust this value? Is this the truth? The answer to these questions is *no*, the value reported by the sensor is an *estimate* of the true temperature of the room. This estimate faces uncertainty because it relies on a physical device (e.g., a thermocouple) that might be affected by random effects (e.g., wind or molecular vibrations). Under a statistical view of the world, a more appropriate way to report temperature would be to provide a range of possible values that capture the inherent uncertainty of the value reported. Now, if the sensor is accurate/precise such that the range of possible values is narrow, then a deterministic representation might be an adequate approximation. However, we do not know if our deterministic assumption is correct until we actually characterize the range of possible values of the sensor.

The fact that a deterministic viewpoint is so pervasive in society and engineering practice *is not an accident*. Characterizing uncertainty of variables requires *appropriate amounts of data* (which can be time-consuming to obtain) and requires techniques to analyze such data. We now proceed to discuss concepts and techniques that will allow us to characterize/model uncertainty based on data. Random variables are the fundamental mathematical construct of statistics, and we will thus spend some time talking about what they are and how we can interpret them.

An RV X is a *model* that contains the following elements:

- Set of realizations $\omega \in \Omega$ with associated values $x_\omega \in \mathcal{D}_X$.
- Domain $\mathcal{D}_X$ under which realizations x_ω live.
- Probability measure $\mathbb{P}$.
- Cumulative density function (cdf) F_X.
- Probability density function (pdf) f_X.

The interpretation of these elements is extremely important and can be explained in simple terms as follows (for illustration purposes, consider the example in Figure 1.5).

An RV has a set of possible realizations Ω; think about this as all the possible *data points (observations)* that you can collect for the RV (which can be infinitely many). For instance, in the example, we show a set Ω containing 14 observations. Each observation $\omega \in \Omega$ has an associated value that the RV takes at such observation; this value is denoted as x_ω (the notation $x(\omega)$ is also often used). In the example provided, we have that realization ω takes a value $x_\omega = a_2$. The value x_ω has

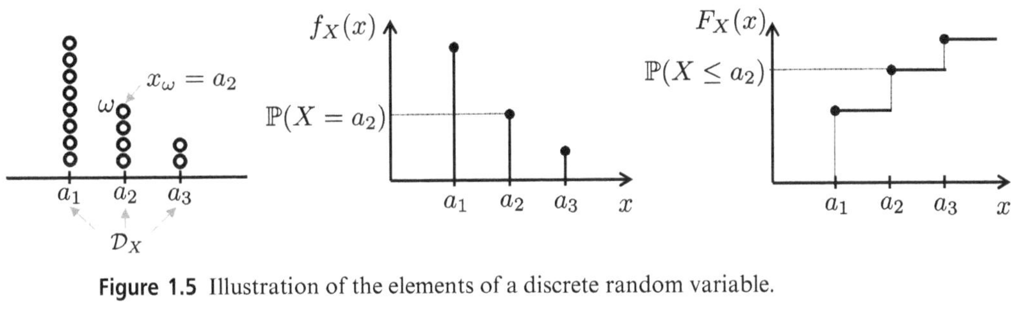

Figure 1.5 Illustration of the elements of a discrete random variable.

to "live" in the domain $\mathcal{D}_X$; therefore, the domain restricts what values the RV can take. In the example, we have that the domain contains three possible values $\mathcal{D}_X = \{a_1, a_2, a_3\}$. The domain of an RV plays a key role, as this tells us the range of values that the RV can take.

Imagine that each realization ω is a pinball that is dropped in the domain (each pinball carries an associated value x_ω); the pinball evolves randomly and lands in a specific location of the domain (think about such point as a "bin" over which the realization falls). As the realizations land randomly in the domain, they will begin to accumulate in specific locations of the domain $\mathcal{D}_X$.

The probability measure assigns probabilities to realizations. Specifically, it tells us how likely (probable) it is to find realizations in certain locations of the domain. For instance, the notation $\mathbb{P}(\omega \in \Omega \mid x_\omega = x)$ tells us the probability that X has realizations satisfying the condition that $x_\omega = x$. Because this notation is verbose, it is often preferred to use the short-hand notation $\mathbb{P}(X = x)$.

The probability measure also allows us to determine probabilities of *events*; we can think of an event as a set of conditions that the RV needs to obey. For instance, for the example provided, consider the probability that X has realizations satisfying the conditions that $x_\omega = a_1$ *and* that $x_\omega = a_2$; we can write this down as $\mathbb{P}(X = a_1 \,\&\, X = a_2)$. As we will see later, we can also consider events with more sophisticated logic such as $\mathbb{P}(a_1 \leq X \leq a_2)$ (probability that X has realizations between the values a_1 and a_2).

The probability measure can be computed by keeping track of the individual realizations of X (e.g., recording all data points x_ω observed). However, because such points can be many (even infinitely many), it is preferable to think about how *densely the realizations accumulate* in certain locations of the domain. The probability density function (pdf) does just that; specifically, $f_X(x)$ tells us the density of realizations at location $x \in \mathcal{D}_X$. Think about the pdf as a function that *summarizes* the behavior of all our realizations. Density is a quantity that is relative to the number of observations; this is a fraction of all the realizations (pinballs) that land in a specific location of the domain.

How likely it is to find realizations in a location depends on the density at such location; the probability measure is thus closely related to the pdf, and this allows us to compute probabilities in a convenient form. In some settings, we can directly relate the pdf and the probability measure as $P(X = x) = f_X(x)$. For instance, in the example shown, we have that $\mathbb{P}(X = a_2) = f_X(a_2)$ (the probability of finding realizations of X landing in a_2 equals the density at such location). Therefore, if we have the pdf of the RV, we do not need to keep track of all the observations to compute probabilities for events! We will see later that the pdf is the essential modeling object that allows us to calculate probabilities of complex events.

The pdf allows us to think about events as regions (or subdomains) of the domain $\mathcal{D}_X$. Specifically, consider that $\mathcal{A}$ is a subdomain of $\mathcal{D}_X$ (we write this as $\mathcal{A} \subseteq \mathcal{D}_X$). For instance, in the example provided, we can define $\mathcal{A} = \{a_1, a_2\}$ (this is a subset of $\mathcal{D}_X = \{a_1, a_2, a_3\}$). The probability that realizations of X land in region $\mathcal{A}$ (denoted as $\mathbb{P}(X \in \mathcal{A})$) can be computed as the *accumulated density* $f_X(x)$ for all values x

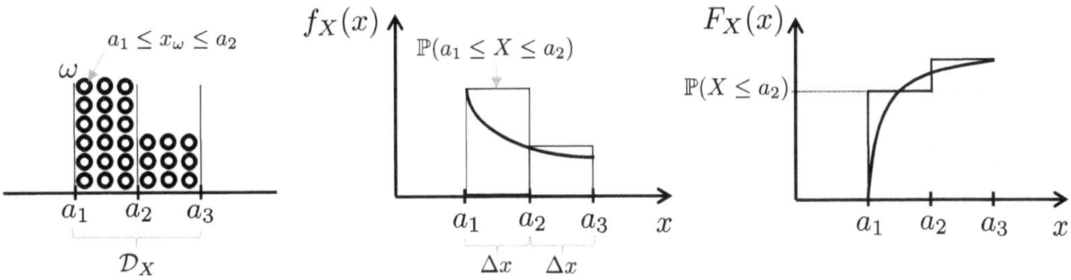

Figure 1.6 Illustration of the elements of a continuous random variable.

in $\mathcal{A}$. For instance, in the example provided, we have that $\mathbb{P}(X = a_1 \,\&\, X = a_2) = f_X(a_1) + f_X(a_2)$. As the name suggests, the cumulative density function (cdf) is a function that keeps track of how density accumulates in a region/subdomain of the domain $\mathcal{D}_X$. For the cdf, the subdomain is defined as an event of the form $X \le x$ (where $x \in \mathcal{D}_X$ is an upper bound or threshold). Specifically, we have that $F_X(x) = \mathbb{P}(X \le x)$ tells us the probability of all realizations that take a value below the threshold x. For instance, in the example provided, we have that $F_X(a_2) = \mathbb{P}(X \le a_2) = f_X(a_1) + f_X(a_2)$.

The previous discussion applies to discrete RVs (those for which the domain $\mathcal{D}_X$ can only take a finite number of values). The discussion for *continuous RVs* (those for which $\mathcal{D}_X$ can take an infinite number of points in the domain) is a bit more intricate but similar in spirit. As an illustration, consider the setting presented in Figure 1.6. Here, we consider a domain that contains any values in the range $[a_1, a_3]$.

For convenience, we discretize/partition the domain in a couple of regions $[a_1, a_2]$ and $[a_2, a_3]$; here, we define the width of the regions as $\Delta x = (a_3 - a_1)/2$. This will allow us to make an analogy with the discrete case; specifically, as before, consider dropping pinballs (realizations) that land on either one of the bins. The probability measure will allow us to determine how likely it is for realizations to land in the bins and to calculate probabilities of events.

The key difference arises when accounting for the density of the realizations. The discrete approximation of the domain allows us to define the density of points in a given bin (subdomain) of size Δx. We therefore can compute probabilities for such subdomains. But now imagine that we begin to decrease the size of the bins Δx, and we start increasing the number of bins (so as to refine our approximation). In the limit $\Delta x \to 0$, the density function will begin forming a continuous function. However, note that when the bin size is close to zero, it will not be possible for any realization to land in the bin. As such, the interpretation of $f_X(x)$ as the probability that $X = x$ breaks. This indicates that, in the continuous case, we need to think about probabilities on subdomains and not about specific locations of the domain. This is why, in the continuous case, the cdf $F_X(x)$ becomes easier to work with, as this tells us the probability that realizations will have values *below* a certain value x (as opposed to take values equal to x). Now, recall that the cumulative density is precisely the accumulated density up to a certain threshold; as such, we will see that the cdf is the integral of the pdf (and therefore the pdf is the derivative of the cdf).

As such, for continuous RVs, it is often convenient to think about the pdf as how rapidly the cdf is increasing.

We will see later that, in practice, one often resorts to discrete approximations of continuous RVs. This will allow us to compute approximations of the pdf and cdf directly from data (observations), and then, we will try to find continuous functions for the pdf/cdf that match those approximations.

1.2.1 Types of Random Variables

In our previous discussion, we introduced the basic elements of RVs and their interpretation. We now proceed to formalize their definition and properties from a mathematical standpoint.

Random variables can be categorized as univariate and multivariate.

- A *multivariate* RV $X = (X_1, X_2, \ldots, X_n)$ has realizations that are vector values $x_\omega = (x_{\omega,1}, x_{\omega,2}, \ldots, x_{\omega,n}) \in \mathbb{R}^n$ for $\omega \in \Omega$. Think about a pinball/realization ω carrying information about multiple variables (contained in the vector x_ω), such as temperature, pressure, and conversion. For a multivariate RV, we have a multidimensional domain $\mathcal{D}_X \subseteq \mathbb{R}^n$.
- A *univariate* RV X is a multivariate RV with $n = 1$ and has realizations that are scalar values $x_\omega = x_{\omega,1} \in \mathbb{R}$. Think about a pinball/realization ω carrying information about only one variable (contained in the scalar x_ω), such as pressure. For a univariate RV, we have a unidimensional domain $\mathcal{D}_X \subseteq \mathbb{R}$.

In the above definitions, we note that $\mathbb{R}$ is the set of real numbers (positive and negative); as such, we have that $\mathbb{R} = (-\infty, \infty)$. We also note that $\mathbb{R}^n$ denotes the n-dimensional generalization of this, which indicates that all the elements of the n-dimensional vector x_ω are values that are real numbers; one denotes this as $\mathbb{R} = (-\infty, \infty)^n$. The notation $\mathcal{D}_X \subseteq \mathbb{R}$ indicates that $\mathcal{D}_X$ is a subdomain of the domain $\mathbb{R}$. This notation might appear to be unnecessary at this point, but we will see that it is essential that we keep track of the types of numbers that we are dealing with.

Random variables can also be categorized as continuous and discrete.

- A *continuous* RV X has a continuous domain $\mathcal{D}_X$. For example, a multivariate continuous RV $X = (X_1, X_2)$ has realizations satisfying $x_{\omega,1} \in [0, 1]$ and $x_{\omega,2} \in [0, 1]$ for $\omega \in \Omega$.
- A *discrete* RV X has a discrete domain $\mathcal{D}_X$. For example, a multivariate discrete RV $X = (X_1, X_2)$ has realizations satisfying $x_{\omega,1} \in \{0, 1\}$ and $x_{\omega,2} \in \{0, 1\}$ for $\omega \in \Omega$.

The type of RV chosen to characterize uncertainty depends on the nature of the system under study. In the reactor of Example 1.2, pressure is a univariate continuous RV because its realizations can take any value in a continuous domain $[50, 250]$ bar, while temperature is a univariate discrete RV since its realizations can only take values $\{513, 613\}$ K. We highlight that, for discrete domains, we use the notation $\{\cdot, \cdot\}$ while, for continuous domains, we use the $[\cdot, \cdot]$ notation.

In the following discussion, we will focus our attention on univariate RVs (a discussion on multivariate RVs is provided in Chapter 3).

1.2.2 Cumulative and Probability Density Functions

The cdf and pdf provide a *complete description* of the behavior of an RV and are thus essential functions to study. In fact, we will see that cdfs and pdfs will be encountered in *every aspect of engineering practice* (e.g., from analyzing data to developing models from data and to making decisions using models). To understand such essential functions, it is necessary to define the domains over which they are valid.

For a *univariate* RV X, its domain $\mathcal{D}_X$ is a subset of the set of real numbers $\mathbb{R}$; we denote this fact by using the notation $\mathcal{D}_X \subseteq \mathbb{R}$. In 1D, a domain $\mathcal{D}_X$ is also called an *interval or range*.

For a continuous X, the domain $\mathcal{D}_X$ is typically given by the intervals $(-\infty, \infty)$, $[0, \infty)$, or $[a, b]$. For a discrete X, the domain $\mathcal{D}_X$ is typically given by the collections of points $\{0, 1, 2, \ldots, \infty\}$, $\{0, 1\}$, or $\{a, a + 1, \ldots, b\}$. Here, we use "(" and ")" to indicate that the value at either extreme of the domain is not included, and we use "[" and "]" or "{" and "}" to indicate that the values are included. Understanding these domains is important because these define the values that an RV can take or not (values that we will actually observe or not). For instance, we know that a flow cannot take negative values and thus its domain cannot be $\mathcal{D}_X = (-\infty, \infty)$; it has to be $\mathcal{D}_X \in [0, \infty)$. This implies that there is a *zero probability* that a flow takes a negative value. In some systems, we might also have physical constraints that prevent variables from taking certain values; for example, in a chemical reactor, it might not be physically possible for temperature to exceed a value of 500°C (this implies that there is a zero probability of finding a temperature that is higher than 500°C). Understanding domains is also important because RVs have inherent variability, and we would like to know if such variability is bounded somehow. Domains might also tell us how extreme the outcomes of a phenomenon might be (e.g., weather and product demands); moreover, understanding domains will be essential in understanding probabilities of events because events are subdomains of the domain.

The *cumulative density function* of a univariate X is a function of the form $F_X : \mathcal{D}_X \to [0, 1]$; this notation indicates that the function F_X maps input values in the domain $\mathcal{D}_X$ into output values in the domain $[0, 1]$. The cdf tells us the probability that X is in the subdomain $(-\infty, t]$, where $t \in \mathbb{R}$ is a given threshold value. This can be stated in the following *mathematically equivalent* forms:

$$\begin{aligned} F_X(t) &= \mathbb{P}(X \in (-\infty, t]) \\ &= \mathbb{P}(-\infty \le X \le t) \\ &= \mathbb{P}(X \le t). \end{aligned} \tag{1.6}$$

We express the fact that the range of the cdf is [0, 1] as:

$$0 \le F_X(t) \le 1, \qquad t \in \mathcal{D}_X. \tag{1.7}$$

A cdf has an associated *probability density function* $f_X : \mathcal{D}_X \to \mathbb{R}_+$. Here, $\mathbb{R}_+$ denotes the set of nonnegative real numbers (zero or positive); consequently, we have that $\mathbb{R}_+ = [0, \infty)$ and:

$$f_X(x) \ge 0, \qquad x \in \mathcal{D}_X. \tag{1.8}$$

We use the subscript X in F_X, f_X, and $\mathcal{D}_X$ to highlight the fact that the cdf, pdf, and domain belong to the RV X. This notation will become convenient when encountering situations in which multiple RVs are present. In the next discussion, we will focus on a single RV; as such, we will drop the subscript to simplify the notation and use F, f, and $\mathcal{D}$. We will use the more verbose notation when necessary.

The nature of the cdf and pdf of an RV is quite different for continuous and discrete RVs.

A *discrete* X has a pdf/cdf that satisfy the following properties:

- $f(x) = \mathbb{P}(X = x), \quad x \in \mathcal{D}$.
- $\sum\limits_{x \in \mathcal{D}} f(x) = 1$.
- $F(t) = \sum\limits_{x \in \mathcal{D} \,|\, x \le t} f(x), \quad t \in \mathbb{R}$.

The first property tells us that the probability of the event in which X takes a specific value $x \in \mathcal{D}$ is given by the pdf $f(x)$. The second property thus tells us that the sum of the probabilities that X takes all possible values x in $\mathcal{D}$ is equal to one. This is equivalent to stating that:

$$\mathbb{P}(X \in \mathcal{D}) = \sum_{x \in \mathcal{D}} \mathbb{P}(X = x) = \sum_{x \in \mathcal{D}} f(x) = 1. \tag{1.9}$$

The third property establishes the relationship between the pdf and the cdf of X. Here, the statement $x \in \mathcal{D} | x \le t$ indicates all values $x \in \mathcal{D}$ satisfying the property that $x \le t$ (the symbol "|" means "such that"). The notation $\sum_{x \in \mathcal{D} \,|\, x \le t}$ is often simplified as $\sum_{x \,|\, x \le t}$.

The pdf is a function that is used to compute the probability of events $\mathcal{A} \subseteq \mathcal{D}$; for a discrete RV, we have that:

$$\mathbb{P}(X \in \mathcal{A}) = \sum_{x \in \mathcal{A}} f(x), \quad \mathcal{A} \subseteq \mathcal{D}. \tag{1.10}$$

Here, we emphasize that an event $\mathcal{A}$ is a subset of the domain $\mathcal{D}$ (an event is a subdomain); this will be discussed in detail in Chapter 3. Note also that, if we have the event $\mathcal{A} = \mathcal{D}$, then $\mathbb{P}(X \in \mathcal{A}) = 1$ (because the probability of finding X in $\mathcal{D}$ is one).

From the previous basic properties, we can derive a number of interesting and useful properties. For instance, consider the subdomain $\mathcal{A} = \{t_1, t_1 + 1, \dots, t_2\}$ with $t_1, t_2 \in \mathcal{D}$ and $t_2 \geq t_1$. We can establish that:

$$\begin{aligned} \mathbb{P}(X \in \{t_1, t_1 + 1, \dots, t_2\}) &= \mathbb{P}(t_1 \leq X \leq t_2) \\ &= F(t_2) - F(t_1). \end{aligned} \tag{1.11}$$

Moreover, we can establish that:

$$\begin{aligned} \mathbb{P}(X > t) &= 1 - \mathbb{P}(X \leq t) \\ &= 1 - F(t). \end{aligned} \tag{1.12}$$

Another useful property is:

$$\begin{aligned} \mathbb{P}(X \in \mathcal{A}) &= \sum_{x \in \mathcal{A}} f(x) \\ &= \sum_{x \in \mathcal{D}} \mathbf{1}[x \in \mathcal{A}] \cdot f(x). \end{aligned} \tag{1.13}$$

Here, we used the *indicator function*:

$$\mathbf{1}[x \in \mathcal{A}] = \begin{cases} 1 & \text{if} \quad x \in \mathcal{A} \\ 0 & \text{if} \quad x \notin \mathcal{A} \end{cases}. \tag{1.14}$$

The indicator function is a useful tool that will allow us to highlight binary logic (e.g., when an event/condition occurs or not).

The pdf/cdf of a discrete RV are *discontinuous* functions. Visually, the pdf is given by a set of points (called points of mass) defined at the different values of the domain $\mathcal{D}$. The cdf takes the form of a staircase, in which the jumps occur at the different values of the domain $\mathcal{D}$.

A discrete RV is *easy to handle computationally* because it involves simple summations and counting operations. For instance, we can compute the pdf and cdf by *counting* how many times the realizations of X (given by x_ω) take a certain value. Specifically, we have that:

$$\begin{aligned} f(x) = \mathbb{P}(X = x) &= \frac{1}{|\Omega|} \sum_{\omega \in \Omega} \mathbf{1}[x_\omega = x] \\ F(t) = \mathbb{P}(X \leq t) &= \frac{1}{|\Omega|} \sum_{\omega \in \Omega} \mathbf{1}[x_\omega \leq t]. \end{aligned} \tag{1.15}$$

Here, $|\Omega|$ denotes the number of realizations (the size of the set Ω). This reveals that the pdf is a function that predicts the number of realizations under which the RV takes a specific value x, relative to the total number of realizations (predicts how probable it is for X to take a specific value x). Similarly, the cdf predicts the number of realizations under which the RV takes a value below t, relative to the total number of realizations (predicts how probable it is for X to take a value below t). Importantly, the above expressions also reveal that we can *directly determine the pdf/cdf of a discrete RV from observations x_ω (data)*.

Example 1.3 Pdf and cdf of a discrete RV

Consider a discrete RV with domain $\mathcal{D} = \{-1, 0, 1, 2\}$ and pdf:

$$f(x) = \begin{cases} 0 & \text{for } x = -1 \\ 0.4 & \text{for } x = 0 \\ 0.6 & \text{for } x = 1 \\ 0 & \text{for } x = 2. \end{cases}$$

Note that the pdf satisfies $\sum_{x \in \mathcal{D}} f(x) = 1$ and $f(x) \geq 0$ for all $x \in \mathcal{D}$.

The RV has a set of 10 possible realizations $\Omega = \{1, 2, 3, \ldots, 10\}$ (thus $|\Omega| = 10$) with associated values x_ω:

$$\{0, 1, 0, 0, 1, 1, 1, 1, 1, 0\}.$$

The RV X takes a value of 0 in four realizations and thus $\mathbb{P}(X = 0) = 4/10 = 0.4$ and takes a value of 1 in six realizations and thus $\mathbb{P}(X = 1) = 6/10 = 0.6$. There are no realizations under which X takes values of -1 or 2 and thus $\mathbb{P}(X = -1) = 0$ and $\mathbb{P}(X = 2) = 0$. We thus see that the pdf predicts how frequently X takes specific values. The pdf values can also be computed using the general formula:

$$f(0) = P(X = 0) = \frac{1}{10} \sum_{\omega \in \Omega} \mathbf{1}[x_\omega = 0] = 0.4$$

$$f(1) = P(X = 1) = \frac{1}{10} \sum_{\omega \in \Omega} \mathbf{1}[x_\omega = 1] = 0.6.$$

We use the pdf to compute the cdf at threshold values $t = -1, 0, 0.5, 1$:

$$F(-1) = \sum_{x | x \leq -1} f(x) = f(-1) = 0$$

$$F(0) = \sum_{x | x \leq 0} f(x) = f(-1) + f(0) = 0.4$$

$$F(0.5) = \sum_{x | x \leq 0.5} f(x) = f(-1) + f(0) = 0.4$$

$$F(1) = \sum_{x | x \leq 1} f(x) = f(-1) + f(0) + f(1) = 1.0.$$

The cdf values are consistent with the fraction of realizations under which X takes a value below the corresponding threshold. The pdf/cdf of the RV are visualized in Figure 1.7. The pdf has points of mass at the domain values $\{-1, 0, 1, 2\}$, and we note the staircase form of the cdf with jumps occurring at the mass points.

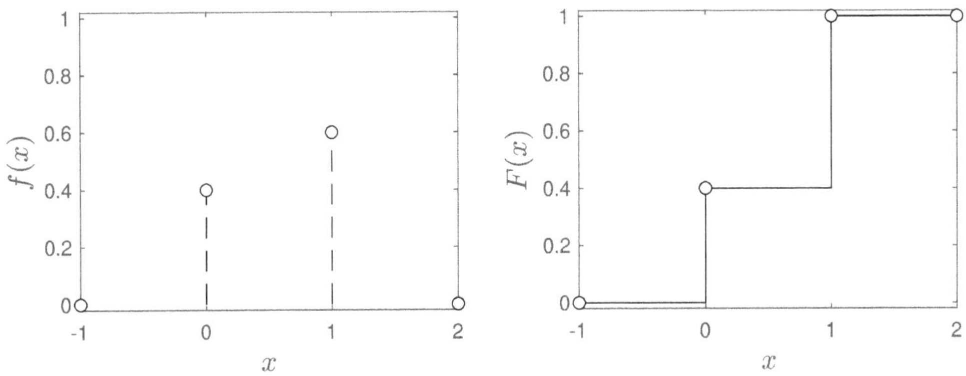

Figure 1.7 Pdf and cdf for simple discrete random variable.

Script to reproduce results: `ch1_pdf_cdf_discrete` □

The previous example is simple and seeks to illustrate some key concepts; in practice, discrete RVs can have a large number of realizations/data, and their domains might contain many values. This is why it is convenient to express how frequently the RV will take certain values by using the pdf/cdf (as opposed to carrying the entire set of data).

A *continuous* RV X has a pdf and cdf that satisfy the properties:

- $f(x) = \frac{dF(x)}{dx}, \quad x \in \mathcal{D}.$
- $\displaystyle\int_{x \in \mathcal{D}} f(x)dx = 1.$
- $F(t) = \int_{x \in \mathcal{D} | x \leq t} f(x)dx.$

From the first property, we can see that the pdf for a continuous RV X is not the probability that this takes a specific value x (as in the discrete case). Instead, we state that the pdf is the derivative of the cdf. This is an important observation and requires some explanation. For a continuous RV, we have that $f(x)$ is the *rate* at which $F(x)$ changes at x and thus:

$$f(x) = \lim_{\Delta x \to 0} \frac{F(x + \Delta x) - F(x)}{\Delta x}. \tag{1.16}$$

From this expression, we can see that:

$$\begin{aligned} f(x)\Delta x &\approx F(x + \Delta x) - F(x) \\ &= \mathbb{P}(X \leq x + \Delta x) - \mathbb{P}(X \leq x) \\ &= \mathbb{P}(x \leq X \leq x + \Delta x). \end{aligned} \tag{1.17}$$

The first equality follows from the fact that $F(x + \Delta x) = \mathbb{P}(x + \Delta x)$ and $F(x) = \mathbb{P}(X \leq x)$. The second equality arises from the observation that:

$$\mathbb{P}(X \leq x + \Delta x) = \mathbb{P}(X \leq x) + \mathbb{P}(x \leq X \leq x + \Delta x). \tag{1.18}$$

Consequently, $f(x)\Delta x$ is *approximately* the probability that X is in the interval $[x, x + \Delta x]$. Here, Δx is the width of the interval. By setting $\Delta x = 0$ in Eq. (1.17), we can see that $\mathbb{P}(X = x) = 0$; this explains why $f(x)$ *should not be interpreted* as $\mathbb{P}(X = x)$ for a continuous RV. To avoid confusion, it is often recommended to think about continuous RVs in terms of their cdf and not of their pdf. Moreover, even if technically $\mathbb{P}(X = x) \neq f(x)$ for a continuous RV, one often uses the notation $f(x) = \mathbb{P}(X = x)$ to state that $f(x)$ is the probability that X is in a *small neighborhood* around x.

The second and third properties of the continuous pdf/cdf are analogous to the discrete case, but we see that summation is replaced with an integral (because the pdf and cdf are continuous functions). For instance, we have that:

$$\mathbb{P}(X \in \mathcal{D}) = \int_{x \in \mathcal{D}} f(x) = 1. \tag{1.19}$$

As in the discrete case, the pdf can be used to compute probabilities of events $\mathcal{A} \subseteq \mathcal{D}_X$ for a continuous RV; this is done by using:

$$\mathbb{P}(X \in \mathcal{A}) = \int_{x \in \mathcal{A}} f(x)dx, \quad \mathcal{A} \subseteq \mathcal{D}_X. \tag{1.20}$$

As in the discrete case, the basic properties can be used to derive a number of interesting properties. For instance, if we consider the subdomain $\mathcal{A} = [t_1, t_2]$ with $t_1, t_2 \in \mathcal{D}$ and $t_2 \geq t_1$, we can establish that:

$$\mathbb{P}(t_1 \leq X \leq t_2) = F(t_2) - F(t_1). \tag{1.21}$$

Moreover, because $\int_{x \in \mathcal{D}} f(x)dx = 1$, we have that:

$$\mathbb{P}(X > t) = 1 - F(t). \tag{1.22}$$

A continuous RV is significantly more difficult to handle computationally because it involves integrals instead of summations (this prevents the use of counting operations). However, from a mathematical standpoint, properties are analogous to those of the discrete case. For instance, we have the following useful property:

$$\begin{aligned} \mathbb{P}(X \in \mathcal{A}) &= \int_{x \in \mathcal{A}} f(x)dx \\ &= \int_{x \in \mathcal{D}} \mathbf{1}[x \in \mathcal{A}] \cdot f(x)dx. \end{aligned} \tag{1.23}$$

Continuous RVs are often *approximated* by using discrete RVs (this is analogous to how continuous time is often approximated using discrete time intervals). Specifically, it is possible to "discretize" the domain $\mathcal{D}$ by defining intervals (also known as bins) of given length Δx and by computing the probability that X is in such intervals. For instance, based on Eq. (1.17), we can construct an approximation of the pdf of the form:

$$\tilde{f}(x) = \frac{F(x + \Delta x) - F(x)}{\Delta x}. \tag{1.24}$$

This discrete approximation is known as a *finite difference* approximation. We will also see how to approximate continuous pdfs and cdfs by using realizations (data) of the RV (as was done for discrete RVs).

Example 1.4 Pdf and cdf of a continuous RV

Consider a continuous RV with domain $\mathcal{D} = (-\infty, \infty)$ and pdf of the form:

$$f(x) = \frac{1}{\sqrt{2\pi}} e^{-\frac{x^2}{2}}. \tag{1.25}$$

It is easy to see that the pdf satisfies $f(x) \geq 0$ for any $x \in \mathcal{D}$ and one can show that $\int_{x \in \mathcal{D}} f(x)dx = 1$. The cdf associated with this pdf is given by:

$$\begin{aligned} F(t) &= \int_{x \mid x \leq t} f(x)dx = \int_{-\infty}^{t} f(x)dx \\ &= \frac{1}{2}\left(1 + \text{erf}\left(\frac{t}{\sqrt{2}}\right)\right). \end{aligned} \tag{1.26}$$

Here, erf$(\cdot)$ is the error function (we will discuss this function in more detail in Chapter 2). The expression for the cdf is obtained by inserting the pdf (1.25) in the integral and conducting integration (analytically). This highlights the complex computations that one encounters when operating with continuous RVs. One can show that the pdf and cdf satisfy $f(x) = dF(x)/dx$ and that the cdf satisfies $0 \leq F(t) \leq 1$ for any $t \in \mathcal{D}$, $F(-\infty) = 0$, and $F(\infty) = 1$.

In Figure 1.8, we present the pdf and cdf of the RV. The cdf tells us that there is virtually no probability that X takes a value below -5 or above 5. In other words, we have that $\mathbb{P}(X \leq -5) \approx 0$ and $\mathbb{P}(X > 5) = 1 - \mathbb{P}(X \leq 5) \approx 0$. We also see that the cdf curve always increases as one moves from the left to the right side of the domain; this is because $f(x) = dF(x)/dx$ and $f(x) \geq 0$. We can also see that the cdf value grows at the fastest rate at a value of $x = 0$ (the derivative of the cdf is the largest at this point). This point corresponds to the peak point of the pdf at $x = 0$ and further highlights the close relationship between the pdf and the cdf.

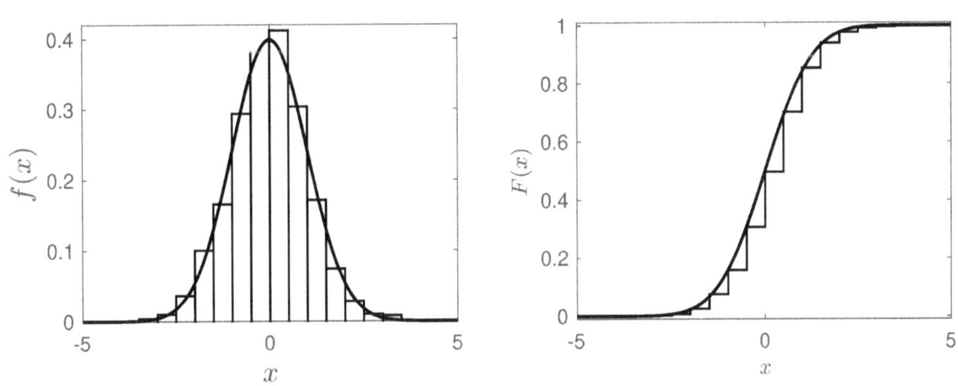

Figure 1.8 Pdf and cdf of a continuous RV and discrete approximations.

In Figure 1.8, we also show the discrete approximations of the pdf and cdf using a bin interval of $\Delta x = 0.5$ (this splits the domain $[-5, 5]$ into 20 bins). We can see that the approximations capture the general features of the pdf/cdf. From these functions, we can reinforce the interpretation of the pdf; for instance, the peak point of the discretized pdf is $\tilde{f}(0) = 0.38$. Here, $\tilde{f}(0)$ is interpreted as the probability of finding X in the interval $[0, 0.5]$ divided by the length of the bin $\Delta x = 0.5$. In other words, from Eq. (1.24), we have that:

$$\begin{aligned}\tilde{f}(0) &= \frac{\mathbb{P}(0 \leq X \leq 0.5)}{0.5} \\ &= \frac{F(0.5) - F(0)}{0.5} = \frac{0.69 - 0.5}{0.5} = 0.38.\end{aligned}$$

This is thus a good approximation of $f(x)$ at $x = 0$ (the peak of the continuous pdf is $f(0) = 0.4$).

Script to reproduce results: `ch1_pdf_cdf_continuous` □

1.3 From Data to Models

In practical situations, we often have access to observations (data) of our system. Our goal is to use such data to gain knowledge (understanding) about our system. Our knowledge will be extracted from the data in the form of *models* (knowledge is embedded in models).

The models will be of a couple of types. The first type is *structural* models (mechanistic or empirical), which provide a characterization of structural dependencies (relationships) between variables in our system and allow us to make predictions. The second type is RV models, which provide a characterization of behavior that cannot be predicted by structural models (RV models are also known as *uncertainty models*). Both types of models are necessary to make predictions and decisions that take uncertainty into account; specifically, quantifying the prediction uncertainty of structural models requires RV models.

1.3.1 Models of Random Variables

We begin our discussion with RV models. Assume that we have observations x_ω, $\omega \in S$; here, $S \subseteq \Omega$ denotes the set of available observations (which is a subset of all the possible observations Ω).

Given the data x_ω, $\omega \in S$, we are interested in answering the following specific questions:

- What type of RV model are the observations following?
- How do we obtain the cdf $F(x)$ and the pdf $f(x)$ for the RV model from data?

In general, it will be difficult to identify trends for random phenomena directly from data. As such, our goal will be to develop procedures to summarize and visualize the data in a way that reveals trends. Our goal will then be to postulate a *theoretical RV* model (denoted as X) that captures such trends. This highlights the fact that our RV X is a *model* of an uncertain phenomenon that we are assuming is generating the observed data.

There are many theoretical RV models that have been proposed to model uncertain phenomena seen in real life (just as there are many mechanistic models that have been proposed to explain physical phenomena seen in real life). These models differ in the type of domain, pdf, and cdf that they use.

A widely used RV model is that of a *Gaussian RV*; this model assumes that X is continuous with domain $\mathcal{D} = (-\infty, \infty)$ and has a pdf of the form:

$$f(x) = \frac{1}{\sqrt{2\pi\sigma^2}} e^{-\frac{(x-\mu)^2}{2\sigma^2}}, \quad x \in \mathcal{D}, \tag{1.27}$$

where $\mu \in \mathbb{R}$ and $\sigma \in \mathbb{R}_+$ are *parameters* of the model. These parameters define the behavior of the RV and can be "tuned" (adjusted) to match the available data. This model might look familiar because it is a generalization of the model used in Example 1.4. The cdf associated with the Gaussian model is:

$$F(x) = \frac{1}{2}\left(1 + \text{erf}\left(\frac{(x-\mu)/\sigma}{\sqrt{2}}\right)\right), \quad x \in \mathcal{D}. \tag{1.28}$$

Another widely used RV model is that of an *exponential RV*; this model assumes that X is continuous, has a domain $\mathcal{D} = [0, \infty)$, and has a pdf of the form:

$$f(x) = \frac{1}{\beta} e^{-x/\beta}, \quad x \in \mathcal{D}, \tag{1.29}$$

where $\beta \in \mathbb{R}_+$ is the parameter of the model. The cdf is:

$$F(x) = 1 - e^{-x/\beta}, \quad x \in \mathcal{D}. \tag{1.30}$$

Another popular theoretical model is the Poisson RV; here, the RV is assumed to be discrete with domain $\mathcal{D} = \{0, 1, 2, \ldots, \infty\}$, and the pdf is:

$$f(x) = \frac{\lambda^x e^{-\lambda}}{x!}, \quad x \in \mathcal{D}, \tag{1.31}$$

where $\lambda \in \mathbb{R}_+$ is the parameter. We will study the behavior of these (and of many other) theoretical models in Chapter 2.

1.3.2 Estimation

Our goal is to investigate whether the observations x_ω, $\omega \in \mathcal{S}$ follow the pdf/cdf of a given theoretical RV model that we postulate. In other words, we hypothesize that the observations follow the pdf/cdf of the theoretical model. To verify the hypothesis, we use the data to construct *empirical* approximations for the pdf $f(x)$ and cdf $F(x)$ of the theoretical model and verify if these match. The empirical

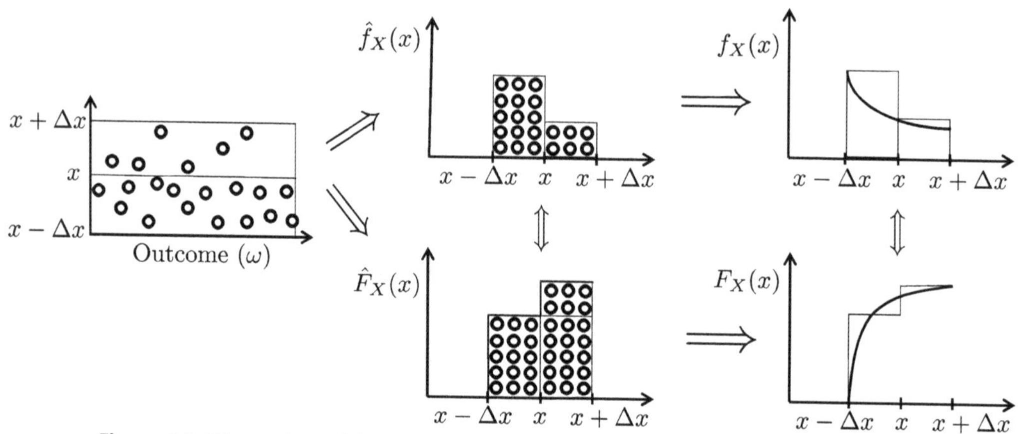

Figure 1.9 Illustration of the procedure used for obtaining empirical estimates of the pdf/cdf from data. The empirical estimates are matched to the pdf/cdf of a theoretical model. Once we have an estimate of the pdf, we can get an estimate of the cdf (and vice versa).

approximations are also known as data-driven approximations and are denoted as $\hat{f}(x)$ and $\hat{F}(x)$; these approximations are used to *estimate* the parameters of $f(x)$ and $F(x)$.

The approach to constructing $\hat{f}(x)$ and $\hat{F}(x)$ for a continuous RV is quite simple; we will discretize the domain into intervals (bins) and count how many observations/pinballs (data points) land on different intervals. This "binning" procedure is illustrated in Figure 1.9 and is summarized below. The binning procedure is significantly simpler for a discrete RV because the bins are the discrete values of the domain.

Empirical Estimates of Domain and Density Functions

1. Construct an *empirical domain* $\hat{\mathcal{D}}$; this is the domain covered by the data points x_ω, $\omega \in \mathcal{S}$. This gives us some initial indication of the actual domain $\mathcal{D}$ of the theoretical RV model (e.g., $(-\infty, \infty)$, $[0, \infty)$, $\{0, 1\}$, or $\{0, 1, \ldots, \infty\}$). For instance, if our observations are appearing as discrete values, then a Gaussian or an exponential model would not be appropriate (because such models assume a continuous domain).
2. If the empirical domain $\hat{\mathcal{D}}$ is continuous, discretize the domain into bins of size Δx and construct *the empirical pdf* as:

$$\hat{f}(x) = \frac{\hat{\mathbb{P}}(x \leq X \leq x + \Delta x)}{\Delta x} \tag{1.32}$$

$$= \frac{\frac{1}{S} \sum_{\omega \in \mathcal{S}} \mathbf{1}[x \leq x_\omega \leq x + \Delta x]}{\Delta x}, \tag{1.33}$$

where $S = |\mathcal{S}|$ is the number of observations. The total number of observations in the bin $[x, x + \Delta x]$ is:

$$\sum_{\omega \in \mathcal{S}} \mathbf{1}[x \leq x_\omega \leq x + \Delta x], \tag{1.34}$$

and is known as the *frequency*. The empirical pdf is the total number of observations x_ω that lie in the interval $[x, x + \Delta x]$ divided by the bin length and the number of observations.

3. If the empirical domain $\hat{\mathcal{D}}$ is discrete, construct *the empirical pdf* as:

$$\hat{f}(x) = \frac{1}{S} \sum_{\omega \in \mathcal{S}} \mathbf{1}[x_\omega = x], \tag{1.35}$$

where:

$$\sum_{\omega \in \mathcal{S}} \mathbf{1}[x_\omega = x], \tag{1.36}$$

is the *frequency*. The empirical pdf is the total number of observations x_ω that equals x divided by the number of observations.

4. Construct *the empirical cdf* as:

$$\hat{F}(x) = \frac{1}{S} \sum_{\omega \in \mathcal{S}} \mathbf{1}[x_\omega \leq x], \tag{1.37}$$

The empirical cdf counts the number of observations x_ω that take a value below x (normalized by S). The total number of observations below $x + \Delta x$ is given by:

$$\sum_{\omega \in \mathcal{S}} \mathbf{1}[x_\omega \leq x], \tag{1.38}$$

and is known as the *cumulative frequency*.

We highlight that the calculation of the empirical pdf for the continuous and discrete case is analogous but different because one needs to ensure consistency with the type of RV model considered. Note also that the computation of the empirical cdf for the continuous and discrete case uses the same formulas, and the only difference is the values of x that one considers (these are defined by the discretized domain in the continuous case and defined by the values of the domain in the discrete case). We also highlight that the empirical cdf $\hat{F}(x)$ is the empirical probability $\hat{\mathbb{P}}(X \leq x)$ (which approximates the theoretical probability $\mathbb{P}(X \leq x)$).

The empirical domain is a subset of the actual (theoretical) domain and thus $\hat{\mathcal{D}} \subseteq \mathcal{D}$; in other words, the actual domain contains observations that we might have not yet encountered. Note also that the empirical pdf/cdf are only defined for values x in the empirical domain $\hat{\mathcal{D}}$ (i.e., they cannot be computed for any values that have not been observed). As such, the empirical functions cannot be used to predict probabilities of outcomes that have not been observed. One of the main goals of having a theoretical pdf/cdf is precisely to be able to *predict the probability of outcomes that we have not observed*. Consequently, after computing the empirical pdf/cdf from data, we now proceed to find theoretical counterparts that match them. To do so, we will *tune* the parameters of the theoretical pdf/cdf to determine if a given value gives a good *agreement* with the empirical counterparts (i.e., the model fits the data well).

The procedure of matching an RV model to data is known as *estimation*. If the theoretical model fits our empirical/data-driven estimates sufficiently well, then we can say that we have sufficient evidence that makes us believe that the postulated RV model is correct (our hypothesis is correct). If the model does not fit the data-driven estimates, then we would need to postulate another model. A more formal and complete discussion of estimation procedures for RV models is presented in Chapter 4.

Example 1.5 Empirical and Theoretical Pdf/Cdf for Gibbs Reactor

Consider the equilibrium reactor discussed in Example 1.2. The pressure of the reactor varies due to random fluctuations of the control system, and we thus consider modeling this as an RV. We record the values of the pressure every hour.

In Figure 1.10 (a), we present 1,000 data points x_ω recorded (we have that S = 1,000). We find that these cover the range $[63.31, 234.91]$ bar, and we define this as our empirical domain $\hat{\mathcal{D}}$. This domain appears to be continuous, as the outcomes seem to take any value in this range. This seems to indicate that our theoretical RV model X should have a continuous domain $\mathcal{D}$. Note also that the pressure measurements gravitate around a middle pressure value of 150 bar and concentrate in a range of $[100, 200]$ (there are few observations that lie outside this range).

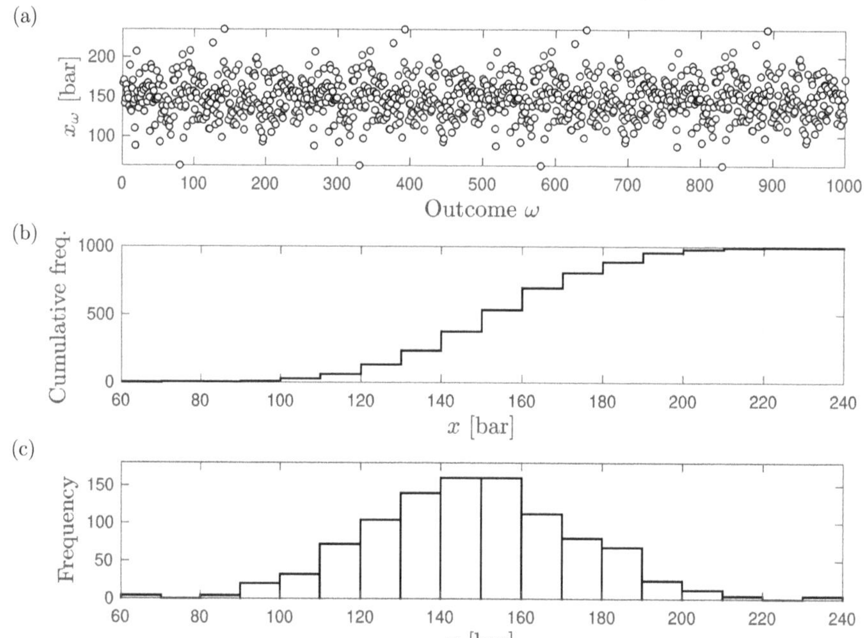

Figure 1.10 Outcomes (a), cumulative frequency (b), and frequency (c) for pressure.

We now proceed to compute the empirical cdf $\hat{F}(x)$; to do so, we define different values for the threshold that cover the empirical domain. We cover the domain $[60, 240]$

and split it into intervals of size $\Delta x = 10$ bar. We compute the cumulative frequency for values $x = 60, 70, \ldots, 240$ by counting the number of outcomes taking a value below a given threshold value. The cumulative frequency is shown in Figure 1.10 (b); from here, we can see that nearly all observations have a pressure value below 210 bar. We also see that few observations have a pressure value below 100 bar and above 210 bar; that is, it is quite unlikely to observe that pressure lies outside the range [100, 210] bar. Note, however, that this range is quite broad, indicating that pressure has quite a high variability. From the cumulative frequency, we compute the empirical cdf $\hat{F}(x)$ (by normalizing using the total number of data points $S = 1{,}000$).

To obtain the empirical pdf $\hat{f}(x)$, we first compute the frequency by using a bin size $\Delta x = 10$ bar; the results are shown in Figure 1.10 (c). This type of plot is known as a *histogram* and tells us how many times a data point falls in a given bin. Note that most outcomes are in the range [140, 160] (with a middle value of 150 bar); specifically, out of the $S = 1{,}000$ outcomes, around 300 of them are in such range. This is consistent with our previous observation that outcomes seem to gravitate around a value of 150 bar. We can see that the number of outcomes decreases rapidly in bins of both sides of this range, and there is a very small number of outcomes in the ranges [210, 240] and [60, 90]. These observations are consistent with those made with the cumulative frequency. The left and right "tails" of the histogram tell us that pressure is unlikely to take such extreme values, but this actually occurs from time to time. This is important because the reactor might face safety issues under such events. This is typical in applications, and extreme values are unlikely (rare) but can have a large impact when they show up. From the frequency, we compute the empirical pdf $\hat{f}(x)$ by normalizing using $S \cdot \Delta x = 1{,}000 \cdot 10$.

The empirical pdf and cdf are shown in Figure 1.11. The shape of the empirical pdf tells us that this is symmetric and has the shape of a "bell" curve. This type of bell shape is an exponential function that is captured by the theoretical Gaussian model (which has a continuous domain). We thus hypothesize that this is the theoretical model that the data follows, and we tune the parameters μ, σ to find the best bell curve that matches the empirical pdf. We find that the best values for the parameters are $\mu = 147.85$ bar and $\sigma = 25.38$ bar.

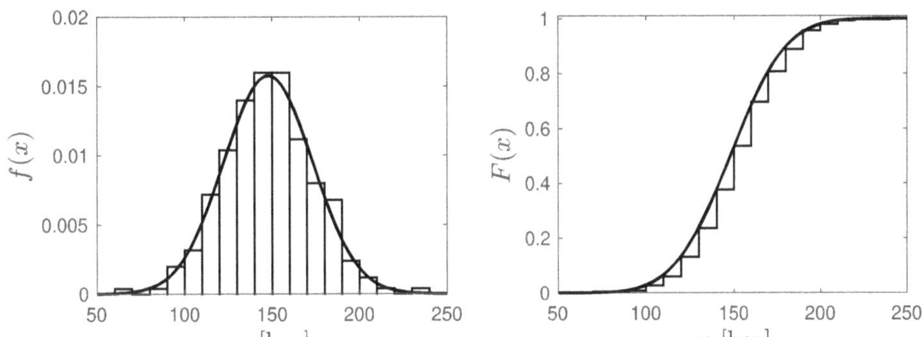

Figure 1.11 Empirical and theoretical pdfs and cdfs for reactor pressure.

In Figure 1.11, we also compare the empirical/theoretical density functions. We can see that the Gaussian model matches the data quite well; we thus conclude that our random phenomenon (pressure) behaves as a Gaussian RV. We can now use the theoretical model to analyze behavior and to make predictions about outcomes at values that we have not observed (e.g., those outside the range covered by the data).

Script to reproduce results: `ch1_gibbs_example` □

The theoretical RV model provides a characterization of our random phenomenon of interest. Having such a model has key implications:

- Having a model that matches the data indicates that the phenomenon under study is not arbitrary but follows a *trend*. Extracting trends (models) help us summarize the data (extract knowledge from data).
- If we know that the phenomenon follows a trend, then we can make predictions about possible outcomes and their respective probabilities. For instance, how likely is that the RV takes a certain value? We can also use the RV model to make predictions about outcomes that have not been observed before (are not recorded in our data). For instance, how likely it is that the reactor takes an extreme temperature that has not been seen before? Answering these types of questions is key in failure and safety analysis.
- Having the ability to predict probabilities of different outcomes is key in making decisions (we use acquired knowledge to make decisions). For instance, should we invest money to modify/retrofit our system so that this can withstand some rare (but extreme) events?
- Note that an RV model predicts the probability of outcomes and does not predict the outcomes themselves. This is a key difference from structural models, which we will explore later.

1.3.3 Summarizing Statistics

The pdf and cdf are *functions* that fully describe the RV X. However, in practice, we might be interested in using simple values (and not functions) to describe X. This is done by using quantities known as *summarizing statistics*. There is a wide range of summarizing statistics available; in fact, we will see that it is not difficult for you to design your own statistic. Different statistics capture different aspects or features of an RV that might be relevant to the application at hand. In other words, there is no universal statistic that is used in all applications. Summarizing statistics are critical quantities that are used for comparing RVs and for making decisions; later on, we will see that summarizing statistics are used as *measures of risk*.

1.3.3.1 Mean, Variance, and Standard Deviation

The mean and variance are popular statistics that are used to characterize (or describe) an RV. The *mean* (also known as expectation) of a discrete RV X is a scalar value $\mathbb{E}[X] \in \mathbb{R}$ (we use $\mathbb{E}_X$ for short) given by:

$$\mathbb{E}_X = \sum_{x \in \mathcal{D}_X} x \cdot f_X(x). \tag{1.39}$$

The *variance* of a discrete RV X is a scalar value $\mathbb{V}[X] \in \mathbb{R}_+$ (we use $\mathbb{V}_X$ for short) given by:

$$\mathbb{V}_X = \sum_{x \in \mathcal{D}_X} (x - \mathbb{E}_X)^2 \cdot f_X(x). \tag{1.40}$$

Associated with the variance is the standard deviation (SD), which is given by $\mathbb{SD}_X = \sqrt{\mathbb{V}_X}$. Note that the units of $\mathbb{E}_X$ and $\mathbb{SD}_X$ are the same as those of X (e.g., bar for pressure), but the units of $\mathbb{V}_X$ are not (e.g., bar^2). Because of this, the SD is often preferred.

The mean (expected value) is a measure of the magnitude of X, while the variance and SD are measures of variability or dispersion. The mean tells us something about how "large" is X by weighting different values x by the pdf $f_X(x)$. The variance and SD tell us how "large" the deviations of X are from its mean by weighting distances $(x - \mathbb{E}_X)^2$ with the pdf $f_X(x)$. The variance and SD are also quantities that are often associated with the amount of uncertainty that X has (i.e., the larger the variance, the more uncertainty that X has). As a result, the variance and the SD are often associated with *risk*.

For a continuous RV X, the mean and variance are given by:

$$\mathbb{E}_X = \int_{x \in \mathcal{D}_X} x \cdot f_X(x)dx \tag{1.41}$$

$$\mathbb{V}_X = \int_{x \in \mathcal{D}_X} (x - \mathbb{E}_X)^2 \cdot f_X(x)dx, \tag{1.42}$$

while the SD is again $\mathbb{SD}_X = \sqrt{\mathbb{V}_X}$.

Note that we require a theoretical pdf $f_X(\cdot)$ to compute the mean, variance, and SD. In other words, such quantities are theoretical (you need to have a model to compute them). But what if we do not have a model available? Moreover, even if you have a model, note that computing the mean and variance for a continuous RV requires conducting integration, which can be difficult or impossible to do. To address these limitations, we compute *empirical estimates for summarizing statistics* directly from available data x_ω, $\omega \in \mathcal{S}$. The empirical estimates for the mean and variance are:

$$\hat{\mathbb{E}}_X = \frac{1}{S} \sum_{\omega \in \mathcal{S}} x_\omega \tag{1.43}$$

$$\hat{\mathbb{V}}_X = \frac{1}{S} \sum_{\omega \in \mathcal{S}} (x_\omega - \hat{\mathbb{E}}_X)^2. \tag{1.44}$$

The empirical SD is $\hat{\mathbb{SD}}_X = \sqrt{\hat{\mathbb{V}}_X}$. The empirical mean and variance are also often referred to as sample or data-driven estimates. Note that the pdf/cdf are not needed to compute these estimates. The name "sample estimate" is used because we can generate realizations of the RV by *sampling an available model.* This sampling

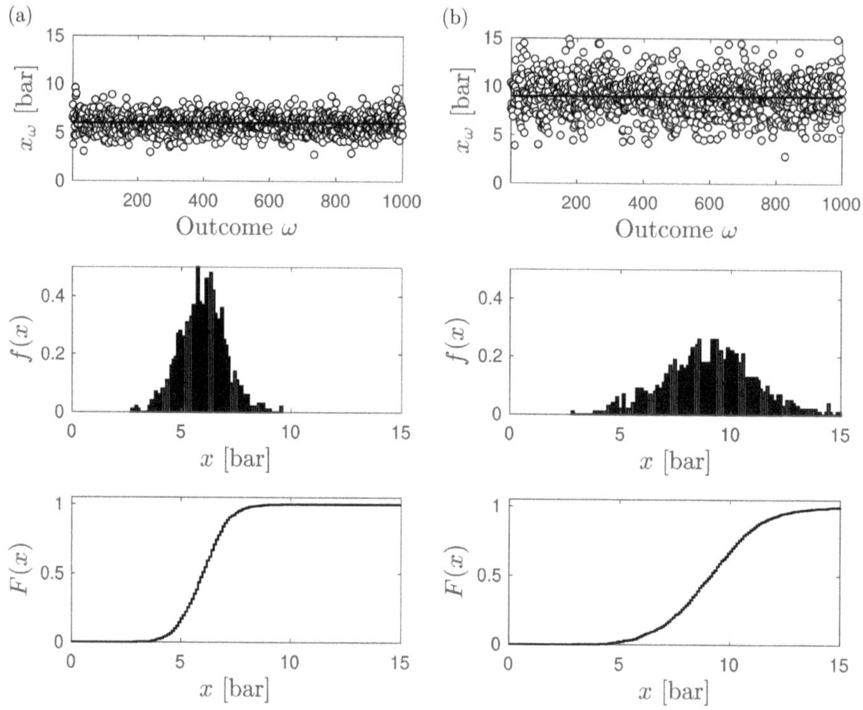

Figure 1.12 Datasets for RVs with $\mathbb{E}_X = 6$ and $\mathbb{V}_X = 1$ (a) and with $\mathbb{E}_X = 9$ and $\mathbb{V}_X = 2$ (b).

process is useful when one seeks to avoid computing integrals to obtain summarizing statistics; samples for an RV model can be obtained by using sampling algorithms implemented in numerical packages. We will discuss this topic in more detail in Chapter 4.

The empirical mean and variance are useful descriptors of data; specifically, these quantities allow us to summarize data and get a quick idea of the behavior of the random phenomenon. For instance, in Figure 1.12, we can see how the mean and variance give us a quick idea of the position and spread of the data. Note also how the mean and variance give us an idea of the shape of the pdf/cdf. For instance, the pdf and cdf spread out more over the domain as the variance increases. We will be exploring different types of descriptors that capture different features of the data and of the associated pdf/cdf.

It seems reasonable to hypothesize that the empirical estimates $\hat{\mathbb{E}}_X$ and $\hat{\mathbb{SD}}_X$ are good approximations of the theoretical statistics $\mathbb{E}_X$ and $\mathbb{SD}_X$. In Chapter 4, we will see that this is indeed the case, provided that our data is collected in appropriate ways (e.g., in a way that it reflects the inherent randomness of the system).

Example 1.6 Empirical and Theoretical Statistics for Reactor Pressure

We use the reactor pressure data to compute empirical estimates for the mean $\hat{\mathbb{E}}_X$ and SD $\hat{\mathbb{SD}}_X$ by using different amounts of data. Specifically, we compute such estimates using $S = 5, 10, 15, \ldots, 1{,}000$ observations. In Figure 1.13, we can see that the empirical mean and SD exhibit high variability when the sample size is small (S is below 100) but quickly converge (settle) at some limiting values. These limiting values coincide with the theoretical values (shown as solid lines). This illustrates that empirical estimates are expected to provide good approximations of the true statistics of the phenomenon.

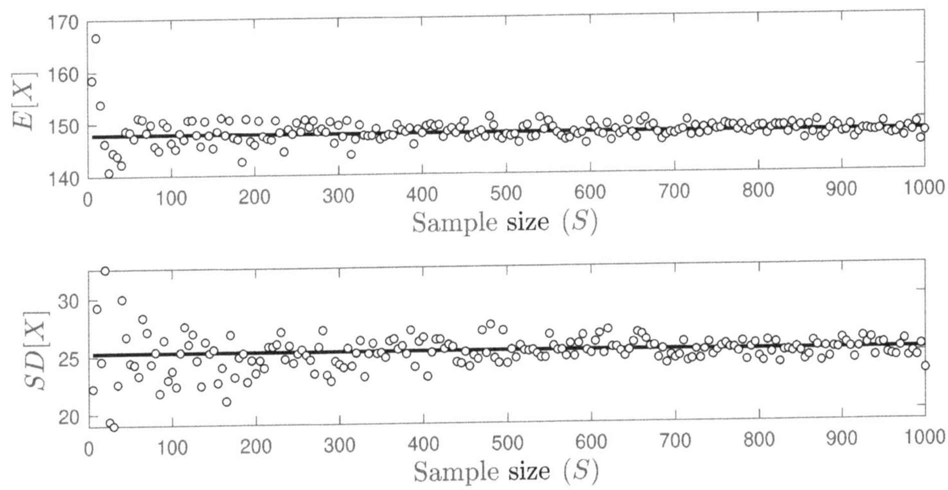

Figure 1.13 Convergence of empirical mean and standard deviation to their theoretical values as the number of observations (sample size) increases.

Script to reproduce results: `ch1_gibbs_example` □

In summary, we can see that data can be used not only to approximate pdfs/cdfs but also to approximate statistics of interest and that such approximations are expected to become increasingly accurate as we collect more data. This principle can be applied to different summarizing statistics such as quantiles and moments, which we discuss next.

1.3.3.2 Quantiles

An important family of summarizing statistics is the *quantiles* (a.k.a. percentiles). The quantile is the inverse function of the cdf, and as such, it is easier to think about quantiles from this perspective. Consider the following equation for a given probability level $\alpha \in [0, 1]$:

$$F_X(x) = \alpha. \tag{1.45}$$

This can also be written as $\mathbb{P}(X \leq x) = \alpha$. A threshold value x that satisfies this equation is known as the α-quantile of the RV X and is given by the scalar value $\mathbb{Q}_X(\alpha) \in \mathbb{R}$. We can thus express the quantile as:

$$\mathbb{Q}_X(\alpha) = F_X^{-1}(\alpha). \tag{1.46}$$

We make the following important observations about quantiles:

- The cdf of an RV can have a "staircase" form, and thus, there might be multiple values of x that satisfy $F_X(x) = \alpha$. Consequently, the α-quantile of X might not be unique; that is, multiple values of x satisfy Eq. (1.45). Typically, the definition of the quantile is refined by taking the smallest or center values of x satisfying $F_X(x) \geq \alpha$.
- Quantiles are related to other statistics of interest. For instance:
  - $\mathbb{Q}_X(1/2)$ is the center value of X; this is the threshold value x at which $\mathbb{P}(X \leq x) = \mathbb{P}(X > x) = \frac{1}{2}$ (that is, it is equally probable to find observations of X below and above this threshold). The center value is known as the *median* of X and is denoted as $\mathbb{M}[X] \in \mathbb{R}$ (or $\mathbb{M}_X$ for short notation).
  - $\mathbb{Q}_X(1)$ is the maximum value of X; in other words, we have that:

$$\mathbb{Q}_X(1) = \max_{x \in \mathcal{D}_X} x. \tag{1.47}$$

    This is the extreme outcome on the right side of the pdf.
  - $\mathbb{Q}_X(0)$ is the minimum value of X; that is, we have that:

$$\mathbb{Q}_X(0) = \min_{x \in \mathcal{D}_X} x. \tag{1.48}$$

    This is the extreme outcome on the left side of the pdf.
- As with other summarizing statistics, we can use data to compute empirical estimates for the quantiles, denoted as $\hat{\mathbb{Q}}_X(\alpha)$. This is done by using the empirical cdf $\hat{F}_X(x)$. We can then use these estimates to tune the parameters for the quantile of a theoretical RV model; this approach is known as quantile regression.

Example 1.7 Quantiles for Reactor Pressure

Empirical and theoretical (Gaussian) quantile functions for the reactor pressure are shown in Figure 1.14. Note that the quantile function is just a "flipped" version of the cdf function. Here, we compute the empirical cdf by using intervals of size $\Delta x = 5$ bar; note that this provides a better approximation of the theoretical cdf than that obtained with $\Delta x = 10$ bar (see Figure 1.11).

From the theoretical curve, we can see that the threshold value for pressure that gives us a cumulative density of 0.9 is 180 bar. Consequently, we say that the quantile $\mathbb{Q}(0.9)$ is 180 bar. We also see that the threshold value that gives us a probability of 0.5 (the median) is 148 bar and thus $\mathbb{M}_X = \mathbb{Q}(1/2) = 148$. It is equally likely to find the pressure below or above this middle value. Specifically, we have that $\mathbb{P}(X \leq 148) = \mathbb{P}(X > 148) = 0.5$. Note that the median coincides with the mean $\mathbb{E}_X = 148$; this is a property of certain RV models such as the Gaussian model used here (but it does not hold in general).

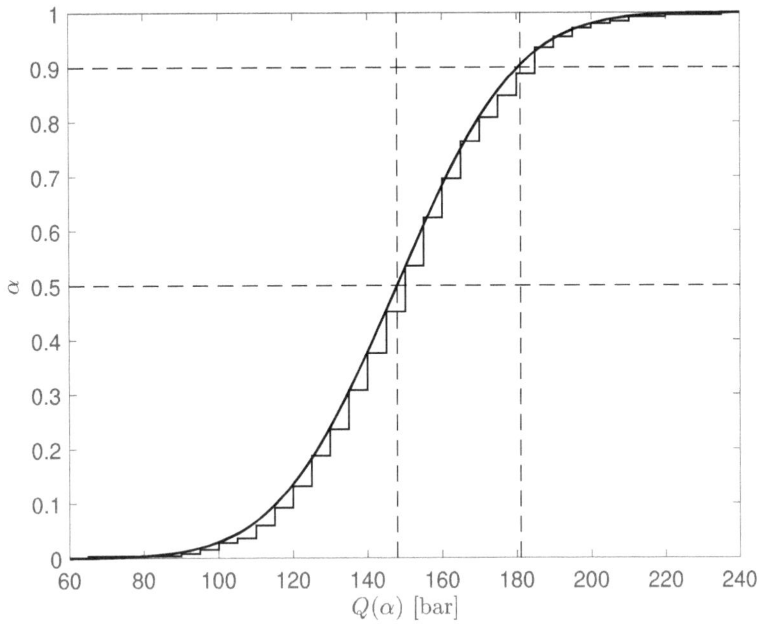

Figure 1.14 Empirical and theoretical quantiles for Gibbs reactor pressure.

Script to reproduce results: `ch1_gibbs_example` □

1.3.3.3 Mode and Moments

The *mode* of an RV is the outcome of maximum probability:

$$\mathbb{MO}[X] \in \underset{x \in \mathcal{D}_X}{\text{argmax}}\, f_X(x). \tag{1.49}$$

Here, the notation $\text{argmax}_{x\in\mathcal{D}_X} f_X(x)$ delivers a value of x in the domain $\mathcal{D}_X$ at which $f_X(x)$ takes its maximum value. For instance, for a Gaussian RV, this maximum value occurs at $x = \mu$; we thus have that the mode of a Gaussian RV is $\mathbb{MO}_X = \mu$. Note that there might be multiple values of x that maximize $f_X(x)$. Random variables that only have one maximum point are known as *unimodal* RVs (e.g., the Gaussian RV is unimodal); *multimodal* RVs are those that have pdfs with multiple maxima (sometimes with the same maximum value, as the peak might be flat).

Another important family of statistics is the *moments*; these quantities provide useful insights into the nature of X. Moments are particularly useful in characterizing the shape of a pdf. Empirical estimates of moments can also be computed from data and are often used to tune parameters of RV models; this approach is known as moment matching (we will discuss this in Chapter 4).

The *raw* moments of X are defined as:

$$\mathbb{RMO}_k[X] = \mathbb{E}[X^k],\ k = 1, 2, 3, 4, \ldots. \tag{1.50}$$

Note that the first raw moment of X (with $k = 1$) is the mean $\mathbb{E}_X$.

The *central* moments of X are given by:

$$\mathbb{CMO}_k[X] = \mathbb{E}[(X - \mathbb{E}_X)^k], \; k = 1, 2, 3, 4, \ldots. \tag{1.51}$$

Note that the second central moment of X (for $k = 2$) is the variance $\mathbb{V}_X$.

The *standardized* moments of X are given by:

$$\mathbb{SMO}_k[X] = \frac{\mathbb{E}[(X - \mathbb{E}_X)^k]}{(\mathbb{SD}[X])^k}, \; k = 1, 2, 3, 4, \ldots. \tag{1.52}$$

The third ($k = 3$) standardized moment is known as *skewness* and the fourth ($k = 4$) standardized moment is known as *kurtosis*. Skewness is a measure of symmetry of the pdf, while kurtosis is a measure of the nature of the tails of the pdf. For instance, a Gaussian RV has a skewness of zero since its pdf is symmetric.

Related to standardized moments is the *coefficient of variation*:

$$\mathbb{CV}[X] = \frac{\mathbb{SD}_X}{\mathbb{E}_X}. \tag{1.53}$$

The coefficient of variation captures a trade-off between the magnitude of X and its variability/uncertainty. This statistic is useful in assessing investment options because, in that setting, we must balance the expected (average) return and uncertainty of the return.

Summarizing statistics allow us to capture different aspects of RVs, which manifest in the shape of the pdf/cdf. For instance, in Figure 1.15, we show how different summarizing statistics capture different aspects of a pdf. In this case, for instance, the pdf is asymmetric (it is skewed right); this manifests in having the mode to the left of the median and the median left to the expected value. A symmetric pdf (such as the Gaussian pdf) has a mode, median, and mean that coincide. An asymmetric pdf that is skewed to the left has a mode that is on the right of the median and a median that is on the right of the mean. The amount of skewness can be quantified using the third central moment. An asymmetric pdf also has a tail that is longer than the other; capturing such tails is important, as they reflect the presence of rare (but extreme) events. The variance captures the overall spread of the pdf (e.g., lower variance has narrower pdfs); we will see later, however, that the variance has difficulty in characterizing the shape of asymmetric pdfs. The quantiles are flexible quantities that capture effects at different probability levels; for instance, as $\alpha \to 1$, the quantile focuses on extreme events on the right tail, and as $\alpha \to 0$, the quantile focuses on extreme events on the left tail. In Chapter 7, we will see that summarizing statistics can be used for quantifying different types of risk and for making decisions. Moreover, we emphasize that all of the empirical estimates for all of these summarizing statistics can be obtained from data (e.g., we obtain empirical quantiles by using the empirical cdf). In Chapter 2, we will see that this is important because empirical estimates of summarizing statistics can be used to determine parameters of a model.

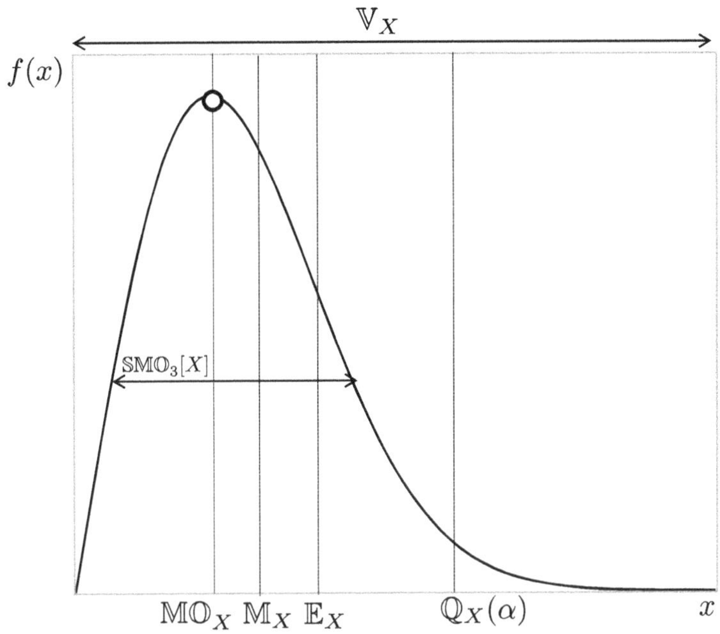

Figure 1.15 Features of the probability density function that different summarizing statistics capture.

1.3.3.4 Event Probabilities

Probabilities of specific events are summarizing statistics and are widely used in decision-making. An event probability can be written in the general form $\mathbb{P}(X \in \mathcal{A})$, where $\mathcal{A} \subseteq \mathcal{D}$ is the event of interest. For instance, imagine that you want to find out the probability that the conversion of a reactor (a continuous RV X) is in the range $\mathcal{A} = [a_1, a_2]$, which is a subdomain of the domain $\mathcal{D}$; the probability is written as $\mathbb{P}(X \in [a_1, a_2])$ or $\mathbb{P}(a_1 \leq X \leq a_2)$. This probability can be computed as:

$$\mathbb{P}(a_1 \leq X \leq a_2) = \int_{x \in [a_1, a_2]} f(x)dx. \tag{1.54}$$

Computing this probability requires us to know the pdf $f(x)$ and requires the computation of an integral (potentially complex). If we have data $x_\omega, \omega \in \mathcal{S}$, we can overcome these limitations by computing an empirical estimate:

$$\hat{\mathbb{P}}(a_1 \leq X \leq a_2) = \frac{1}{S} \sum_{\omega \in \mathcal{S}} \mathbf{1}[a_1 \leq x_\omega \leq a_2]. \tag{1.55}$$

In other words, we simply count how many of our data points satisfy the condition that $x_\omega \in [a_1, a_2]$. In Chapter 3, we will see that we can use a similar procedure to compute empirical probabilities for events $\hat{\mathbb{P}}(X \in \mathcal{A})$ that capture *decision-making logic*. For instance, imagine that you want to determine the probability that conversion is either lower than a_1 or higher than a_2; this can be computed as:

$$\hat{\mathbb{P}}(X \leq a_1 \text{ or } X \geq a_2) = \frac{1}{S} \sum_{\omega \in \mathcal{S}} \mathbf{1}[x_\omega \leq a_1 \text{ or } x_\omega \geq a_2]. \tag{1.56}$$

Event probabilities are widely used as *measures of risk* (e.g., compute the probability that an investment yields a positive profit or the probability that a reactor fails).

1.3.4 Structural Models

Statistics offers tools to *discover* structural relationships between system variables from data. Such structural relationships can be captured by data-driven empirical models, mechanistic models, or combinations of both (known as hybrid models). Relationships captured by structural models embed *knowledge* and are needed in engineering to make predictions and decisions.

Structural models are different from RV models in that they aim to extract predictable behavior from data (as opposed to characterizing purely random/unpredictable behavior). For instance, we might want to develop a model that allows us to predict reactor conversion from temperature and pressure. The ability to make such predictions is hindered by uncertainty; this uncertainty arises from our inability to perfectly measure variables (e.g., sensor errors) or due to exogenous factors that are not under our control (e.g., energy prices and molecular fluctuations). The main statistical tools used for developing structural models from data are *experimental design* and *estimation*. These concepts will be discussed in detail in Chapter 5.

To explain the main concepts behind the development of structural models, we consider a system model of the form:

$$Y = \varphi(u, \theta) + \epsilon. \tag{1.57}$$

Here, $\varphi(u, \theta)$ is the structural model that captures the *relationship* between a deterministic input u (e.g., operating conditions) and a random output Y (e.g., reactor conversion). The structural model allows us to predict the output using input data; this is done by exploiting systematic *trends* that arise because of correlations between input and output variables. We will explore the topic of correlations in Chapter 3. The deterministic input u are variables that we can control and are thus not subject to uncertainty (they are decisions). The output Y is an RV that is given by the model prediction $\varphi(u, \theta)$, which is contaminated by *"noise"* ϵ; the noise captures unknown/unpredictable aspects of the system that cannot be captured by the structural model.

The variables θ are *tunable* parameters of the structural model; we are interested in tuning these parameters in order to extract/separate the trend $\varphi(u, \theta)$ by filtering out random noise ϵ. To do so, we collect data for (u, Y) and consider that such observations are related as:

$$y_\omega = \varphi(u_\omega, \theta) + \epsilon_\omega, \ \omega \in \mathcal{S}. \tag{1.58}$$

The trend is extracted by solving a parameter estimation problem that aims to find parameter values that best match the structural model to the output data; in other words, we want the residual error $\epsilon_\omega = y_\omega - \varphi(u_\omega, \theta)$ to be as small as possible. As

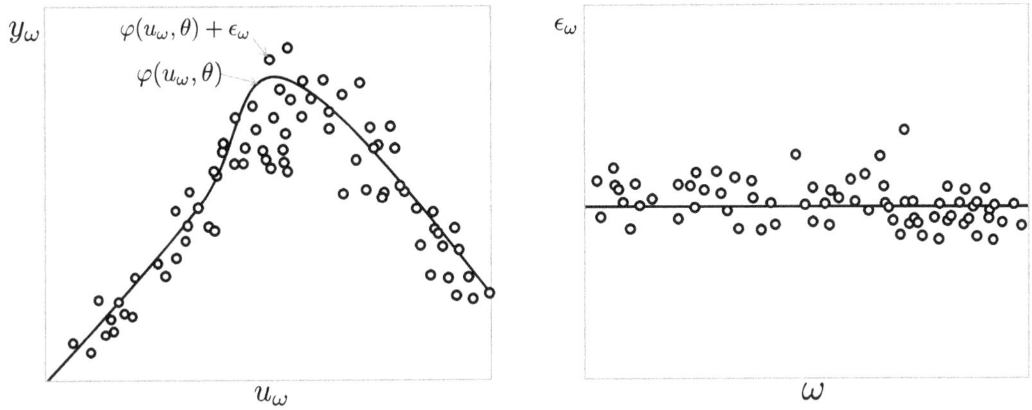

Figure 1.16 Goal of the estimation process is to separate the knowledge (a model that captures predictable trends) embedded in the data from unknown effects (random noise). If the process is effective, the residual should be a purely random effect (with no trend).

such, our goal is to separate the random/unknown effects (captured by the noise ϵ) from the predictable effects (captured by the model). In other words, the estimation process aims to extract knowledge (the model or the trend) from the data. This separation process is illustrated in Figure 1.16.

The selection of input conditions u plays a key role in obtaining response data for Y that helps us separate the model from the noise. Specifically, we want to manipulate our input conditions to *explore* a suitable space over which we want the model to be able to predict behavior. Moreover, we want conditions that sufficiently *excite* the system such that we can uncover variable relationships (the trend) of interest. In other words, the selection of the inputs u can help us *gain useful information* about the system. The selection of inputs that maximize information content is at the core of an area of statistics known as the *design of experiments*. Once we have obtained a structural model, it will be possible for us to *exploit* this knowledge to find decisions u that optimize the output Y (e.g., minimize cost or maximize efficiency). As we will see next, however, we need to be *careful* in capturing the uncertainty of any random (uncontrollable) factors ϵ that affect the system, as such uncertainties will introduce randomness in the output Y (and will thus limit our ability to perfectly predict and control system behavior). The use of decisions u to gain information and/or optimize performance is an important goal of human *decision-making*. These topics are introduced in the next sections and are discussed in detail in Chapters 5 and 7.

1.4 From Models to Decisions

Imagine that we have a characterization of a given random phenomenon (an RV model X) that affects a system of interest. In addition, we have a structural model that captures how this random input (uncontrollable factors such as noise or unpredictable effects) X and a deterministic input (controllable decisions) u impact a system output Y. We use the compact notation $Y = \varphi(X,u)$ to denote the structural model relating the deterministic and random inputs to the output. We would like to exploit this model to make decisions that manipulate Y in desirable ways.

In this context, it is important to make a couple of important observations:

- Uncertainty/variability *propagates* through systems in complex ways; in other words, uncertainty in one variable (e.g., sensor errors and market prices) will result in uncertainty in any variable that depends on it (e.g., conversion and profit).
- Uncertainty/variability can be *mitigated* by manipulating the system (e.g., via design or control decisions).

1.4.1 Uncertainty Propagation

The system model $Y = \varphi(X,u)$ captures how the input uncertainty of X propagates through the system to generate uncertainty in the output Y (see Figure 1.17). We make the following observations:

- The output Y is an RV because X is an RV.
- The nature of Y (its cdf, pdf, and domain) depends on the system φ. Some systems magnify uncertainty (e.g., a sensitive system such as an exothermic reactor), while others dampen it (e.g., a passive system such as a storage tank).
- The nature of Y depends on the decision u; this means that we can use u to control (e.g., damp or magnify) the uncertainty of Y. The decision u is a deterministic variable (because we decide its value); we use lowercase u (as opposed to U) to highlight this fact.

As an example, consider the propagation of uncertainty of wind power (obtained from wind turbines) that is injected into the power grid. The uncertain power inputs X will introduce uncertainty in the operating cost of the grid Y; however, the grid has available variables u (e.g., generation from other power plants) that can be used to counteract wind power uncertainty and keep cost under control.

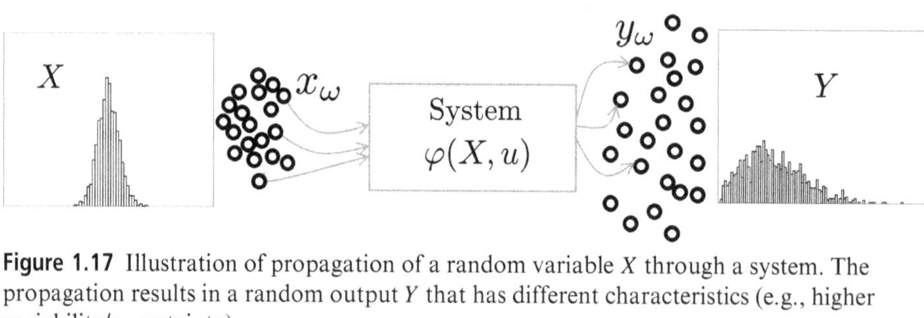

Figure 1.17 Illustration of propagation of a random variable X through a system. The propagation results in a random output Y that has different characteristics (e.g., higher variability/uncertainty).

To make predictions and decisions, it is necessary to characterize the uncertainty of the system output Y, which is influenced by the uncertainty of X and by the decision u. Having data x_ω, $\omega \in \mathcal{S}$ (realizations of X), a system model φ, and a decision u, we can characterize the cdf, pdf, and domain of Y by using *simulations*, as illustrated in Figure 1.17. Specifically, consider the procedure outlined below.

Monte Carlo simulation

1. For a given decision u and for realizations x_ω, $\omega \in \mathcal{S}$, perform simulations of the form:

$$y_\omega = \varphi(x_\omega, u), \; \omega \in \mathcal{S}. \tag{1.59}$$

2. Use the output data y_ω, $\omega \in \mathcal{S}$ to compute empirical approximations for Y such as the empirical mean and variance:

$$\begin{aligned} \hat{\mathbb{E}}_Y &= \frac{1}{S}\sum_{\omega \in \mathcal{S}} y_\omega \\ \hat{\mathbb{V}}_Y &= \frac{1}{S}\sum_{\omega \in \mathcal{S}} (y_\omega - \hat{\mathbb{E}}_Y)^2, \end{aligned} \tag{1.60}$$

and the empirical cdf:

$$\hat{F}_Y(y) = \frac{1}{S}\sum_{\omega \in \mathcal{S}} \mathbf{1}[y_\omega \le y]. \tag{1.61}$$

The above procedure is known as *Monte Carlo (MC) simulation* and is widely used to estimate diverse quantities of interest for RVs. For example, the above procedure computes empirical estimates for the expected value, variance, and cdf of Y, but one can also use it to estimate the pdf, quantiles, moments, variance, probability of events, and so on.

The properties of estimates obtained with MC will be covered in detail in Chapter 4. For example, we will see that the estimates $\hat{\mathbb{E}}_Y$ and $\hat{\mathbb{V}}_Y$ are good approximations of the *actual* expected value and variance of Y; for continuous Y, these are given by:

$$\mathbb{E}_Y = \int_{x \in \mathcal{D}_X} f(x) \cdot \varphi(x,u) dx \tag{1.62a}$$

$$\mathbb{V}_Y = \int_{x \in \mathcal{D}_X} f(x) \cdot (\varphi(x,u) - \mathbb{E}_Y)^2 \, dx. \tag{1.62b}$$

Here note that, in order to compute $\mathbb{E}_Y$ and $\mathbb{V}_Y$, we need to know the pdf $f(x)$ of X; moreover, when the RV is continuous, we need to perform the integration of a potentially complex function $\varphi(X,u)$. MC is a powerful and straightforward approach that overcomes these limitations; specifically, MC can compute estimates of the mean and variance by simply propagating realizations of X through the model $\varphi(X,u)$. Importantly, these realizations can be actual available data for X or can be samples of the RV model of X obtained via a sampling algorithm.

MC is a *highly versatile computational procedure* that can be used to understand how uncertainty propagates through a complex system, but it can also be seen as a method to conduct "sensitivity analysis" or as a method for conducting numerical integration (i.e., we can avoid integration when computing summarizing statistics by using MC). The characterization of the propagation of uncertainty in complex systems is studied in an area of statistics known as *uncertainty quantification*.

Example 1.8 Uncertainty Propagation in Gibbs Reactor

We want to understand how uncertainty in pressure generates uncertainty in the conversion. We thus model pressure as the input RV X and the conversion as the output RV Y and the decision/control u is the temperature at which we operate the reactor. The structural model $\varphi(X,u)$ is the equilibrium model (1.1)–(1.3).

We use the MC simulation procedure to obtain the empirical pdfs/cdfs for the input and output, assuming that the reactor is operated at high temperature. The results are shown in Figure 1.18; as we have seen, X behaves as a Gaussian RV, but note that Y does not seem to behave as a Gaussian RV. Specifically, note the emergence of a second peak in the pdf of the conversion (indicating that the pdf is no longer unimodal and does not have a "bell" shape). The complex behavior of the conversion pdf is the result of the strong nonlinear behavior of the reactor (i.e., the conversion depends on the pressure in a complex manner). This also highlights that RVs can have complex pdfs.

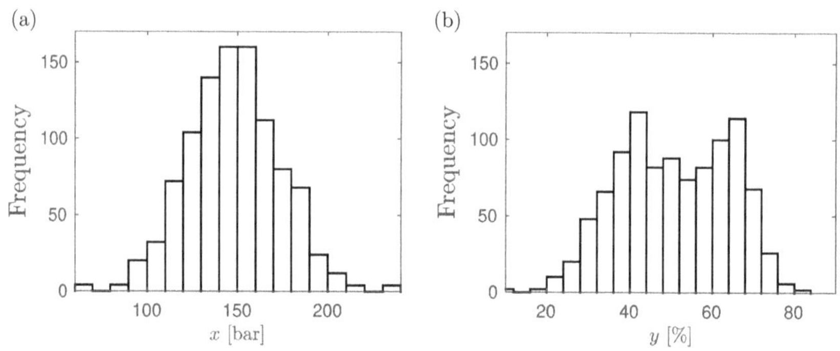

Figure 1.18 Empirical pdfs for reactor pressure (a) and conversion (b).

Script to reproduce results: `ch1_gibbs_example` □

1.4.2 Decision-Making under Uncertainty

Consider now that we would like to find a decision u that manipulates the output $Y = \varphi(X,u)$ in some desirable way. For convenience, we highlight the dependence of Y on the deterministic decision u explicitly by using the simplified notation $Y(u)$. Recall that $Y(u)$ is an RV for any choice of u because the output depends on X. Our goal is to answer the following questions:

- If we have candidate decisions u and u' with associated random outputs $Y(u)$ and $Y(u')$, how can we tell which decision is better?
- How can we find the best decision u that maximizes/minimizes the random output $Y(u)$?

These types of questions are easy to answer under a *deterministic view of the world* but are notoriously difficult to answer under a statistical view of the world; specifically:

- In a *deterministic* setting (which ignores uncertainty), the outputs $Y(u)$ and $Y(u')$ will each take a single known value (denoted as $y(u)$ and $y(u')$). Consequently, we would select, unambiguously, the one with a larger/smaller value. Specifically, if we are trying to minimize the output (e.g., cost), then we simply select u over u' if $y(u) \leq y(u')$.
- In a *statistical* setting (which accounts for uncertainty), the previous reasoning is no longer possible because $Y(u)$ and $Y(u')$ have multiple possible outcomes and probabilities. Consequently, the concept of "better" when comparing decisions is not clearly defined (it is ambiguous). In fact, the mathematical statement $Y(u) \leq Y(u')$ does not make sense! For instance, does $Y(u) \leq Y(u')$ mean that all the outcomes of $Y(u)$ are lower than those of $Y(u')$? Does it mean that the outcomes of $Y(u)$ are almost always lower than those of $Y(u')$?

Ideally, we would like to select a decision by using all the information contained in the pdfs/cdfs of $Y(u)$ and $Y(u')$ because these fully characterize the RVs. This paradigm is known as *stochastic dominance*. However, in practice, decision-makers prefer to deal with values (and not functions) when comparing RVs. These values are also known as *risk measures*, and these are nothing but *summarizing statistics* of $Y(u)$ (e.g., mean, variance, and quantiles). Risk measures are key in comparing RVs and in making decisions; for instance, such values are used to determine the uncertainty associated with a given decision. In Chapter 7, we will see that there are many risk measures that one can use to guide decisions (as we have seen, we have many types of summarizing statistics). The reason for this is that decision-makers (humans) take different attitudes toward uncertainty. For instance, some humans are less tolerant of risk than others (while some others love risk). This makes the definition of risk subjective (just like fairness or happiness). We will discuss strategies to define risk measures that enable *consistent* (proper) comparisons between RVs.

Example 1.9 Decision-Making under Uncertainty for Gibbs Reactor

Consider again the Gibbs reactor model as $Y(u) = \varphi(X,u)$ with pressure as input X and conversion as output Y. In this reactor, we have the ability of controlling the reactor temperature u. Specifically, the temperature can be set at a low level (583 K) or high level (613 K) to counteract the uncertainty (variability) in pressure X.

In Figure 1.19, we compare the empirical pdf/cdf for the conversion at low $Y(u)$ and high $Y(u')$ temperatures. The question that we address here is: Should we operate at low or high temperature? This is not necessarily an easy question because it depends on what we want to achieve. For instance, perhaps we are interested in maximizing the conversion or perhaps we are interested in controlling its variability (i.e., we want to stabilize conversion).

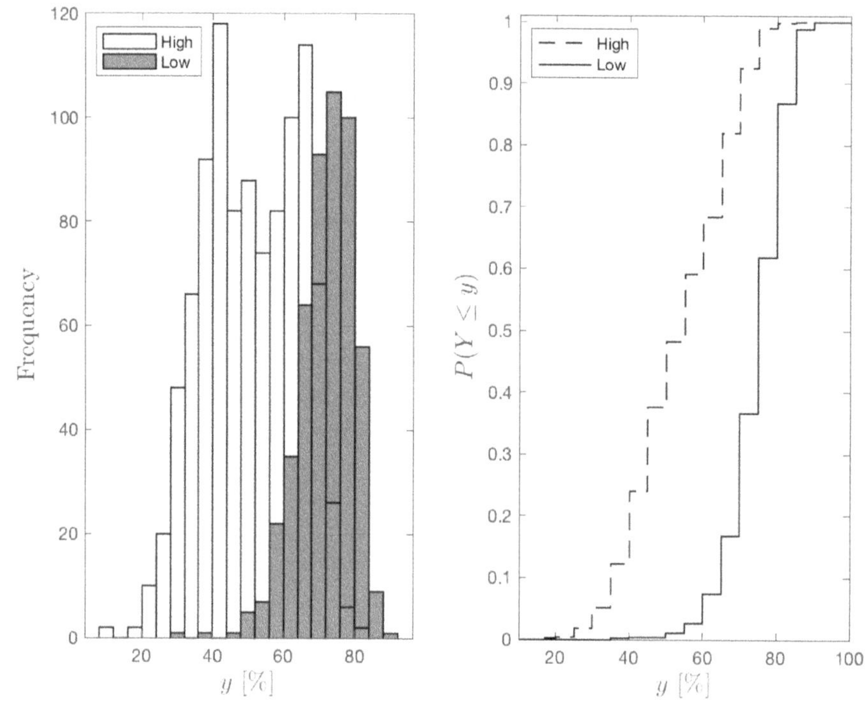

Figure 1.19 Effect of reactor operating temperature on conversion.

If we are interested in maximizing the conversion, it is natural to select the temperature that gives the highest average conversion. If we follow this logic, we should operate at a low temperature. This makes physical sense because the reaction is exothermic and thus operating at a low temperature increases conversion. As we can see for the pdf, lowering the temperature has the effect of shifting the pdf to the right (toward larger values of conversion). The mean conversion at low temperature is around 70%, while at high temperature, it is around 50%. Note also how the pdf at low temperature is skewed. From the cdf, we see that lowering the temperature also shifts the cdf to the right; this indicates that there is a higher probability of finding the conversion at a higher value if we operate at low temperature. For instance, the cdf tells us that there is virtually a zero probability of seeing a conversion below 40% if we operate at low temperature, while the probability of seeing a conversion below 40% at high temperature is 25%.

If we are interested in stabilizing/controlling the conversion, it is natural to select the temperature that gives the smallest variance (or SD). From the pdf, it is obvious that the pdf for low temperature covers a narrower range (40–85%) than for high temperature (10–85%), indicating that the conversion sees significantly lower variability. In other words, by operating at low temperature, we are more certain about where the conversion will lie. By comparing the cdfs, we can see that the lower variability for the conversion at low temperature results in a sharper increase in the cdf (compared to the cdf at high temperature). Specifically, for low temperature, the probability moves from zero to one over a range of 45–85%, while, for high temperature, the probability moves from zero to one over a range of 20–80%. We thus conclude that operating at low temperature both increases the mean conversion and reduces its uncertainty/variability. In more complicated examples, we will see that making a decision is less obvious. For instance, it might be possible that an action increases the mean but also increases uncertainty.

Script to reproduce results: `ch1_gibbs_example` □

The previous example illustrates how one can control uncertainty by making strategic decisions (e.g., operating a system differently). Specifically, decisions have the effect of shaping the pdf/cdf of the output of interest, which can make the system less susceptible to uncertainty (e.g., it protects the system). For instance, we might want to select operating conditions for a chemical process that reduce variance.

In the context of decision-making, we might not only be interested in comparing decisions but might actually want to find the best possible decision (see Figure 7.1). This is a topic that is explored in the field of *stochastic optimization*; we will study this in Chapter 7. To decide what is "best" or "optimal," we need to select a proper characterization of the output $Y(u)$ (a summarizing statistic) that we denote as $\rho(Y(u))$. With this, we can aim to shape the pdf of the output $Y(u) = \varphi(X,u)$ by solving the optimization problem:

$$\min_{u \in \mathcal{U}} \rho(Y(u)). \tag{1.63}$$

Here, $\mathcal{U}$ is the set of possible decisions that we can choose from, and this can be continuous or discrete. For instance, in the Gibbs reactor, the set is discrete because we can only operate at a couple of temperatures and is given by $\mathcal{U} = \{583, 613\}$. If we select $\rho(Y(u)) = \mathbb{SD}[Y(u)]/\mathbb{E}[Y(u)]$, the optimization problem (1.63) aims to find the best decision u^* such that the output $Y(u^*)$ minimizes the *coefficient of variation*. This will have the effect of minimizing the variability of $Y(u)$ (making the numerator small reduces $\rho(Y(u))$) while at the same time maximizing the average magnitude of $Y(u)$ (making the denominator large reduces $\rho(Y(u))$). For instance, in the Gibbs reactor, operating at a low temperature yields a coefficient of variation for the conversion of 0.07, while operating at a high temperature yields 0.10; we thus conclude that the optimal decision is $u^* = 583$ K.

In Chapter 7, we will use different summarizing statistics as risk measures that aim to shape the pdf of the output in different ways; for instance, the coefficient of variation aims to simultaneously shift the mean and reduce variance. It is possible to propose different optimization problems that use different risk measures to manipulate the shape of $Y(u)$ (e.g., quantiles, moments, or event probabilities). For instance, one might want to minimize the probability that the conversion is below a certain value (minimize the risk of low conversion). We will also see how, in some situations, one might want to make decisions that intentionally put the system at risk (in order to gain information). This mimics how humans tend to "push the boundaries" in order to learn.

1.5 Summary

This chapter provides an end-to-end introduction to statistics; the following are important take-home messages:

- Statistics is a branch of mathematics that provides tools for characterizing random behavior of systems (behavior that cannot be fully predicted and therefore that is unknown).
- Random variables are the fundamental modeling abstraction that statistics uses to model random behavior; the key elements of an RV model are its domain and its probability and cumulative density functions.
- There are different types of RVs (univariate, multivariate, discrete, and continuous) that can be used to model different types of phenomena.
- Summarizing statistics are quantities that allow us to capture key features of an RV; some examples include the mean, variance, quantiles, and the mode. These summarizing statistics facilitate decision-making (e.g., comparing RVs) and are often used as risk measures.
- We can compute empirical estimates for the density functions and for the summarizing statistics of RVs directly from available data; these estimates approximate the theoretical values of an RV and become increasingly accurate as we accumulate data. Empirical estimates can also be used to determine parameters for RV models.
- Statistics provides tools to understand connections between variables and to develop structural models that capture such connections; these structural models capture predictable behavior (knowledge) about a system.
- Uncertainty/variability contained in an RV will propagate through a system to generate uncertainty/variability in system outputs. One can often mitigate the output uncertainty via decisions (e.g., system design or control); shaping the density function of system outputs in desirable ways is at the core of decision-making under uncertainty.
- This chapter presents a *data–models–decision pipeline* (using data to derive models and using models to make decisions), which is typically encountered in engineering practice. We often start from data, and from these data, we develop

models to characterize predictable behavior (structural models) and unpredictable behavior (uncertainty models). We use a combination of these models to make decisions to obtain outcomes of interest (e.g., minimize risk or gain information).

1.6 Exercises

Exercise 1.1. What Is Known and Unknown about a Heat Exchanger?
Consider the countercurrent heat exchanger shown below. Here, a "hot" stream with inlet temperature T_{in} is cooled down to reach outlet temperature T_{out}; this is achieved by using a "cold" stream that is fed into the outer shell of the exchanger at inlet temperature t_{in}, and this agent leaves the system at temperature t_{out}.

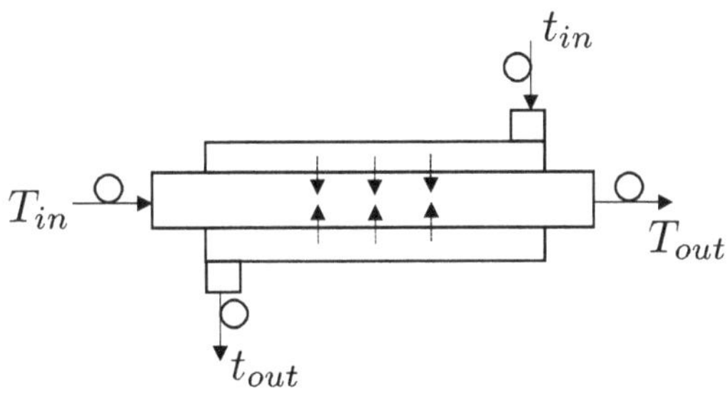

Address the following questions:

(a) Describe what is known about this system based on your intuition. Describe your knowledge as general logical descriptions (no equations). For example, we know that if a given variable increases, then a given output variable decreases, we know that a given temperature cannot be higher/lower than another one, we know that these factors affect the heat transferred between fluids, and so on.
(b) Describe what is known about this system as a physical model that connects the inputs and outputs of the system to the total heat transferred between the streams. Describe *all* variables present in the physical model (not only the input and output variables but also any intermediate variables or parameters). Explain what assumptions are made by the proposed model.
(c) Are the values for all variables and parameters of your model known? If not, how would you get their values (e.g., sensor, experiments, and formulas)? If you need formulas to get some variables or parameters, state them and describe their origin (e.g., fundamental laws or empirical correlations).
(d) Is this model mechanistic or empirical or a combination of both? If a combination, describe which aspects are mechanistic and which are empirical.
(e) Imagine that you have a real heat exchanger in a laboratory or a manufacturing facility. Could your model perfectly predict its behavior? Describe what aspects of the system are known and unknown (captured and not captured by your model).

(f) If your model cannot perfectly predict behavior, how can statistics help you characterize unknown behavior?

Exercise 1.2. Energy Retrofit of a Building.
The air handling unit (AHU) of a building is a piece of equipment that is responsible for conditioning ambient air that is supplied to a building (it controls air humidity and delivery temperature). The AHU of an engineering building (let's call this Engineering Hall) is at least 30 years old; its efficiency has degraded and is consuming too much energy.

You hire an energy company to replace the AHU; after analyzing data on the weekly energy consumption of the old AHU for the last year, the company promises that the new unit will yield energy savings. To verify that this is indeed the case, the company records the weekly energy consumption for an entire year after installing the new AHU.

Energy use data can be found in `energy_retrofit.dat`; the first column contains the weekly energy consumption for the old AHU and the second column is the weekly consumption for the new AHU (in MWh). Address the following:

(a) Use the data to determine summarizing statistics for energy use (mean, variance, SD, and mode); report the units of these quantities. By comparing the statistics for the old and new AHUs, determine if the new AHU is performing better (or not) and in what sense.
(b) Obtain the frequency and cumulative frequency functions for both AHUs and compare them. From these functions, determine if the investment in the new unit is justified.
(c) Obtain the empirical pdf and cdf for both units. Verify that these follow theoretical Gaussian models and tune the parameters for such models so that the theoretical pdf/cdf match the empirical pdf/cdf. Verify that the mean, variance, SD, and mode for the theoretical models are close to the previously computed values from data.
(d) Due to weekly budget limitations, it is unacceptable that the AHU consumes more than 20 MWh per week. What are the probabilities that the old and new AHUs have an acceptable energy consumption? Which AHU has the most frequent violations of budget constraints?

Exercise 1.3. Exponential RV Model.
The exponential RV model is widely used in survival analysis (e.g., the RV represents the time to wait until an asset fails). This model has a pdf of the form:

$$f(x) = \frac{1}{\beta} e^{-\frac{x}{\beta}},$$

with domain $\mathcal{D}_X = [0, \infty)$ and with parameter $\beta \in \mathbb{R}_+$.

Address the following:

(a) Determine the cdf $F(x)$ associated with the pdf $f(x)$.
(b) Verify that the pdf and cdf satisfy the following properties (which are required for a continuous RV model):
  - $f(x) \geq 0$ for all $x \in \mathcal{D}_X$.
  - $0 \leq F(x) \leq 1$ for all $x \in \mathcal{D}_X$.
  - $F(\infty) = 1$.
  - $f(x) = \frac{dF(x)}{dx}$.

Exercise 1.4. A la Mode.
The mode $\mathbb{MO}[X]$ of an RV X is an important summarizing statistic that tells us the most likely value(s) that the RV can take. The mode can be determined from the pdf $f(x)$ by finding value(s) of x that maximize it. From basic calculus, we know that a value x^* that maximizes $f(x)$ satisfies the conditions:

$$\frac{df(x^*)}{dx} = 0 \quad \text{and} \quad \frac{d^2 f(x^*)}{dx^2} \leq 0.$$

Moreover, if $\frac{d^2 f(x^*)}{dx^2} < 0$, we know that x^* is the unique value that maximizes $f(x)$; if $\frac{d^2 f(x^*)}{dx^2} = 0$, we have that x^* is not unique (there are other values that maximize the pdf). These conditions allow us to determine the mode analytically from the pdf.

Determine the mode for the following RVs analytically; for each RV, determine conditions under which the mode is unique (there is only one value maximizing the pdf). Comment on how the mode is related to the parameters of the RV. Plot the pdfs (select arbitrary values of the parameters) and verify the location of the mode based on the value determined analytically.

Hint: A point x^* that maximizes $f(x)$ also maximizes $\log f(x)$.

(a) Gaussian:

$$f(x) = \frac{1}{\sqrt{2\pi\sigma^2}} e^{\frac{-(x-\mu)^2}{2\sigma^2}},$$

with domain $\mathcal{D}_X = (-\infty, \infty)$ and parameters $\mu \in \mathbb{R}$, $\sigma \in \mathbb{R}_+$.

(b) Gamma:

$$f(x) = \frac{1}{\beta^\alpha \Gamma(\alpha)} \cdot e^{-x/\beta} \cdot x^{\alpha-1},$$

with domain $\mathcal{D}_X = [0, \infty)$ and parameters $\alpha \in \mathbb{R}_+$, $\beta \in \mathbb{R}_+$.

(c) Exponential:

$$f(x) = \frac{1}{\beta} e^{-x/\beta},$$

with domain $\mathcal{D}_X = [0, \infty)$ and parameter $\beta \in \mathbb{R}_+$.

Exercise 1.5. Random Orders.
Modular chemical plants are small-scale plants that can be preassembled and that can be easily transported to their final destination (thus enabling mass manufacturing and reducing engineering and deployment costs) [27]. The number of monthly orders for modular electrolysis plants at a manufacturing assembly facility is random. You have available historical order data recorded over the last 4 years (see `orders_data.dat`). Determine the following:

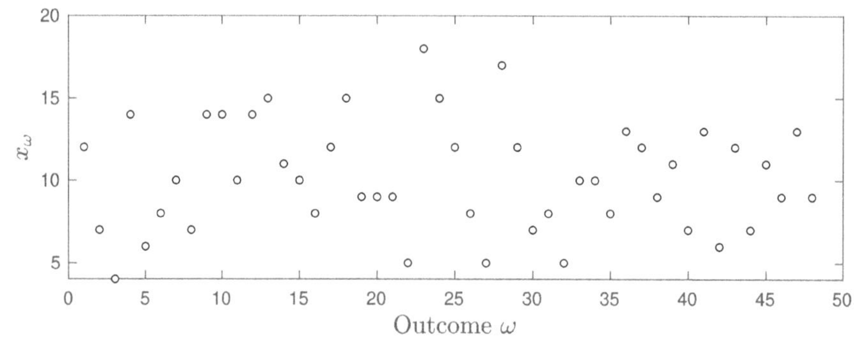

(a) Use the data to determine the frequency and probability that the number of orders is 10, 15, 20.
(b) Determine if the data matches a Poisson RV model.
(c) Use the model to determine the probability that the number of orders is 10, 15, 20.
(d) Use the model to determine the probability that the number of orders is above the monthly capacity of the manufacturing assembly line (which is 25). Should you worry about not having enough capacity?

Exercise 1.6. Assessing Teaching Performance.
A chemical engineering professor (call him Professor Z) calculated final grades for his statistics course last semester and was unhappy with student performance. As a result, he implemented a new teaching strategy that he thinks can lead to better grades this semester. He would like to compare student performance for the previous and current semester by using the 90% percentile of the grades. Use the data provided in `grades_data.dat` (the first column is data for the previous semester and the second column is data for the current semester) to address the following:

(a) Compare the percentile curves (obtained for different % values) for both semesters. Determine the 90% percentiles for the previous and new semester

based on these curves. Do not use any built-in functions from software packages to compute percentiles (do this yourself directly from data).
(b) Determine whether the new teaching strategy of the professor led to improvements in student performance.
(c) Assess whether the conclusion will change if different percentile values are used to make the comparison.

Exercise 1.7. Clinical Trials.
A pharmaceutical company is testing a new drug that aims to reduce high blood pressure. To test the drug, the company conducted a clinical trial for a small group of 10 people and computed their average pressure; the cost associated with the test of each person is approximately 200 USD. To ensure that the study is robust, they repeated the trial five times (with five different groups); the data is reported in `clinical_smallgroup.dat`. The average pressure across trials is often called the "mean estimate" and the SD across trials is a measure of its robustness or reproducibility (known as precision). A widely used metric of precision is the coefficient of variation (ratio of the SD and the mean estimate); this coefficient relates variability to overall magnitude. The company deems that the study is consistent if the coefficient of variation is less than 1%.

Address the following:

(a) Determine the average pressure for each trial. How much variability do you observe across trials?
(b) Determine the coefficient of variation. Is the study robust? What is the overall cost of this study?
(c) Repeat the procedure using data for larger groups of 100 people (and five trails); the data is reported in `clinical_largegroup.dat`. Is this study robust? What is the overall cost of this study? Is the cost justified?

Exercise 1.8. Wind Power Scaling.
The amount of power produced by a wind turbine is known to scale cubically with wind speed [28]. This is one of the reasons why the height of wind turbines has been increasing over the years (higher wind speeds are obtained at higher altitudes). For instance, in 2000, the tallest wind turbine in the world was 90 m tall, while, in 2018, it was 160 m tall.

Over 1,000 data points of produced power and wind speed have been collected by an engineer for an experimental turbine; see `wind_data.dat` and the figure shown below. We can see that there is a clear growing trend in the data, but there is also significant variability.

Address the following:

(a) An engineer is seeking to confirm that the cubic model $p_\omega = \theta \cdot u_\omega^3 + x_\omega,\ \omega \in S$ explains the trend observed in the data; here, p_ω is the measured power, u_ω is the corresponding wind speed, θ is a tunable parameter, and x_ω is any source

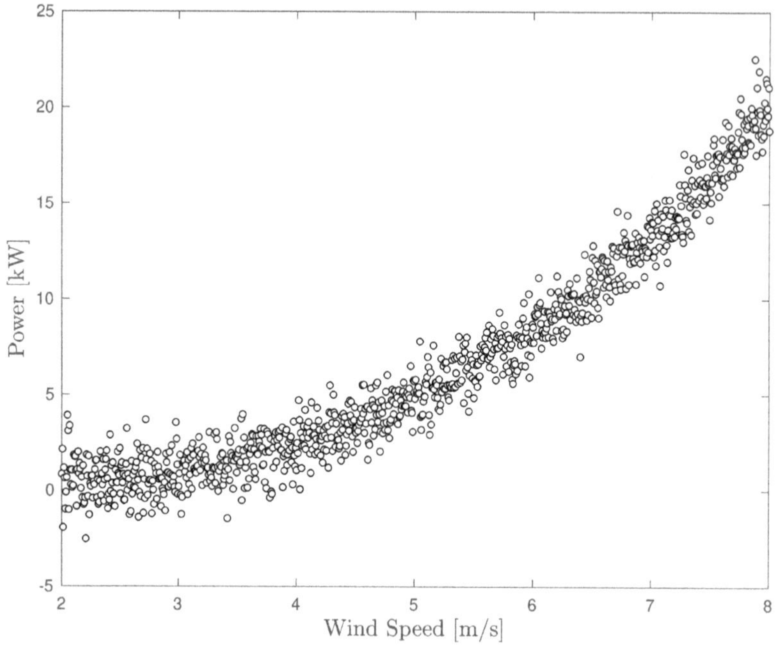

of randomness that cannot be explained by the model. To conduct the analysis, the engineer tries different values of the parameter to determine the one that best fits the data. Determine whether the residual errors $x_\omega = p_\omega - \theta \cdot u_\omega^3$, $\omega \in \mathcal{S}$ obtained with the best fit are purely random; to do so, verify that the residuals do not follow any particular trend.

(b) To confirm that the cubic model is appropriate, the engineer also tested the quadratic model $p_\omega = \theta \cdot u_\omega^2 + x_\omega$, $\omega \in \mathcal{S}$ and tested different parameter values to find the best fit. Determine if the model with the best fit can explain the trend and lead to residuals that are purely random.

Exercise 1.9. Uncertainty Propagation and Time Value of Money.
The present value PV of a cash flow C (revenue) obtained at a future time t is given by:

$$PV = \frac{C}{(1+i)^t},$$

where i is the interest rate; assume that this $i = 0.1$ (10%). The term $\alpha = \frac{1}{(1+i)^t}$ is known as the discount factor, and this factor relates the present and future value of money. Note that the discount factor α is less than or equal to one and thus the present value is always less than or equal to the future value. In other words, the value of money always increases over time; to see this, we can also write the above expression as:

$$C = \frac{PV}{\alpha}.$$

The cash flow C that you expect from an investment is uncertain, and a financial analyst has obtained a number of possible scenarios for you (see data in `cash_flow.dat`).

Address the following:

(a) Visualize the pdf of the cash flow C and use MC simulations to obtain the pdf for the present value PV for times $t = 1, 5, 10$ (in years). Use MC simulations also to obtain the mean and standard deviation (SD) of the PV as a function of time. Discuss how these summarizing statistics and the pdf vary over time.

(b) Assume that C is a continuous RV and derive a formula for the uncertainty (SD) of the PV as a function of the discount factor α and of the SD of the cash flow.

(c) Use the formula to determine how does uncertainty of a cash flow obtained in the future affect the uncertainty of its corresponding PV. Specifically, what happens to the uncertainty of the PV when time is long ($t \to \infty$)?

(d) Use the formula to determine what happens to the uncertainty of the PV when the interest rate is small ($i \to 0$).

(e) Imagine now the situation in which you have uncertainty in the PV and you want to analyze how this propagates to uncertainty in the future cash flow. How are the uncertainty of the cash flow and of the present value related? What happens to the uncertainty of the cash flow when time is long ($t \to \infty$)?

Exercise 1.10. Uncertainty Mitigation.

The effect of pressure and temperature on the equilibrium conversion for the reaction:

$$CO + 2H_2 \longleftrightarrow CH_3OH,$$

can be predicted by using a mechanistic (physical) model. This model has the following components:

- Conservation conditions (assuming CO is the limiting component):

$$\mu_{CO}^{out} = \mu_{CO} - \xi \cdot \mu_{CO}$$
$$\mu_{H_2}^{out} = \mu_{H_2} - 2 \cdot \xi \cdot \mu_{CO}$$
$$\mu_{CH_3OH}^{out} = \mu_{CH_3OH} + \xi \cdot \mu_{CO},$$

  where ξ is the conversion (value between 0 and 1), $\mu_{CO}, \mu_{H_2}, \mu_{CH_3OH}$ are the input molar flows, and $\mu_{CO}^{out}, \mu_{H_2}^{out}, \mu_{CH_3OH}^{out}$ are the output molar flows.

- Equilibrium condition:

$$K_{eq} = \frac{(P \cdot \mu_{CH_3OH}^{out}/\mu_{tot})}{(P \cdot \mu_{H_2}^{out}/\mu_{tot})^2 \cdot (P \cdot \mu_{CO}^{out}/\mu_{tot})},$$

  where $\mu_{tot} = \mu_{CO}^{out} + \mu_{H_2}^{out} + \mu_{CH_3OH}^{out}$ is the total molar output flow and pressure P is in bar.

- Van't Hoff equation:

$$\log K_{eq} = -12.275 + 4938 \cdot \frac{1}{T + 273.15}.$$

- The input flows are fixed to $\mu_{CO} = 100$, $\mu_{H_2} = 600$, and $\mu_{CH_3OH} = 0$ (kmol/hr).

Simulating this mechanistic model requires advanced numerical techniques. A computer implementation of this model has been provided to you; this simulation model can predict conversion as a function of pressure and temperature. Your goal is to use the model to identify temperature conditions that can help counteract pressure variability (which generate variability in the conversion). The pressure data is `gibbs_pressure.dat`; the model is `ch1_gibbs_solve`.

Address the following:

(a) Use MC simulations to obtain the empirical pdf and cdf of the conversion when operating at $T = 300°C$. Determine the SD of conversion and the probability of conversion being greater than 0.8.

(b) Show that the variability of the conversion can be nearly completely mitigated by operating at $T = 200°C$. Determine the SD of conversion and the probability of conversion being greater than 0.8.

2 Univariate Random Variables

A wide range of random variable (RV) models can be used to capture trends observed in real-life, random phenomena (e.g., diffusion, mixing, extreme events, particle sizes, and waiting times). As we saw in Chapter 1, RV models are uncertainty models and can be categorized as discrete/continuous and as univariate/multivariate. As expected, the type of model to be used depends on the nature of the phenomenon under study. This is directly analogous to how we select physical models on the basis of the nature of the phenomenon under study (e.g., an ideal or nonideal gas model).

In this chapter, we will discuss different types of RV models and highlight connections between them. In particular, we will see that many RVs result from algebraic transformations (e.g., logarithmic, exponential, and quadratic) that are applied to other RVs. We will also see that some RV models are generalizations of other models. We will show how some RV models can help characterize random phenomena observed in physical systems (e.g., material failures and motion of particles). This will aim to highlight how *physical systems can naturally exhibit random behavior* at certain scales, while their behavior might appear to be deterministic at other scales. Finally, we will highlight how properties of models can be used to obtain estimates for their parameters from data (e.g., from empirical approximations of summarizing statistics).

Our focus in this chapter will be on univariate RV models, while in Chapter 3, we will discuss multivariate RV models. We will focus our attention on some of the most used RV models and explain their emergence and relevance in applications. In Chapter 4, we will develop techniques to *optimally* tune the parameters of an RV model and to select a model type that best explains available data. In Chapters 5–7, we will see that the ability to model uncertainty will be critical in developing structural models, quantify their prediction uncertainty, and use them to make decisions.

2.1 Discrete Models

2.1.1 Uniform

The uniform RV is the simplest model that one can use to model a random phenomenon. This model is inspired by the fact that, in many applications, we might not have much information available to infer behavior. Specifically, imagine that all that we know is that our RV can take discrete values $a, a+1, a+2, \ldots, b-1, b$. In the absence of any other information, it seems reasonable to assume that it is equally likely (probable) that the RV X takes any of these values. Accordingly, the pdf of this model is:

$$\mathbb{P}(X = x) = f(x) = \frac{1}{n}, \; x \in \mathcal{D}. \tag{2.1}$$

with domain $\mathcal{D} = \{a, a+1, a+2, \ldots, b-1, b\}$ and domain size $n = b - a + 1$.

The probability that X takes any value outside of the domain $\mathcal{D}$ is zero. The pdf is a set of points of equal height (this is a discontinuous function). We express the fact that X follows a uniform RV model as $X \sim \mathcal{U}(a, b)$, and we note that $a, b \in \mathbb{Z}$ are the parameters of the model; here, $\mathbb{Z}$ is the set of integer numbers.

We recall that the parameters of the model are values that can be adjusted to match the model to available data. For this model, the parameters are simply the extreme points of the domain (a and b), and thus, these can be obtained by simply using the smallest and largest values observed.

The cdf for the uniform RV model is given by:

$$\mathbb{P}(X \leq x) = F(x) = \frac{x - a + 1}{n}, \; x \in \mathcal{D}. \tag{2.2}$$

The cdf has a staircase form and is discontinuous.

The uniform model has the property that its mean (expected value) is given by:

$$\begin{aligned}
\mathbb{E}_X &= \sum_{x \in \mathcal{D}} x \cdot f(x) \\
&= \sum_{x=a}^{b} \frac{1}{n} \cdot x \\
&= \frac{1}{n}(a + (a+1) + (a+2) + \cdots + (b-2) + (b-1) + b) \\
&= \frac{1}{2}(a + b).
\end{aligned} \tag{2.3}$$

This makes sense, as it indicates that the mean is the middle of the domain. The median of this RV is the value at which $F(x) = 1/2$; as such, we find the value of x that satisfies:

$$\frac{1}{2} = \frac{x - a + 1}{n}. \tag{2.4}$$

By isolating x, we find that the median is $\mathbb{M}_X = \frac{1}{2}(a + b)$. This indicates that, for the uniform model, the mean and the median coincide (this is a distinctive property of this model).

The variance of this RV is:

$$\mathbb{V}_X = \sum_{x \in \mathcal{D}} f(x) \cdot (x - \mathbb{E}_X)^2 = \frac{1}{n} \sum_{x=a}^{b} \left(x - \frac{1}{2}(a + b) \right)^2 = \frac{1}{12}(n^2 - 1). \tag{2.5}$$

This indicates that the variance increases with the domain size n (which makes sense). The uniform model has many applications and is particularly used in sampling (data collection) schemes. Specifically, in such schemes, we say that it is equally likely to select a sample over another (there is no preference).

Example 2.1 Pdf and Cdf of Discrete Uniform RV

In Figure 2.1, we show 100 observations and the pdf/cdf of the uniform RV $\mathcal{U}(-1,1)$. We can see that its domain is $\mathcal{D} = \{-1,0,1\}$ and its pdf is $f(x) = 1/3$ for $x = \{-1,0,1\}$. We recall that the pdf tells us how densely the outcomes/pinballs concentrate in particular values of the domain; in this case, the density is the same for the three values. There is no particular trend that we can spot in the observations, but we know that the outcomes will land in each value with equal probability.

Script to reproduce results: `ch2_compare_uniformd` □

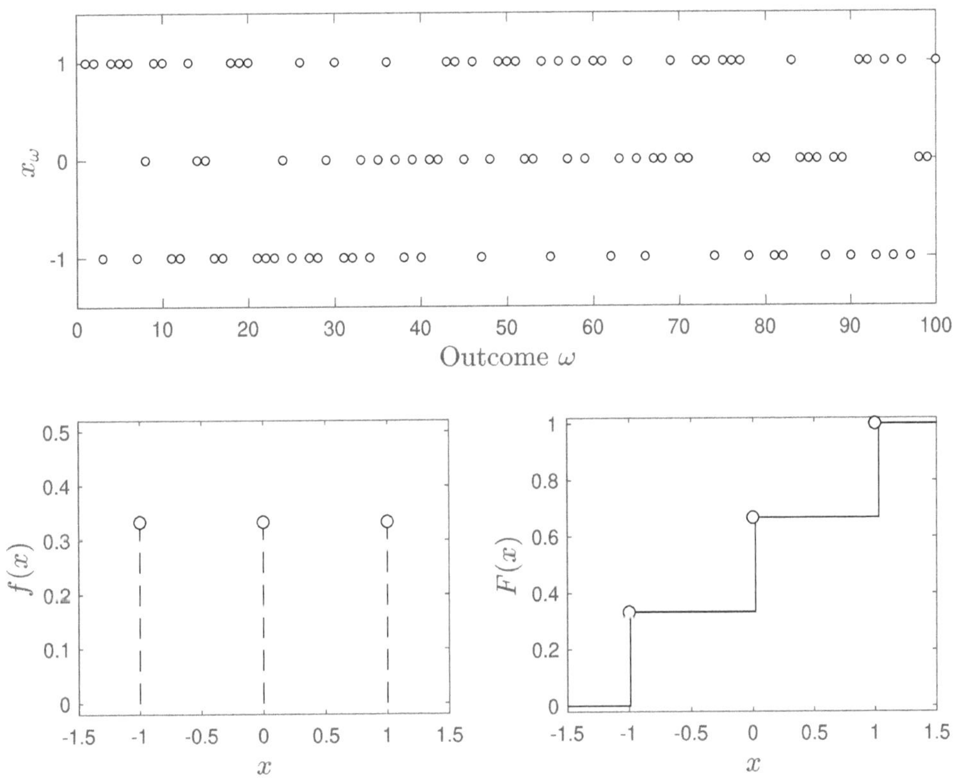

Figure 2.1 Pdf and cdf for a uniform random variable with $a = -1$ and $b = 1$.

Example 2.2 Machine Status as Uniform RV

You have a machine that appears to fail randomly and you have collected weekly data for over 3 years (156 data points) to try to identify a pattern (see Figure 2.2). You have not identified any particular pattern, but you have noticed that the machine has not failed in 79 out of 156 weeks.

- Is it suitable to model the machine status as a uniform RV?
- If so, what would be its parameters?

We define the machine status as $X = 0$ when it fails and $X = 1$ when it works. With this, we have that the domain of the RV is $\mathcal{D} = \{0, 1\}$ and thus $a = 0$, $b = 1$, and $n = 2$. We recall that, for a uniform RV, we have that $\mathbb{P}(X = x) = 1/n$ for all $x \in \mathcal{D}$. This would indicate that, if the machine status is indeed a uniform RV, we must have that $\mathbb{P}(X = 0) = 1/2$ and $\mathbb{P}(X = 1) = 1/2$. From our data, we can see that the empirical pdf is:

$$\hat{\mathbb{P}}(X = 0) = 79/156 = 0.5064,$$
$$\hat{\mathbb{P}}(X = 1) = (156 - 79)/156 = 0.4936.$$

We thus have that the data suggests that the machine status seems to follow the behavior of a uniform RV of the form $\mathcal{U}(0, 1)$.

Script to reproduce results: `ch2_example_uniform` □

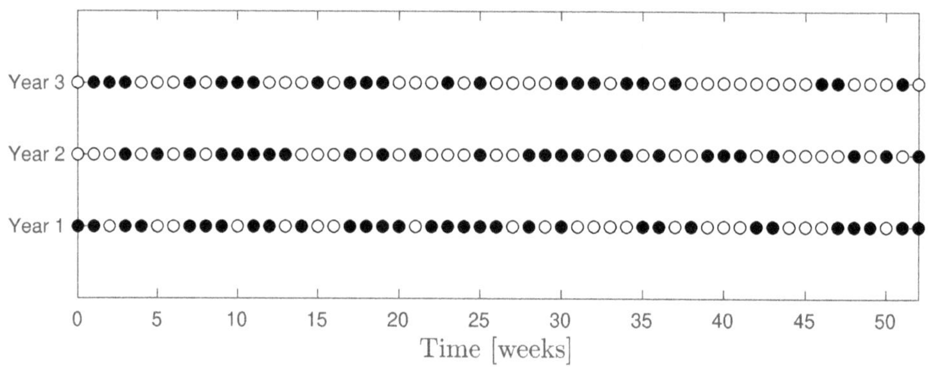

Figure 2.2 Machine failures (dark dots) recorded over 3 years.

2.1.2 Bernoulli

Another simple RV model is the *Bernoulli RV*. Consider that you run an experiment and that the experiment only has a couple of possible outcomes: Y (yes) or N (no). We define $p \in [0, 1]$ as the probability of outcome Y, and we thus have that the probability of outcome N is $(1 - p)$. The RV X takes a value of 1 if the outcome is Y or a value of 0 if the outcome is N. We thus have that the domain of X is $\mathcal{D} = \{0, 1\}$ and the pdf is given by:

$$f(x) = \begin{cases} (1 - p) & \text{if } x = 0 \\ p & \text{if } x = 1. \end{cases} \tag{2.6}$$

The pdf can be written in compact form using indicator functions as:

$$f(x) = p^{\mathbf{1}[x=1]}(1 - p)^{\mathbf{1}[x=0]}. \tag{2.7}$$

As we will see in Chapter 6, this form is used in binary classification models for statistical learning.

We express the fact that X follows a Bernoulli model as $X \sim \mathcal{B}(p)$, where p is the parameter. The mean of this RV is given by:

$$\mathbb{E}_X = \sum_{x \in \mathcal{D}} x \cdot f(x) = 0 \cdot (1-p) + 1 \cdot p = p, \tag{2.8}$$

and one can show that its variance is:

$$\mathbb{V}_X = \sum_{x \in \mathcal{D}} (x - \mathbb{E}_X)^2 \cdot f(x) = p \cdot (1-p). \tag{2.9}$$

The Bernoulli RV has applications in data collection schemes and forms the basis of more sophisticated RV models.

Example 2.3 Pdf and Cdf of Discrete Bernoulli RV

In Figure 2.3, we show 100 samples and the pdf/cdf of $\mathcal{B}(0.3)$. We can see that its domain is $\mathcal{D} = \{0, 1\}$ and its pdf is $f(x) = 0.3$ for $x = 1$ and $f(x) = 0.7$ for $x = 0$. Note how the observations land unevenly across domain values.

Script to reproduce results: `ch2_compare_bernoulli` □

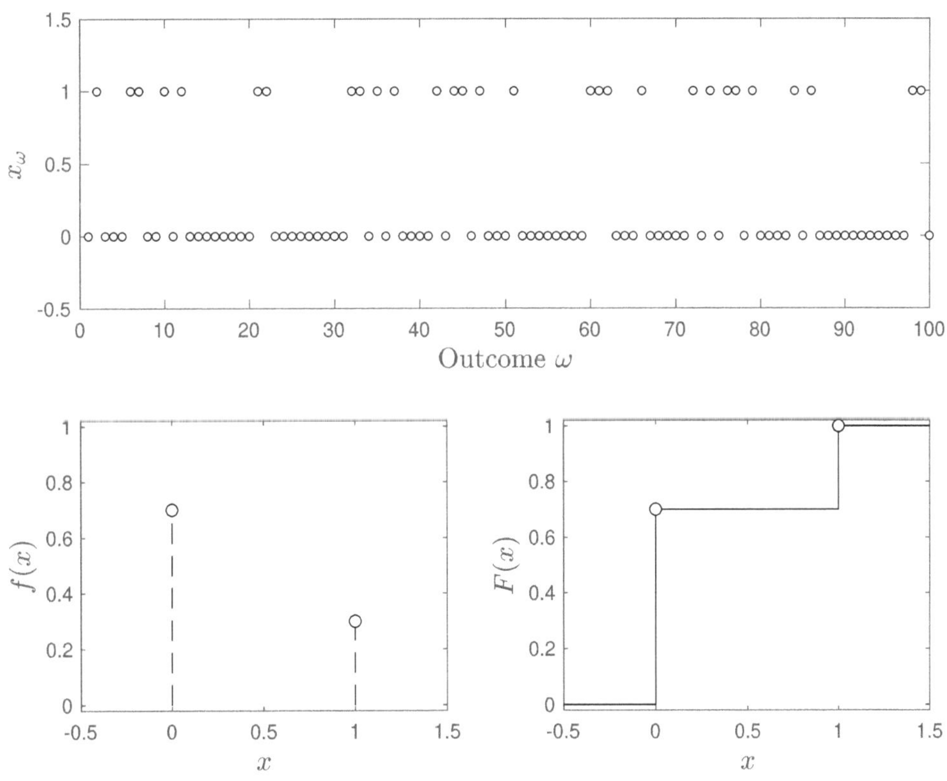

Figure 2.3 Pdf and cdf for Bernoulli random variable with $p = 0.3$.

Example 2.4 Machine Status as Bernoulli RV

Consider the machine status problem of Example 2.2 and consider the following questions:

- Is it suitable to model the machine status as a Bernoulli RV?
- If so, what would be its parameters?
- What makes this different from a uniform RV?

We define the machine status as $X = 0$ when it fails and $X = 1$ when it works. We thus have that the domain of the RV is $\mathcal{D} = \{0, 1\}$. We recall that, for a Bernoulli model, we must have that $\mathbb{P}(X = 0) = p$ and $\mathbb{P}(X = 1) = 1 - p$. From our data, we have seen that the empirical pdf is:

$$\hat{\mathbb{P}}(X = 0) = 79/156 = 0.5064,$$
$$\hat{\mathbb{P}}(X = 1) = (156 - 79)/156 = 0.4936.$$

We thus have that the machine status can be modeled as a Bernoulli RV $\mathcal{B}(0.5)$. Note that, in this particular example, the uniform and Bernoulli RV models are identical. □

In the previous example, we encountered the situation in which the uniform and Bernoulli models are identical. However, note that the uniform RV is a more general model because it can capture multiple possible values for the RV, while the Bernoulli can only capture a couple of possible values. In other words, the uniform model is a generalization of the Bernoulli model; this highlights the fact that RV models can be related.

2.1.3 Hypergeometric

The *hypergeometric RV* arises in applications in which one draws random samples from a finite population. For instance, in a manufacturing facility, we choose a set of products to inspect for quality control. Similarly, imagine that you poll a subset of the US population to identify their political affiliation. The fundamental issue here is that the population can be *extremely large*, and therefore, it is too expensive (or impossible) to inspect it all. As a result, we want to select a subset and try to infer from this the behavior of the general population. This type of situation is also typically encountered in manufacturing; here, we might need to infer the behavior of a large population of products from a small sample that is inspected.

Assume that the population is composed of a couple of mutually exclusive groups (Y or N). The total units in the population is N_t, and we define the total number of units of type Y as N_y. We thus have that $(N_t - N_y)$ are of type N. The fraction of Y items is thus $p = N_y/N_t$.

We draw a sample of n items from the population and check how many items are of type Y. The RV X is the number of items of type Y in our sample. Now note that the total number of ways of choosing n items from the population is:

$$N_\Omega = \binom{N_t}{n} = \frac{N_t!}{n! \cdot (N_t - n)!}. \tag{2.10}$$

Here, $n!$ denotes the *factorial* of the integer number n. The notation $\binom{N_t}{n}$ is the *binomial coefficient* or the *choose* function. The total number of distinct ways in which we can select n items from a population of N_t items is N_Ω. This is the size of the sample set Ω and is, in general, an extremely large number. For instance, for $N_t = 100$ and $n = 10$, we have $N_\Omega = 1.73 \times 10^{13}$. Recall that X is the number of items of type Y in our sample of size n.

The probability of obtaining a given number x of items of type Y in the sample is given by the pdf:

$$f(x) = \frac{\binom{N_y}{x} \cdot \binom{N_t - N_y}{n - x}}{\binom{N_t}{n}}, \quad x \in \mathcal{D}, \tag{2.11}$$

where the domain is $\mathcal{D} = \{\max(0, n + N_y - N_t), \ldots, \min(n, N_y)\}$ (only values in this domain can be used to evaluate the choose functions in $f(x)$). We say that X is a hypergeometric RV by using the notation $X \sim \mathcal{H}(N_y, N_t, n, p)$.

Example 2.5 Pdf and Cdf of Hypergeometric RV

In Figure 2.4, we show 1,000 realizations and the pdf/cdf of a hypergeometric RV for $N_t = 500$, $N_y = 50$, and $n = 50$ (the faction of items of type Y in the population is $p = 0.1$). We can see that the pdf grows quickly with the x, reaches a peak (mode) at $x = 5$, and then decays rapidly to nearly zero probabilities for $x > 10$. This indicates that it is unlikely to find more than 10 items of type Y in the sample and that it is highly likely to find exactly 5 items of type Y. Interestingly, it is much more likely to find exactly 5 items of type Y than exactly 1 item of type Y in our sample. This is because the sample size is quite large (we would have to be lucky to find just 1 item of type Y in our sample). From the cdf, we see that the probability of finding less than 10 items of type Y is nearly one. These insights are confirmed when we analyze the realizations; note how these concentrate around a value of $x = 5$ and decay quickly (in an asymmetric manner) as we move away from this value.

Script to reproduce results: `ch2_compare_hypergeometric` □

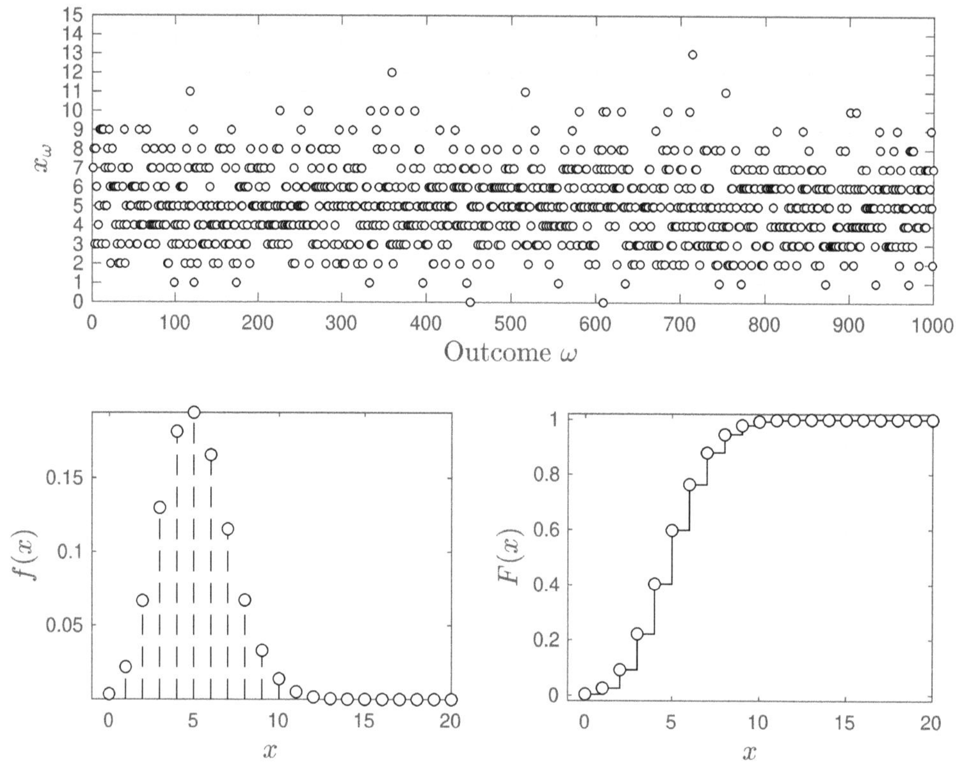

Figure 2.4 Pdf/cdf for hypergeometric RV with $N_t = 500, N_y = 50$, and $n = 50$.

The expected value of X (the average number of items of type Y in our sample) is:

$$\mathbb{E}_X = n \cdot p. \tag{2.12}$$

Note that the value n is known (we control this), but $p = N_y/N_t$ is generally not known because it would be extremely time-consuming or expensive to determine how many Y items N_y we actually have in the entire population. Consequently, p is usually estimated from empirical (data-driven) estimates of $\mathbb{E}_X$. For instance, assume that you pick S different samples (each of size n) and record the number of items of type Y in each sample; this gives the observations of X (x_ω, $\omega \in \mathcal{S}$). The empirical estimate of the expected value is $\hat{\mathbb{E}}_X = \frac{1}{S}\sum_{\omega \in \mathcal{S}} x_\omega$, and we can use (2.12) to estimate $\hat{p}$ as:

$$\hat{p} = \hat{\mathbb{E}}_X / n. \tag{2.13}$$

The variance of the hypergeometric RV is given by:

$$\mathbb{V}_X = n \cdot p \cdot (1 - p) \cdot \frac{N_t - n}{n - 1}. \tag{2.14}$$

From this expression, we can see that the variance is zero if $n = N_t$. In other words, if we pick all items $n = N_t$ each time, then there is no variability in the number of items of type Y that we observe. Generally speaking, the variance will decrease as n approaches N_t. This highlights how the RV model exhibits consistent behavior. However, note that the inspection cost increases with increasing n; this highlights how reducing variance/uncertainty can be expensive or cost-prohibitive. In other words, there is a *price to pay for certainty*.

Note also that the parameter p can be estimated using either the empirical mean or the variance, but it is easier to do it via the mean (as the relationship is linear). This highlights how some summarizing statistics might be more convenient to use than others for obtaining parameters.

Example 2.6 Monitoring Machine Status

Consider again the machine status problem and data of Example 2.2. You cannot afford to monitor the status of the machine every week for the next 3 years (this is time-consuming and expensive), and therefore, you would only want to inspect it randomly. You assume that the number of failures observed over the next 3 years will be roughly the same as those observed over the last 3 years. Consider the following questions:

- Consider that you inspect the machine 5 times (selected at random). What is the probability that the machine will be in failure mode 5 out of 5 times that you inspected it? What is the probability that the machine will fail in 0 out of 5 times that you inspected it?
- Consider that you inspect the machine 10 times (selected at random). What is the probability that the machine will be in failure mode 10 out of 10 times that you inspected it?

In Example 2.2, we observed that the machine failed 79 times (weeks) out of 156 weeks. Consider the case in which we will inspect the machine a total of 5 times. Because we will be choosing the inspection times at random, it is appropriate to model the number of failures observed as a hypergeometric RV $X \sim \mathcal{H}(N_y, N_t, p, n)$. Here, we define $N_t = 156$, $N_y = 79$, $p = N_y/N_t = 0.50$, and $n = 5$. The probability that the machine will be in failure mode 5 times out of the 5 times inspected is:

$$\mathbb{P}(X = 5) = f(5) = 0.1312.$$

The probability that the machine will be in failure mode 0 times out of the 5 times inspected is:

$$\mathbb{P}(X = 0) = f(0) = 0.0274.$$

We can thus see that it is significantly less likely that we will find no failures in our 5 inspections than that we will find 5 failures in our 5 inspections. Note that this behavior is asymmetric (it is significantly more likely that all inspections are failures than no inspections are failures).

To compute the probabilities for 10 inspections chosen at random, we set $n = 10$; the probability of observing 10 failures is:

$$\mathbb{P}(X = 10) = f(10) = 8 \times 10^{-4}.$$

It is thus highly unlikely that we will find 10 failures in our 10 inspections.

Script to reproduce results: `ch2_example_hypergeometric` □

2.1.4 Binomial

We now generalize the setting of the Bernoulli RV; in the Bernoulli setting, a trial has a couple of possible outcomes: Y (yes) or N (no). Imagine now that we perform a *sequence* of n Bernoulli trials and record the number of times that the outcomes are of type Y. Importantly, each trial is independent of the others (they are not influenced by each other); we will formally define what we mean by "independent" in Chapters 3 and 4.

The *binomial RV* X counts the number of trials of type Y in our n trials. We thus have that the domain is $\mathcal{D} = \{0, 1, \ldots, n\}$. The probability that the number of Y trials is x is given by:

$$f(x) = \binom{n}{x} \cdot p^x \cdot (1-p)^{(n-x)}, \quad x \in \mathcal{D}. \tag{2.15}$$

We say that X is a Binomial RV by using the notation $X \sim \text{Bi}(n, p)$.

The expected value and variance are given by:

$$\begin{aligned} \mathbb{E}_X &= n \cdot p \\ \mathbb{V}_X &= n \cdot p \cdot (1-p). \end{aligned} \tag{2.16}$$

Note that these statistics are just n times the expected value and variance of the Bernoulli RV. This seems to highlight the fact that the binomial RV is a generalization of the Bernoulli RV. This is indeed the case, specifically, if we consider a sequence of independent Bernoulli RVs $X_1, X_2, \ldots, X_n$ (each with probability p), then the RV:

$$X = \sum_{i=1}^{n} X_i, \tag{2.17}$$

is a binomial RV of the form $X \sim \text{Bi}(n, p)$. This is consistent with the way that we constructed the binomial setting based on a sequence of independent Bernoulli trials. We thus conclude that $\text{Bi}(1, p) = \mathcal{B}(p)$.

We also note that the binomial RV is a limiting case of a hypergeometric RV; specifically:

$$\lim_{N_t \to \infty} \mathcal{H}(N_y, N_t, n, p) = \text{Bi}(n, p). \tag{2.18}$$

In other words, a binomial setting is equivalent to a hypergeometric setting when the size of the population is large. Here, it is important to highlight the difference in the setting of the hypergeometric and binomial RVs. In the hypergeometric setting, we choose n items at once, while in the binomial setting, we choose n items by choosing one item at a time (resetting the system every time we chose a new item).

Intuitively, when the population N_t is large, the size of the sample does not matter and picking n items at once or as an independent sequence is close to equivalent.

The connections between the binomial, Bernoulli, and hypergeometric RVs highlight that understanding the relationships between models can be interesting and useful.

Example 2.7 Pdf and Cdf of Binomial RV

In Figure 2.5(a) we can see the pdf and cdf of a binomial RV with $n = 10$ and $p = 0.3$. We can see that, after $n = 10$ trials, it is most likely to find three trials with outcome Y. We can also see that there is a small (but non-zero) probability of finding zero trials with outcome Y and that there is virtually no chance of finding eight trials or more with outcome Y.

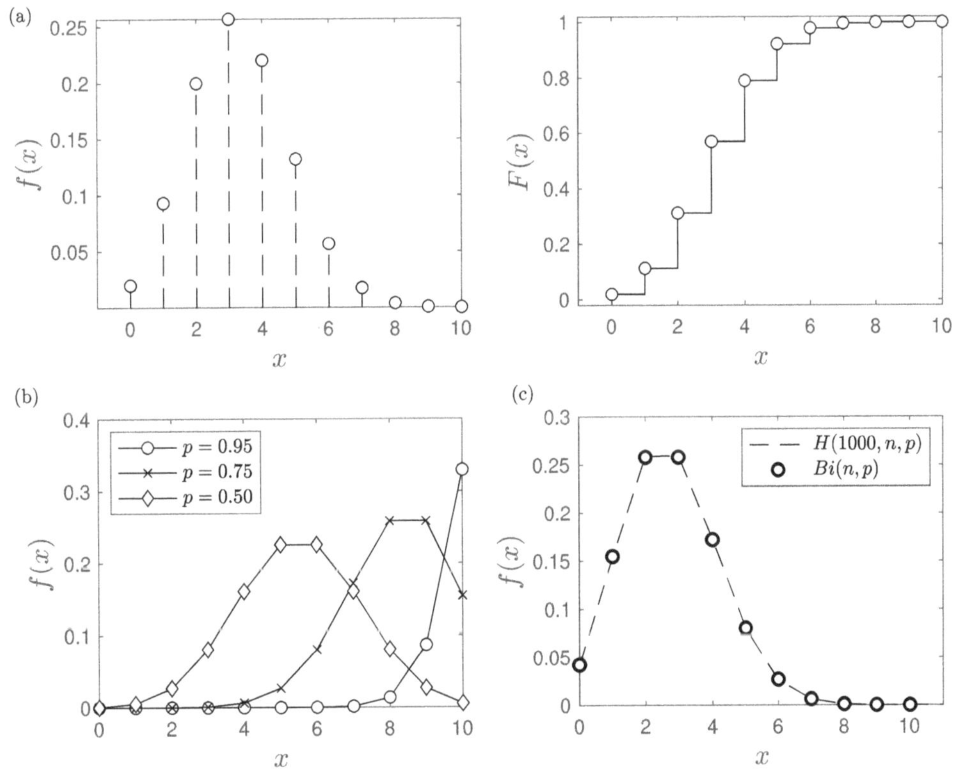

Figure 2.5 Pdf and cdf for binomial RV with $n = 10$ and $p = 0.3$ (a). Binomial pdf for different values of p (b). Convergence of hypergeometric pdf to binomial pdf for $N = 1{,}000$ (c).

In Figure 2.5(b), we compare the pdfs for $n = 10$ and different probability values p. Note that a value of $p = 0.95$ indicates that it is highly likely that an item of type Y will show up; consequently, there is a high probability that 10 trials will have outcome Y, and it is unlikely that less than 8 trials will have an outcome of Y.

In Figure 2.5(c), we compare the pdf of the RV $\mathcal{H}(N_y, N, n, p)$ for $N = 1{,}000, n = 10, p = 0.25$ and the pdf of $\text{Bi}(n,p)$ for $n = 10, p = 0.25$. We see that the pdfs are nearly identical, demonstrating that the hypergeometric RV converges to a binomial RV when the population is large.

Script to reproduce results: `ch2_compare_binomial` □

Example 2.8 Monitoring Machine Status

Consider the machine status problem and data of Example 2.2. Recall that in Example 2.4, we established that the status of the machine can be modeled as a Bernoulli RV with $p = 0.5$. Imagine now that you are interested in determining how likely it is to see a given sequence of failures. Specifically, consider the following questions:

- What is the probability of finding 2 failures in a sequence of 10 times?
- What is the probability of finding 10 failures in a sequence of 10 times?
- What is the average number of failures in a sequence of 10 times? What is the probability of finding the average number of failures?

Since the status of the machine is a Bernoulli RV with p, the number of failures in a sequence of length n can be modeled as a binomial RV $\text{Bi}(n,p)$. We thus have that the probability of finding 2 failures in a sequence of $n = 10$ times is:

$$\mathbb{P}(X = 2) = f(2) = 0.0439.$$

The probability of finding 10 failures is:

$$\mathbb{P}(X = 10) = f(10) = 9.76 \times 10^{-4}.$$

The average is:

$$\mathbb{E}_X = n \cdot p = 10 \cdot 0.5 = 5,$$

with probability $\mathbb{P}(X = 5) = 0.24$. Note that it is less likely to see 2 failures than to see 5 failures. At first, this might seem counterintuitive, but this makes sense because we recall that it is equally likely that the machine is in failure or normal mode at any given time ($p = 0.5$), and it is thus unlikely that it will be in either mode all the n times. Specifically, note that the probability of finding 0 failures is $\mathbb{P}(X = 0) = 9.76 \times 10^{-4}$, which is the same probability as that of finding 10 failures.

Script to reproduce results: `ch2_example_binomial` □

2.1.5 Poisson

The *Poisson RV* is an important generalization of the binomial RV. In the binomial setting, we considered a sequence of n Bernoulli trials and recorded the number of occurrences of type Y. Imagine now that the trials are taken at consecutive times $z_1, z_2, \ldots, z_n$, and we define Δz as the time between trials (Δz is the sampling interval). If we have a continuous time interval $[0, z]$ of total length z, then the number of trials in the interval is $n = z/\Delta z$. As the sampling interval becomes small ($\Delta z \to 0$), the number of trials increases ($n \to \infty$). The setting of a Poisson RV is the limiting case of $\Delta z \to 0$ (equivalently, $n \to \infty$); specifically, we are interested in understanding the probability of observing occurrences over infinitesimally small sampling intervals and with this find out probabilities of occurrences over an interval of fixed length.

We note that the interval does not have to be time but can also be space. For instance, imagine that you are inspecting a long pipeline (say z =100 km) in trying to find defects; you want to inspect this carefully, and thus, you need to examine small intervals of pipeline (say Δz =1 m). Here, every Δz, we inspect the pipeline and check if there is a defect or not (a Bernoulli trial is of type Y or N).

A Poisson RV gives the total number of occurrences of type Y in an interval of length z. Note that the domain of this RV is $\mathcal{D} = \{0, 1, 2, \ldots, \infty\}$, which highlights the fact that we are looking at the limit $n \to \infty$. The pdf of this RV is given by:

$$f(x) = \frac{\lambda^x e^{-\lambda}}{x!}, \quad x \in \mathcal{D}. \tag{2.19}$$

We state that X is a Poisson RV using $X \sim \mathcal{P}(\lambda)$ with parameter $\lambda \in \mathbb{R}_+$.

In Figure 2.6, we show the pdf/cdf of $\mathcal{P}(\lambda)$ with different values of λ, and we show 1,000 realizations for $\mathcal{P}(1/2)$. Here, we show the functions as continuous for visualization purposes but highlight that these functions are discontinuous because

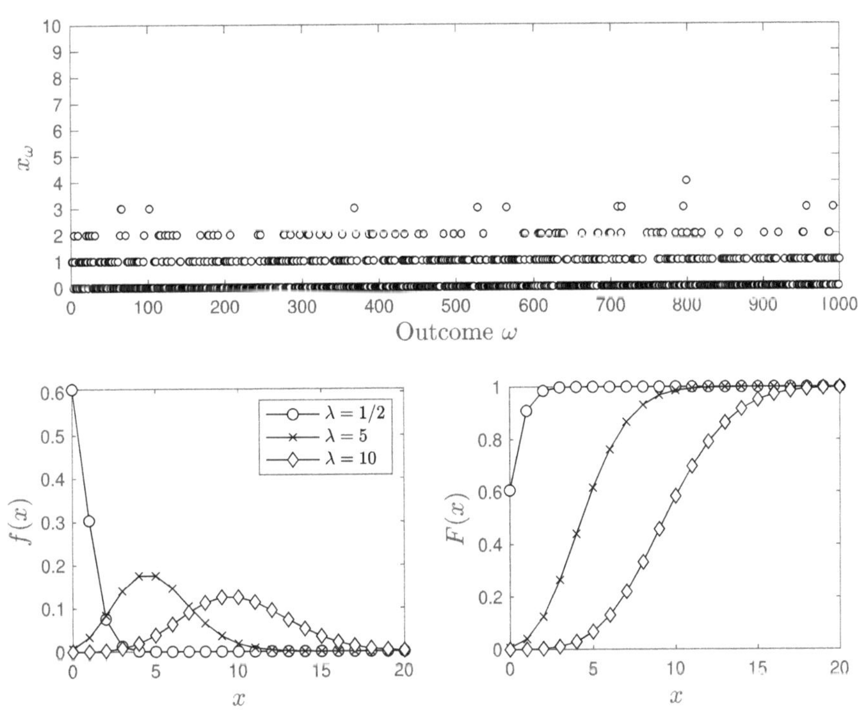

Figure 2.6 Pdf/cdf for Poisson random variable and realizations of $\mathcal{P}(1/2)$.

they are only defined at the discrete values in $\mathcal{D}$. Note that the location of the peak of the pdf (the mode) is close to the value of λ. This is consistent with the fact that the mode of a Poisson RV is given by:

$$\mathbb{MO}[X] = \{\lceil \lambda \rceil - 1, \lfloor \lambda \rfloor\}. \tag{2.20}$$

Here, $\lceil \cdot \rceil$ is the ceiling function and $\lfloor \cdot \rfloor$ is the floor function. This reveals that the mode can only be an integer value because it must be in the domain $\mathcal{D}$ (which only includes integer values). For $\lambda = 5$, the mode is $\{4, 5\}$ (there are two modes), and for $\lambda = 10$, we have $\{9, 10\}$ (two modes). In other words, the mode is not unique. This is the reason why we see a flat peak in the pdfs at these values. For $\lambda = 1/2$, on the other hand, the mode is $\{0, 0\}$ (there is only one mode). For this value, we also see that the pdf behaves as a purely exponential function (this can be confirmed from the realizations, which show a high concentration at a value of 0 and quick decay in concentration as we move away from zero). We can anticipate the emergence of a peak in the pdf from the shape of the function (2.19). Specifically, for $\lambda \geq 1$, the terms λ^x increase with x, but the term $1/x!$ decreases with x. We thus have that there is a competing effect in this terms, giving rise to a peak. The cdf is a complicated function that will not be discussed here, but we can see that its behavior is exponential in x.

The expected value and variance of this RV are given by:

$$\begin{aligned} \mathbb{E}_X &= \lambda, \\ \mathbb{V}_X &= \lambda. \end{aligned} \tag{2.21}$$

The expected value reveals that the parameter λ can be interpreted as the average number of occurrences (events) found in interval. Interestingly, note that the mean and variance are the same for this RV (this is a characteristic property of the Poisson model). This behavior can also be seen in the pdfs in Figure 2.6; for instance, a small value of λ tells us that the average number of occurrences is small and that the variance is small. This is why the pdf decays quickly and becomes zero at $x = 3$ (the cdf rises sharply and reaches 1 at $x = 3$); in other words, if the variability is small, the range of possible values of x is small. In contrast, when $\lambda = 10$, we have that the average number of occurrences is 10, but the variability is large; this is manifested as a pdf/cdf that expands over a wide range (from a value of 0 to 20).

The parameter λ is also often expressed as:

$$\lambda = \eta \cdot z, \tag{2.22}$$

where $\eta \in \mathbb{R}_+$ is known as the intensity (or event rate); this parameter is the average number of occurrences per unit of interval (e.g., per unit of time).

The Poisson model is widely used in queueing theory; for instance, it can characterize the number of customers that arrive during a time interval (which is random) (see Exercise 2.2). The Poisson model can also be used to characterize the occurrence of events in spatial domains (see Exercise 2.1).

Example 2.9 Failure of Control System for Reactor

The experimental reactor of Example 1.2 is equipped with an automated cooling system that is used to manipulate the temperature (keep it at a low or high level). Unfortunately, the pump of the cooling agent is not reliable (cavitates quite often). The pump failures have been recorded daily for the last 3 months; the records are shown in Figure 2.7.

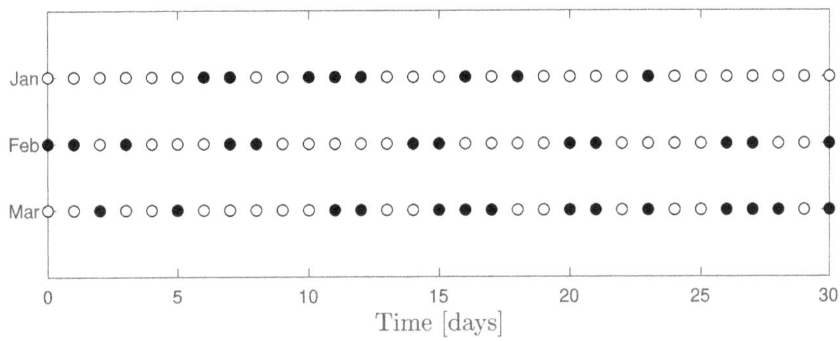

Figure 2.7 Pump failures (dark dots) recorded over 3 months.

Assuming that the number of failures per month is a Poisson RV, determine the following:

- Is it more probable that the pump does not fail or that the pump fails daily?
- Probability that the pump fails an average number of times.

From the data recorded, the number of failures per month were $x_{Jan} = 8$, $x_{Feb} = 12$, and $x_{Mar} = 14$. From the values, we can obtain the empirical estimate of the expected value $\mathbb{E}_X$:

$$\hat{\mathbb{E}}_X = \frac{1}{3}(8 + 12 + 14) = 11.3. \tag{2.23}$$

For a Poisson RV, we know that $\mathbb{E}_X = \lambda$, and we thus estimate that $\hat{\lambda} = 11.3$. Using this estimate and Eq. (2.19), we obtain:

$$\begin{aligned} \mathbb{P}(X = 0) &= \frac{\lambda^0 e^{-\lambda}}{0!} = 1.23 \times 10^{-5}, \\ \mathbb{P}(X = 30) &= \frac{\lambda^{30} e^{-\lambda}}{30!} = 1.82 \times 10^{-6}. \end{aligned} \tag{2.24}$$

We thus conclude that it is more probable that the pump does not fail. The probability that the pump fails an average number of times is:

$$\mathbb{P}(X = 11) = \frac{\lambda^{11} e^{-\lambda}}{11!} = 0.12. \tag{2.25}$$

Note that we evaluated this at $x = 11$ (and not at $x = 11.3$) because the domain of $f(x)$ only contains integer numbers. Specifically, the factorial function is not defined for non-integer numbers.

Script to reproduce results: `ch2_example_poisson` □

2.2 Continuous Models

2.2.1 Uniform

As with the discrete case, this model is inspired by the fact that in most applications, we might not have data to determine an actual trend. Specifically, imagine that we only know a typical range $[a, b]$ that our variable of interest covers. Consequently, it makes sense to assume that it is equally likely (probable) that X is anywhere inside this domain. Accordingly, the pdf of a uniform RV is given by:

$$f(x) = \frac{1}{b-a}, \ x \in \mathcal{D}, \tag{2.26}$$

where the domain of the RV is $\mathcal{D} = [a, b]$. We have that the probability that X takes any value outside of $\mathcal{D}$ is 0 and thus the pdf is a rectangular function. We express the fact that X follows a uniform RV model as $X \sim \mathcal{U}_c(a, b)$ (we introduce the subscript c to indicate this is a continuous RV). The cdf is given by:

$$F(x) = \frac{x-a}{b-a}, \ x \in \mathcal{D}, \tag{2.27}$$

and note that $F(x) = 0$ for $x = a$ and $F(x) = 1$ for $x = b$. Note that the pdf is precisely the derivative of the cdf; i.e., $f(x) = dF(x)/dx$.

In Figure 2.8, we sketch the pdf/cdf with $a = -1$ and $b = 1$ and show 1,000 realizations. We can see that the pdf is a rectangular function with a probability $1/(b-a) = 1/2$ over the domain $[a, b]$. We note that the pdf is the derivative of the cdf (it is the slope of the cdf); consequently, because the pdf is constant in $\mathcal{D}$, we have that the cdf is a line with constant slope $1/(b-a)$. From the realizations, note how these concentrate evenly in the domain $[-1, 1]$, and there is a zero probability of finding them outside this domain.

The mean of the uniform RV is given by:

$$\begin{aligned} \mathbb{E}_X &= \int_a^b x \cdot \frac{1}{b-a} dx, \\ &= \frac{1}{b-a} \cdot \frac{1}{2}(b^2 - a^2), \\ &= \frac{1}{2}(a+b), \end{aligned} \tag{2.28}$$

this is the center point of the domain. For the example shown in Figure 2.8, we have that the mean is 0 (because $a = -1$ and $b = 1$).

From the cdf, we note that the α-quantiles are the values x satisfying:

$$\alpha = F(x) = \frac{x-a}{b-a}, \tag{2.29}$$

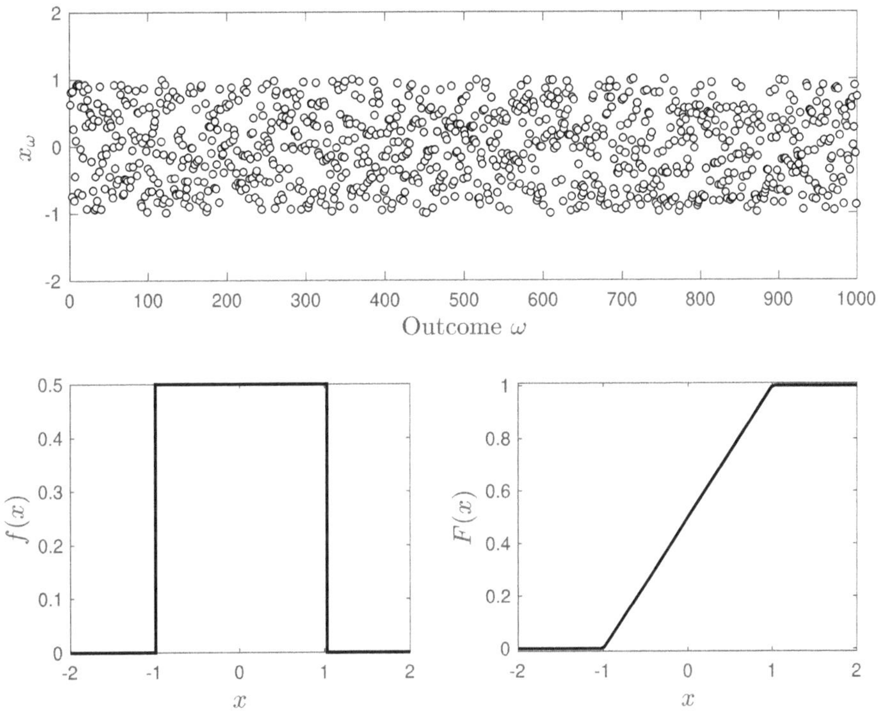

Figure 2.8 Pdf and cdf for continuous uniform RV with $a = -1$ and $b = 1$.

we thus have that $\mathbb{Q}_X(\alpha) = a + \alpha \cdot (b - a)$. Consequently, we have that $\mathbb{Q}_X(1) = b$ and $\mathbb{Q}_X(0) = a$ (the maximum and minimum values of $x \in \mathcal{D}$). Moreover, the median is $\mathbb{M}_X = \mathbb{Q}_X(1/2) = \frac{1}{2}(a + b)$ (the mean and median coincide for this RV).

Note that the mode for this RV (the point at which $f(x)$ is maximum) is not unique because the pdf is a constant (flat) function. The mode can be any value in the domain $x \in \mathcal{D}$. We also note that the pdf of this RV is symmetric. Finally, we observe that the variance for this RV is given by:

$$\mathbb{V}_X = \frac{1}{b-a} \int_a^b \left(x - \frac{1}{2}(a+b) \right)^2 dx = \frac{1}{12}(b-a)^2. \tag{2.30}$$

This reveals a connection between the variance and the range that the RV covers. Specifically, when the range $[a, b]$ is wide, we have that $(b - a)^2$ is large and the variance (uncertainty) is large.

The uniform distribution has many applications and is particularly used in developing sampling schemes. All that we need to know to specify this model are the boundaries of the domain, given by the parameters a and b. These can be estimated empirically (from data) by using $\hat{\mathcal{D}}$ with boundaries $\hat{a}, \hat{b}$ and with this we can estimate all needed statistics (e.g., $\hat{\mathbb{E}}_X$). Note, however, that the boundaries observed from available data might not be the true boundaries; in other words, it is possible that our variable lies outside the estimated domain. As such, it would be necessary to expand the empirical domain as we observe more data.

Example 2.10 Setting up a Safety Alarm

Consider the reactor of Example 1.2. Consider the situation in which there is no data recorded for the reactor pressure; all that you know is that, historically, the pressure has remained in the range 63–234 bar. Consider the questions:

- Is it adequate to model pressure as a uniform RV?
- Thermodynamics tells us that we will have low conversions at pressures below 100 bar. What is the probability that the reactor will exhibit low conversions?
- The reactor vessel is 30 years old and you are unsure that it can tolerate high pressures for much longer. To be safe, you want to set up a threshold value to trigger a safety alarm system. You do not want the threshold value to be too low because the alarm will be triggered often and you do not want it too high because this will put the system at risk. You decide that it is reasonable to set a threshold that triggers the alarm only 1% of the time. What is such threshold value?

In the absence of any evidence other than the fact that pressure stays in the continuous range $[63, 234]$ bar, we have no reason to doubt that it is equally likely to find pressure anywhere in the range. Consequently, the uniform RV is an adequate model with $b = 234$ and $a = 63$. Given this model, the probability that the reactor exhibits low conversion is:

$$\mathbb{P}(X \leq 100) = \frac{100 - 63}{234 - 63} = 0.21.$$

We want to find a threshold value x such that $\mathbb{P}(X > x) = 0.01$. This is equivalent to $\mathbb{P}(X \leq x) = 0.99$, and thus, we want to find x such that:

$$0.99 = \frac{x - 63}{234 - 63}$$

By solving for x, we have that the threshold is 232.3 bar. Note that this threshold value is the quantile:

$$\mathbb{Q}_X(0.99) = 63 + 0.99 \cdot (234 - 63) = 232.3.$$

□

2.2.2 Gaussian

Many random phenomena observed in real-life applications follows the behavior of a normal RV model (also known as the *Gaussian RV*). In fact, in Chapter 4, we will see that *all* random phenomena, *on average*, behave as a Gaussian RV. Because of this, it is important that we spend a bit of time understanding the behavior and properties of Gaussian RVs. Here, we will focus on univariate Gaussian RVs; in Chapter 3, we will generalize the analysis to multivariate Gaussian RVs.

A Gaussian RV X has a continuous domain of the form $\mathcal{D} = (-\infty, \infty)$ and has a pdf given by an exponential function of the form:

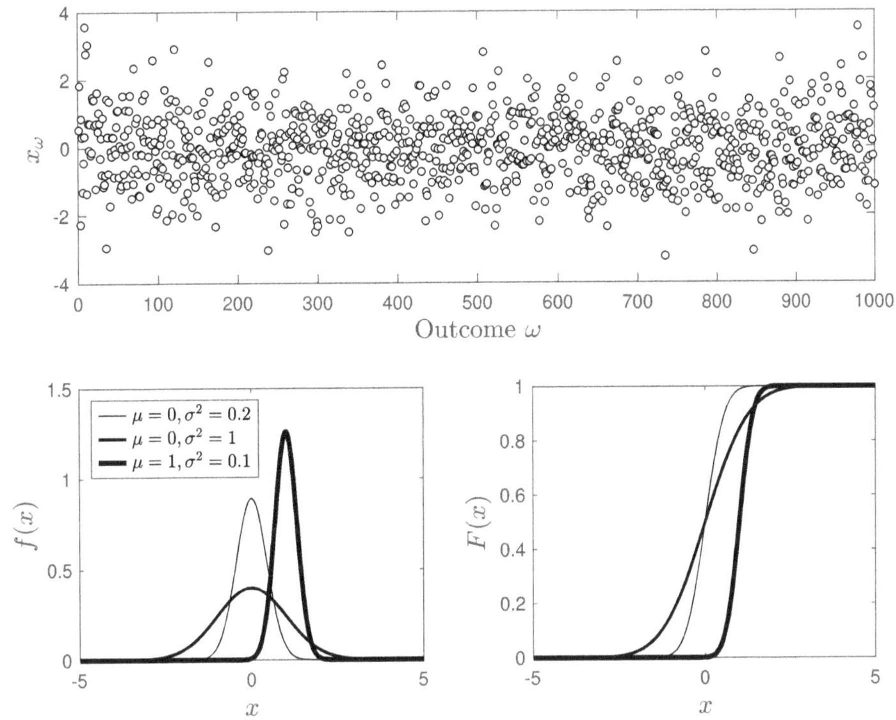

Figure 2.9 Pdf and cdf of Gaussian RV models and realizations for $\mathcal{N}(0,1)$.

$$f(x) = \frac{1}{\sqrt{2\pi\sigma^2}} \exp\left[-\frac{(x-\mu)^2}{2\sigma^2}\right], \quad x \in \mathcal{D}. \tag{2.31}$$

The scalars $\mu \in \mathbb{R}$ and $\sigma \in \mathbb{R}_+$ are the parameters of the model. We express the fact that an RV is Gaussian by using the notation $X \sim \mathcal{N}(\mu, \sigma^2)$. The cdf is given by:

$$F(x) = \frac{1}{2}\left(1 + \text{erf}\left(\frac{(x-\mu)/\sigma}{\sqrt{2}}\right)\right), \quad x \in \mathcal{D}, \tag{2.32}$$

where erf : $\mathbb{R} \to \mathbb{R}$ is the error function and is given by:

$$\text{erf}\,(x) = \frac{2}{\sqrt{\pi}} \int_0^x e^{-t^2} dt. \tag{2.33}$$

In Figure 2.9, we visualize the pdf/cdf of the Gaussian model for different values of its parameters and show 1,000 realizations for $\mathcal{N}(0,1)$. From here, we can make the following interesting observations:

- Recall that $f(x)$ is the derivative of $F(x)$; we thus note that the point of fastest growth in probability (fastest growth in $F(x)$) is given at the point that maximizes $f(x)$:

$$\max_{x \in \mathcal{D}} f(x). \tag{2.34}$$

  This maximum point occurs at $x = \mu$ and thus $\mathbb{MO}_X = \mu$.
- Because the pdf is *symmetric* around $x = \mu$, we have that $\mathbb{P}(X \leq \mu) = \mathbb{P}(X \geq \mu) = \frac{1}{2}$. In other words, μ is the median of X ($\mathbb{M}_X = \mu$).

- The probability $\mathbb{P}(X = x)$ (technically $\mathbb{P}(x \leq X \leq x + \Delta x)$ because this is a continuous RV) decays exponentially fast as we move away from μ. The speed of decay is dictated by σ (the smaller σ, the faster the decay). Note also that the range of possible outcomes of X decreases as σ decreases. Consequently, σ can be interpreted as a measure of uncertainty and variability; this can be visualized in the realizations shown in Figure 2.9 (note how the realizations spread around the mean, but the spread in this case decays quickly). In the limit when $\sigma \to 0$, the range becomes infinitesimally narrow and X becomes certain with a value of μ (becomes a deterministic variable).
- The domain of the Gaussian RV is $\mathcal{D} = (-\infty, \infty)$ and thus its outcomes can take negative or positive values and potentially very large or small values. This can represent a problem in some applications because many phenomena involve variables that cannot take negative values (e.g., mass) or that cannot take values above or below a certain threshold (e.g., temperatures). However, in practice, it is often assumed that probabilities are negligibly small in undesired portions of the domain.

Gaussian RVs naturally emerge from diffusion-like phenomena. An insightful visualization of this behavior can be obtained using a device known as the Galton board (see Figure 2.10). In this device, we have a set of pinballs (think of particles) that fall through a system of obstacles. The obstacles generate *random walks* for the pinballs. One observes that the pinballs accumulate in the bins at the bottom and that these bins form a bell-like shape that is characteristic of a Gaussian RV. This device also establishes a connection between the Gaussian RV and the binomial RV, which characterizes the probability of pinballs bouncing one way or another in the sequence of obstacles (sequence of Bernoulli events).

The Galton board also helps illustrate the *multi-scale* behavior of systems. Specifically, despite the fact that there is random (unpredictable) behavior occurring at small (micro) scales, the observed behavior at a large (macro) scale might be predictable. For instance, even if we cannot predict the exact trajectory that each individual pinball (particle) follows in the Galton board, their aggregated/bulk behavior *always* leads to a bell-like bin (concentration) profile.

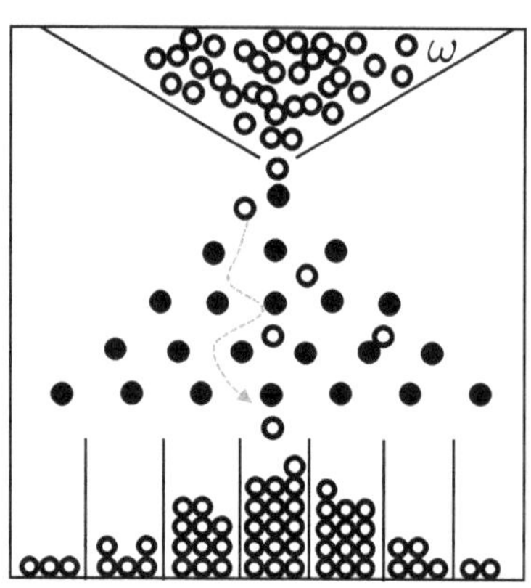

Figure 2.10 Illustration of a Galton board. This is a device that generates random walks for pinballs; the pinballs accumulate in the bins at the bottom by following a Gaussian distribution.

The connection between Gaussian RVs and diffusion-like phenomena can be formalized mathematically; for instance, consider a particle that diffuses randomly through a medium. We thus consider that the position of the particle is random and given by the RV X. Specifically, consider the question:

> What is probability of finding a particle at a location x in the spatial domain $\mathcal{D} = (-\infty, \infty)$ and at a given time t in $[0, T]$?

One can show that such probability, denoted as the function $f(x,t)$, is the solution of the diffusion equation:

$$\frac{\partial f(x,t)}{\partial t} = D\frac{\partial^2 f(x,t)}{\partial x^2}, \tag{2.35}$$

with boundary conditions:

$$\begin{aligned} f(x,t) &= 0, \quad x \in \mathcal{D},\ t = 0, \\ \int_{-\infty}^{\infty} f(x,t)dx &= 1, \quad t \in [0,T]. \end{aligned} \tag{2.36}$$

The solution of the diffusion equation is:

$$f(x) = \frac{1}{\sqrt{2\pi\sigma^2}} \exp\left[-\frac{x^2}{2\sigma^2}\right], \quad x \in \mathcal{D}, \tag{2.37}$$

with $\sigma^2 = 2Dt$. We thus have that, at any time t, the particle position X is a Gaussian RV of the form $X \sim \mathcal{N}(0, \sigma^2)$. Note the connection between σ and the diffusivity D and time t. As diffusivity increases, our uncertainty of where the particle is increases; similarly, as time increases, our uncertainty on where the particle is increases. Gaussian RVs also naturally emerge in residence time distributions (RTDs) for tubular reactors due to axial dispersion [29]. These examples highlight how random RVs can emerge from natural phenomena and how one can provide a physical interpretation to their behavior. A beautiful and detailed derivation of the diffusion equation based on statistical arguments is provided by Ogunnaike in [8].

The parameters of the Gaussian RV have a direct connection with the expected value and variance. Specifically, one can show that:

$$\mathbb{E}_X = \int_{-\infty}^{\infty} x \frac{1}{\sqrt{2\pi\sigma^2}} \exp\left[-\frac{(x-\mu)^2}{2\sigma^2}\right] dx = \mu. \tag{2.38}$$

Obtaining this result is nontrivial, see [30] for a detailed derivation. We can thus conclude that the mean, mode, and median of a Gaussian RV coincide. Moreover, one can show that the variance is:

$$\mathbb{V}_X = \int_{-\infty}^{\infty} (x-\mu)^2 \frac{1}{\sqrt{2\pi\sigma^2}} \exp\left[-\frac{(x-\mu)^2}{2\sigma^2}\right] dx = \sigma^2. \tag{2.39}$$

We thus have that $\sigma = \mathbb{SD}_X$, and thus, we can see that σ is indeed a measure of uncertainty.

Equations (2.38) and (2.39) indicate that we can estimate the parameters μ and σ directly from the empirical approximations for the mean and variance $\hat{\mathbb{E}}_X$ and $\hat{\mathbb{V}}_X$, respectively. We denote the parameter estimates as $\hat{\mu}$ and $\hat{\sigma}^2$. In Chapter 4, we will see that these empirical estimates converge to the true expected value and variance of X as our number of observations increases ($\hat{\mu} \to \mathbb{E}_X$ and $\hat{\sigma}^2 \to \mathbb{V}_X$). Moreover, we will see that the empirical mean and variance are the best possible estimates for the parameters.

Gaussian RVs have the extremely useful property that they remain Gaussian under *linear transformations*. Specifically, consider the propagation of $X \sim \mathcal{N}(\mu,\sigma^2)$ through a linear system of the form $Y = \varphi(X)$ with $\varphi(X) = a + b \cdot X$ (where a and b are scalar values). One can show that:

$$Y \sim \mathcal{N}(a + b \cdot \mu, b^2 \cdot \sigma^2). \tag{2.40}$$

In other words, Y is a Gaussian RV with expected value $\mathbb{E}_Y = a+b\cdot\mathbb{E}_X$ and variance $\mathbb{V}_Y = b^2 \cdot \mathbb{V}_X$.

Think about some of the implications and applications of this basic result; for instance, it tells us that a linear transformation $\varphi(X) = a + bX$ will *shrink/damp* uncertainty if $b < 1$ and will *magnify* uncertainty if $b > 1$. Note also that, if $Z \sim \mathcal{N}(0,1)$ and $X = \mu + \sigma \cdot Z$, then $X \sim \mathcal{N}(\mu,\sigma^2)$; conversely, we have that $Z = (X - \mu)/\sigma \sim \mathcal{N}(0,1)$. The Gaussian RV $Z \sim \mathcal{N}(0,1)$ is known as the standard normal RV; we will see that this RV is useful in computing quantities for general Gaussian RVs.

Example 2.11 Gaussian Mixing Problem

In this example, we illustrate how we can use the properties of Gaussian RVs to infer properties of other RVs derived from them. Consider the system in Figure 2.11; here, we have an input flow $f_1 = 10$ gpm that can be measured with high accuracy (so it is appropriate to assume this to be deterministic). However, we have a problematic input flow F_2 (also in gpm) that cannot be measured with the existing sensor to high accuracy, and it is thus uncertain; historical data indicates that this flow follows a Gaussian RV $\mathcal{N}(20,1)$. Even if the flow F_2 is uncertain, we can control its impact by using a valve with coefficient $\kappa \in [0,1]$ ($\kappa = 1$ is fully open and $\kappa = 0$ is fully closed).

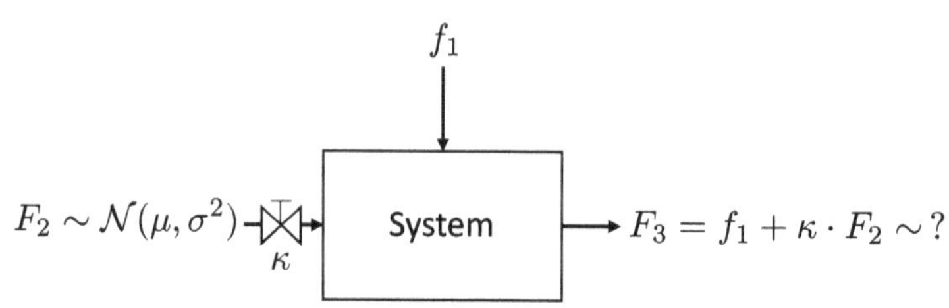

Figure 2.11 Mixing a certain flow f_1 with an uncertain flow F_2.

We have the following questions:

- What type of an RV is the output flow $F_3 = f_1 + \kappa \cdot F_2$? What is its mean and standard deviation?
- How is the uncertainty of F_3 affected as we close the valve $\kappa \to 0$ and as we open the valve $\kappa \to 1$? Why?

To answer these questions, we note that $F_3 = 10 + \kappa \cdot F_2$ and thus F_3 is a linear transformation of the Gaussian F_2. Consequently, we have that F_3 is a Gaussian RV and $F_3 \sim \mathcal{N}(10 + \kappa \cdot 20, \kappa^2 \cdot 1)$; in other words, its expected value and variance are $\mathbb{E}[F_3] = 10 + \kappa \cdot 20$ and $\mathbb{SD}[F_3] = \kappa$.

From the standard deviation, note that the uncertainty of F_3 is directly proportional to the opening of the valve. The standard deviation becomes zero when we close the valve ($\kappa = 0$) because $F_3 = f_1$ (the effect of the uncertain flow disappears and thus F_3 becomes certain/deterministic). The uncertainty of F_3 is maximum when the valve is fully open ($\kappa = 1$). □

Consider again the Gaussian $X \sim \mathcal{N}(\mu, \sigma^2)$ and the linear transformation $Y = a + b \cdot X$. One can show that the cdfs of X and Y satisfy:

$$F_Y(y) = F_X(x) \quad \text{for all} \quad y = a + b \cdot x. \tag{2.41}$$

This is equivalent to state that

$$F_Y(a + b \cdot x) = F_X(x), \tag{2.42}$$

or $F_Y(y) = F_X((y - a)/b)$. This also implies that:

$$\mathbb{Q}_Y(\alpha) = a + b \cdot \mathbb{Q}_X(\alpha), \tag{2.43}$$

or $\mathbb{Q}_X(\alpha) = (\mathbb{Q}_Y(\alpha) - a)/b$.

These properties tell us that the cdfs and quantiles of Y and X are scaled versions of one another, and this has interesting implications; for instance, the cdf of $Z \sim \mathcal{N}(0, 1)$ is given by:

$$F_Z(z) = \frac{1}{2}\left(1 + \text{erf}\left(\frac{z}{\sqrt{2}}\right)\right), \tag{2.44}$$

and this does not depend on any parameters. Consider now the transformation $X = \mu + \sigma \cdot Z$, substituting $z = (x - \mu)/\sigma$:

$$\begin{aligned} F_Z((x - \mu)/\sigma) &= \frac{1}{2}\left(1 + \text{erf}\left(\frac{(x - \mu)/\sigma}{\sqrt{2}}\right)\right) \\ &= F_X(x). \end{aligned} \tag{2.45}$$

Consequently, we confirm that $F_X(x) = F_Z(z)$ with $z = (x - \mu)/\sigma$ holds. Moreover, we have that

$$\mathbb{Q}_X(\alpha) = \mu + \sigma \cdot \mathbb{Q}_Z(\alpha). \tag{2.46}$$

This result establishes that all that we need to know to compute the cdf and quantiles of $X \sim \mathcal{N}(\mu, \sigma^2)$ are the cdf and quantiles for $Z \sim \mathcal{N}(0,1)$. The α-quantiles (also known as critical values) of $Z \sim \mathcal{N}(0,1)$ are often denoted as $z_\alpha = \mathbb{Q}_Z(\alpha)$.

Calculations with standard normal RVs form the basis behind calculations for general Gaussian RVs in statistical software packages. The scaling of Gaussian variables as $(X - \mu)/\sigma$ is standard practice; this *scalarization* procedure is used, for instance, to put different Gaussian RVs on an equal basis. Note also that scalarization eliminates units because the mean and the standard deviation have the same units.

Example 2.12 Quantiles for Mixing Problem

Recall that, for the mixing problem of Figure 2.11, we have that $F_3 \sim \mathcal{N}(10+\kappa \cdot 20, \kappa^2)$. Assume that the valve is fully open ($\kappa = 1$) and thus $F_3 \sim \mathcal{N}(30, 1)$.

We compute the quantile function for $\mathcal{N}(30,1)$ and of $\mathcal{N}(0,1)$ in order to see how the quantile functions are related. The functions are shown in Figure 2.12; notice how the shapes of the functions are exactly the same, but the function of $\mathcal{N}(30,1)$ is displaced 30 units to the right. In other words, the quantile function of $\mathcal{N}(30,1)$ is a rescaled version of that of $\mathcal{N}(0,1)$.

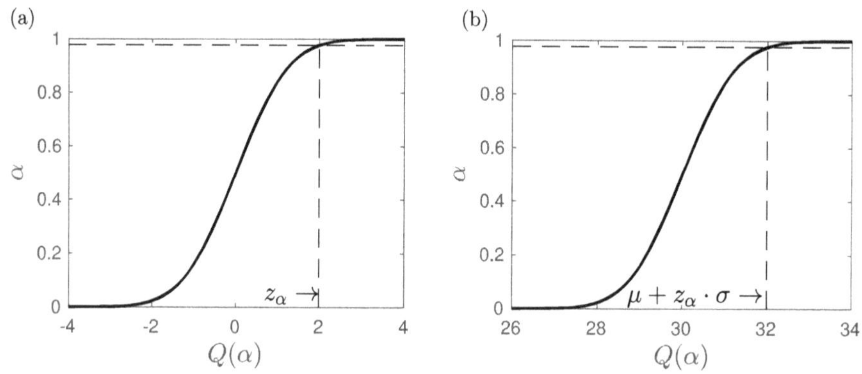

Figure 2.12 Quantile functions for $\mathcal{N}(0,1)$ (a) and $\mathcal{N}(\mu, \sigma^2)$ with $\mu = 30$ and $\sigma = 1$ (b).

We know that the quantile of $Z \sim \mathcal{N}(0,1)$ at $\alpha = 0.977$ is $z_\alpha = Q_Z(\alpha) = 2$. From this result, we can directly obtain that the quantile for $F_3 \sim \mathcal{N}(30,1)$ is $\mu + 2 \cdot \sigma = 30 + 2 \cdot 1 = 32$. From the quantile functions, we can confirm that this is true. This result indicates that $\mathbb{P}(F_3 \leq 32) = 0.977$; in other words, we are 97.7% confident that the flow F_3 is below a value of 32 gpm. By comparison, we can say that we are only 50% confident that F_3 is below 30 gpm and that we are only 84.1% confident that it is below 31 gpm; we can see this from the quantile for $\mathcal{N}(30,1)$.

Script to reproduce results: `ch2_mixing_gaussians` □

The cdfs and critical values of the standard normal $Z \sim \mathcal{N}(0,1)$ also have an interesting interpretation. Consider the cdf $F_Z(k) = \mathbb{P}(Z \leq k)$ evaluated at the integer values $k = 0, 1, 2, 3, \ldots,$. The cdf values for the Gaussian $X \sim \mathcal{N}(\mu, \sigma^2)$ satisfy:

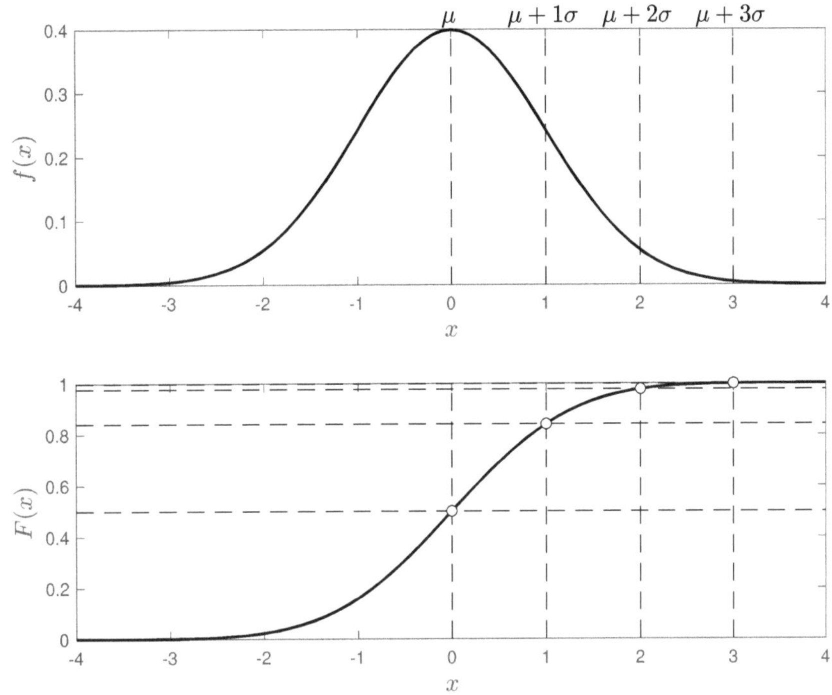

Figure 2.13 Critical values of $\mathcal{N}(0,1)$ at $\mu, \mu + 1\sigma, \mu + 2\sigma$, and $\mu + 3\sigma$ along with corresponding probabilities.

$$\begin{aligned}
\mathbb{P}(Z \leq 0) = 50.0\% &\Longleftrightarrow \mathbb{P}(X \leq \mu) = 50.0\% \\
\mathbb{P}(Z \leq 1) = 84.1\% &\Longleftrightarrow \mathbb{P}(X \leq \mu + \sigma) = 84.1\% \\
\mathbb{P}(Z \leq 2) = 97.7\% &\Longleftrightarrow \mathbb{P}(X \leq \mu + 2\sigma) = 97.7\% \\
\mathbb{P}(Z \leq 3) = 99.9\% &\Longleftrightarrow \mathbb{P}(X \leq \mu + 3\sigma) = 99.9\%.
\end{aligned}$$

In other words, the probability that X is below μ is always 50% (regardless of the value of μ), the probability that it is below $\mu + \sigma$ is always 84.1% (regardless of μ and σ), and so on. You can thus remember these general values to do *back-of-the-envelope* calculations of the probability levels for a given value of μ and σ. These levels are illustrated in Figure 2.13. Compare these values against the quantiles computed in Example 2.12 (compared for $z_\alpha = 0, 1, 2$).

The cdfs and critical values of the standard normal are also used to compute *confidence regions* (a.k.a. confidence intervals) for $X \sim \mathcal{N}(\mu, \sigma^2)$. Assume a probability level $\alpha \in [0, 1]$; it can be shown that the critical value z that satisfies the equation:

$$\mathbb{P}(-z \leq Z \leq z) = 1 - \alpha, \tag{2.47}$$

is given by $z = \mathbb{Q}_Z(1 - \frac{\alpha}{2})$ (this critical value is often denoted using the notation $z_{\alpha/2}$). In other words, we have that:

$$\mathbb{P}(-z_{\alpha/2} \leq Z \leq z_{\alpha/2}) = 1 - \alpha. \tag{2.48}$$

Using our linear transformation properties for $X \sim \mathcal{N}(\mu, \sigma^2)$, we have that the critical value for X corresponding to $z_{\alpha/2}$ is $\mu + \sigma z_{\alpha/2}$ and for $-z_{\alpha/2}$ is $\mu - \sigma z_{\alpha/2}$; consequently:

$$\mathbb{P}(\mu - z_{\alpha/2}\sigma \leq X \leq \mu + z_{\alpha/2}\sigma) = 1 - \alpha. \tag{2.49}$$

This expression indicates that the probability of finding $X \sim \mathcal{N}(\mu, \sigma^2)$ in the region $\mu \pm z_{\alpha/2} \cdot \sigma$ is $1 - \alpha$. This gives us an idea of how confident we are of finding X in a given range. Confidence intervals are widely used in engineering and science, as a way to express uncertainty associated with different random quantities; for example, you can express that an investment will yield a return in the range $\mu \pm z_{\alpha/2} \cdot \sigma$ with certainty (probability) of $1 - \alpha$.

Example 2.13 Confidence Region for Gaussian Mixing Problem

Consider again the mixing problem of Example 2.11. We are interested in answering the following question:

- In what range do we expect the output flow $F_3 \sim \mathcal{N}(30, 1)$ to be with 90% confidence?

Applying the calculations that we have discussed, we have $1 - \alpha = 0.9$ and thus $\alpha = 0.1$ and $\mathbb{Q}(1 - \frac{\alpha}{2}) = 1.644$. We thus have F_3 is in the interval:

$$[\mu - z_{\alpha/2}, \mu + z_{\alpha/2}] = [30 - 1.64 \cdot 1, 30 + 1.64 \cdot 1] = [28.36, 31.64],$$

with 90% probability. In other words, we have that $\mathbb{P}(28.36 \leq F_3 \leq 31.64) = 0.9$; this indicates that we are fairly certain that F_3 is within 28.36 and 31.64 gpm. The confidence region is depicted in Figure 2.14.

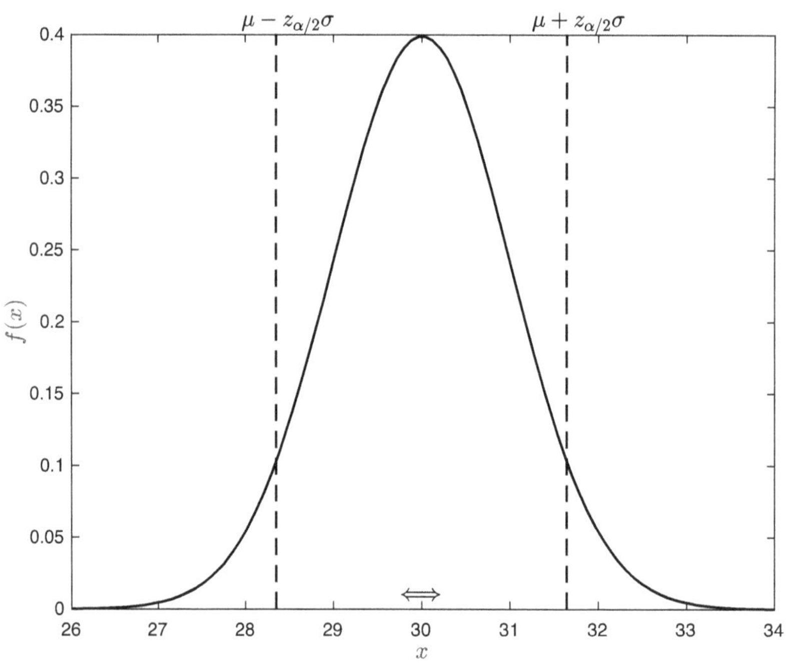

Figure 2.14 90% confidence region for flow $F_3 \sim \mathcal{N}(30, 1)$

Script to reproduce results: `ch2_mixing_gaussians` □

2.2.3 Lognormal

The *lognormal RV* is important in the description of natural phenomena that might span *multiple scales*. For instance, natural growth processes are often driven by the accumulation of multiplicative effects, and these become additive on a log scale. Lognormal RVs are used to model the economic income of the population of a country (to measure inequality) and to model particle size distributions (PSDs) in soil and powder products [31, 32].

As the name suggests, the lognormal RV is related to the normal RV (it is a logarithmic transformation); this can be easily seen from the form of its pdf:

$$f(x) = \frac{1}{x\sqrt{2\pi\sigma^2}} \exp\left[-\frac{(\log(x)-\mu)^2}{2\sigma^2}\right], \; x \in \mathcal{D}, \tag{2.50}$$

with hyperparameters $\mu \in \mathbb{R}$ and $\sigma \in \mathbb{R}_+$. The domain of this RV is $\mathcal{D} = (0, \infty)$, which is different from that of the normal RV because the logarithm is not defined for zero and for negative values. This is a crucial difference from an application standpoint because the lognormal model is more appropriate to capture variables that cannot be negative. The pdf of this RV is presented in Figure 2.15.

We state the fact that X is a lognormal RV by writing $X \sim \log\mathcal{N}(\mu, \sigma)$. We can make the connection between the normal and lognormal more explicit; specifically, note that if $X \sim \log\mathcal{N}(\mu, \sigma)$, then $\log(X) \sim \mathcal{N}(\mu, \sigma^2)$.

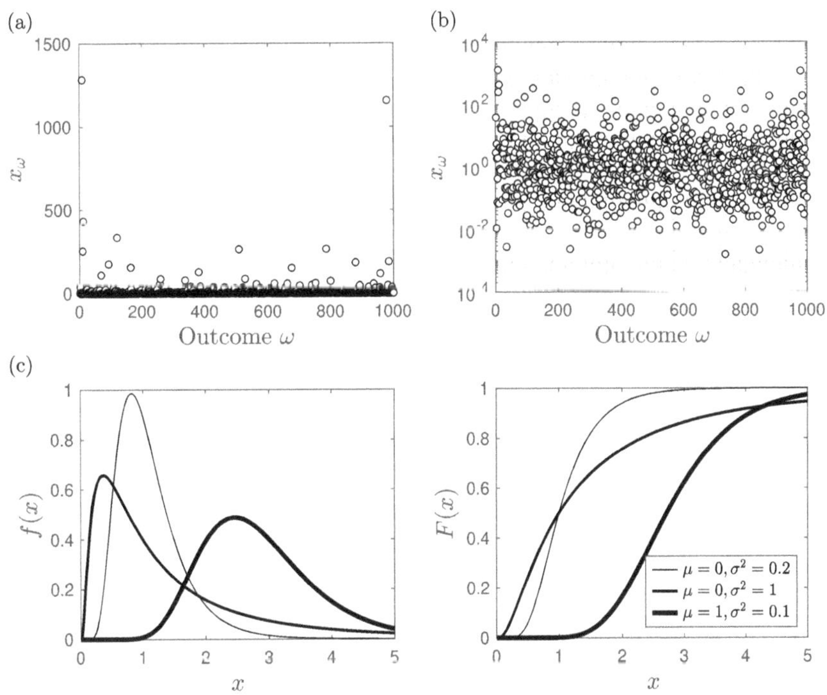

Figure 2.15 Realizations visualized in regular scale (a) and logarithmic scale (b). Pdf/cdf for lognormal random variable (c).

The cdf of the lognormal has the form:

$$F(x) = \frac{1}{2}\left(1 + \text{erf}\left(\frac{(\log(x) - \mu)/\sigma}{\sqrt{2}}\right)\right), \quad x \in \mathcal{D}. \tag{2.51}$$

The α-quantiles for this RV are given by:

$$\mathbb{Q}_X(\alpha) = \exp\left(\mu + \sigma \cdot \sqrt{2} \cdot \text{erf}^{-1}(2\alpha - 1)\right). \tag{2.52}$$

From here, we can see that the median of this RV is $\mathbb{M}_X = \exp(\mu)$. The mean and variance of the lognormal RV are given by:

$$\begin{aligned} \mathbb{E}_X &= \exp\left(\mu + \frac{1}{2}\sigma^2\right) \\ \mathbb{V}_X &= \exp(\sigma^2 - 1)\exp(2\mu + \sigma^2). \end{aligned} \tag{2.53}$$

Interestingly, the mean for a lognormal RV does not coincide with the median, and the mean is not given by μ. Note also that both the mean and variance are functions of μ and σ, and thus, the interpretation of the parameters is different from that in the normal RV case. Moreover, because the mean and variance depend on both parameters, it might be more convenient to estimate μ from an empirical approximation of the median (because this only depends on μ) and then estimate σ from either the empirical approximation of the mean or variance. The mode of the lognormal RV is given by $\mathbb{MO}_X = \exp(\mu - \sigma^2)$; we thus have that $\mathbb{MO}_X < \mathbb{M}_X < \mathbb{E}_X$. This indicates that the pdf is skewed to the right (it is not symmetric).

In Figure 2.15, we present 1,000 realizations for $\log\mathcal{N}(0,4)$; in Figure 2.15(a), we see how the realizations appear to concentrate around small values, but this is only a visual effect because there are also realizations with very large values. This manifests in a pdf that has a long "tail." When we visualize the realizations using a logarithmic transformation (Figure 2.15(b)), we can see that they follow a normal-like distribution.

Example 2.14 Particle Size Distribution in Pharma Products

In the pharmaceutical (pharma) industry, the particle characterization of powder materials is crucial in drug product development and quality control of solid oral dosage forms. Specifically, the particle size distribution (PSD) of a drug substance may have significant effects on final drug product performance (e.g., dissolution and bioavailability) [33]. Furthermore, the PSDs can affect manufacturability of the final drug product (e.g., flowability and blend uniformity), which can impact processing [34]. A laboratory characterization method known as laser diffraction was used to characterize the PSD of 1,000 kg of a batch of a drug substance. The data provided by lab method is in terms of a cumulative distribution for particle sizes, as shown in Figure 2.16.

Address the following questions:

- Does the PSD follow a lognormal RV?
- If so, what are the parameters of the RV? What is the PSD?

- The product is considered to be acceptable if the particle sizes are in the range [250, 400] μm. How many kilograms of the batch are within the acceptable range?

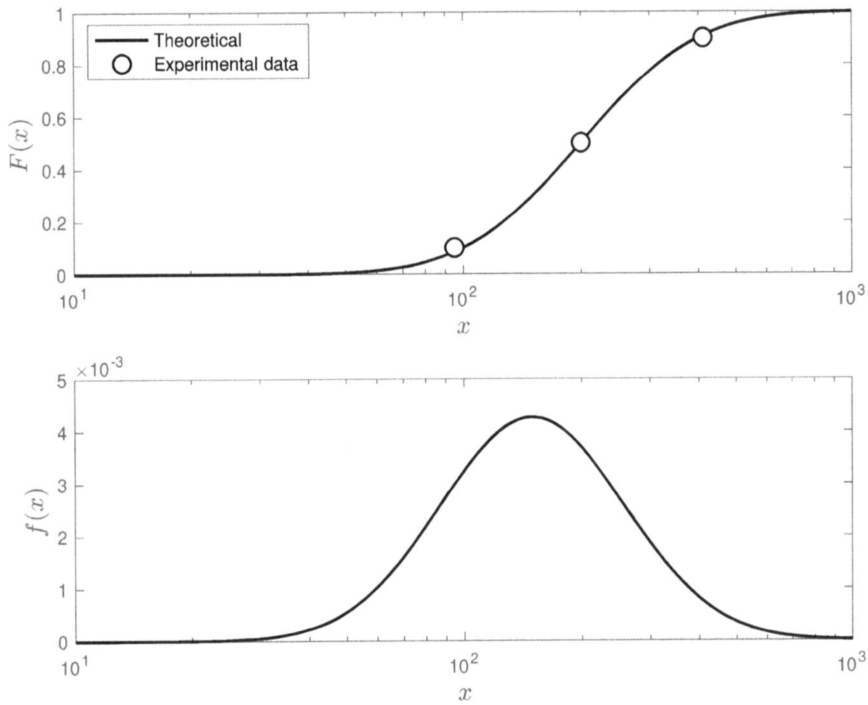

Figure 2.16 Cdf and pdf for particle size distribution of a pharma product.

Note that the particle sizes span three orders of magnitude (from 10 μm to 1,000 μm); this already tells us that a lognormal RV model might be appropriate. Our goal is to demonstrate that the three experimental data points correspond to those of a lognormal RV. We thus hypothesize that the particle size is indeed a lognormal RV.

From the experimental data point at particle size of 200 μm, we have that the probability of finding a particle with size below this value is 50%. We thus have that this particle size is the median of the PSD. Under our hypothesis, we have that the median of the distribution must be $\mathbb{M}_X = \exp(\mu) = 200$, and from here, we estimate that $\mu = \log 200$. From the data, we also observe that the probability of finding a particle size below 400 μm is 90%. Under our hypothesis, we thus have that:

$$\mathbb{Q}_X(\alpha) = \exp\left(\mu + \sigma \cdot \sqrt{2} \cdot \text{erf}^{-1}(2 \cdot \alpha - 1)\right).$$

Solving for σ, we have that

$$\sigma = \frac{\log \mathbb{Q}_X(\alpha) - \mu}{\sqrt{2} \cdot \text{erf}^{-1}(2 \cdot \alpha - 1)}.$$

Substituting terms, we find that:

$$\sigma = \frac{\log \mathbb{Q}_X(0.9) - \mu}{\sqrt{2} \cdot \text{erf}^{-1}(2 \cdot 0.9 - 1)}$$

$$= \frac{\log 400 - \log 200}{\sqrt{2} \cdot \text{erf}^{-1}(2 \cdot 0.9 - 1)}$$
$$= 0.54.$$

In Figure 2.16, we show the cdf of the RV model $\log \mathcal{N}(\mu, \sigma)$ and see that this matches the experimental data. This seems to validate our hypothesis that the particle size is a lognormal RV. Here, we also present the pdf for the particle size. From there, it is clear that the particle sizes cover a wide range. The probability that the particle size is in the acceptable range is:

$$\mathbb{P}(250 \leq X \leq 400) = F(400) - F(250) = 0.24.$$

This means that only 24% of our product is acceptable and indicates that a total mass of $1{,}000 \cdot 0.24 = 240$ kg is acceptable. The rest of the product will have to be separated (e.g., using sieves) and recycled. Note that estimating the amount of product in the desired range requires a theoretical RV model because the experimental data is not collected at the cdf values needed. This highlights how theoretical models can be used to make predictions.

Script to reproduce results: `ch2_example_lognormal` □

2.2.4 Exponential

An important RV that often appears in applications is the *exponential RV*. An exponential RV has a pdf of the form:

$$f(x) = \frac{1}{\beta} \exp\left(-x/\beta\right), \quad x \in \mathcal{D}, \tag{2.54}$$

with domain $\mathcal{D} = [0, \infty)$. The only parameter of this model is $\beta \in \mathbb{R}_+$ (this is known as the scale value). The reciprocal $\eta = 1/\beta$ is known as the intensity; the pdf can also be written as $f(x) = \eta \cdot e^{-\eta \cdot x}$.

We express the fact that X is exponential as $X \sim \text{Exp}(\beta)$. The cdf of the exponential RV is given by:

$$F(x) = 1 - \exp\left(-x/\beta\right), \quad x \in \mathcal{D}. \tag{2.55}$$

For this RV, it is rather easy to establish that the pdf is the derivative of the cdf. In Figure 2.17, we show the behavior the exponential RV for different values of β.

From the structure of the pdf and cdf, we can observe that the probability of finding X away from zero decays exponentially fast at rate η. Note also that this RV never takes negative values. Note also that the pdf is asymmetric and one can think of an exponential RV as a one side of a Gaussian RV.

The expected value and variance of an exponential RV are given by $\mathbb{E}_X = \beta$ and $\mathbb{V}_X = \beta^2$. We thus have that we can compute an empirical estimate for the parameter by using empirical estimates of either the mean or variance; however, there is also an easy way to do this by using the cdf. The exponential RV has the interesting property that $F(\beta) = 0.63$ (this can be easily confirmed from the cdfs

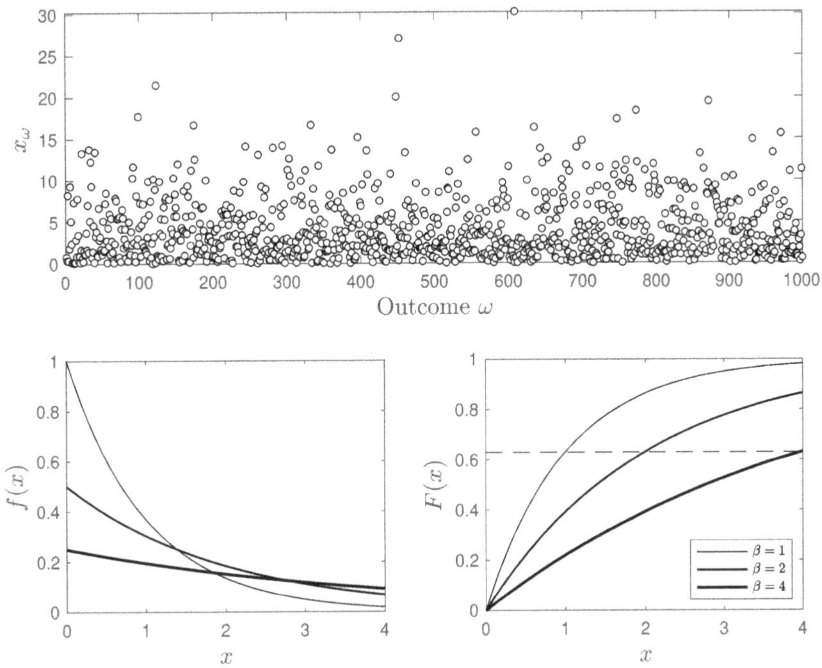

Figure 2.17 Pdf/cdf for exponential RV and realizations for Exp(4).

shown in Figure 2.17). Consequently, one can infer the value of β directly from the empirical cdf.

For this RV, it is easy to see that the quantile function is given by:

$$\mathbb{Q}(\alpha) = -\beta \log(1 - \alpha). \tag{2.56}$$

Note also that the cdf is strictly increasing and thus the quantiles are unique. From the quantile function, we can also see that the median is simply $\mathbb{M}_X = \beta \cdot \log(2)$. We can also observe that the mode is

$$\mathbb{MO}_X = \underset{x \in \mathcal{D}}{\operatorname{argmax}} f(x) = 0. \tag{2.57}$$

Consequently, the median, mean, and mode do not coincide; finally, we observe that the pdf evaluated at the mode is $f(0) = 1/\beta$.

The exponential RV is often used to model time-dependent phenomena (e.g., *failures* in reliability analysis) [35, 36]. In this context, it is important to highlight the relationship between the exponential and Poisson RVs. For instance, an exponential RV can be used to model the time that we have to wait between Poisson events; conversely, if the time between events follows an exponential RV, then the number of events in a given interval $[0, z]$ is a Poisson RV. The average waiting time between events is $\mathbb{E}_X = \beta$ and the intensity $\eta = 1/\beta$ is the average event rate (average rate at which Poisson events occur). We recall that the average number of Poisson events in an interval $[0, z]$ is $\lambda = \eta \cdot z$.

Due to the use of the exponential RV in failure analysis, it is often useful to think of this RV in terms of the so-called *survival function* $\mathbb{P}(X > x) = S(x) = 1 - F(x)$.

In failure analysis, this characterizes the probability that a system survives after time x.

Exponential RVs arise in time-dependent applications because they naturally characterize the response time of a dynamical system after an instantaneous excitation (perturbation). Specifically, the exponential cdf naturally arises from the solution of the first-order differential equation:

$$\beta \frac{dF(t)}{dt} + F(t) = 1, \quad t \in [0, \infty), \tag{2.58}$$

with $F(0) = 0$. The solution of this differential equation is simply $F(t) = 1 - e^{-t/\beta}$. We can also see that β can be interpreted as the time constant of the response. A wide range of dynamical systems exhibit first-order responses, and thus, the exponential RV has many applications. These observations are analogous to the diffusion interpretation of the Gaussian RV. One can also interpret the exponential RV as spatial distance (instead of time); in such a case, we would be interested in characterizing the distance that we need to travel until we see a certain event.

Example 2.15 Residence Time in a Continuously Stirred Tank Reactor (CSTR)

To provide an example on how exponential RVs naturally emerge in physical systems, consider a CSTR with volume V and input and output flow Q. This system is illustrated in Figure 2.18. At time $t = 0$, we inject a pulse of particles (e.g., a tracer) so that the concentration of the input flow is C_0. The particle concentration inside the vessel at time t is $C(t)$.

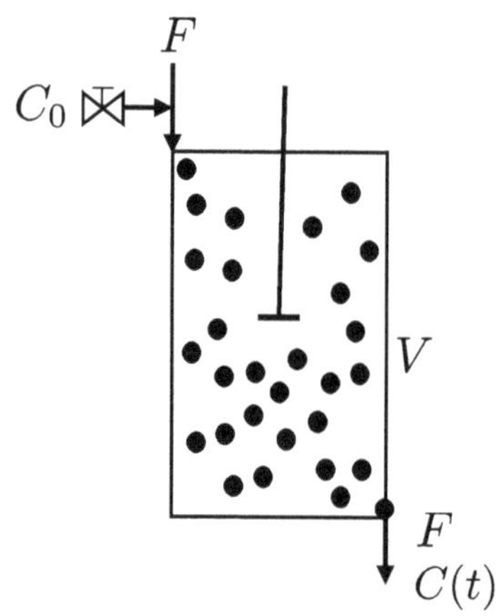

Figure 2.18 Particle residence time in a perfectly mixed system.

Consider the questions:

- For how long will a given particle reside in the system?
- What factors influence this time?

We expect that some particles will have a longer residence time (denoted as T) than others and thus the residence time is an RV. The material balance for this system is:

$$f(t) = \frac{1}{\tau}e^{-t/\tau}, \quad t \in [0, \infty) \tag{2.59}$$

with $\tau = V/Q$ and $f(t) = \frac{C(t)}{C_0}$. Consequently, the residence time of the particles can be modeled as an exponential random variable $T \sim \text{Exp}(\tau)$. Now let's use the properties of exponential RVs to see the implications of this:

- The pdf of the residence time is $f(t) = \frac{1}{\tau}e^{-t/\tau}$. This tells us that the probability that a particle resides in the system decays exponentially with time; in other words, few particles will have a long residence time.
- The mean residence time is $\mathbb{E}[T] = \tau = V/Q$. This indicates that the mean residence time increases as the volume of the vessel increases and decreases as the flow increases.
- The cdf of the residence time is:

$$F(t) = \int_0^t f(t')dt = \int_0^t (1/\tau)e^{-t'/\tau}dt' = (1 - e^{-t/\tau}), \tag{2.60}$$

  and thus $F(0) = 0$ and $F(\infty) = 1$. From a physical viewpoint, the cdf $F(t)$ is interpreted as the fraction of particles that have exited up to time t (have residence time T less than t). At $t = \infty$, we have that all particles have exited, and at $t = 0$, we have that none have exited. From a statistical viewpoint, the cdf $F(t)$ is interpreted as the probability that a particle entering at time $t = 0$ has left before time t. This interpretation is the basis behind the experimental characterization of residence times in reactors, as has been widely studied in chemical engineering (see the book of Hill and Root [29] and Fogler [5] for detailed discussions on the subject).
- The fraction of particles that remains (survive) in the system at time t is given by $\mathbb{P}(T > t) = 1 - F(t)$. □

Example 2.16 Product Failure and Guarantee

Your company manufactures adhesive strips; after performing experimental tests for a new commercial product (over 1,000), it has been observed that the time to failure can be modeled as an exponential RV. In Figure 2.19, we can see the outcomes of the experimental tests, the pdf, and the cdf.

Address the following questions:

- What is the average time to failure of the product?
- The adhesive is commercialized with a "guarantee" that it will last more than 1 year. What is the probability that the adhesive satisfies this guarantee?
- How frequently will the adhesive last more than 5 years?
- You have been tasked with improving the average lifetime of the product by investigating a new chemistry. What needs to be the target average lifetime of the new product so that the 1-year guarantee is satisfied with 95% probability?

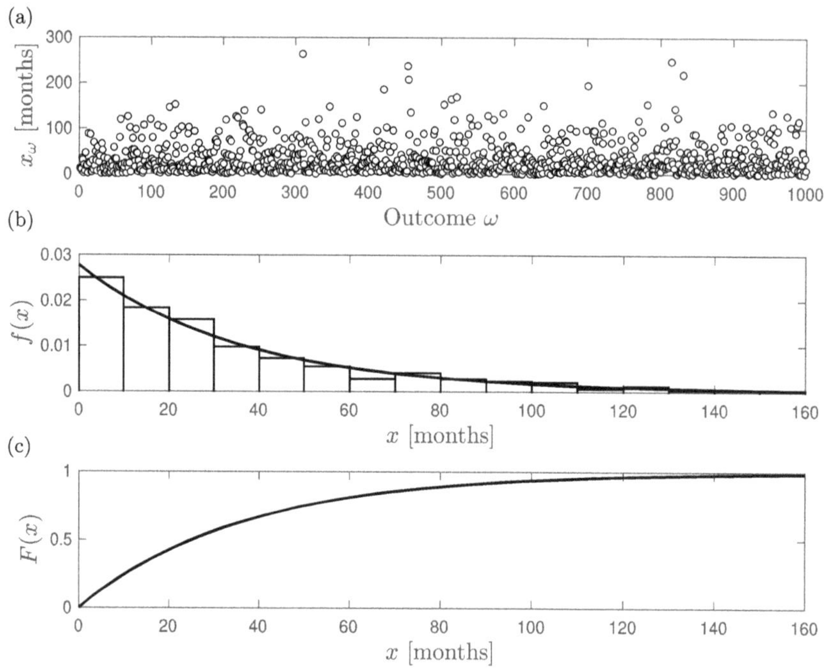

Figure 2.19 Experimental tests (a), empirical and theoretical pdf for exponential RV (b), and theoretical cdf (c).

From the cdf, it is easy to observe that $F(x) = 0.63$ occurs at $x = \beta = 36$ months. The probability that the adhesive will last longer than 12 months (i.e., survives after 12 months) is given by the survival function:

$$\mathbb{P}(X > 12) = S(12) = 1 - F(12) = e^{-12/36} = 0.716.$$

The adhesive will last more than 5 years, $S(60) = 0.19$ (19% of the time). To address the last question, we need to find the average lifetime that makes the adhesive survive after 12 months with 0.95 probability. This is obtained from the survival function as:

$$0.95 = e^{-12/\beta},$$

which gives $\beta = -12/\log 0.95 = 233$ months. You need to improve the average lifetime by a factor of $233/36 \approx 6.5$. Good luck!

Script to reproduce results: `ch2_example_exponential` □

2.2.5 Gamma

The *Gamma RV* is a generalization of the exponential RV and has pdf:

$$f(x) = \frac{1}{\beta^\alpha \Gamma(\alpha)} \cdot e^{-x/\beta} \cdot x^{\alpha-1}, \quad x \in \mathcal{D}. \tag{2.61}$$

The pdf has shape and scale parameters $\alpha, \beta \in \mathbb{R}_+$ and domain $\mathcal{D} = [0, \infty)$. The function $\Gamma(\alpha)$ is the Gamma function; for integer $\alpha \geq 1$, we have $\Gamma(\alpha) = (\alpha - 1)!$ but for non-integer α, this function is quite complicated. A Gamma RV with integer α is also known as the Erlang RV.

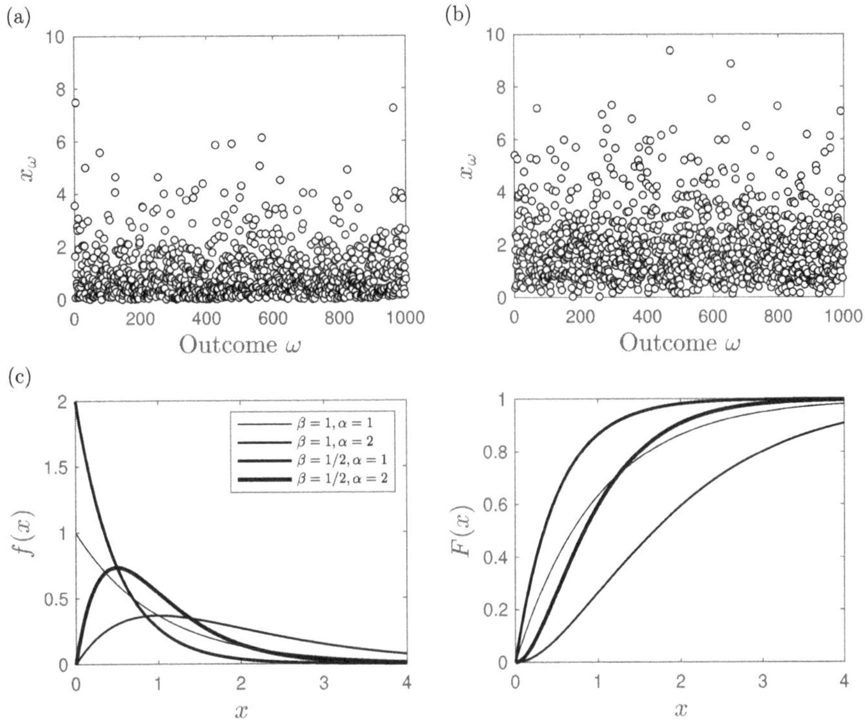

Figure 2.20 Realizations obtained with Gamma(1, 1) (a) and Gamma(2, 1) (b). Pdfs and cdfs for Gamma random variable (c).

We express the fact that X is a Gamma RV as $X \sim \text{Gamma}(\alpha, \beta)$. The behavior of the Gamma RV is illustrated in Figure 2.20. The pdf tells us that one recovers an exponential RV when $\alpha = 1$. Note also the emergence of peaks in the pdf due to competing effects for $\alpha = 2$. The term $x^{\alpha-1}$ in the pdf introduces a competing (opposite) effect for the exponential decay, giving rise to peak in the pdf. The location of this peak is the mode and is given by $\mathbb{MO}_X = (\alpha - 1)\beta$.

The expected value and variance for this RV are $\mathbb{E}_X = \alpha\beta$ and $\mathbb{V}_X = \alpha\beta^2$. We thus have that, if we have empirical approximations for mean and variance, the parameters can be estimated by solving a set of two coupled equations.

In the context of time-dependent phenomena, the Gamma RV generalizes the exponential RV in that it models the waiting time until we observe the α-th occurrence of Poisson events. For $\alpha = 1$, X is the time we wait until the occurrence of the first event (as in the exponential RV). The realizations shown in Figure 2.20, for instance, show how setting a value of $\alpha = 2$ leads to longer waiting times (compared to realizations with $\alpha = 1$). The parameter β tells us the average waiting time between events. The parameter $1/\beta$ tells us the rate at which the events occur; this is the event rate and corresponds to λ in the Poisson RV (recall that λ is the mean number of occurrences in an interval). Consequently, when the event rate $\lambda = 1/\beta$ is higher, we have to wait a shorter time. This can be seen from the shape of the pdf in Figure 2.20, note how the pdf is shifted to the left when $\beta = 1/2$ (this corresponds to an event rate of $\lambda = 1/\beta = 2$). Note that the mean value of the Gamma RV is the number of occurrences times the average waiting time between events.

For integer α, the cdf of the Gamma RV takes the form:

$$F(x) = 1 - \sum_{k=0}^{\alpha-1} \frac{1}{k!} \left(\frac{x}{\beta} \right)^k e^{-x/\beta}. \tag{2.62}$$

The cdf provides a useful interpretation of the Gamma RV. Specifically, $\mathbb{P}(X > x) = 1 - F(x)$ tells us the probability that the waiting time until the next α-th event is greater than x. This probability is equivalent to the probability of observing less than α Poisson events in a time interval $[0, x]$. Recall that in the exponential case, $\alpha = 1$ and thus $\mathbb{P}(X > x)$ is the probability of observing zero Poisson events. We emphasize that the Poisson RV tells us the number of events occurring in an interval (but tells us nothing about the timing of such events). The exponential RV provides information on the timing between consecutive events, and the Gamma RV provides information about the timing of event queues.

The Gamma RV has many applications not only in modeling time-dependent phenomena (e.g., queuing theory and service times) but also in modeling spatial phenomena (e.g., inspection of a pipeline or of sequences). For instance, it can be used to model the distance traveled until we find α-th occurrence of certain type of atom type in a chain.

Example 2.17 Service Time

You run the warehouse logistics for your company and are responsible for making sure that the queue of electric forklift trucks at the charging stations remains under control. The average charging rate at a station is 0.25 trucks/hr; that is, the average charge time is 4 hr. The charge time is random because there is variability in the age of the batteries.

You are interested in addressing the following questions:

- If the current queue of a station has two trucks, what is the probability that the next truck will wait more than 3 hrs?
- What is the probability if the queue has three trucks?
- What is the probability of waiting more than 4 hr if the queue has one truck?
- You buy brand new batteries, and this allows you to cut down the average charging time to 2 hr. What is the probability of waiting more than 4 hr if the queue has one truck?

We use X to denote the waiting time of the next truck (time to wait until two charging events are observed). We thus have that $\mathbb{P}(X > x)$ is the probability that the time to wait until we observe $\alpha = 2$ events is more than $x = 3$ hr. The average waiting time per event is $1/0.25 = 4$ hr. We thus use:

$$P(X > x) = 1 - F(x) = \sum_{k=0}^{\alpha-1} \frac{1}{k!} \left(\frac{x}{\beta} \right)^k e^{-x/\beta},$$

with $x = 3$ (hr), $\alpha = 2$ and $\beta = 4$ (hr). Substituting:

$$\begin{aligned} P(X > 3) &= \sum_{k=0}^{1} \frac{1}{k!}\left(\frac{3}{4}\right)^k e^{-3/4}, \\ &= \frac{1}{0!}\left(\frac{3}{4}\right)^0 e^{-3/4} + \frac{1}{1!}\left(\frac{3}{4}\right)^1 e^{-3/4}, \\ &= e^{-3/4} + (3/4)e^{-3/4} = 0.826. \end{aligned}$$

It is quite likely that we will have to wait more than 3 hr. If there are three trucks in the queue, we would expect an even higher probability. Indeed, if we set $\alpha = 3$, we obtain:

$$P(X > 3) = \sum_{k=0}^{2} \frac{1}{k!}\left(\frac{3}{4}\right)^k e^{-3/4} = 0.959.$$

If there is only one truck in the queue, then $\alpha = 1$, the probability of waiting more than 4 hr is:

$$P(X > 4) = \sum_{k=0}^{0} \frac{1}{k!}\left(\frac{4}{4}\right)^k e^{-4/4} = 0.368.$$

Improving the service rate decreases the mean waiting time to $\beta = 2$ hours. The probability that the waiting time is more than 4 hr (for a queue of one truck) is decreased from 0.368 to 0.135. □

2.2.6 Chi-Squared

The *chi-squared RV* has a pdf of the form:

$$f(x) = \frac{1}{2^{r/2}\Gamma(r/2)} e^{-x/2} x^{r/2-1}, \; x \in \mathcal{D}. \tag{2.63}$$

The domain is $\mathcal{D} = [0, \infty)$, and the pdf has only one parameter, given by a positive integer value $r \in \mathbb{Z}_+$ (this is also known as the degrees of freedom). We express the fact that X is a chi-squared RV by using the notation $X \sim \chi^2(r)$.

The chi-squared RV has interesting connections to the Gamma and to the Gaussian RV. Specifically, note that the chi-squared RV is a Gamma RV with $\beta = 2$ and $\alpha = r/2$ (for any positive integer r). We can exploit this connection to obtain summarizing statistics for a chi-squared RV such as the expected value and variance.

One can also show that, if $Z \sim \mathcal{N}(0, 1)$, then $Z^2 \sim \chi^2(1)$. This property has many interesting implications; for instance, if we have that $Z \sim \mathcal{N}(0, 1)$ and $Y = Z^2 \sim \chi^2(1)$, then:

$$\mathbb{Q}_Z\left(1 - \frac{\alpha}{2}\right) = \sqrt{\mathbb{Q}_Y(1 - \alpha)}, \tag{2.64}$$

this can also be stated as $\mathbb{Q}_Y(1 - \alpha) = \mathbb{Q}_Z(1 - \frac{\alpha}{2})^2$. This relationship holds because:

$$\mathbb{P}(-z \leq Z \leq z) = \mathbb{P}(Z^2 \leq z^2). \tag{2.65}$$

As a result, we can also obtain the critical value $z_{\alpha/2} = \mathbb{Q}_Z(1 - \frac{\alpha}{2})$ to construct confidence regions for Gaussians by using the quantiles of a chi-squared RV. Another interesting property is that, for a set of independent $X_i \sim \mathcal{N}(0,1)$, $i = 1, \ldots, n$, we have:

$$\sum_{i=1}^{n} X_i^2 \sim \chi^2(n). \tag{2.66}$$

In Chapter 3, we will see that this property will allow us to compute confidence regions for multivariate Gaussian RVs.

Example 2.18 Computing Confidence Intervals for Gaussian Flow

Consider the output flow of Example 2.12 $F_3 \sim \mathcal{N}(30, 1)$. Compute the 90% confidence interval for F_3 using the quantile of a chi-squared RV.

We know that the confidence interval is given by $\mu \pm z_{\alpha/2} \cdot \sigma$ with $z_{\alpha/2} = \mathbb{Q}_Z(1-\alpha/2)$ and $Z \sim \mathcal{N}(0,1)$. From Eq. (2.64), we observe that we can compute $z_{\alpha/2}$ as:

$$z_{\alpha/2} = \sqrt{\mathbb{Q}_Y(1-\alpha)},$$

with $Y \sim \chi^2(1)$. We use a numerical package to compute $\mathbb{Q}_Y(1 - 0.1) = 2.71$, and we find that $z_{\alpha/2} = \sqrt{2.71} = 1.64$. We thus have that the confidence interval is $[30 \pm 1 \cdot 1.64] = [28.36, 31.64]$ gpm. This is the same confidence interval computed in Example 2.12. □

Example 2.19 Circle of Trust

We have established that we can find the boundaries of the confidence region for $Z \sim \mathcal{N}(0,1)$ by finding z such that $\mathbb{P}(-z \leq Z \leq z) = 1 - \alpha$. We have also seen that the same boundary can be found by finding z such that $\mathbb{P}(Y \leq z^2) = 1 - \alpha$ and that the boundary is given by $z_{\alpha/2} = \sqrt{\mathbb{Q}_Y(1-\alpha)}$ for $Y \sim \chi^2(1)$. From this, we conclude that Z^2 is in the region defined by $Z^2 \leq z_{\alpha/2}$ with 90% probability. This is equivalent to say that Z is in the region $-z_{\alpha/2} \leq Z \leq z_{\alpha/2}$ with 90% probability.

The confidence region of $Y = Z_1^2 + Z_2^2$ with $Z_1, Z_2 \sim \mathcal{N}(0,1)$ defines a ball (a circle) in 2D that is centered at $(0,0)$. In this case, the boundary z_r satisfying $\mathbb{P}(Y \leq z_r^2) = 1 - \alpha$ is the radius of the ball. Specifically, the confidence ball is defined by the equation:

$$Z_1^2 + Z_2^2 \leq z_r^2.$$

Address the following questions:

- What is the radius z_r of the 90% confidence ball? What is the interpretation of this ball?
- Assume that we know that $Z_2 = 0$, what is the interpretation of the region $Z_1^2 \leq z_r^2$? What is the 90% confidence region of Z_1?

- Assume that we know that $Z_1 = 0$, what is the interpretation of the region $Z_2^2 \le z_r^2$? What is the 90% confidence region of Z_2?
- Sketch the confidence ball and the regions.

From Eq. (2.66), we know that $Y \sim \chi^2(2)$, and thus, the threshold value z that satisfies $\mathbb{P}(Y \le z^2) = 1 - \alpha$ is given by $z_r = \sqrt{\mathbb{Q}_Y(0.9)}$. From a numerical package, we obtain that $z_r^2 = \mathbb{Q}_Y(0.9) = 4.58$ and thus the radius is $z_r = 2.14$ and thus the confidence ball is defined by the equation $Z_1^2 + Z_2^2 \le 4.58$. The confidence ball indicates that there is a 90% probability of finding (Z_1, Z_2) in the ball. In other words, 90% of all the outcomes $(z_{1,\omega}, z_{2,\omega})$ will lie in the ball.

The region $Z_1^2 \le 4.58$ tells us that, provided that $Z_2 = 0$, there is 90% of finding Z_1 in the region $-2.14 \le Z_1 \le 2.14$. The equation $Z_2^2 \le 4.58$ tell us that, provided that $Z_1 = 0$, there is 90% of finding Z_2 in the region $-2.14 \le Z_2 \le 2.14$. Note that these regions are segments that lie along the horizontal and vertical axes of the ball (Figure 2.21).

Script to reproduce results: `ch2_example_chisquare` □

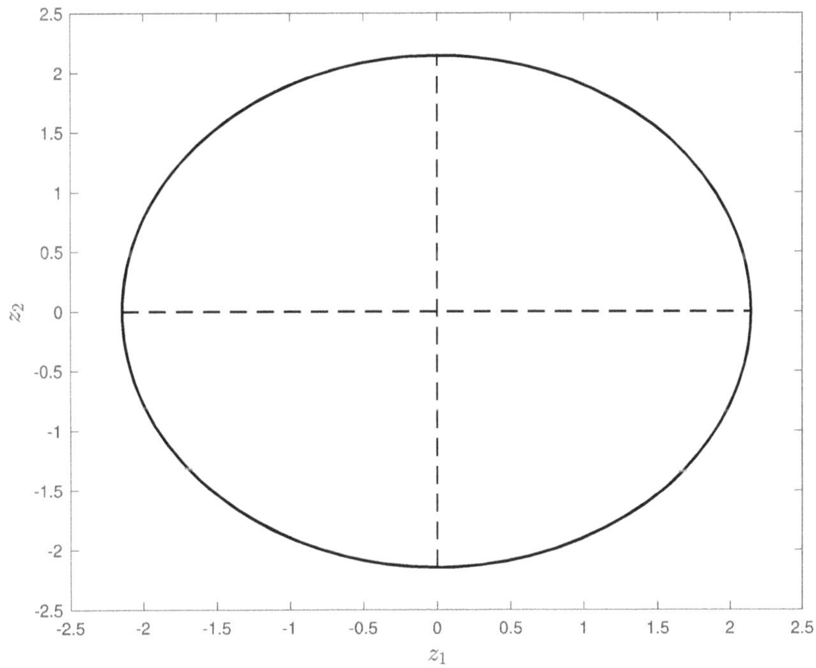

Figure 2.21 Confidence ball $Z_1^2 + Z_2^2 \le 4.58$ and intervals defined by $Z_1^2 \le 4.58$ and $Z_2^2 \le 4.58$ along the axes.

2.2.7 Weibull

The *Weibull RV* is a generalization of the exponential RV with pdf:

$$f(x) = \frac{\xi}{\beta}\left(\frac{x}{\beta}\right)^{\xi-1} \exp\left(-\left(\frac{x}{\beta}\right)^{\xi}\right), \; x \in \mathcal{D}. \tag{2.67}$$

The parameters of this model are $\beta, \xi \in \mathbb{R}_+$ (these are known as the scale and shape), and the domain is $\mathcal{D} = [0, \infty)$. We express the fact that X is a Weibull RV as $X \sim \text{Weibull}(\xi, \beta)$ and note that one recovers an exponential RV when $\xi = 1$. We can express this fact as $\text{Weibull}(1, \beta) \sim \text{Exp}(\beta)$.

The expected value and variance for this RV are given by:

$$\mathbb{E}_X = \beta \cdot \Gamma(1 + 1/\xi), \tag{2.68a}$$

$$\mathbb{V}_X = \beta^2 \cdot \left(\Gamma(1 + 2/\xi) - \Gamma(1 + 1/\xi)^2\right). \tag{2.68b}$$

The dependence of these statistics on the Gamma function makes it difficult to compute estimates for the parameters from the empirical mean and variance ($\hat{\mathbb{E}}_X$ and $\hat{\mathbb{V}}_X$).

Fortunately, the cdf of the Weibull has a simple form, and we can leverage this property to estimate the model parameters. Specifically, the cdf is:

$$F(x) = 1 - e^{-(x/\beta)^{\xi}}, \; x \in \mathcal{D}. \tag{2.69}$$

The behavior of the pdf and cdf of a Weibull RV is shown in Figure 2.22. From the cdf, we can see that $\mathbb{P}(X \leq \beta) = 0.63$ (as for the exponential RV), but, interestingly, this holds for any value of ξ. This property can be used to estimate β directly from the empirical cdf. We can also use the empirical cdf (or quantiles) to estimate ξ.

Because the cdf of the Weibull has an exponential form (function can be easily inverted), it is easy to see that the α-quantiles are:

$$\mathbb{Q}_X(\alpha) = \beta\left(-\log(1 - \alpha)\right)^{1/\xi}. \tag{2.70}$$

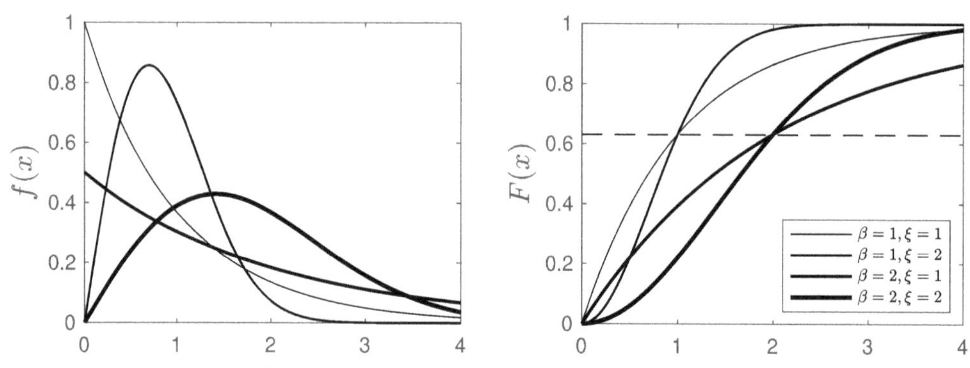

Figure 2.22 Pdfs and cdfs for a Weibull random variable.

From here, we can determine that the median is $\mathbb{M}_X = \beta(\log 2)^{1/\xi}$. One can show that the mode is given by:

$$\mathbb{MO}_X = \begin{cases} \beta\left(\frac{\xi-1}{\xi}\right)^{1/\xi} & \xi > 1 \\ 0 & \xi \leq 1. \end{cases} \tag{2.71}$$

This reveals the influence of the shape parameter ξ; when $\xi \leq 1$, the pdf behaves as that of an exponential RV and the peak is at $x = 0$. When $\xi > 1$, we see the emergence of a peak at $x > 0$.

The Weibull RV is the de facto model used in the failure/reliability analysis [35]. This RV was extensively studied by Waloddi Weibull (a materials scientist) to study the failure of materials [37]. It turns out that, as in the case of the Gaussian RV, many phenomena have the Weibull RV as a limiting case. Specifically, extreme values of RVs tend to behave as a Weibull-like RVs (while average values behave as Gaussian RVs). We will explore this topic in Chapter 4. Due to applications in failure analysis (which tend to have time-dependent behavior), the survival function of the Weibull RV is often used to characterize phenomena. We recall that this function is:

$$S(x) = \mathbb{P}(X > x) = 1 - F(x) = \exp\left(-\left(\frac{x}{\beta}\right)^{\xi}\right), \; x \in \mathcal{D}. \tag{2.72}$$

Example 2.20 Modeling Wind Conditions

Wind is known to be rather unpredictable, but it is also known that the 10-min wind speed (time average over 10 minutes) tends to follow a Weibull distribution [38]. In Figure 2.23, we present the empirical and theoretical pdf and cdf for the 10-min wind speed captured by an anemometer at a particular geographical location. Information on wind speed distribution is used to determine how much wind power can potentially be extracted using wind turbines (e.g., some locations have more favorable wind conditions than others).

Use the pdf and cdf to determine the following:

- Scale and shape parameters for the 10-min wind speed.
- Average and most likely wind speed at this location. Which of these statistics would you use to recommend or not recommend this location?
- Wind turbines extract a large amount of power when the wind speed is in the 10–20 m/s range. How often will the turbine operate in this high-power range at this location?

From pure visual inspection, we have that $F(x) = 0.63$ occurs at approximately x =6 m/s (this gives the scale parameter β, and its units are m/s). Now note that $F(x) = 0.5$ occurs at approximately $x = 5$ m/s. We thus have that $\mathbb{M}_X = 5$ and thus $5 = \beta(\log 2)^{1/\xi}$. Applying logarithms, we obtain:

$$\log 5 = \log(6) + (1/\xi)\log(\log 2).$$

Solving for ξ we obtain:

$$\xi = \frac{\log(\log 2)}{\log 5 - \log(6)} = 2.$$

Note that the shape parameter is adimensional. The average 10-min windspeed (in m/s) is:

$$\mathbb{E}_X = 6 \cdot \Gamma(1 + 1/2) = 5.3,$$

while the most likely value (the mode, also in m/s) is:

$$\mathbb{MO}_X = 6 \cdot \left(\frac{2-1}{2}\right)^{1/2} = 4.2.$$

The mode has a more intuitive interpretation here than the average because it tells us the most likely wind speed value. The average value only gives us an indication of the magnitude of the wind speed. The probability that the 10-min wind speed is in the [10–20] m/s range is given by:

$$\begin{aligned}\mathbb{P}(10 \le X \le 20) &= F(20) - F(10)\\ &= (1 - e^{-(20/\beta)^{\xi}}) - (1 - e^{-(10/\beta)^{\xi}})\\ &= 0.06.\end{aligned}$$

We thus have that the wind turbine will only operate 6% of the time in the high-power range at this location.

Script to reproduce results: `ch2_example_weibull` □

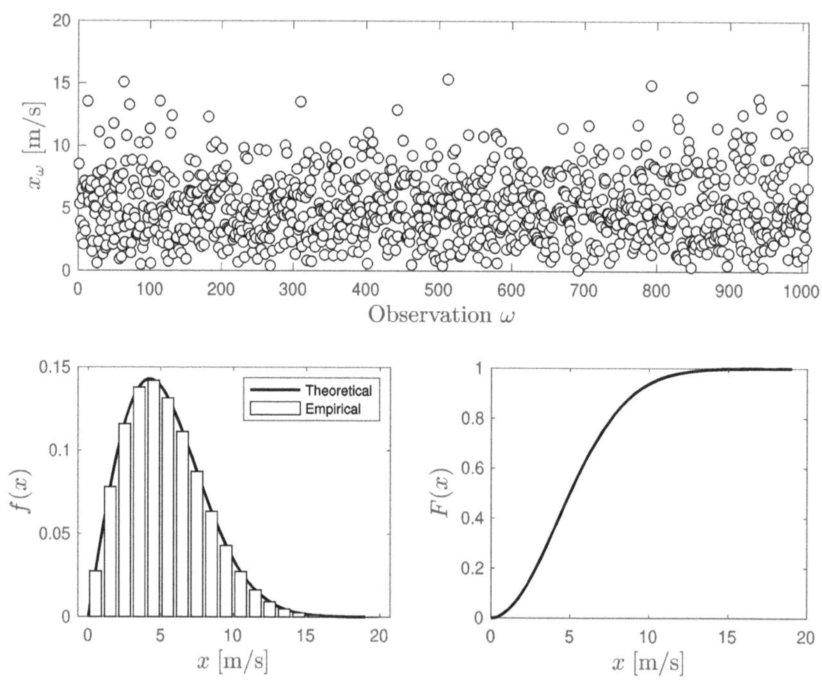

Figure 2.23 Pdf and cdf for 10-min average wind speed.

2.3 Summary

This chapter provides an overview of different theoretical RV models that can be used to model random phenomena encountered in applications.

The following are important take-home messages:

- Our goal is to identify a suitable RV model that explains available data for a random phenomenon.
- An RV model is defined by its domain, pdf, cdf, and parameters.
- An RV model captures random behavior that is specific to different types of applications; it is important to understand what is the setting that gives rise to your data and to select a model appropriately. For instance, it is not appropriate to use a Gaussian RV to model data that is discrete; similarly, it is not appropriate to use a Poisson RV to model data that is continuous.
- An RV model has characteristic behavior that can be helpful when comparing against available data. For instance, a characteristic of a Poisson model is that its mean and variance coincide.
- An RV model can be derived from basic principles, and in some cases, they have physical origins (e.g., the Weibull model has origins in materials science). RV models can also be used to explain physical phenomena that appears in applications (e.g., diffusion and residence times); this is not a coincidence, one can show that some RV models naturally emerge from random behavior observed in physical systems.
- It is important to emphasize that the list of RV models discussed in this chapter is by no means complete, and there are a broad range of models available. However, some of the ideas discussed in this chapter are generally applicable. A detailed discussion of the properties of all existing RVs and their connections and transformations is beyond the scope of this work. One should be aware, however, that many models and relationships between them exist.
- Random variables are interrelated in the sense that one is the generalization of another one (e.g., Weibull and Gamma are generalizations of the exponential). This is not an accident, as many RVs are part of general families that share common features. For instance, the exponential, Gamma, chi-squared, and Weibull RVs are special cases of the generalized Gamma model, which has the pdf:

$$f(x) = \frac{1}{\beta^{\alpha\cdot\xi}\cdot\Gamma(\alpha)}\cdot\exp\left[-\left(\frac{x-\delta}{\beta}\right)^{\xi}\right]\cdot\xi\cdot(x-\delta)^{\alpha\cdot\xi-1}, \quad x\in\mathcal{D}.$$

  with domain $\mathcal{D} = [0,\infty)$ and parameters α,β, and ξ. RVs derived from this generalized model are known as the Gamma family.
- We have seen that RVs are related to each other through algebraic transformations; for example, the Gaussian and lognormal RVs are connected via a logarithmic transformation, and the Gaussian and chi-squared RVs are connected via a squared transformation. We have also seen that certain transformations do not

alter the nature of the RVs; for instance, a linear transformation of a Gaussian yields a Gaussian.

- We have seen that some RVs are limiting cases of other RVs; for instance, we have seen that a hypergeometric RV converges to a binomial RV.
- It is important to emphasize that we can, in principle, design our own RV model (nothing prevents us from doing so). In defining such a model, however, we need to ensure that it satisfies fundamental properties of pdfs and cdfs. All the models discussed here are consistent in this way.
- The RV models discussed in this chapter have been widely studied in the literature. Our goal here is to provide a simple and concise explanation of their relevance, particularly with the goal of highlighting connections to engineering applications and to physical phenomena. A detailed exposition of RV models is provided in [39]. A detailed derivation of RV models (from basic principles) along with discussions on connections to physical phenomena is provided in [8].

2.4 Exercises

Exercise 2.1. Fluorescent Microscopy.
The Poisson RV can be used to characterize the occurrence of rare, random events that are distributed in time or space. This typically arises in settings where you have many observations but a small number of events. Scientists have found that Poisson behavior is encountered in small signals commonly found in analytical instrumentation such as fluorescence detection. In this type of analysis, one is interested in characterizing the appearance of individual particles, molecules, and atoms in microscopy images.

Consider the image shown below; this type of image is typically obtained in single-molecule characterization experiments using florescent microscopy [6]. Here, the white pixels indicate the presence of a molecule. You have been provided with data (see `fluorescence_image.dat`) that contains a 200×200 matrix that represents this image; each matrix entry is a pixel and the entry value is either 1 (molecule present, white) or 0 (molecule not present, gray).

Address the following:

(a) Partition the matrix into small boxes of size 5×5 and count how many molecules you see in each box. Obtain the empirical probability density for the number of molecules per box.
(b) Obtain the average number of molecules per box and use this as a parameter for the theoretical Poisson RV model. Verify that the model matches your empirical pdf.
(c) An interesting property of the Poisson RV is that it converges to a Gaussian RV when the probability of events increases. A way to verify this is to increase the size of our boxes; partition your original matrix into larger boxes of size

20 × 20 and count the number of molecules in the boxes. Obtain the empirical probability density and the theoretical Gaussian model. What do you observe?

(d) Can you think of another application in which you can use the Poisson RV to characterize the occurrence of events in a spatial context?

Exercise 2.2. Barista Planning.

A coffee shop in Coyoacán, México, is open from 6 am to 8 pm every weekday; the shop records the number of customers that arrive every hour, and they have data for a total of 200 days. The figure below provides a visualization of the data; the y-axis is the operating hours in a day, the x-axis is the number of days, and a darker color indicates more customers.

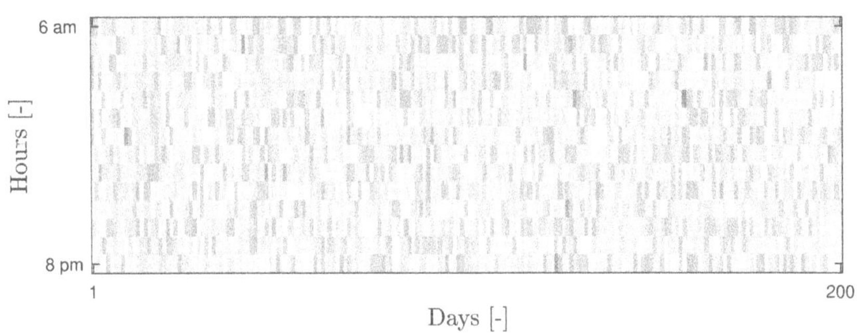

Keeping track of customer arrivals is important to ensure that the coffee shop has enough staff to attend the customers. Use the available data (see `barista_data.dat`) to address the following:

(a) Determine an empirical probability density for the number of consumers per hour. Determine a suitable model to capture this behavior.

(b) It is estimated that the shop needs one staff member to handle the orders of eight customers per hour (this prevents long waiting lines). How many staff members would you recommend that the shop has available? (Use the theoretical model to conduct this calculation.)

(c) If, due to budget limitations, you need to operate with two staff members, what fraction of the time will you be understaffed? (Use the theoretical model to conduct this calculation.)

Exercise 2.3. Quality Inspection and AQL.
A company produces approximately 100,000 batteries every day. As usual, inspecting every item is cost-prohibitive. The Acceptance Quality Limit (AQL) is a sampling standard that is used in industry [40]. For a production lot of 100,000 units, the AQL standard recommends that you should sample 500 units. The standard suggests that you should accept the batch if you encounter ≤ 21 defects in your sample and suggests that you should reject the batch if you encounter ≥ 22 defects.

(a) If there are 1,000 defective items in your lot, determine the probability that a batch will be accepted and determine the probability that it will be rejected. Comment on what does this say about the quality of your production process.

(b) If there are 5,000 defective items in your lot, determine the probability that a batch will be accepted and determine the probability that it will be rejected. If you will be rejecting a lot of batches, what should you do?

Exercise 2.4. Uncertainty Compounding.
Consider a random investment $X \sim \mathcal{N}(\mu, \sigma^2)$ (in USD) that is put in a fund that generates a rate of return r; at the end of the first year, the value of the investment is:

$$Y_1 = X + r \cdot X = (1 + r) \cdot X.$$

At the end of the nth year, the investment value is:

$$Y_n = (1 + r)^n \cdot X.$$

(a) What is the pdf of the future value Y_n? Obtain the parameters of this pdf.

(b) Assume $n = 30$, $\mu = 10{,}000$, $\sigma = 1{,}000$, and $r = 0.08$. What is the mean $\mathbb{E}[Y_n]$ and standard deviation $\mathbb{SD}[Y_n]$? If the goal is to reach 100,000 USD, does the investment meets the target (on average)?

(c) Does Y_n reach the target of 100,000 USD with high probability? Specifically, verify that $\mathbb{P}(Y_n > 100{,}000) \geq 0.99$. Why is it (or not) possible to meet the required probability?

(d) How many years should elapse to ensure that $\mathbb{P}(Y_n > 100{,}000) \geq 0.99$? What is the mean and standard deviation after these many years? Why is it (or not) possible to meet the required probability?

Exercise 2.5. Persistent Noise.
A deterministic signal x containing key temperature sensor data (in °F) for a feedback controller goes through a noisy communication channel. The channel contaminates the signal by adding random noise $\epsilon \sim \mathcal{N}(0,\sigma^2)$; this noise can be naturally occurring or can be the result of an intentional cyberattack. The addition of noise is troublesome because this can distort the actual data that the controller receives, and this can thus affect the action of the controller.

(a) What is the pdf of the contaminated output channel signal and what are its parameters?
(b) A tolerable level of noise for the output signal requires that its standard deviation is at most 1% of the value of the input signal x; this guarantees that the signal received by the controller is useful (there is a high signal-to-noise ratio). What is the limit value of σ if we have $x = 20\,°\text{F}$? Report the 95% confidence interval of the output signal using this parameter value.

Exercise 2.6. Energy Use.
A company has been recording the daily energy demand of a facility, and this has been estimated to be 15 ± 1 MWh. The daily energy demand is assumed to be Gaussian and the confidence intervals correspond to a probability level of 95%. The company is concerned of having enough capacity to meet the energy demand. What is the probability that the energy demand will exceed 16 MWh?

Exercise 2.7. Weibull and Stress Failure Analysis.
Waloddi Weibull was a Swedish civil engineer, materials scientist, and mathematician who conducted extensive research on strength of materials, fatigue, and rupture in solids. In some of his most influential work, he postulated that a stressed material under uniaxial tension has failure statistics that are described by the now-called Weibull model:

$$F(x) = 1 - \exp\left(-\left(\frac{x}{\beta}\right)^{\xi}\right).$$

Due to the physical origins of this model, this has wide applicability in analyzing the failure of materials. Material failures are often the result of inherent defects that arise in manufacturing (some of these defects can be small and difficult to control). In the context of stress failure, X is an RV that captures the stress level observed at fracture (in MPa), $F(x)$ is the cumulative failure probability at stress level x, and the parameters β and ξ must be determined experimentally (as these depend on the type of material and of its preparation). Experiments conducted on similar samples of glass (grade zinc sulfide) found stresses at fracture that vary significantly [3]. The collected experimental data was used to construct the plot shown below.

(a) Determine the parameters of the Weibull model for the glass material.
(b) Use the model to determine the stress levels that give cumulative failure probabilities of 1×10^{-6}, 1×10^{-1}, and 0.99.
(c) Use the model to determine the probability that the glass withstands a stress of 126 MPa or higher.
(d) Use the model to determine the most likely stress level that leads to failure.

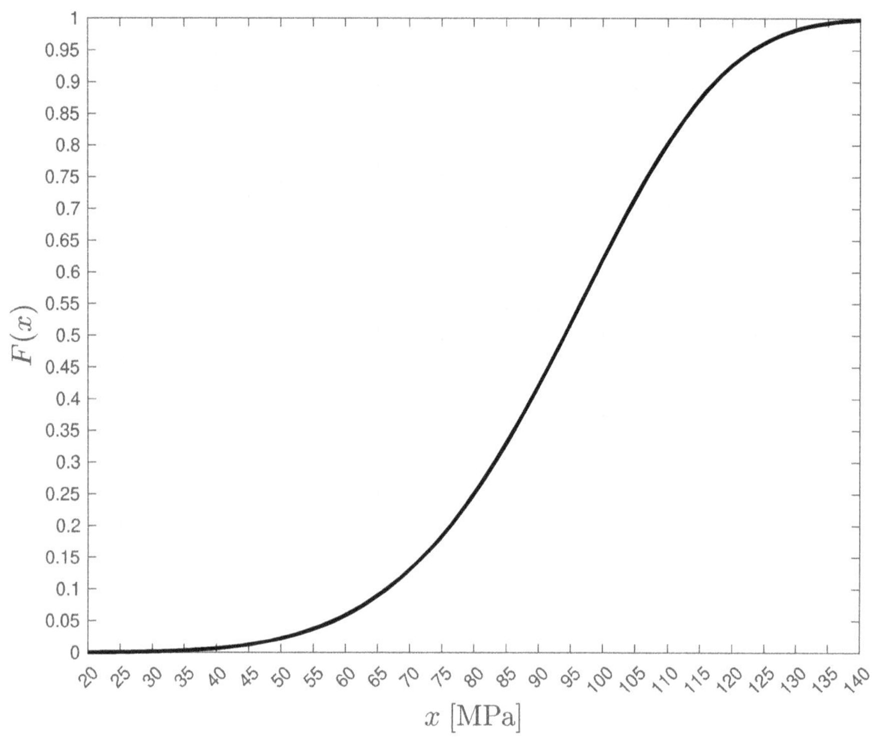

Exercise 2.8. Wisconsin Lakes.
You have recently been hired by the Department of Natural Resources of the State of Wisconsin to study pollution of the lakes in the state. It is estimated that Wisconsin has nearly 15,000 lakes and, in order to prioritize, you would like to characterize the distribution of lake sizes. Load the available data in `WI_lake_areas.csv` [41] (which contains the lake areas in km^2) and address the following:

(a) Determine the summarizing statistics (mean, standard deviation, mode, median, min, and max) of the lake areas.
(b) Obtain the empirical cdf of the lake areas and determine a suitable model.
(c) Because your dataset does not contain all the lakes in the state, use the model to determine the probability that a lake has an area larger than Lake Mendota ($40\ \text{km}^2$).

Exercise 2.9. Sieve Analysis.
A pharmaceutical powder product has particles with sizes in the μ-meter range and thus its particle size distribution (PSD) can be characterized using sieve analysis. Here, a given mass of product M_0 is put on a sieve stack that contains eight sieves. The sieves have increasing aperture sizes (from bottom to top) and cover the range of interest. The particles of the product naturally separate in the stacks according to their size. This device provides data on the total mass of product that is in each sieve (with corresponding particle size ranges). A representative dataset is [2]:

R (μm)	0–45	45–63	63–150	150–250	250–500
$M(R)$ (g)	4.5	6	15	22	48
R (μm)	500–1,000	1,000–2,000	2,000–4,000	4,000–6,000	
$M(R)$ (g)	22	15	6	3	

Here, R is the particle size (diameter) and $M(R)$ is the corresponding mass of material (in grams). The PSD that is characterized in terms of weight/mass is known as a mass-based distribution; this can be expressed as:

$$f(R) = \frac{M(R)}{\int_0^\infty M(R)dR}.$$

Here, the numerator can be interpreted as the mass of material of a specific size, while the denominator can be interpreted as the total mass of material.

Use the available data to address the following:

(a) Obtain the empirical size distribution and cumulative size distribution.
(b) Show that scaling the particle sizes as $\log(R)$ yields a distribution that resembles a Gaussian. What does this tell you about what type of distribution is $f(R)$?

Exercise 2.10. Managing Packages.
A delivery company needs to manage tens of thousands of packages every week in its inventory; these packages can vary significantly in size, and such variability poses a challenge from a transportation and storage perspective (e.g., homogeneous package sizes are easier to manage and store than heterogeneous ones). As such, the company collects data for all package sizes and carefully monitors the variability/spread of the sizes; they have collected the data in `packages_data.dat` . The first column is the package size v (measured in terms of volume); the second column is the number of packages $N(v)$ that have size in the interval $[v - \Delta v, v]$. To model the inherent variability of the packages, the company models the size as a random variable V.

Because the number of packages is significant; the company prefers to keep track of the fraction of packages of difference sizes; as such, they normalize the data to obtain a size distribution function of the form:

$$f(v) = \frac{N(v)}{\int_0^\infty N(v)dv}.$$

The company also likes to use a simple metric that captures the variability of the size of the packages; they use the polydispersity index, which is defined as:

$$\mathbb{PD}[V] = \frac{\mathbb{E}[V^2]}{\mathbb{E}[V]^2} = 1 + \frac{\mathbb{V}[V]}{\mathbb{E}[V]^2}.$$

Ideally, it is desired that the polydispersity index is as close as possible to one. This index is closely related to the coefficient of variation; in particular:

$$\mathbb{PD}[V] = 1 + \left(\frac{\mathbb{SD}[V]}{\mathbb{E}[V]}\right)^2.$$

Use the available data to address the following:

(a) Obtain the empirical size distribution function of the packages from data and comment on its shape. You can approximate any integrals using a numerical integration method such as the trapezoidal rule.
(b) Obtain the average size and the polydispersity index.
(c) Use the empirical estimates for the average size and for the polydispersity index to obtain the parameters of a lognormal model. Show that the model explains the data.

Exercise 2.11. Residence Time Distribution from Data.
A mass of tracer is suddenly injected at time $t = 0$ (in one shot) to the feedstream of a sequence of $n = 4$ CSTRs. The tracer outlet concentration $c(t)$ at different times t is recorded after the injection. The residence time distribution (RTD) of the sequence of CSTRs can be obtained from $c(t)$ data using:

$$f(t) = \frac{v \cdot c(t)}{\int_0^\infty v \cdot c(t)dt}, \qquad t \in [0, \infty).$$

Assume that the flow v is constant and that the units of time are min and units of concentration are g/m^3. After conducting an experiment in a laboratory reactor, you find the data shown in the table.

t (min)	0	1	2	3	4	5
$c(t)$ (g/m^3)	0	6.13	18.045	22.404	19.537	14.037
t (min)	6	7	8	9	10	11
$c(t)$ (g/m^3)	8.923	5.213	2.862	1.499	0.756	0.3705

Address the following:

(a) What is the random variable that we are studying in this system?
(b) Obtain a plot for $f(t)$ and $F(t) = \int_0^t f(t')dt'$ (you can approximate this using a numerical integration technique such as the trapezoidal approximation). Comment on the shape of $f(t)$ and $F(t)$ and provide a physical interpretation of their shape. Report the units of t, $f(t)$, $f(t)dt$, and $F(t)$.
(c) What is the physical and statistical interpretation of $f(t)$?
(d) What is the physical and statistical interpretation of $f(t)dt$?
(e) What is the physical and statistical interpretation of $F(t) = \int_0^t f(t')dt'$?

(f) What is the fraction of tracer that remains in the reactor after 6 min?
(g) What is the fraction of tracer that has exited the reactor up to time 10 min?
(h) What is the average residence time of the sequence of CSTRs (denote this as $\mathbb{E}$)?
(i) Show that the RTD $f(t)$ follows the pdf of a Gamma RV with $\alpha = n$ and $\beta = \mathbb{E}/n$.
(j) Use the Gamma model to show that the RTD $f(t)$ of a single CSTR is an exponential pdf. What is the average residence time of this CSTR?

3 Multivariate Random Variables

Our previous discussions have focused on univariate random variables (RVs); these RVs have realizations that are scalars (single numbers). In this chapter, we will extend our discussion to account for multivariate RVs; these RVs have realizations that are vectors (collections of numbers). These types of RVs are extremely important in applications because we usually have to deal with systems that generate data for different variables. For instance, a chemical process can contain hundreds to thousands of sensors that provide data for flows, temperatures, pressures, and so on. As such, every time we observe/sample the process, we obtain a realization that contains data for all the sensors. Our discussion on multivariate RVs will also give rise to important concepts such as correlation, independence, and joint events; these concepts will become important in advanced topics covered in the book.

To formalize our discussion, we recall that the realization $\omega \in \Omega$ of a univariate RV generates a scalar value $x_\omega \in \mathcal{D}_X \subseteq \mathbb{R}$. This implies that the domain $\mathcal{D}_X$ is unidimensional (e.g., a line segment). We now consider multivariate RVs $X = (X_1, X_2, \ldots, X_n)$; for multivariate RVs, we have that a realization $\omega \in \Omega$ generates a vector $x_\omega = (x_{\omega,1}, x_{\omega,2}, \ldots, x_{\omega,n}) \in \mathcal{D}_X \subseteq \mathbb{R}^n$. This implies that the domain $\mathcal{D}_X$ is multidimensional.

Specific questions that we would like to answer in a multivariate setting are as follows:

- Are there any connections (dependencies) between the RVs? Is there a clear trend that suggests that they vary together? Do they vary independently of one another?
- If the RVs are related, how strong are these connections?
- If the RVs are related, how does knowledge of one resolves the uncertainty of the other?
- How to analyze connections between *many* RVs? (e.g., when the number of RVs n is in the hundreds)
- How to generalize previous tools from a univariate case to a multivariate case?

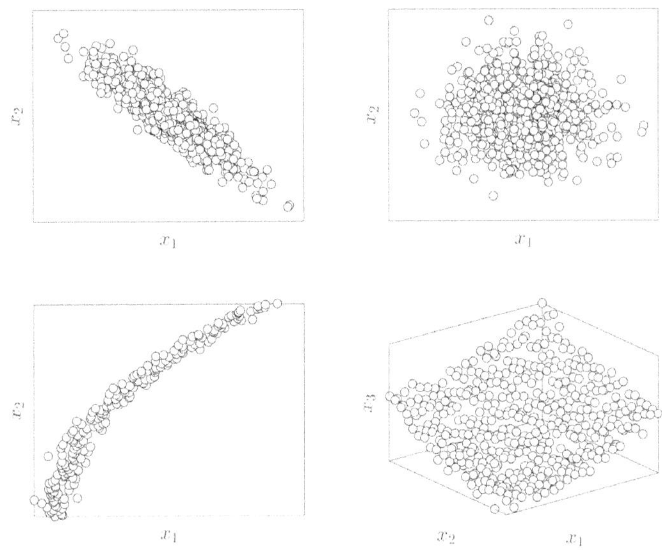

Figure 3.1 Realizations of multivariate RVs highlighting possible relationships between them.

These questions are important by themselves but also point at questions that are relevant from a *decision-making* standpoint:

- If data suggests that the variables are related, is there a hidden *mechanistic* relationship that explains such connections?
- Can we exploit such relationships to develop *predictive* models and make decisions?

Some example datasets that we might encounter and that motivate such questions are presented in Figure 3.1. Note how, in some cases, the RVs tend to vary together (e.g., as one increases, the other one tends to increase as well); this suggests that there is a trend that allows us to predict one variable as a function of others. For instance, if we have data that suggests that reactor conversion increases with temperature, then we might be able to develop a model that predicts conversion from temperature; this will be the topic of Chapter 5. If variables vary together, this implies that information about one variable carries information about the others (if I know how one behaves, I can infer how the other behaves); in Chapter 6, we will see that this principle can be used to derive techniques that reduce large datasets and that eliminate data redundancies.

The elements of a multivariate RV are similar to those of a univariate RV, but the concepts of *dependence* and *independence* introduce a number of new concepts that need to be carefully explained. As in the univariate case, our goal will also be to understand if our data seems to follow the behavior of a specific multivariate RV model.

We have seen that the pdf and cdf fully characterize an RV model and that these are necessary to determine the probability of events associated with the RV. The pdf and cdf of a multivariate RV are *multidimensional* functions that are more complex than their univariate counterparts. Specifically, a challenging aspect that we will encounter is that data and density functions live in high dimensions that are difficult to analyze and visualize. Moreover, in the multivariate case, dependencies between variables give rise to the concepts of covariance and correlations; we will see that these are generalizations of the concept of variance in the univariate case. Dependencies between variables also give rise to the concepts of joint, marginal, and conditional probabilities; we will see that these types of probabilities are extremely useful in determining the likelihood of complex events.

Unfortunately, there are not that many established theoretical models for multivariate RVs, which is a significant difference from the univariate case. However, we will see that many useful calculations for multivariate RVs can be conducted directly from data (without having an actual theoretical model). We will also explore the multivariate Gaussian RV model and see that this model has many interesting properties and applications.

3.1 Domains and Subdomains

Multivariate RVs give rise to realizations (data) and functions that live in multiple dimensions. Specifically, domains of multivariate RVs are no longer lines or points along a line (as in the univariate case) but can take complex shapes (e.g., cubes, ellipses, or polyhedra). Because of this, it will be necessary to develop the ability to visualize and analyze data and functions over high dimensions. For simplicity in our discussion, we will focus on the case of the bivariate RV $X = (X_1, X_2)$ (i.e., $n = 2$), but we will see that the concepts presented can be generalized to arbitrary n.

For a bivariate RV, any realization $\omega \in \Omega$ of $X = (X_1, X_2)$ generates an observation pair $x_\omega = (x_{\omega,1}, x_{\omega,2}) \in \mathbb{R}^2$. In other words, the observation x_ω is a data point in a 2D plane with coordinates $(x_{\omega,1}, x_{\omega,2})$. For example, an experiment can generate an observation containing pressure and conversion data.

The domain of X is given by $\mathcal{D}_X \subseteq \mathbb{R}^2$; this domain results from the cartesian product of the individual domains of the univariate RVs X_1 and X_2:

$$\mathcal{D}_X = \mathcal{D}_{X_1} \times \mathcal{D}_{X_2}, \tag{3.1}$$

where $\mathcal{D}_{X_1} \subseteq \mathbb{R}$ is the domain of X_1 and $\mathcal{D}_{X_2} \subseteq \mathbb{R}$ is the domain of X_2. When appropriate, we will use the compact notation $\mathcal{D} = \mathcal{D}_X$, $\mathcal{D}_1 = \mathcal{D}_{X_1}$, and $\mathcal{D}_2 = \mathcal{D}_{X_2}$.

For *discrete* X, both X_1 and X_2 are discrete RVs (both $\mathcal{D}_1$ and $\mathcal{D}_2$ are discrete). Here, we will assume that the domains are given by $\mathcal{D}_1 = \{l_1, u_1\}$ and $\mathcal{D}_2 = \{l_2, u_2\}$. In other words, the domain $\mathcal{D}_1$ only contains a couple of possible values $l_1 \in \mathbb{R}$, and $u_1 \in \mathbb{R}$, and we will assume $-\infty \leq l_1 < u_1 \leq \infty$. For example, if we set $l_1 = 0$ and $u_1 = 1$, we have that X_1 can model on/off behavior. We use the same logic for

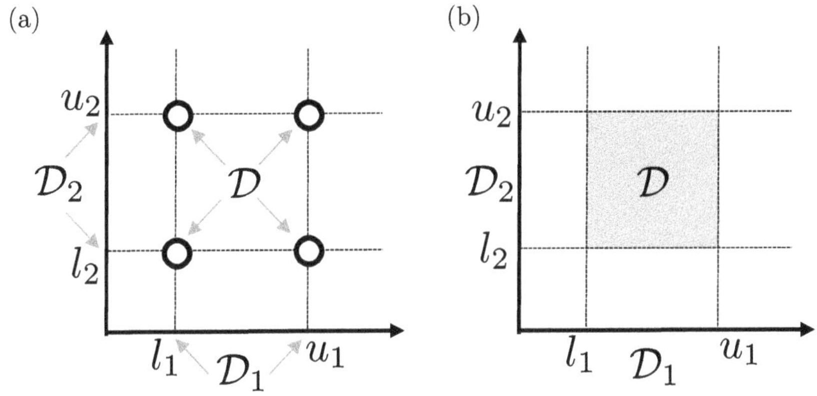

Figure 3.2 Domains for 2D discrete (a) and continuous (b) RVs.

domain $\mathcal{D}_2$. Based on these definitions of the univariate discrete domains, we have that the domain of X is:

$$\mathcal{D} = \mathcal{D}_1 \times \mathcal{D}_2 = \{(l_1,l_2),(l_1,u_2),(u_1,l_2),(u_1,u_2)\}. \tag{3.2}$$

This indicates that any observation of X must satisfy $(x_{\omega,1},x_{\omega,2}) \in \mathcal{D}$ and thus can only take four possible values. Note how the cartesian product of discrete domains is simply all the combinations of the elements. The domain of the discrete X is illustrated in Figure 3.2(a). We interpret the domain $\mathcal{D}$ as the set of values that the observation x_ω is *restricted* to take. As such, we can interpret the domain $\mathcal{D}$ as a restriction (a constraint) of the domain $\mathbb{R}^2$.

For a *continuous* RV, both X_1 and X_2 are continuous RVs (both the domains $\mathcal{D}_1$ and $\mathcal{D}_2$ are continuous). Here, we will assume that the domains are given by the line intervals $\mathcal{D}_1 = [l_1,u_1]$ and $\mathcal{D}_2 = [l_2,u_2]$. In other words, the domain $\mathcal{D}_1$ contains all possible values along the segment $[l_1,u_1]$, and we will assume $-\infty \leq l_1 < u_1 \leq \infty$ (we use the same logic to define $\mathcal{D}_2$), and with this, we obtain the domain:

$$\mathcal{D} = \mathcal{D}_1 \times \mathcal{D}_2 = [l_1,u_1] \times [l_2,u_2]. \tag{3.3}$$

This indicates that any observation x_ω must satisfy $(x_{\omega,1},x_{\omega,2}) \in \mathcal{D}$ and thus can take any value in the region $\mathcal{D}$. Note how the cartesian product for continuous domains is the region that covers all the possible values of the individual domains. This region is illustrated in Figure 3.2(b). By comparing the domains of the continuous and discrete RVs, we note that a continuous domain is less restrictive (x_ω is allowed to take a wider range of values in the continuous case). In other words, the discrete domain is more constrained than the continuous one. The importance of the notion of how restricted a domain is will become apparent soon; specifically, we will see that subdomains of a domain can be defined by specifying conditions and logic (events) that the RVs need to satisfy.

The definition of $\mathcal{D}$ in terms of the cartesian product is intuitive for discrete domains (it gives all combinations of points), but it is not particularly intuitive in continuous domains. To obtain a more intuitive notation, we note that we can also express $\mathcal{D}_1$ in the following equivalent forms:

$$\begin{aligned}\mathcal{D}_1 &= \{X_1 \in [l_1, u_1]\}\\ &= \{l_1 \le X_1 \le u_1\}\\ &= \{l_1 \le X_1 \; \& \; X_1 \le u_1\}\\ &= \{l_1 \le X_1 \le \infty \; \& \; -\infty \le X_1 \le u_1\}\\ &= \{-\infty \le X_1 \; \& \; l_1 \le X_1 \; \& \; X_1 \le u_1 \; \& \; X_1 \le \infty\}.\end{aligned} \tag{3.4}$$

From the second expression, we see that any value of X_1 in $\mathcal{D}_1$ must satisfy $l_1 \le X_1 \le u_1$. From the third expression, we recall that the "&" symbol inside the brackets indicates that any value of X_1 in $\mathcal{D}_1$ must *simultaneously* satisfy the restriction $l_1 \le X_1$ *and* the restriction $X_1 \le u_1$. The fourth and fifth expressions arise from the "hidden" fact that $-\infty \le l_1$ and $u_1 \le \infty$. These forms are rarely used in practice because they are often obvious from context; we use them here to highlight the important fact that a domain is constructed by *restricting* a larger domain. The use of inequalities of the form $l_1 \le X_1 \le u_1$ is common in the field of optimization; in such field, these inequalities are known as *constraints* (conditions that a variable needs to satisfy). We can use similar logic to express the domain of X_2 in terms of constraints as $\mathcal{D}_2 = \{l_2 \le X_2 \le u_2\}$.

Based on these definitions, we have that the domain for continuous X can be written in terms of constraints.

$$\mathcal{D} = \mathcal{D}_1 \times \mathcal{D}_2 = \{l_1 \le X_1 \le u_1 \; \& \; l_2 \le X_2 \le u_2\}. \tag{3.5}$$

This indicates that the domain is a *box* region under which $l_1 \le X_1 \le u_1$ holds *and* $l_2 \le X_2 \le u_2$ holds.

The *empirical approximation* of the (theoretical) domain $\mathcal{D}$ is denoted as $\hat{\mathcal{D}} \subseteq \mathcal{D}$ and is given by the region covered/occupied by the available observations (data) x_ω, $\omega \in \mathcal{S}$ for X. This is illustrated in Figure 3.3. Note that a large number of observations might be needed to cover a domain; in fact, note that you need an infinite number of observations to *perfectly* cover a continuous domain. Covering a domain using observations becomes more difficult when the number of dimensions increases; for instance, in 3D, the domain $\mathcal{D}$ of the RV $X = (X_1, X_2, X_3)$ can be a cube or a sphere that will be difficult to cover. We also note that the empirical domain is a subdomain of the theoretical domain; this is important because it implies that large portions of the actual domain might remain unobserved/unexplored.

Domains and subdomains are important in statistics and decision-making because we are interested in computing *probabilities of events*. As we discussed in Chapter 1, events can be represented as subdomains of $\mathcal{D}$. For example, $\mathbb{P}(X \in \mathcal{A})$ is the probability of the event that X is in the subdomain $\mathcal{A} \subseteq \mathcal{D}$. In the following section, we will discuss how to compute probabilities associated with subdomains; we will see that this can be done by using the pdf of the RV, and we will see that we can compute empirical approximations of the probabilities directly from

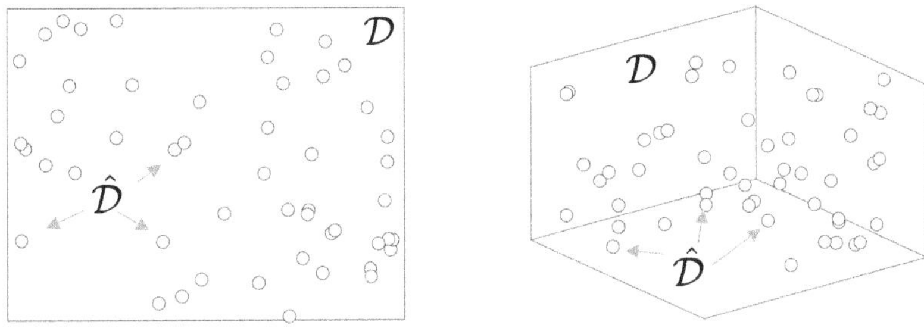

Figure 3.3 Empirical (observed) and theoretical domains in 2D and 3D.

observations (data). For now, we will just highlight the fact that the probability of an event is a measure of the size of its associated subdomain; this is because, the larger the domain, the more likely it will be to find observations of X landing in it.

Using the definitions that we have already established, we can construct events (with associated subdomains $\mathcal{A}$) that capture *logical constraints*. For instance, imagine that we want to compute the probability that X is in the domain $\mathcal{R} = \{X_1 \geq 0 \,\&\, X_2 \geq 0\}$ (i.e., the probability that X takes nonnegative values). To construct our subdomain $\mathcal{A} \subseteq \mathcal{D}$, we note that any observation $x_\omega \in \mathcal{A}$ must satisfy the fact that $x_\omega \in \mathcal{R}$ *and* $x_\omega \in \mathcal{D}$. Consequently, we have that:

$$\begin{aligned}\mathcal{A} &= \mathcal{R} \cap \mathcal{D} \\ &= \{\{X_1 \geq 0 \,\&\, X_2 \geq 0\} \,\&\, \{l_1 \leq X_1 \leq u_1 \,\&\, l_2 \leq X_2 \leq u_2\}\} \\ &= \{X_1 \geq 0 \,\&\, X_2 \geq 0 \,\&\, l_1 \leq X_1 \leq u_1 \,\&\, l_2 \leq X_2 \leq u_2\}.\end{aligned} \tag{3.6}$$

The symbol "$\cap$" in the first expression denotes the *intersection of sets* and captures the "&" logic (this is also represented as $\wedge$ in propositional logic). Specifically, the intersection tells us that any observation satisfying $x_\omega \in \mathcal{A}$ must be, simultaneously, in the region $\mathcal{R}$ and in the region $\mathcal{D}$. In other words, the event must satisfy all the constraints imposed by $\mathcal{R}$ and $\mathcal{D}$; this means that the domain $\mathcal{A}$ is more restrictive than the domain $\mathcal{D}$ (we say that $\mathcal{A}$ has more constraints than $\mathcal{D}$), and this is why $\mathcal{A} \subseteq \mathcal{D}$. Note that this also implies that events are simply conditions/constraints that the RV needs to satisfy. These concepts are illustrated in Figure 3.4.

We can envision building many different types of subdomains $\mathcal{A}$ to capture different types of constraints associated with X. For instance, we can build an event $\mathcal{A}$ that enforces the constraint that $\{X_1 = X_2\}$ (to compute the probability that $X_1 = X_2$). We can also build a subdomain $\mathcal{A}$ that enforces the constraint that $\{X_1^2 + X_2^2 \leq 1\}$ (to compute the probability that X is in a circular region). These domains are illustrated in Figure 3.5. We will use circular (ellipsoidal) regions when discussing multivariate Gaussian models (we will see that realizations of a multivariate Gaussian accumulate in regions that form ellipsoids).

It is also often of interest to construct events that capture "*or*" logic. For example, imagine that we want to compute the probability that X is in $\mathcal{R}_1 = \{X_1 \geq 0 \,\&\, X_2 \geq$

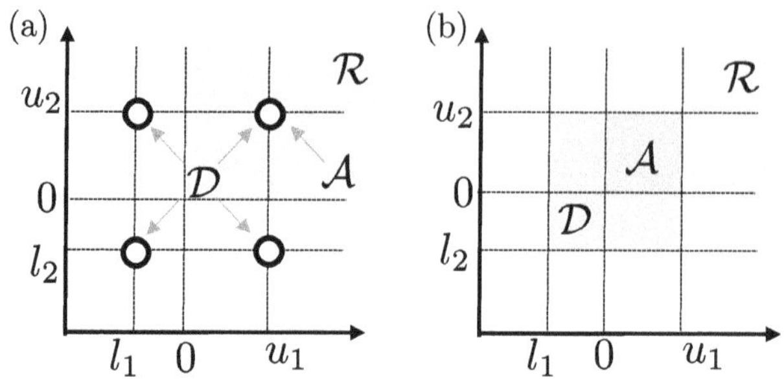

Figure 3.4 Subdomain $\mathcal{A}$ of $\mathcal{D}$ for discrete (a) and continuous (b) RV.

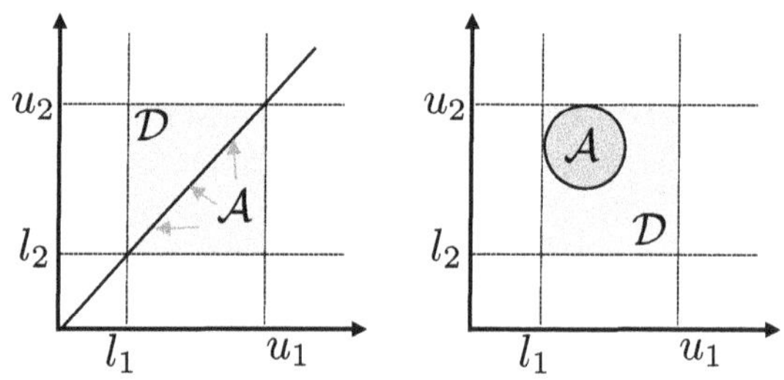

Figure 3.5 Examples of subdomains $\mathcal{A} \subseteq \mathcal{D}$ capturing different constraints.

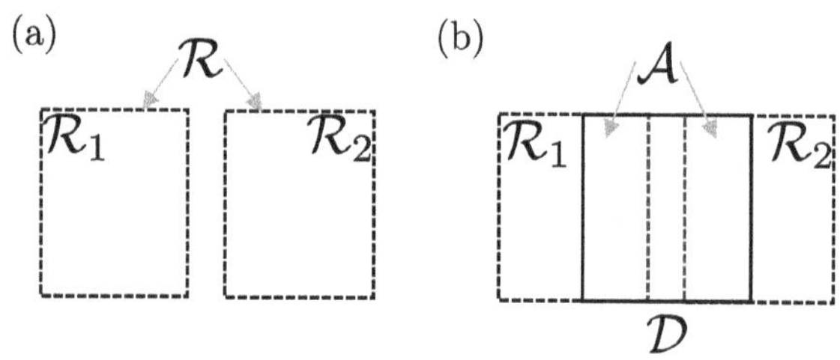

Figure 3.6 Union of $\mathcal{R}_1$ and $\mathcal{R}_2$ to obtain $\mathcal{R} = \mathcal{R}_1 \cup \mathcal{R}_2$ (a). Intersection of $\mathcal{R}$ with $\mathcal{D}$ to obtain $\mathcal{A} = (\mathcal{R}_1 \cup \mathcal{R}_2) \cap \mathcal{D}$ (b).

0} *or* that it is in $\mathcal{R}_2 = \{X_1 \leq -1 \,\&\, X_2 \leq -1\}$. Here, we build our subdomain $\mathcal{A}$ by first defining:

$$\mathcal{R} = \mathcal{R}_1 \cup \mathcal{R}_2 = \{\{X_1 \geq 0 \,\&\, X_2 \geq 0\} \cup \{X_1 \leq -1 \,\&\, X_2 \leq -1\}\}, \tag{3.7}$$

where the symbol $\cup$ represents *set union* and captures *or* logic (also represented as $\vee$). The expression $\mathbb{P}(X \in \mathcal{R})$ is thus the probability that $X \in \mathcal{R}_1$ or that $X \in \mathcal{R}_2$. We thus see that set union has the effect of *expanding* the range of values that X can take (provides flexibility). The event $\mathcal{A}$ must be a subdomain of $\mathcal{D}$ (it must satisfy $\{l_1 \leq X_1 \leq u_1 \,\&\, l_2 \leq X_2 \leq u_2\}$); we achieve this by enforcing:

$$\begin{aligned} \mathcal{A} &= \mathcal{R} \cap \mathcal{D} \\ &= (\mathcal{R}_1 \cup \mathcal{R}_2) \cap \mathcal{D} \\ &= (\mathcal{R}_1 \cap \mathcal{D}) \cup (\mathcal{R}_2 \cap \mathcal{D}). \end{aligned} \tag{3.8}$$

The event $X \in \mathcal{A}$ thus indicates that X is in $\mathcal{R}_1$ and $\mathcal{D}$ or that it is in $\mathcal{R}_2$ and $\mathcal{D}$. The construction of this event is illustrated in Figure 3.6.

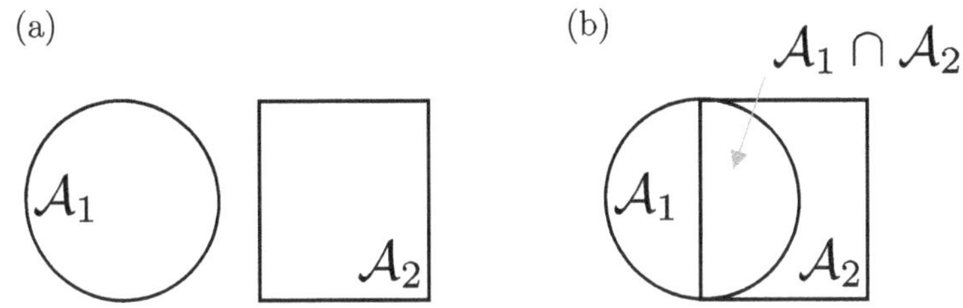

Figure 3.7 Disjoint events (a) and joint events (b).

From the previous discussion, we see that set union (set expansion) is associated with events that include *or* logic, while set intersection (set restriction) is associated with events that include *and* logic. As one can imagine, it is possible to construct highly complex events that involve combinations of "or" and "and" logic (set union and intersection). This allows us to convey complex logic that a person might follow when making a decision.

Imagine that we have events $X \in \mathcal{A}_1$ and $X \in \mathcal{A}_2$, and we want to compute their joint probability (probability that both occur simultaneously); this is given by $\mathbb{P}(X \in \mathcal{A})$ with $\mathcal{A} = \mathcal{A}_1 \cap \mathcal{A}_2$. Now note that it might be impossible for X to simultaneously satisfy $X \in \mathcal{A}_1$ and $X \in \mathcal{A}_2$; in such a case, we have that the domain $\mathcal{A}$ is empty (denoted as $\mathcal{A} = \mathcal{A}_1 \cap \mathcal{A}_2 = \emptyset$). An empty domain $\mathcal{A}$ indicates that the associated event is *not realizable* and thus $\mathbb{P}(X \in \mathcal{A}) = 0$ (i.e., it is impossible for the RV to satisfy the conditions of $\mathcal{A}$). Moreover, this indicates that the events $\mathcal{A}_1$ and $\mathcal{A}_2$ are *mutually exclusive* (also called disjoint events). For instance, the events $X \geq 0$ and $X \leq -1$ are disjoint (cannot be satisfied simultaneously and thus their intersection is empty), while $0 \leq X \leq 1$ and $0 \leq X \leq 2$ are joint events (their intersection is non-empty and given by $0 \leq X \leq 1$). The concepts of joint and disjoint events are illustrated in Figure 3.7.

Now imagine that we want to compute the probability that either event $X \in \mathcal{A}_1$ or event $X \in \mathcal{A}_2$ occurs; this is given by $\mathbb{P}(X \in \mathcal{A})$ with $\mathcal{A} = \mathcal{A}_1 \cup \mathcal{A}_2$. Now note that:

$$\mathbb{P}(\mathcal{A}_1 \cup \mathcal{A}_2) = \mathbb{P}(\mathcal{A}_1) + \mathbb{P}(\mathcal{A}_2) - \mathbb{P}(\mathcal{A}_1 \cap \mathcal{A}_2). \tag{3.9}$$

This is because the domain covered by $\mathcal{A}_1$ and $\mathcal{A}_2$ ($\mathcal{A}_1 \cup \mathcal{A}_2$) equals the sum of the individual domains minus the overlapping domain (where $\mathcal{A}_1$ and $\mathcal{A}_2$ occur simultaneously). Here, the overlapping domain $\mathcal{A}_1 \cap \mathcal{A}_2$ is subtracted because this is "double-counted" when adding the individual domains $\mathcal{A}_1$ and $\mathcal{A}_2$. Note that we have eliminated X from the definition of probabilities (this is standard and helpful in simplifying notation). When the events are disjoint, we have an empty overlapping region $\mathcal{A}_1 \cap \mathcal{A}_2 = \emptyset$, Eq. (3.9) tells us that $\mathbb{P}(\mathcal{A}_1 \cup \mathcal{A}_2) = \mathbb{P}(\mathcal{A}_1) + \mathbb{P}(\mathcal{A}_2)$. These properties are useful for visualizing and interpreting event logic.

It is worth emphasizing that the *or* logic induced by domain union is *inclusive or* logic (events can occur simultaneously). *Exclusive or* logic is captured using the concept of disjunctive domain union; this not discussed here, as this is a more complex topic.

Of interest in decision-making will also be the situation in which we want to determine the probability that an event $\mathcal{A}_1$ occurs *given knowledge* that another event $\mathcal{A}_2$ has occurred. This is captured by the *conditional probability* $\mathbb{P}(\mathcal{A}_1|\mathcal{A}_2)$, which is defined as:

$$\mathbb{P}(\mathcal{A}_1|\mathcal{A}_2) = \frac{\mathbb{P}(\mathcal{A}_1 \cap \mathcal{A}_2)}{\mathbb{P}(\mathcal{A}_2)}. \tag{3.10}$$

Here, $\mathbb{P}(\mathcal{A}_1 \cap \mathcal{A}_2)$ is the *joint* probability and $\mathbb{P}(\mathcal{A}_2)$ is known as the *marginal* probability of $\mathcal{A}_2$. Intuitively, if we know that $\mathcal{A}_2$ has occurred and the overlapping region $\mathcal{A}_1 \cap \mathcal{A}_2$ is large, then there is a high probability that $\mathcal{A}_1$ also occurs. In other words, event $\mathcal{A}_2$ carries information about event $\mathcal{A}_1$.

The expression for the conditional probability can also be written in the following useful form:

$$\mathbb{P}(\mathcal{A}_1 \cap \mathcal{A}_2) = \mathbb{P}(\mathcal{A}_1|\mathcal{A}_2) \cdot \mathbb{P}(\mathcal{A}_2). \tag{3.11}$$

If the event $\mathcal{A}_2$ does not affect occurrence of $\mathcal{A}_1$ (the events are said to be *independent*), then we have that: $\mathbb{P}(\mathcal{A}_1|\mathcal{A}_2) = \mathbb{P}(\mathcal{A}_1)$; we thus have that:

$$\mathbb{P}(\mathcal{A}_1) = \frac{\mathbb{P}(\mathcal{A}_1 \cap \mathcal{A}_2)}{\mathbb{P}(\mathcal{A}_2)}, \tag{3.12}$$

and

$$\mathbb{P}(\mathcal{A}_1 \cap \mathcal{A}_2) = \mathbb{P}(\mathcal{A}_1) \cdot \mathbb{P}(\mathcal{A}_2). \tag{3.13}$$

That is, if the events $\mathcal{A}_1$ and $\mathcal{A}_2$ are independent, their joint probability equals the product of the marginal probabilities of $\mathcal{A}_1$ and $\mathcal{A}_2$.

One can use the fact that $\mathbb{P}(\mathcal{A}_1 \cap \mathcal{A}_2) = \mathbb{P}(\mathcal{A}_1|\mathcal{A}_2) \cdot \mathbb{P}(\mathcal{A}_2)$ and $\mathbb{P}(\mathcal{A}_1 \cap \mathcal{A}_2) = \mathbb{P}(\mathcal{A}_2|\mathcal{A}_1) \cdot \mathbb{P}(\mathcal{A}_1)$ to establish the relationship:

$$\mathbb{P}(\mathcal{A}_1|\mathcal{A}_2) = \frac{\mathbb{P}(\mathcal{A}_2|\mathcal{A}_1)\mathbb{P}(\mathcal{A}_1)}{\mathbb{P}(\mathcal{A}_2)}. \tag{3.14}$$

This is an important result in statistics that is known as *Bayes' rule*. This result has many applications; in Chapter 5, we will use this result to estimate parameters for predictive models. Bayes' rule is also widely used in the field of uncertainty quantification.

One is often interested in computing the probability that an event $\mathcal{A} \subseteq \mathcal{D}$ *does not occur*. The event that $\mathcal{A}$ does not occur has associated subdomain $\mathcal{D} \setminus \mathcal{A}$, where the symbol "\" eliminates (slices off) the region $\mathcal{A}$ from the domain $\mathcal{D}$. The domain $\mathcal{D} \setminus \mathcal{A}$ is known as the complement of $\mathcal{A}$. The probability of $\mathcal{A}$ not occurring is:

$$\mathbb{P}(\mathcal{D} \setminus \mathcal{A}) = \mathbb{P}(\mathcal{D}) - \mathbb{P}(\mathcal{A}) = 1 - \mathbb{P}(\mathcal{A}). \tag{3.15}$$

The previous discussion allows us to see that events (and associated subdomains) convey complex logic and conditions/constraints about our application. Understanding the interplay between logic, events, and constraints is an important topic in *decision-making under uncertainty*, which we will cover in Chapter 7.

Specifically, *knowledge* and decisions are often expressed in terms of logic and constraints. For example, we might decide to make an investment only if certain conditions/constraints are satisfied. Such conditions are important in restricting the space of possible outcomes and thus can facilitate decision-making. Visualizing how domains arise from logic is also useful when analyzing complex datasets (as it helps us narrow down the space over which we search for insight).

Example 3.1 Subdomains and Logic

Consider the Gibbs reactor with random pressure P and conversion C. The associated domain of these variables is $\mathcal{D} = \{P \geq 0 \ \& \ 0 \leq C \leq 1\}$. Determine the events (subdomains $\mathcal{A}$) associated with the following probabilities.

- Probability that pressure is in the range $[100, 200]$ bar.
- Probability that pressure is in the range $[100, 200]$ bar *and* that conversion is below 0.5.
- Probability that pressure is below 100 bar *or* above 200 bar *and* that conversion is above 0.5.
- Probability that pressure is below 100 bar *and* that pressure is above 200, and that conversion is above 0.5.

To address the first probability, we need to build a subdomain that restricts pressure to be $100 \leq P \leq 200$. Since we do not need to restrict conversion, we can simply require that $0 \leq C \leq 1$ (as in the domain $\mathcal{D}$). We thus build the event $\mathcal{R} = \{100 \leq P \leq 200 \ \& \ 0 \leq C \leq 1\}$, and we build our subdomain $\mathcal{A} = \mathcal{R} \cap \mathcal{D} = \{100 \leq P \leq 200 \ \& \ 0 \leq C \leq 1 \ \& \ P \geq 0 \ \& \ 0 \leq C \leq 1\}$. The subdomain can be simplified as $\mathcal{A} = \{100 \leq P \leq 200 \ \& \ 0 \leq C \leq 1\}$ by removing redundancies.

To address the second probability, we simply define the restricted domain $\mathcal{R} = \{100 \leq P \leq 200 \ \& \ C \leq 0.5\}$. After simplification, the subdomain is $\mathcal{A} = \mathcal{R} \cap \mathcal{D} = \{100 \leq P \leq 200 \ \& \ 0 \leq C \leq 0.5\}$.

For the third probability, we define events $\mathcal{R}_1 = \{P \leq 100 \ \& \ C \geq 0.5\}$ and $\mathcal{R}_2 = \{P \geq 200 \ \& \ C \geq 0.5\}$, and we need that either of them holds. We thus build $\mathcal{R} = \mathcal{R}_1 \cup \mathcal{R}_2 = \{\{P \leq 100 \ \& \ C \geq 0.5\} \cup \{P \geq 200 \ \& \ C \geq 0.5\}\}$. From this, we have that $\mathcal{A} = (\mathcal{R}_1 \cup \mathcal{R}_2) \cap \mathcal{D} = (\mathcal{R}_1 \cap \mathcal{D}) \cup (\mathcal{R}_2 \cap \mathcal{D})$.

For the fourth probability, we define events $\mathcal{R}_1 = \{P \leq 100 \ \& \ C \geq 0.5\}$ and $\mathcal{R}_2 = \{P \geq 200 \ \& \ C \geq 0.5\}$, and we need that both of them hold. We thus build $\mathcal{R} = \mathcal{R}_1 \cap \mathcal{R}_2$ and $\mathcal{A} = (\mathcal{R}_1 \cap \mathcal{R}_2) \cap \mathcal{D} = \mathcal{R}_1 \cap \mathcal{R}_2 \cap \mathcal{D}$. The subdomain is given by $\mathcal{A} = \{P \leq 100 \ \& \ C \geq 0.5 \ \& \ P \geq 200 \ \& \ C \geq 0.5 \ \& \ P \geq 0 \ \& \ 0 \leq C \leq 1\}$. We note that it is impossible for P to simultaneously satisfy $P \leq 100$ and $P \geq 200$ and thus $\mathcal{A} = \emptyset$.

In Figure 3.8, we illustrate the subdomains (in gray) associated with the four events discussed. Here, we also show realizations of P and C. In order to compute the probabilities of each of the four events, we simply count the number of realizations that land on the subdomain of interest. Note that the probability of the first event is quite high, while the probability of the last event is zero because the subdomain $\mathcal{A}$ is empty (the event cannot occur).

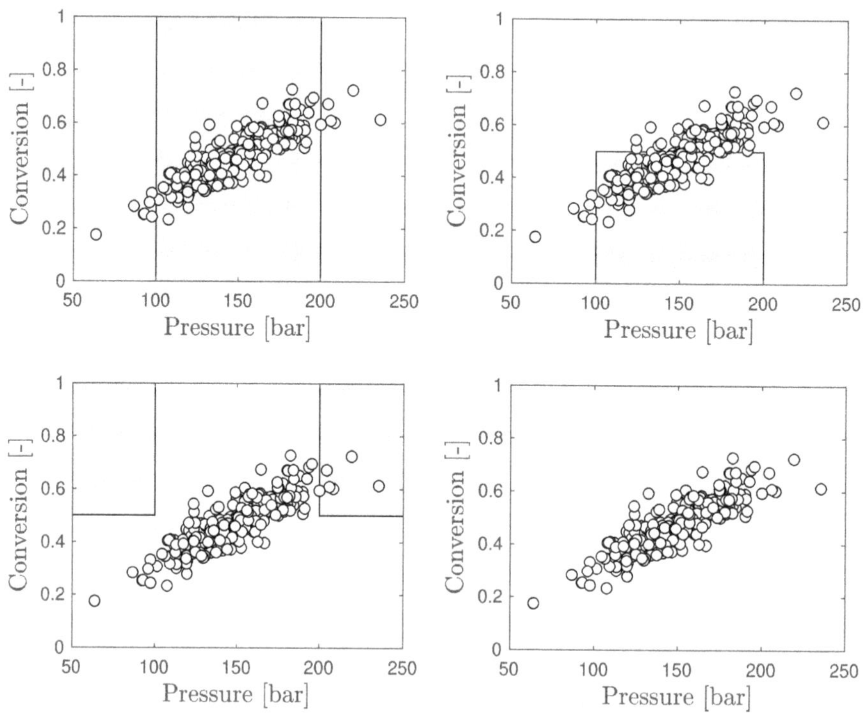

Figure 3.8 Subdomains associated with logic of Example 3.1.

Script to reproduce results: `ch3_gibbs_subdomains` □

3.2 Joint Density Functions

The *joint pdf* is the essential function needed to quantify the probability of events associated with the multivariate RV X. The joint pdf is the generalization of the pdf of a univariate RV to a multivariate setting. The joint pdf for $X = (X_1, X_2)$ is defined as:

$$f(x_1, x_2) = \mathbb{P}(X_1 = x_1 \,\&\, X_2 = x_2), \quad (x_1, x_2) \in \mathcal{D}. \tag{3.16}$$

For a discrete RV X, $f(x_1, x_2)$ is interpreted as the probability of the joint event in which X_1 takes value x_1 *and* X_2 takes value x_2. For continuous RVs, as we discussed for the univariate case, the pdf $f(x_1, x_2)$ is to be interpreted as the probability that (X_1, X_2) is in an infinitesimal neighborhood around (x_1, x_2) (we will explain this in more detail shortly). We will use notation $f(x_1, x_2) = \mathbb{P}(X_1 = x_1 \,\&\, X_2 = x_2)$ for both discrete and continuous RVs in order to simplify the discussion. Just remember the special significance in the continuous case.

For discrete RVs, the joint pdf is a discontinuous function, while for continuous RVs, it is continuous. For continuous X, the joint pdf must satisfy:

$$\begin{aligned} &f(x_1,x_2) \geq 0,\ (x_1,x_2) \in \mathcal{D} \\ &\int_{(x_1,x_2)\in\mathcal{D}} f(x_1,x_2)dx_1dx_2 = 1. \end{aligned} \tag{3.17}$$

For discrete RV, we replace the integral with summation $\sum_{(x_1,x_2)\in\mathcal{D}}$.

The joint pdf allows us to compute the probability of events $\mathcal{A} \subseteq \mathcal{D}$. For a continuous RV, this is done as:

$$\mathbb{P}(X \in \mathcal{A}) = \int_{(x_1,x_2)\in\mathcal{A}} f(x_1,x_2)dx_1dx_2. \tag{3.18}$$

For the discrete case, we simply replace integral with summation. The joint pdf gives us the *density* of points at a location in the subdomain $\mathcal{A}$, while the probability $\mathbb{P}(X \in \mathcal{A})$ gives us the *total density* of points in $\mathcal{A}$. We thus have that $\mathbb{P}(X \in \mathcal{A})$ measures the size (in terms of density) of the subdomain $\mathcal{A}$ (i.e., the larger the subdomain, the higher the probability that realizations land there). Note that, in order to compute the probabilities, we need to know the joint density function. We will see later that we can also obtain empirical estimates for the probabilities directly from data by counting observations that land on the subdomains (as illustrated in Example 3.1). It is important to highlight that analytic operations using integrals are rarely used for computation; however, such manipulations are useful for conceptual understanding and for deriving properties of RVs. In the discrete case, these computations are much easier to carry out.

We are often interested in building events associated with *box* subdomains $\mathcal{A} \subseteq \mathcal{D}$ of the form:

$$\begin{aligned} \mathcal{A} &= \{a_1 \leq X_1 \leq b_1 \ \&\ a_2 \leq X_2 \leq b_2\} \\ &= \{X_1 \in [a_1,b_1] \ \&\ X_2 \in [a_2,b_2]\}. \end{aligned} \tag{3.19}$$

For continuous RVs, we have that the probability of box events is given by:

$$\begin{aligned} \mathbb{P}(X \in \mathcal{A}) &= \int_{(x_1,x_2)\in\mathcal{A}} f(x_1,x_2)dx_1dx_2 \\ &= \int_{a_2}^{b_2}\int_{a_1}^{b_1} f(x_1,x_2)dx_1dx_2. \end{aligned} \tag{3.20}$$

To explain how we can compute empirical probabilities from data, we note that the event probability can also be written as:

$$\mathbb{P}(X \in \mathcal{A}) = \mathbb{E}[\mathbf{1}[X \in \mathcal{A}]]. \tag{3.21}$$

As we discuss in more detail in Chapter 4, we can use Monte Carlo simulations to compute an empirical approximation of the probability. Specifically, we consider x_ω, $\omega \in \mathcal{S}$ samples of X and compute:

$$\hat{\mathbb{P}}(X \in \mathcal{A}) = \frac{1}{S} \sum_{\omega \in \mathcal{S}} \mathbf{1}[x_\omega \in \mathcal{A}]. \tag{3.22}$$

In simple terms, the empirical probability is obtained by counting the realizations that satisfy the conditions (the constraints) of the event $\mathcal{A}$. As we discuss in Section 3.7.3, this approach can be combined with properties of probabilities to compute diverse types of event probabilities.

Example 3.2 Joint Pdf for Thermostat

The operation of a thermostat relies on a sensor (thermocouple) and a computing processor that processes the sensor data and sends feedback signals to the air conditioning unit. The thermostat fails to operate if either of these components fails. The time to failure of the components is (X_1, X_2) and has a joint pdf of the form:

$$f(x_1, x_2) = \frac{1}{\mu_1 \cdot \mu_2} e^{-(x_1/\mu_1 + x_2/\mu_2)}, \tag{3.23}$$

with $\mathcal{D} = \{0 \le x_1 \le \infty \;\&\; 0 \le x_2 \le \infty\}$ and $\mu_1 = 1$ and $\mu_2 = 10$ years.

- Verify that the joint pdf satisfies the required properties.
- Determine the probability that the components last less than 1 year.
- Determine the probability of event that $X_1 \ge 1$ and $X_2 \ge 2$.

It is clear that $f(x_1, x_2) \ge 0$ for all $(x_1, x_2) \in \mathcal{D}$. Now note that:

$$\begin{aligned}
&\int_{(x_1,x_2)\in\mathcal{D}} f(x_1, x_2) dx_1 dx_2 \\
&= \int_0^\infty \int_0^\infty f(x_1, x_2) dx_1 dx_2 \\
&= \int_0^\infty \int_0^\infty \left(\frac{1}{\mu_1 \cdot \mu_2} e^{-(x_1/\mu_1 + x_2/\mu_2)} \right) dx_1 dx_2 \\
&= \frac{1}{\mu_1 \cdot \mu_2} \left(\int_0^\infty e^{-x_1/\mu_1} dx_1 \right) \left(\int_0^\infty e^{-x_2/\mu_2} dx_2 \right) \\
&= \frac{\mu_1 \cdot \mu_2}{\mu_1 \cdot \mu_2} \left(e^{-0/\mu_1} - e^{-\infty/\mu_1} \right) \left(e^{-0/\mu_2} - e^{-\infty/\mu_2} \right) \\
&= (1 - 0) \cdot (1 - 0) \\
&= 1.
\end{aligned}$$

To determine the probability that the components last less than 1 year, we note that this corresponds to the event $\mathcal{A} = \{0 \le x_1 \le 1 \;\&\; 0 \le x_2 \le 1\}$. Clearly, $\mathcal{A} \subseteq \mathcal{D}$, and we thus expect to see that $\mathbb{P}(X \in \mathcal{A}) \le 1$. The probability is given by:

$$\begin{aligned}\mathbb{P}(X \in \mathcal{A}) &= \int_0^1 \int_0^1 f(x_1, x_2) dx_1 dx_2 \\ &= \left(e^{-0/1} - e^{-1/1}\right)\left(e^{-0/10} - e^{-1/10}\right) \\ &= (1 - 0.36) \cdot (1 - 0.90) = 0.064.\end{aligned}$$

To obtain the second probability, we construct the event:

$$\mathcal{A} = \{1 \le x_1 \le \infty \;\&\; 2 \le x_2 \le \infty\},$$

and compute:

$$\begin{aligned}\mathbb{P}(X \in \mathcal{A}) &= \int_1^\infty \int_2^\infty f(x_1, x_2) dx_1 dx_2 \\ &= \left(e^{-1/1} - e^{-\infty/1}\right)\left(e^{-2/10} - e^{-\infty/10}\right) \\ &= (0.36 - 0) \cdot (0.82 - 0) = 0.30.\end{aligned}$$

This illustrates how we can manipulate subdomains and events to obtain different probabilities of interest. □

Example 3.3 Flow Cytometer Data

A microemulsion containing 10,000 droplets is passed through a flow cytometer [7]. The device reports the side scatter (SSC) and front scatter (FSC) values for the droplets. For simplicity, the SSC and FSC values are assumed to belong to three possible categories – low (L), medium (M), and high (H) – as shown in Figure 3.9. In other words, the FSC and SSC are treated as discrete RVs. Use the

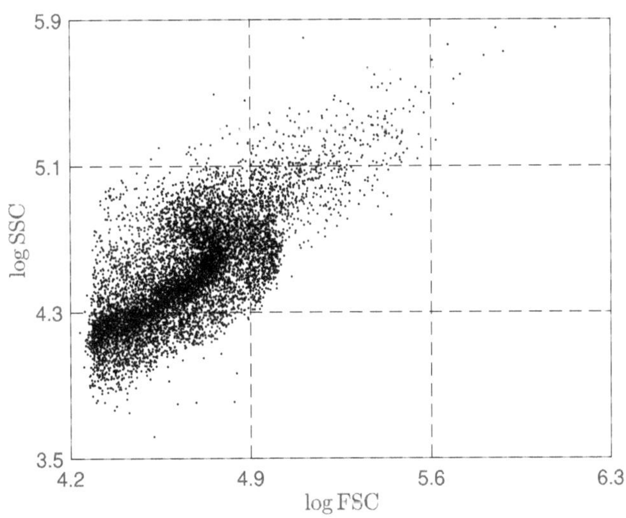

Figure 3.9 Flow cytometer data categorized in three FSC and SSC values (low, medium, and high).

data to obtain the joint pdf for the FSC and SSC. Determine what is the most likely category that the droplets belong to and determine the probability that FSC takes a low value and that SSC takes a low, medium, and high value.

We define the RV $X = (X_1, X_2)$ with X_1 = FSC and X_2 = SSC; the individual domains of the RVs are $\mathcal{D}_1 = \{L, M, H\}$ and $\mathcal{D}_2 = \{L, M, H\}$, and the joint domain is:

$$\begin{aligned}\mathcal{D} &= \mathcal{D}_1 \times \mathcal{D}_2 \\ &= \{(L,L),(L,M),(L,H),(M,L),(M,M),(M,H),(H,L),(H,M),(H,H)\}.\end{aligned}$$

We thus have that the droplets can belong to nine possible categories.

To obtain the joint pdf (empirical), we count the number of droplets (observations) in each value of the domain and normalize them by the total number of observations $S = 10{,}000$. The joint pdf can be expressed as a table (matrix) of the form:

$\hat{f}(x_1, x_2) =$

	L	M	H
H	0.0028	0.0239	0.0012
M	0.5994	0.1114	0
L	0.2613	0	0

Here, the rows correspond to the SSC values and the columns correspond to the FSC values. One can confirm that all the entries in the matrix add up to one. The most likely combination of values for the droplets is (L, M), which has a joint probability $\hat{\mathbb{P}}(X_1 = L, X_2 = M) = 0.5994$. This means that 5,994 droplets lie in this category.

To determine the second probability, we note that this corresponds to the event $\mathcal{A} = \{x_1 = L \ \& \ x_2 \in \{L, M, H\}\}$. Consequently,

$$\begin{aligned}\mathbb{P}(X \in \mathcal{A}) &= \sum_{x_1 = L} \sum_{x_2 \in \{L,M,H\}} f(x_1, x_2) \\ &= 0.2613 + 0.5994 + 0.0028 = 0.8635.\end{aligned}$$

We thus have that 8,635 droplets take a low value of FSC. Note that, technically speaking, the values that an RV can take should be numbers, but in this case, we use symbols (because these have a natural interpretation that allows us to draw comparisons).

Script to reproduce results: `ch3_flow_cytometer_discrete` □

The *joint cdf* of a discrete RV is a discontinuous function, while for a continuous RV, it is a continuous function. In the continuous case, we have that the joint pdf and cdf are related as:

$$f(x_1, x_2) = \frac{\partial^2 F(x_1, x_2)}{\partial x_1 \partial x_2}, \quad (x_1, x_2) \in \mathcal{D}. \tag{3.24}$$

That is, the joint pdf is the mixed partial derivative of the joint cdf (this is a generalization of the univariate case, in which the pdf is the derivative of the cdf). The interpretation of the derivative as the limit of an infinitesimal small domain is a bit more complicated here than in the univariate case. The limit can be written as:

$$f(x_1,x_2) = \lim_{\Delta x_2 \to 0} \frac{1}{\Delta x_2}\left(\frac{\partial F(x_1,x_2+\Delta x_2)}{\partial x_1} - \frac{\partial F(x_1,x_2)}{\partial x_1}\right), \tag{3.25}$$

where:

$$\begin{aligned}
\frac{\partial F(x_1,x_2+\Delta x_2)}{\partial x_1} &= \lim_{\Delta x_1 \to 0} \frac{F(x_1+\Delta x_1,x_2+\Delta x_2) - F(x_1,x_2+\Delta x_2)}{\Delta x_1} \\
\frac{\partial F(x_1,x_2)}{\partial x_1} &= \lim_{\Delta x_1 \to 0} \frac{F(x_1+\Delta x_1,x_2) - F(x_1,x_2)}{\Delta x_1}.
\end{aligned} \tag{3.26}$$

These expressions help us derive an interesting interpretation of the pdf of a multivariate RV. Recall that for the univariate (1D) case, we have that:

$$\begin{aligned}
f(x)\Delta x &\approx F(x+\Delta x) - F(x) \\
&= \mathbb{P}(X \le x+\Delta x) - \mathbb{P}(X \le x) \\
&= \mathbb{P}(x \le X \le x+\Delta x).
\end{aligned} \tag{3.27}$$

The pdf thus gives the probability that X is in the infinitesimal domain $[x, x+\Delta x]$. We recall that the last equation is obtained by noticing that the probability $\mathbb{P}(X \le x+\Delta x)$ can be split as $\mathbb{P}(X \le x) + \mathbb{P}(x \le X \le x+\Delta x)$ and thus the term $\mathbb{P}(X \le x)$ cancels out. For the 2D case, we use the approximation (3.25) to note that:

$$\begin{aligned}
f(x_1,x_2) \approx \frac{1}{\Delta x_1 \Delta x_2}\Big[&F(x_1+\Delta x_1,x_2+\Delta x_2) - F(x_1,x_2+\Delta x_2) \\
&+ F(x_1,x_2) - F(x_1+\Delta x_1,x_2)\Big].
\end{aligned} \tag{3.28}$$

This can be rewritten as:

$$\begin{aligned}
f(x_1,x_2)\Delta x_1\Delta x_2 \approx\ & \mathbb{P}(X_1 \le x_1+\Delta x_1 \ \&\ X_2 \le x_2+\Delta x_2) \\
&- \mathbb{P}(X_1 \le x_1 \ \&\ X_2 \le x_2+\Delta x_2) \\
&+ \mathbb{P}(X_1 \le x_1 \ \&\ X_2 \le x_2) \\
&+ \mathbb{P}(X_1 \le x_1+\Delta x_1 \ \&\ X_2 \le x_2).
\end{aligned} \tag{3.29}$$

By splitting probabilities, we can cancel out terms to obtain:

$$f(x_1,x_2)\Delta x_1\Delta x_2 \approx \mathbb{P}(x_1 \le X_1 \le x_1+\Delta x_1 \ \&\ x_2 \le X_2 \le x_2+\Delta x_2). \tag{3.30}$$

The pdf thus gives us the probability that $X = (X_1, X_2)$ is in an infinitesimal domain of the form $[x_1, x_1+\Delta x_1]\times[x_2, x_2+\Delta x_2]$. This domain is illustrated in Figure 3.10. As in the 1D case, the above approximation indicates that, in the limit $\Delta x_1, \Delta x_2 \to 0$, we have that $\mathbb{P}(X_1 = x_1 \ \&\ X_2 = x_2) = 0$. This, again, reveals why one has to be careful when drawing direct connections between $f(x_1,x_2)$ and $\mathbb{P}(X_1 = x_1 \ \&\ X_2 = x_2)$ for continuous RVs.

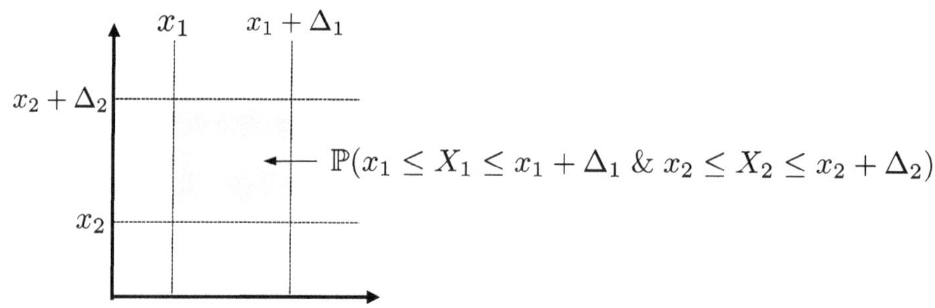

Figure 3.10 Interpretation of joint pdf for a bivariate RV X. This captures the probability that realizations land in a small box of the domain. Here, we write $\Delta_1 = \Delta x_1$ and $\Delta_2 = \Delta x_2$ for simplicity.

Example 3.4 Joint Cdf of Thermostat

- Determine the cdf for the thermostat example.
- Use this to determine the probability that the lifetime of the sensor and the processor is less than 1 year.
- Determine the probability that the lifetime of the sensor is 1 year and that the lifetime of the processor is 1 year.

The domain of the RV is $\mathcal{D} = \{0 \leq x_1 \leq \infty \;\&\; 0 \leq x_2 \leq \infty\}$. We thus have that the domain of the cdf is:

$$\begin{aligned}\mathcal{F} &= \mathcal{D} \cap \{x_1 \leq t_1 \;\&\; x_2 \leq t_2\}\\ &= \{0 \leq x_1 \leq t_1 \;\&\; 0 \leq x_2 \leq t_2\}.\end{aligned}$$

The cdf is thus:

$$\begin{aligned}F(t_1,t_2) &= \int_0^{t_2} \int_0^{t_1} f(x_1,x_2)dx_1dx_2\\ &= \left(e^{-0/\mu_1} - e^{-t_1/\mu_1}\right)\left(e^{-0/\mu_2} - e^{-t_2/\mu_2}\right) \qquad (3.31)\\ &= \left(1 - e^{-t_1/\mu_1}\right)\left(1 - e^{-t_2/\mu_2}\right).\end{aligned}$$

We thus have that $\mathbb{P}(X_1 \leq t_1 \;\&\; X_2 \leq t_2) = F(t_1,t_2) = 0.06$ (the probability that the thermostat fails before 1 year is rather small).

To determine the probability that the component lifetimes are exactly 1 year, we note that the pdf is a continuous function and thus we should think of $\mathbb{P}(X_1 = t_1 \;\&\; X_2 = t_2)$ in an infinitesimal sense. If we select $\Delta x_1 = \Delta x_2 = 0.1$ years, from Eq. (3.30), we have that $\mathbb{P}(x_1 \leq X_1 \leq x_1 + \Delta x_1 \;\&\; x_2 \leq X_2 \leq x_2 + \Delta x_2)$ can be approximated as $f(x_1,x_2) \cdot \Delta x_1 \cdot \Delta x_2$. Consequently, we find that:

$$
\begin{aligned}
&\mathbb{P}(x_1 \le X_1 \le x_1 + \Delta x_1 \;\&\; x_2 \le X_2 \le x_2 + \Delta x_2) \\
&= \frac{1}{\mu_1 \cdot \mu_2} e^{-(x_1/\mu_1 + x_2/\mu_2)} \cdot \Delta x_1 \cdot \Delta x_2 \\
&= \frac{1}{1 \cdot 10} e^{-(1/1+1/10)} \cdot 0.1 \cdot 0.1 \\
&= 3 \times 10^{-4}.
\end{aligned} \tag{3.32}
$$

From this, we confirm that the probability that X_1 is in $[1, 1.1]$ years and that X_2 is in $[1, 1.1]$ years is indeed small. The probability becomes smaller as the interval narrows down $\Delta x_1, \Delta x_2 \to 0$, confirming that $\mathbb{P}(X_1 = x_1 \;\&\; X_2 = x_2) = 0$. □

As in the univariate case, we can construct empirical approximations of the pdf/cdf. Given data x_ω, $\omega \in \mathcal{S}$, the cdf approximation is:

$$
\begin{aligned}
\hat{F}(t_1, t_2) &= \hat{\mathbb{P}}(X_1 \le t_1 \;\&\; X_2 \le t_2) \\
&= \frac{1}{S} \sum_{\omega \in \mathcal{S}} \mathbf{1}[x_{\omega,1} \le t_1 \;\&\; x_{\omega,2} \le t_2].
\end{aligned} \tag{3.33}
$$

As in the univariate case, we can obtain an empirical pdf by using a finite-difference approximation of the cdf. There are many ways to do this and one has to be careful with how accurate different approximations are. A straightforward approximation arises from the limit Eq. (3.25):

$$
\begin{aligned}
\hat{f}(x_1, x_2) = \frac{1}{\Delta x_1 \Delta x_2} \Big[&\hat{F}(x_1 + \Delta x_1, x_2 + \Delta x_2) - \hat{F}(x_1, x_2 + \Delta x_2) \\
&\hat{F}(x_1, x_2) - \hat{F}(x_1 + \Delta x_1, x_2) \Big].
\end{aligned} \tag{3.34}
$$

This is equivalent to:

$$
\hat{f}(x_1, x_2) \approx \frac{1}{\Delta x_1 \Delta x_2} \hat{\mathbb{P}}(x_1 \le X_1 \le x_1 + \Delta x_1 \;\&\; x_2 \le X_2 \le x_2 + \Delta x_2), \tag{3.35}
$$

where:

$$
\begin{aligned}
&\hat{\mathbb{P}}(x_1 \le X_1 \le x_1 + \Delta x_1 \;\&\; x_2 \le X_2 \le x_2 + \Delta x_2) \\
&= \frac{1}{S} \sum_{\omega \in \mathcal{S}} \mathbf{1}[x_1 \le x_{\omega,1} \le x_1 + \Delta x_1 \;\&\; x_2 \le x_{\omega,2} \le x_2 + \Delta x_2].
\end{aligned} \tag{3.36}
$$

Note that this indicates that we can approximate the joint pdf by using a binning procedure in which we split the domain into small boxes and then count how often the observations land in the boxes.

Example 3.5 Empirical Joint Pdf of a Flow Cytometer

In Example 3.3, we treated FSC/SSC as discrete RVs in order to facilitate analysis. However, it is more appropriate to treat these as continuous RVs because their values can lie anywhere in a continuous range. Use the flow cytometer data to obtain an empirical approximation of the joint pdf of FSC/SSC.

To construct the approximation, we first determine the empirical domain covered by the observations. From the data, we find that the minimum and maximum values (after logarithmic transformation) for FSC define the domain $\hat{\mathcal{D}}_1 = [4.2365, 6.0834]$ and that the values of SSC define the domain $\hat{\mathcal{D}}_2 = [3.6191, 5.8543]$. From this, we find that the empirical joint domain is defined by the box:

$$\hat{\mathcal{D}} = \{4.2365 \leq X_1 \leq 6.0834 \ \& \ 3.6191 \leq X_2 \leq 5.8543\}.$$

We partition (discretize) this box by using bins of size $\Delta x_1 = \Delta x_2 = 0.05$, which gives a total of 46 bins in the SSC domain and 38 bins in the FSC domain (giving a total of $46 \times 38 = 1{,}748$ bins).

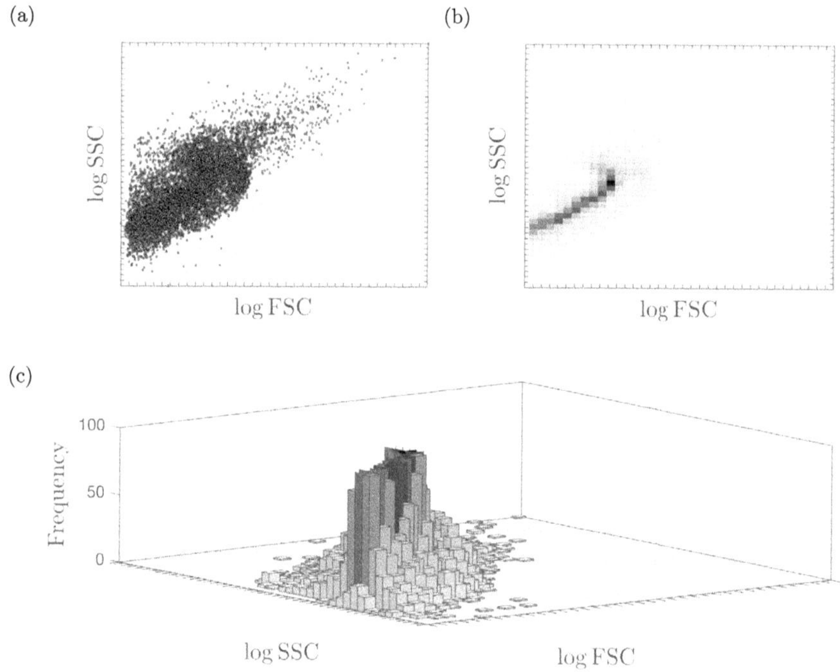

Figure 3.11 Empirical joint pdf for FSC and SSC. (a) Scatter plot showing observations. (b) Density of observations in individual bins (darker is more dense). (c) Empirical pdf shown as a 2D histogram.

To construct our empirical pdf approximation, we use Eq. (3.35) and count the number of observations in each of the bins defined. We visualize these counts in Figure 3.11. In Figure 3.11(a), we see the $S = 10{,}000$ observations and show the mesh created by the binning procedure. In Figure 3.11(b), we show the count of observations in each bin (darker color indicates a higher count). Note that this visualization better reveals the regions of FSC and SSC that are more dense (have more observations). In Figure 3.11(c), we show a 2D histogram that shows the frequency (number of counts) in each bin. We can get the empirical value of the pdf by normalizing the frequency by the total number of observations S and by the bin sizes $\Delta x_1 \cdot \Delta x_2$. We can see that the joint pdf of a bivariate RV has a complex shape that is difficult to analyze.

Script to reproduce results: `ch3_flow_cytometer_empirical` □

Generalizing the concept of the joint pdf for an n-dimensional RV is straightforward. Here, the RV is defined as $X = (X_1, X_2, \ldots, X_n)$, and the joint pdf is a function of the form $f(x) = f(x_1, x_2, \ldots, x_n)$, which is associated with $\mathbb{P}(X_1 = x_1 \;\&\; X_2 = x_2 \;\ldots\; \&\; X_n = x_n)$. Similarly, the joint cdf is a function of the form $F(x) = F(x_1, x_2, \ldots, x_n)$, which is associated with the probability $\mathbb{P}(X_1 \leq x_1 \;\&\; X_2 \leq x_2 \;\ldots\; \&\; X_n \leq x_n)$.

3.3 Marginal Density Functions

The joint pdf $f(x_1, x_2)$ tells us the probability that X_1 takes value x_1 and X_2 takes value x_2. Imagine now we want to know the probability that X_1 takes value x_1 *regardless* of what value X_2 takes (ignoring all knowledge of X_2). Similarly, imagine that we want to know the probability that X_2 takes value x_2 *regardless* of what value X_1 takes (ignoring all knowledge of X_1). These probabilities can be obtained from the *marginal pdfs*.

For a continuous RV, the marginals are obtained from the joint pdf as:

$$\begin{aligned} f_1(x_1) &= \int_{x_2 \in \mathcal{D}_2} f(x_1, x_2) dx_2, \; x_1 \in \mathcal{D}_1 \\ f_2(x_2) &= \int_{x_1 \in \mathcal{D}_1} f(x_1, x_2) dx_1, \; x_2 \in \mathcal{D}_2, \end{aligned} \tag{3.37}$$

while for a discrete RV, these are obtained as:

$$\begin{aligned} f_1(x_1) &= \sum_{x_2 \in \mathcal{D}_2} f(x_1, x_2), \; x_1 \in \mathcal{D}_1 \\ f_2(x_2) &= \sum_{x_1 \in \mathcal{D}_1} f(x_1, x_2), \; x_2 \in \mathcal{D}_2. \end{aligned} \tag{3.38}$$

In other words, the marginal pdfs are obtained by "integrating out" the effect of the RV we ignore (for discrete RV, we sum out).

To verify the above properties, consider that we want to obtain $\mathbb{P}(X_1 \in \mathcal{A}_1)$ for a continuous RV $X = (X_1, X_2)$. We have that:

$$\begin{aligned} \mathbb{P}(X_1 \in \mathcal{A}_1) &= \int_{x_1 \in \mathcal{A}_1} \overbrace{\int_{x_2 \in \mathcal{D}_2} f(x_1, x_2) dx_2}^{f_1(x_1)} dx_1 \\ &= \int_{x_1 \in \mathcal{A}_1} f_1(x_1) dx_1. \end{aligned} \tag{3.39}$$

Marginal pdfs also have associated marginal cdfs $F_1(x_1) = \mathbb{P}(X_1 \leq x_1)$ and $F_2(x_2) = \mathbb{P}(X_2 \leq x_2)$. The marginals of the RV are the pdfs or cdfs of the individual (univariate) RVs X_1 and X_2 and thus have the same properties that we discussed in Chapter 2. For example, we can use the same techniques to compute empirical approximations for marginal pdfs and cdfs. We also note that generalizing the concept of marginals to the n-dimensional case is straightforward (based on the observations previously made for the conditional pdf).

Example 3.6 Marginal Pdfs and Cdfs for Thermostat

Consider the thermostat example and determine the following:

- Marginal pdf and cdf of component lifetimes X_1, X_2.
- Determine the probability that component X_1 survives after 1 year.
- Determine the probability that component X_2 survives after 1 year.
- Compare these probabilities against the probability that both components survive after 1 year.

The marginal pdf for X_1 is obtained by integrating X_2 out over its domain $\mathcal{D}_2 = \{0 \leq X_2 \leq \infty\}$.

$$
\begin{aligned}
f_1(x_1) &= \int_0^\infty f(x_1, x_2) dx_2 \\
&= \int_0^\infty \left(\frac{1}{\mu_1 \cdot \mu_2} e^{-(x_1/\mu_1 + x_2/\mu_2)} \right) dx_2 \\
&= \frac{1}{\mu_1 \cdot \mu_2} e^{-x_1/\mu_1} \left(\int_0^\infty e^{-x_2/\mu_2} dx_2 \right) \\
&= \frac{1}{\mu_1} e^{-x_1/\mu_1} \left(e^{-0/\mu_2} - e^{-\infty/\mu_2} \right) \\
&= \frac{1}{\mu_1} e^{-x_1/\mu_1}.
\end{aligned}
$$

Applying a similar procedure, we obtain that:

$$f_2(x_2) = \frac{1}{\mu_2} e^{-x_2/\mu_2}.$$

We note that the marginals are pdfs of exponential RVs with corresponding mean lifetimes $\mu_1 = 1$ and $\mu_2 = 10$ years. From this, we can infer that the cdfs are:

$$
\begin{aligned}
F_1(x_1) &= 1 - e^{-x_1/\mu_1} \\
F_2(x_2) &= 1 - e^{-x_2/\mu_2}.
\end{aligned}
$$

The probabilities that the components survive are given by the survival functions evaluated at $x_1 = 1$ and $x_2 = 1$:

$$\mathbb{P}(X_1 \geq x_1) = 1 - F_1(x_1) = e^{-1/1} = 0.37$$
$$\mathbb{P}(X_2 \geq x_2) = 1 - F_2(x_2) = e^{-1/10} = 0.90.$$

Clearly, the second component has a much higher probability of surviving (its mean lifetime is much longer). The probability that both components survive after 1 year is:

$$\mathbb{P}(X_1 \geq 1 \ \&\ X_2 \geq 1) = \int_1^\infty \int_1^\infty f(x_1, x_2) dx_1 dx_2, \tag{3.40}$$

and by evaluating we obtain $\mathbb{P}(X_1 \geq 1 \ \&\ X_2 \geq 1) = 0.33$. We note that this is close to that of the lifetime of the first component (which makes sense because this is the limiting component). □

Example 3.7 Empirical Marginal Pdfs of a Flow Cytometer

Use the flow cytometer data to obtain empirical marginal pdfs of FSC/SSC.

To construct the approximation, we partition the individual domains $\hat{\mathcal{D}}_1 = [4.2365, 6.0834]$ (for FSC) and $\hat{\mathcal{D}}_2 = [3.6191, 5.8543]$ (for SSC) by using bins of size $\Delta x_1 = \Delta x_2 = 0.05$, which gives a total of 46 bins in the SCC domain and 38 bins in the FSC domain. The empirical marginals are:

$$\hat{f}_1(x_1) = \frac{1}{S}\frac{1}{\Delta x_1} \sum_{\omega \in \mathcal{S}} \mathbf{1}[x_1 \leq x_{\omega,1} \leq x_1 + \Delta x_1]$$
$$\hat{f}_2(x_2) = \frac{1}{S}\frac{1}{\Delta x_2} \sum_{\omega \in \mathcal{S}} \mathbf{1}[x_2 \leq x_{\omega,2} \leq x_2 + \Delta x_2].$$

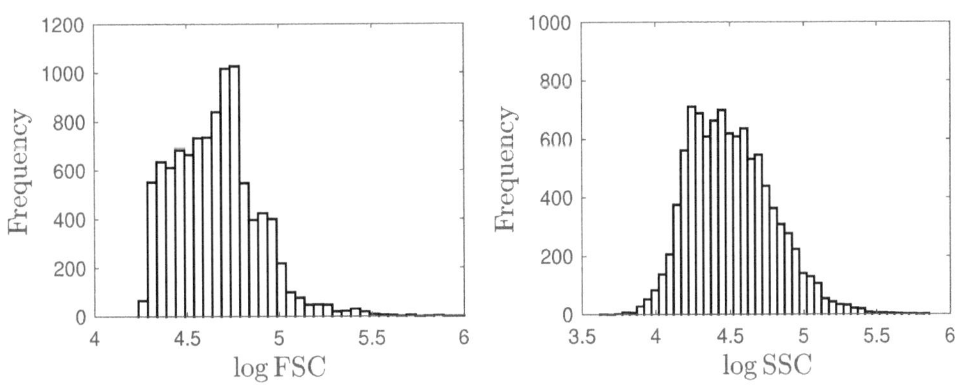

Figure 3.12 Empirical marginal pdfs for FSC/SSC using data.

Note that these are the same expressions used to compute empirical pdfs for univariate RVs. The marginal pdfs are shown in Figure 3.12 (we show these in terms of frequency to facilitate interpretability). From the marginal pdf for FSC, we can see that the marginal is nonsymmetric and has a high density of observations (peak) in the range $[4.5, 5]$. From the marginal pdf for SSC, we can see that the density is symmetric and Gaussian-like.

Script to reproduce results: `ch3_flow_cytometer_empirical` □

3.4 Conditional Density Functions

Imagine now that we want to know the probability that X_1 takes value x_1 *given knowledge* that X_2 takes value x_2 (or other way around). These probabilities are known as *conditional probabilities* and are obtained by using the *conditional pdfs*:

$$\begin{aligned} f(x_1|x_2) &= \frac{f(x_1,x_2)}{f_2(x_2)}, \; x_1 \in \mathcal{D}_1 \\ f(x_2|x_1) &= \frac{f(x_1,x_2)}{f_1(x_1)}, \; x_2 \in \mathcal{D}_2. \end{aligned} \tag{3.41}$$

Here, $f(x_1|x_2)$ gives us the probability $\mathbb{P}(X_1 = x_1|X_2 = x_2)$ and $f(x_2|x_1)$ gives us the probability $\mathbb{P}(X_2 = x_2 \mid X_1 = x_1)$; for discrete RVs, this interpretation is direct, while for continuous, this is interpreted in an infinitesimal sense.

The function $f_1(x_1)$ is the marginal pdf of X_1 and $f_2(x_2)$ is the marginal pdf of X_2. We recall that "|" indicates "given that" or "such that." The conditional probability $\mathbb{P}(X_1 = x_1|X_2 = x_2)$ can also be written in a compact form as $\mathbb{P}(X_1 = x_1|x_2)$. It is important to highlight that the statement $\mathbb{P}(X_1 = x_1|x_2)$ implicitly assumes that X_2 has a known value of x_2 (i.e., it assumes that X_2 is a deterministic variable). Similarly, the statement $\mathbb{P}(X_2 = x_2|x_1)$ implicitly assumes that X_1 is a deterministic variable.

As with joint pdfs, conditional pdfs can be continuous (for continuous RVs) or discontinuous (for discrete RVs). For discrete RVs, we can write the definition of the conditional pdf $f(x_1|x_2)$ in terms of probabilities as:

$$\mathbb{P}(X_1 = x_1|X_2 = x_2) = \frac{\mathbb{P}(X_1 = x_1 \;\&\; X_2 = x_2)}{\mathbb{P}(X_2 = x_2)}. \tag{3.42}$$

For the continuous case, it is more appropriate to interpret the conditional pdf as the ratio of the joint pdf (defined over a domain around (x_1, x_2)) and the marginal pdf (defined over a domain around x_2). This can be easily seen if we consider the empirical approximation for the conditional pdf. This is obtained by using empirical approximations for the joint and marginal pdfs; for instance:

$$\begin{aligned} \hat{f}(x_1|x_2) &= \frac{\hat{f}(x_1,x_2)}{\hat{f}_2(x_2)} \\ &= \frac{\frac{1}{\Delta x_1 \Delta x_2}\hat{\mathbb{P}}(x_1 \le X_1 \le x_1 + \Delta x_1 \;\&\; x_2 \le X_2 \le x_2 + \Delta x_2)}{\frac{1}{\Delta x_2}\hat{\mathbb{P}}(x_2 \le X_2 \le x_2 + \Delta x_2)}. \end{aligned} \tag{3.43}$$

Example 3.8 Conditional Pdf and Cdf for Thermostat

Consider the thermostat example and determine the following:

- Conditional pdf and cdf of component lifetimes X_1, X_2.
- Compute the probability that component X_1 survives after 1 year, given that X_2 survives 1 year.
- Compute the probability that component X_2 survives after 1 year, given that X_1 survives 1 year.
- Compare these probabilities against the marginal probabilities of survival.

The conditional pdfs are obtained from joint and marginal pdfs as:

$$\begin{aligned}
f(x_1|x_2) = \frac{f(x_1,x_2)}{f_2(x_2)} &= \frac{\frac{1}{\mu_1 \cdot \mu_2} e^{-(x_1/\mu_1 + x_2/\mu_2)}}{\frac{1}{\mu_2} e^{-x_2/\mu_2}} \\
&= \frac{\left(\frac{1}{\mu_1} e^{-x_1/\mu_1}\right)\left(\frac{1}{\mu_2} e^{-x_2/\mu_2}\right)}{\frac{1}{\mu_2} e^{-x_2/\mu_2}} \\
&= \frac{1}{\mu_1} e^{-x_1/\mu_1}.
\end{aligned}$$

We apply a similar procedure to obtain:

$$f(x_2|x_1) = \frac{1}{\mu_2} e^{-x_2/\mu_2}.$$

Note that, in this particular example, the conditional $f(x_1|x_2)$ does not depend on x_2 and that the conditional $f(x_2|x_1)$ does not depend on x_1 (this is not always the case). This means that knowing the failure time of X_2 does not influence the probability of failure of X_1 (and the other way around). Note also that the conditional pdfs are the marginal pdfs. We thus have that the probability that X_1 survives after 1 year knowing that X_2 survives 1 year is the same as the probability that X_1 survives after 1 year by ignoring X_2 (the marginal survival). The same applies to the opposite case. This seems to indicate that X_1 and X_2 do not depend on one another (we will address this concept in Section 3.5). We thus have that from our previous example that $\mathbb{P}(X_1 \geq 1|X_2 = 1) = \mathbb{P}(X_1 \geq 1) = 0.37$ and $\mathbb{P}(X_2 \geq 1|X_1 = 1) = \mathbb{P}(X_2 \geq 1) = 0.90$. □

Example 3.9 Conditional Pdf for a Flow Cytometer

Revisit the flow cytometer problem and assume that SSC and FSC are discrete RVs that can take low (L), medium (M), and high (H) values. We recall that the joint pdf (obtained from the observations) is given by:

$f(x_1, x_2) =$

	L	M	H
H	0.0028	0.0239	0.0012
M	0.5994	0.1114	0
L	0.2613	0	0

Use the data to determine the empirical marginal and conditional pdfs for FSC and SSC.

The marginal pdf for FSC is obtained by counting the number of observations that are in the L, M, and H categories. In Figure 3.13(a), we show the regions corresponding to each category. The marginal pdf for SSC is obtained similarly (Figure 3.13(b)).

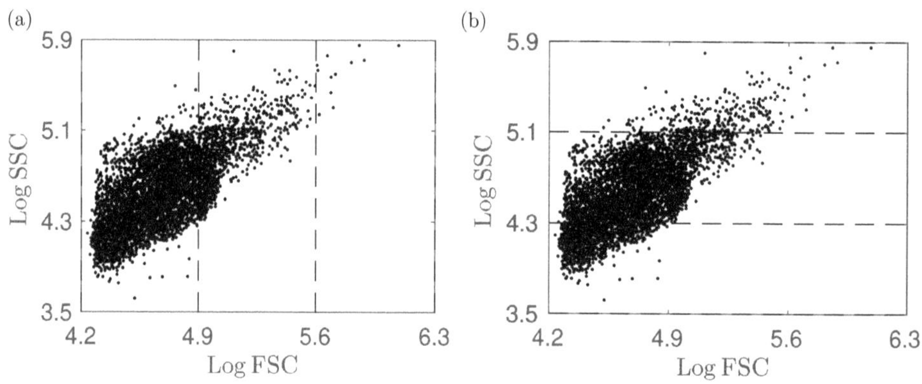

Figure 3.13 Regions for obtaining marginal pdf of FSC (a) and marginal pdf of SSC (b).

The procedure gives the marginal pdfs for FSC (X_1) and SSC (X_2):

$f_1(x_1) =$

L	M	H
0.8635	0.1353	0.0012

$f_2(x_2) =$

L	M	H
0.2613	0.7108	0.0279

From these, we can see that the entries of the marginals add up to one (the marginals are consistent). These also reveal that FSC is more concentrated in the L category, while SSC is more concentrated in the M category (this can be verified visually in Figure 3.13).

The joint pdf is computed from $f(x_1|x_2) = f(x_1, x_2)/f_2(x_2)$ and gives:

$f(x_1|x_2) =$

	L	M	H
L	1.0000	0	0
M	0.8433	0.1567	0
H	0.1004	0.8566	0.0430

The first row gives $f(x_1|x_2 = L)$ for $x_1 \in \mathcal{D}_1$, the second row gives $f(x_1|x_2 = M)$ for $x_1 \in \mathcal{D}_1$, and the third column gives $f(x_1|x_2 = H)$ for $x_1 \in \mathcal{D}_1$. Note that the summation of the entries in every row equals one; for instance, $\sum_{x_1 \in \mathcal{D}_1} f_1(x_1|x_2 = H) = 1$. Consequently, the joint pdf is consistent. The joint pdf $f(x_2|x_1) = f(x_1,x_2)/f_1(x_1)$ is given by:

$f(x_2|x_1) =$

	L	M	H
L	0.3026	0	0
M	0.6942	0.8234	0
H	0.0032	0.1766	1.0000

The first column gives $f(x_2|x_1 = L)$ for $x_2 \in \mathcal{D}_2$, the second column gives $f(x_2|x_1 = M)$ for $x_2 \in \mathcal{D}_2$, and the third column gives $f(x_2|x_1 = H)$ for $x_2 \in \mathcal{D}_2$. The summation of the entries in every column equals one.

The joint pdfs can be used to quantify how knowledge in one RV resolves uncertainty in the other; for instance, the probability that SSC takes a medium value (given knowledge that FSC takes a high value) is $\mathbb{P}(X_2 = M|X_1 = H) = f(x_2 = M|x_1 = H) = 0$ (there is zero chance of seeing an observation with a medium SSC value if the corresponding FSC value is high). This is consistent with the scattered data shown in Figure 3.9. Note also that the probability that FSC takes a low value is $f_1(x_1 = L) = 0.8635$, while the probability that it takes a low value given that SSC takes a low value is $\mathbb{P}(X_1 = L|X_2 = L) = 1.000$. In other words, if we know that a given droplet has an SSC value of L, then we can be sure that it will also have an FSC value of L. This highlights that the FSC and SSC values are somehow related (they vary together).

Script to reproduce results: `ch3_flow_cytometer_discrete` □

Generalizing the concept of the conditional pdf for n-dimensional RVs $X = (X_1, X_2, \ldots, X_n)$ gives rise to interesting concepts. Specifically, imagine that we partition our RV vector into subvectors $X_I = (X_1, \ldots, X_m)$ and $X_{II} = (X_{m+1}, \ldots, X_n)$. For example, if we have the collection:

$$X = (X_1, X_2, X_3, X_4, X_5),$$

we can define the partitions $X_I = (X_1, X_2)$ and $X_{II} = (X_3, X_4, X_5)$. Based on our variable partition, we define our conditional expectation as $f(x_I|x_{II})$ or, in an expanded form, $f(x_1, \ldots, x_m \,|\, x_{m+1}, \ldots, x_n)$. As we can infer, the joint pdf tells us the probability that $\mathbb{P}(X_I = x_I \,|\, X_{II} = x_{II})$, which in an expanded form is:

$$\mathbb{P}(X_1 = x_1, \ldots, X_m = x_m \,|\, X_{m+1} = x_{m+1}, \ldots, X_n = x_n). \tag{3.44}$$

The conditional pdf is given by:

$$f(x_I|x_{II}) = \frac{f(x)}{f_{II}(x_{II})} = \frac{f(x_1, \ldots, x_n)}{f_{II}(x_{m+1}, \ldots, x_n)}, \quad x_I \in \mathcal{D}_I, \tag{3.45}$$

where $f(x_1, \ldots, x_n)$ is the joint pdf of X and $f_{II}(x_{II}) = f_{II}(x_{m+1}, \ldots, x_n)$ is the joint pdf for X_{II} (all the variables that we assume to be known). The density

$f_{II}(x_{II})$ plays the role of the marginal density in the $n = 2$ case. In the general n case, $f_{II}(x_{II})$ is the joint pdf of all the variables that are marginalized (assumed known).

3.5 Independence

Conditional density functions tell us how knowledge in one RV *resolves uncertainty* in another. In other words, they tell us *how much knowledge of one is embedded in the other*. But, what if knowledge of one does not resolve any uncertainty in the other? In other words, what if one conveys no knowledge of the other? This gives rise to the concept of *independence*.

The RVs X_1 and X_2 are said to be *independent* if:

$$\begin{aligned} f(x_1|x_2) &= f_1(x_1),\ x_1 \in \mathcal{D}_1 \\ f(x_2|x_1) &= f_2(x_2),\ x_2 \in \mathcal{D}_2. \end{aligned} \tag{3.46}$$

That is, if X_1 and X_2 are independent, we have that the conditional pdfs equal the marginal pdfs. This implies that:

$$f(x_1,x_2) = f_2(x_2)\cdot f_1(x_1), \quad (x_1,x_2) \in \mathcal{D}. \tag{3.47}$$

Equivalently:

$$\mathbb{P}(X_1 = x_1\ \&\ X_2 = x_2) = \mathbb{P}(X_1 = x_1)\cdot \mathbb{P}(X_2 = x_2). \tag{3.48}$$

That is, the joint pdf of independent X_1 and X_2 equals the product of its marginal pdfs. Note that this is analogous to the notion of independent events discussed in Section 3.1. The concept of independence is important in statistics because it tells us that RVs vary independently of one another, and therefore, there is no connection between them. This would mean, for instance, that using one to predict another would not make sense. In Section 3.7, we will introduce the concept of correlation; we will see that, if RVs are independent, it means that they are not correlated.

Example 3.10 Component Independence in Thermostat

Consider the thermostat example and address the following:

- Prove that the lifetimes X_1 and X_2 are independent and provide a physical explanation of why this might be the case.

The condition for independence is that $f(x_1,x_2) = f_1(x_1)\cdot f_2(x_2)$. Notice that the joint pdf is:

$$\begin{aligned} f(x_1,x_2) &= \frac{1}{\mu_1\cdot\mu_2} e^{-(x_1/\mu_1 + x_2/\mu_2)} \\ &= \left(\frac{1}{\mu_1} e^{-x_1/\mu_1}\right)\left(\frac{1}{\mu_2} e^{-x_2/\mu_2}\right), \end{aligned}$$

and notice that the product of the marginals is:

$$f_1(x_1)f_2(x_2) = \left(\frac{1}{\mu_1}e^{-x_1/\mu_1}\right)\left(\frac{1}{\mu_2}e^{-x_2/\mu_2}\right).$$

We thus have that X_1 and X_2 are independent. The failure of the components might be independent if they do not interact physically or if their interaction is weak. For example, the sensor might send signals to the computer processor, but such signals might not influence the degradation of the processor. The processor might influence the sensor by releasing heat (computation releases heat), but such interaction might not be strong enough to influence the degradation of the sensor. This might not be the case in other systems; for example, the sensor might "flood" the processor with data and might overwork the processor (thus decreasing its lifetime). In such a case, we expect that the lifetimes of the components are dependent (as opposed to independent). □

We generalize the concept of independence to the n-dimensional $X = (X_1, \ldots, X_n)$ case by grouping RVs. Specifically, we say that group $X_I = (X_1, \ldots, X_m)$ is independent from group $X_{II} = (X_{m+1}, \ldots, X_n)$ if:

$$\mathbb{P}(X_I = x_I \,\&\, X_{II} = x_{II}) = \mathbb{P}(X_I = x_I) \cdot \mathbb{P}(X_{II} = x_{II}). \tag{3.49}$$

Here, $\mathbb{P}(X_I = x_I \,\&\, X_{II} = x_{II})$ is computed using $f(x) = f(x_1, .., x_n)$, $\mathbb{P}(X_I = x_I)$ is computed using $f_I(x_I) = f_I(x_1, \ldots, x_m)$, and $\mathbb{P}(X_{II} = x_{II})$ is computed from $f_{II}(x_{II}) = f_{II}(x_{m+1}, \ldots, x_n)$. Here, note that the RVs *within* a group might not need to be independent from one another, but they might still be independent from the RVs of the other group. For example, X_I might be independent from X_{II}, while X_1 and X_m might not.

This *grouping* approach can become useful when analyzing highly complex systems that can naturally be partitioned into groups. For example, a chemical process can be partitioned into unit operations; here, X_I can be the variables in a reactor (e.g., temperature and conversion), while X_{II} can be the variables in a separator (e.g., input and output concentration). If the unit operations are not connected, then X_I and X_{II} will be independent; however, the variables of the reactor are expected to strongly depend on one another and might therefore not be independent (similarly for the separator). Independence (lack of connectivity) and interdependence (connectivity) are key concepts in statistics.

The concept of independence is key in establishing the statistical properties of RVs. In Chapter 2, we established properties for univariate RVs that rely on the assumption that a set of RVs $X_i, i = 1, \ldots, n$ are independent. For example, we saw that, if we have a set of RVs $X_i \sim \mathcal{N}(0, 1), i = 1, \ldots, n$ that are independent, then $\sum_{i=1}^n X_i^2 \sim \chi^2(n)$. The concept of independence is also key when establishing properties of sequences of RVs. For example, in Chapter 2, we saw that a Poisson process is constructed based on a sequence of independent Bernoulli trials. In Chapter 4, we will see that constructing sequences of independent RVs will help establish *unbiased* data collection (sampling) procedures. Specifically, we will see that unbiased sampling procedures are needed to ensure that empirical approximations of summarizing statistics converge to the true statistics as data is accumulated over time.

3.6 Conditional Events

In our previous discussion, we focused on computing probabilities that are conditional to the occurrence of *elementary* (simple) events such as $X_1 = x_1$. Probabilities that are conditional to the occurrence of general events can be derived from the domain properties established previously. The joint pdf of X plays a key role here, as this is the basic tool that enables the computation of probabilities over domains. The joint pdf, together with general logic and constraints captured by subdomains, allows us to compute probabilities of complex events.

To generalize our concepts, we return to the definition of conditional events discussed in Section 3.1:

$$\mathbb{P}(X \in \mathcal{A}_1 | X \in \mathcal{A}_2) = \frac{\mathbb{P}(X \in \mathcal{A}_1 \ \&\ X \in \mathcal{A}_2)}{\mathbb{P}(X \in \mathcal{A}_2)}, \tag{3.50}$$

where $\mathcal{A}_1$ and $\mathcal{A}_2$ are subdomains of $\mathcal{D}$ that capture specific events. Here, $\mathbb{P}(X \in \mathcal{A}_1 | X \in \mathcal{A}_2)$ is the probability that event $\mathcal{A}_1$ occurs given knowledge that $\mathcal{A}_2$ has occurred, $\mathbb{P}(X \in \mathcal{A}_1 \ \&\ X \in \mathcal{A}_2)$ is the joint probability of $\mathcal{A}_1$ and $\mathcal{A}_2$, and $\mathbb{P}(X \in \mathcal{A}_2)$ is the marginal probability of event $\mathcal{A}_2$.

For continuous X, the probability $\mathbb{P}(X \in \mathcal{A}_1 | X \in \mathcal{A}_2)$ can be computed by using the joint pdf as:

$$\mathbb{P}(X \in \mathcal{A}_1 | X \in \mathcal{A}_2) = \frac{\displaystyle\int_{x \in \mathcal{A}_1 \cap \mathcal{A}_2} f(x)dx}{\displaystyle\int_{x \in \mathcal{A}_2} f(x)dx}. \tag{3.51}$$

Here, we recall that the event $\{X \in \mathcal{A}_1 \ \&\ X \in \mathcal{A}_2\}$ is equivalent to $X \in \mathcal{A}_1 \cap \mathcal{A}_2$. The conditional probability can be computed for discrete X by replacing integrals with summations. Given that computing integrals is difficult, in practice, we can also obtain empirical approximations of probabilities by using data, as we see in the next example.

Example 3.11 Conditional Events in Gibbs Reactor

Consider the situation in which we monitor the pressure and conversion of the reactor, consider the events

$$\mathcal{A} = \{P \leq 170 \,\&\, C \geq 0.5\}$$
$$\mathcal{B} = \{P \geq 120 \,\&\, C \geq 0.5\},$$

and consider the data provided in Figure 3.14.

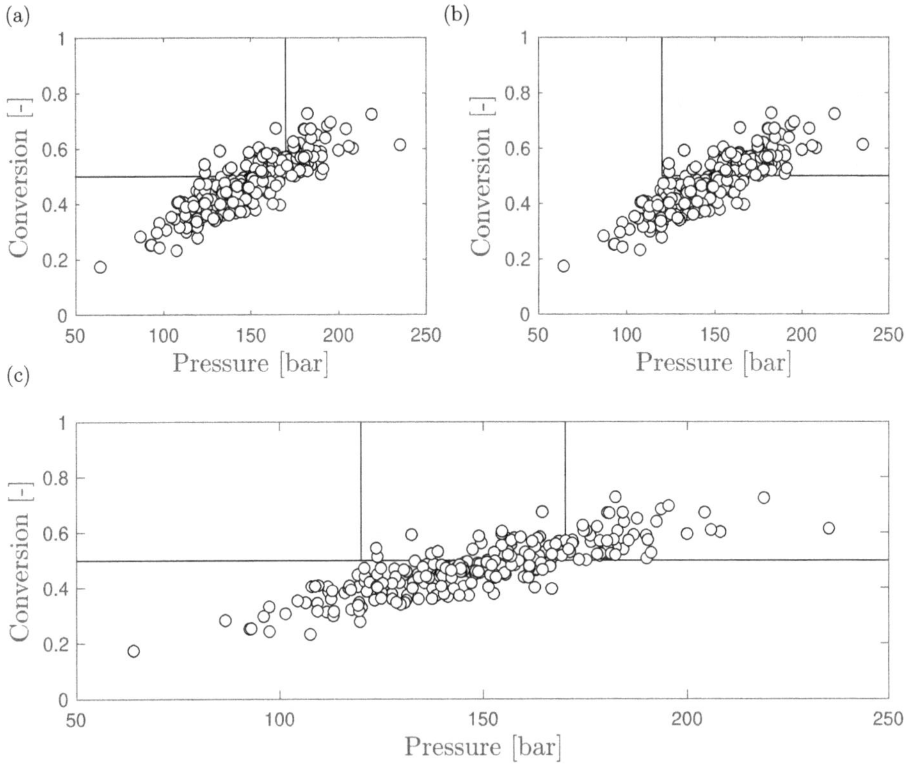

Figure 3.14 Subdomains associated with events $\mathcal{A}$ (a), $\mathcal{B}$ (b), and $\mathcal{A} \cap \mathcal{B}$ (c).

- What is the probability that $\mathcal{A}$ occurs, given that we know that $\mathcal{B}$ has occurred?
- What is the probability that $\mathcal{B}$ occurs, given that we know that $\mathcal{A}$ has occurred?

In Figure 3.14, we show the subdomains of the individual domains (associated with the marginal probabilities) and their intersection (associated with the joint probability). If we count the number of data points that land in each subdomain and divide by the total number of data points, we can estimate that:

$$\hat{\mathbb{P}}(X \in \mathcal{A}) = 0.16$$
$$\hat{\mathbb{P}}(X \in \mathcal{B}) = 0.35$$
$$\hat{\mathbb{P}}(X \in \mathcal{A} \,\&\, X \in \mathcal{B}) = 0.16.$$

Note that every point that lands on $\mathcal{A}$ also lands on $\mathcal{B}$, but not every point that lands on $\mathcal{B}$ lands on $\mathcal{A}$. This explains why the joint probability $\hat{\mathbb{P}}(X \in \mathcal{A} \,\&\, X \in \mathcal{B})$ equals the marginal probability of $\mathcal{A}$. Using the joint and marginals, we can estimate the conditional probabilities:

$$\hat{\mathbb{P}}(X \in \mathcal{A} \,|\, X \in \mathcal{B}) = 0.16/0.35 = 0.45$$
$$\hat{\mathbb{P}}(X \in \mathcal{B} \,|\, X \in \mathcal{A}) = 0.16/0.16 = 1.0.$$

The joint probability $\hat{\mathbb{P}}(X \in \mathcal{A} \,|\, X \in \mathcal{B})$ indicates that knowing that a data point lands on $\mathcal{B}$ increases the probability that it also lands on $\mathcal{A}$ (relative to the marginal probability $\hat{\mathbb{P}}(X \in \mathcal{A})$). As such, event $\mathcal{B}$ conveys information about event $\mathcal{A}$.

Now, from the joint probability $\hat{\mathbb{P}}(X \in \mathcal{B} \,|\, X \in \mathcal{A})$, we note that knowing that a point lands on $\mathcal{A}$ indicates that the probability that it is in $\mathcal{B}$ is one. This results because every point in $\mathcal{A}$ is also in $\mathcal{B}$; therefore, event $\mathcal{A}$ carries full information about event $\mathcal{B}$.

Script to reproduce results: `ch3_gibbs_conditional` □

The concept of conditioning can be applied *sequentially*; this is particularly relevant in applications in which one progressively accumulates knowledge of event occurrences (e.g., investment over uncertain market conditions or planning a trip under uncertain weather). Imagine that you want to determine the probability of event $\mathcal{A}_1$, given that you know that event $\mathcal{A}_2$ has occurred. This probability is given by Eq. (3.50). Now imagine you want the probability of $\mathcal{A}_1$, given that $\mathcal{A}_2$ and $\mathcal{A}_3$ have occurred. This conditional probability is given by:

$$\mathbb{P}(X \in \mathcal{A}_1 | X \in \mathcal{A}_2 \;\&\; X \in \mathcal{A}_3) = \frac{\displaystyle\int_{x \in \mathcal{A}_1 \cap \mathcal{A}_2 \cap \mathcal{A}_3} f(x)dx}{\displaystyle\int_{x \in \mathcal{A}_2 \cap \mathcal{A}_3} f(x)dx}. \tag{3.52}$$

We can use a similar reasoning to derive an expression to compute the conditional probability:

$$\mathbb{P}(X \in \mathcal{A}_1 | X \in \mathcal{A}_2 \;\&\; X \in \mathcal{A}_3 \;\&\; \cdots \;\& X \in \mathcal{A}_n).$$

We recall that, if $\mathcal{A}_1$ and $\mathcal{A}_2$ are *independent* events, we have that:

$$\mathbb{P}(X \in \mathcal{A}_1 \;\&\; X \in \mathcal{A}_2) = \mathbb{P}(X \in \mathcal{A}_1) \cdot \mathbb{P}(X \in \mathcal{A}_2). \tag{3.53}$$

This implies that:

$$\int_{x \in \mathcal{A}_1 \cap \mathcal{A}_2} f(x)dx = \left(\int_{x \in \mathcal{A}_1} f(x)dx\right) \cdot \left(\int_{x \in \mathcal{A}_2} f(x)dx\right). \tag{3.54}$$

We recall that the probability that event $\mathcal{A}_1$ or $\mathcal{A}_2$ occurs is given by:

$$\mathbb{P}(X \in \mathcal{A}_1 \cup \mathcal{A}_2) = \mathbb{P}(X \in \mathcal{A}_1) + \mathbb{P}(X \in \mathcal{A}_2) - \mathbb{P}(X \in \mathcal{A}_1 \cap \mathcal{A}_2). \tag{3.55}$$

This is computed by using the joint pdf of X as:

$$\mathbb{P}(X \in \mathcal{A}_1 \cup \mathcal{A}_2) = \int_{x \in \mathcal{A}_1} f(x)dx + \int_{x \in \mathcal{A}_2} f(x)dx - \int_{x \in \mathcal{A}_1 \cap \mathcal{A}_2} f(x)dx. \tag{3.56}$$

We can thus see that the pdfs and general event properties can be used to compute the probabilities of complex events. In Section 3.7.3, we discuss how to compute empirical approximations for event probabilities using data.

Example 3.12 Probability and Logic for Thermostat

Consider the thermostat example and compute the following:

- Probability that the sensor or processor last more than a year.
- Probability that the sensor and processor last more than a year or that the sensor and processor last more than 2 years.

The first probability is determined from Eq. (3.56) with events $\mathcal{A}_1 = \{x_1 \geq 1 \ \& \ x_2 \in \mathcal{D}_2\}$ and $\mathcal{A}_2 = \{x_1 \in \mathcal{D}_1 \ \& \ x_2 \geq 1\}$, which give $\mathcal{A}_1 \cap \mathcal{A}_2 = \{x_1 \geq 1 \ \& \ x_2 \geq 1\}$, and $\mathcal{A}_1 \cup \mathcal{A}_2 = \{x_1 \geq 1 \text{ or } x_2 \geq 1\}$. Computing probabilities, we have:

$$\int_{x \in \mathcal{A}_1} f(x)dx = \int_0^{\infty} \int_1^{\infty} f(x_1, x_2)dx_1 dx_2 = 0.37$$

$$\int_{x \in \mathcal{A}_2} f(x)dx = \int_1^{\infty} \int_0^{\infty} f(x_1, x_2)dx_1 dx_2 = 0.90$$

$$\int_{x \in \mathcal{A}_1 \cap \mathcal{A}_2} f(x)dx = \int_1^{\infty} \int_1^{\infty} f(x_1, x_2)dx_1 dx_2 = 0.33,$$

and thus $\mathbb{P}(X \in \mathcal{A}_1 \cup \mathcal{A}_2) = 0.37 + 0.90 - 0.33 = 0.94$. We can see that there is a high probability that either one of them lasts more than a year.

The second probability is $\mathbb{P}(X \in \mathcal{B}_1 \cap \mathcal{B}_2)$ with events $\mathcal{B}_1 = \mathcal{A}_1 \cap \mathcal{A}_2 = \{x_1 \geq 1 \ \& \ x_2 \geq 1\}$ and $\mathcal{B}_2 = \mathcal{A}_3 \cap \mathcal{A}_4$ with $\mathcal{A}_3 = \{x_1 \geq 2 \ \& \ x_2 \in \mathcal{D}_2\}$ and $\mathcal{A}_4 = \{x_1 \in \mathcal{D}_1 \ \& \ x_2 \geq 2\}$ to give $\mathcal{B}_2 = \mathcal{A}_3 \cap \mathcal{A}_4 = \{x_1 \geq 2 \ \& \ x_2 \geq 2\}$. Note that $\mathcal{B}_1 \cap \mathcal{B}_2 = \mathcal{A}_1 \cap \mathcal{A}_2 \cap \mathcal{A}_3 \cap \mathcal{A}_4 = \{x_1 \geq 2 \ \& \ x_2 \geq 2 \ \& \ x_1 \geq 1 \ \& \ x_2 \geq 1\}$, which can be simplified to $\mathcal{B}_1 \cap \mathcal{B}_2 = \{x_1 \geq 2 \ \& \ x_2 \geq 2\}$ (the other constraints become redundant). The desired probability can be obtained from:

$$\mathbb{P}(X \in \mathcal{B}_1 \cup \mathcal{B}_2) = \int_{x \in \mathcal{B}_1} f(x)dx + \int_{x \in \mathcal{B}_2} f(x)dx - \int_{x \in \mathcal{B}_1 \cap \mathcal{B}_2} f(x)dx.$$

Computing probabilities:

$$\mathbb{P}(X \in \mathcal{B}_1) = \int_1^{\infty} \int_1^{\infty} f(x_1, x_2)dx_1 dx_2 = 0.33$$

$$\mathbb{P}(X \in \mathcal{B}_2) = \int_2^{\infty} \int_2^{\infty} f(x_1, x_2)dx_1 dx_2 = 0.11$$

$$\mathbb{P}(X \in \mathcal{B}_1 \cap \mathcal{B}_2) = \int_2^{\infty} \int_2^{\infty} f(x_1, x_2)dx_1 dx_2 = 0.11.$$

Thus, $\mathbb{P}(X \in \mathcal{B}_1 \cup \mathcal{B}_2) = 0.33 + 0.11 - 0.11 = 0.33$. Interestingly, note that $\mathbb{P}(X \in \mathcal{B}_2) = \mathbb{P}(X \in \mathcal{B}_1 \cap \mathcal{B}_2)$ because the restrictions imposed by event $\mathcal{B}_1$ become redundant under the constraints imposed by event $\mathcal{B}_2$. □

3.7 Summarizing Statistics

As with univariate RVs, we are often interested in characterizing multivariate RVs by using summarizing statistics (quantities) as opposed to using density functions. Computing such statistics in the multivariate case is similar to what we have already established for the univariate case, but a few new concepts will emerge in this discussion. For instance, the concept of variance will be expanded to consider covariance and correlation (which capture connectivity between variables).

3.7.1 Expectation

The *joint expectation* of $X = (X_1, X_2)$ is a vector $\mathbb{E}[X] \in \mathbb{R}^2$ of the form:

$$\mathbb{E}[X] = \mathbb{E}\begin{bmatrix} X_1 \\ X_2 \end{bmatrix}. \tag{3.57}$$

For a continuous RV, this vector is computed as:

$$\mathbb{E}[X] = \int_{x \in \mathcal{D}} x \cdot f(x) dx. \tag{3.58}$$

If we write $x = (x_1, x_2)$, we can express the expected value as:

$$\mathbb{E}[X] = \int_{(x_1,x_2) \in \mathcal{D}} \begin{bmatrix} x_1 \\ x_2 \end{bmatrix} \cdot f(x_1, x_2) dx_1 dx_2. \tag{3.59}$$

We can thus see that we can write the expectation as the vector:

$$\mathbb{E}[X] = (\mathbb{E}[X_1], \mathbb{E}[X_2]), \tag{3.60}$$

with entries:

$$\begin{aligned} \mathbb{E}[X_1] &= \int_{(x_1,x_2) \in \mathcal{D}} x_1 \cdot f(x_1, x_2) dx_1 dx_2 \\ \mathbb{E}[X_2] &= \int_{(x_1,x_2) \in \mathcal{D}} x_2 \cdot f(x_1, x_2) dx_1 dx_2. \end{aligned} \tag{3.61}$$

It is important to highlight the difference between the joint expectations $\mathbb{E}[X_1]$ and $\mathbb{E}[X_2]$ and the marginal expectations of X_1 and X_2. The *marginal expectations* are given by:

$$\begin{aligned} \mathbb{E}_m[X_1] &= \int_{x_1 \in \mathcal{D}_1} x_1 \cdot f_1(x_1) dx_1 \\ \mathbb{E}_m[X_2] &= \int_{x_2 \in \mathcal{D}_2} x_2 \cdot f_2(x_2) dx_2. \end{aligned} \tag{3.62}$$

The joint expectation of a scalar function $\varphi : \mathbb{R}^2 \to \mathbb{R}$ of X is a scalar value given by:

$$\mathbb{E}[\varphi(X)] = \int_{(x_1,x_2)\in\mathcal{D}} \varphi(x_1,x_2)\cdot f(x_1,x_2)dx_1dx_2. \tag{3.63}$$

The *conditional* expectation of X_1 (given knowledge that $X_2 = x_2$) is:

$$\mathbb{E}[X_1|X_2 = x_2] = \int_{x_1\in\mathcal{D}_1} x_1\cdot f(x_1|x_2)dx_1. \tag{3.64}$$

The conditional expectation of a function $\varphi : \mathbb{R}^2 \to \mathbb{R}$ of X (given knowledge $X_2 = x_2$) is:

$$\mathbb{E}[\varphi(X)|X_2 = x_2] = \int_{x_1\in\mathcal{D}_1} \varphi(x_1,x_2)\cdot f(x_1|x_2)dx_1. \tag{3.65}$$

Conditional expectations are useful in decision-making; for instance, we might be interested in determining the expected revenue of an investment X_1, given that we already know that another investment X_2 takes a specific value x_2. Expressions for discrete RVs are analogous (replace integrals with summations).

Example 3.13 Expected Lifetime of Thermostat

Consider the thermostat example (Example 3.2) and address the following:

- Determine the expected lifetime of the components.
- Determine the expected lifetime of the processor, given that the lifetime of the sensor has reached 1 year.

The expected lifetimes are:

$$\begin{aligned}\mathbb{E}[X] &= \int_{(x_1,x_2)\in\mathcal{D}} \begin{bmatrix} x_1 \\ x_2 \end{bmatrix}\cdot f(x_1,x_2)dx_1dx_2 \\ &= \begin{bmatrix} \int_{(x_1,x_2)\in\mathcal{D}} x_1\cdot f(x_1,x_2)dx_1dx_2 \\ \int_{(x_1,x_2)\in\mathcal{D}} x_2\cdot f(x_1,x_2)dx_1dx_2 \end{bmatrix}.\end{aligned}$$

Now note that:

$$\begin{aligned}&\int_{(x_1,x_2)\in\mathcal{D}} x_1\cdot f(x_1,x_2)dx_1dx_2 \\ &= \int_{(x_1,x_2)\in\mathcal{D}} x_1\left(\frac{1}{\mu_1}e^{-x_1/\mu_1}\right)\left(\frac{1}{\mu_2}e^{-x_2/\mu_2}\right)dx_1dx_2 \\ &= \int_0^\infty x_1\left(\frac{1}{\mu_1}e^{-x_1/\mu_1}\right)dx_1 \int_0^\infty \left(\frac{1}{\mu_2}e^{-x_2/\mu_2}\right)dx_2 \\ &= e^{-0/\mu_1}(\mu_1 + 0) - e^{-\infty/\mu_1}(\mu_1 + \infty) \\ &= \mu_1.\end{aligned}$$

Here, we use the fact that $\int_0^\infty \left(\frac{1}{\mu_2}e^{-x_2/\mu_2}\right)dx_2 = 1$. Using similar logic, we find that:

$$\int_{(x_1,x_2)\in\mathcal{D}} x_2 \cdot f(x_1,x_2)dx_1dx_2 = \mu_2.$$

We thus have that $\mathbb{E}[X] = (\mu_1, \mu_2)$.

To address the second point, we need to compute $\mathbb{E}[X_2|X_1 = x_1]$ with $x_1 = 1$. This is obtained from:

$$\begin{aligned}\mathbb{E}[X_2|X_1 = x_1] &= \int_{x_2\in\mathcal{D}_2} x_2 f(x_2|x_1)dx_2\\ &= \int_{x_2\in\mathcal{D}_2} x_2\frac{1}{\mu_2}e^{-x_2/\mu_2}dx_2\\ &= \mu_2.\end{aligned}$$

We thus have that the conditional expected lifetime is equal to the expected lifetime. This makes sense because we have already established that the lifetime of the processor and sensor are independent. □

The joint expectation of an n-dimensional RV $X = (X_1, X_2, \ldots, X_n)$ is a vector $\mathbb{E}[X] \in \mathbb{R}^n$ of the form:

$$\mathbb{E}[X] = \begin{bmatrix} \mathbb{E}[X_1] \\ \mathbb{E}[X_2] \\ \vdots \\ \mathbb{E}[X_n] \end{bmatrix}, \tag{3.66}$$

where:

$$\mathbb{E}[X_i] = \int_{x\in\mathcal{D}} x_i \cdot f(x)dx, \; i = 1, \ldots, n, \tag{3.67}$$

and where $f(x) = f(x_1, x_2, \ldots, x_n)$ is the joint pdf of X. Similarly, we use the joint pdf $f(x)$ to compute the expected value of any function $\varphi(X)$.

As with conditional probabilities, it is also possible to compute expectations that are conditional on events (known as conditional expectations). For example:

$$\mathbb{E}[\varphi(X)|X \in \mathcal{A}] = \int_{x\in\mathcal{A}} \varphi(x) \cdot f(x)dx, \tag{3.68}$$

is the expectation of $\varphi(X)$, given that event $\mathcal{A}$ has occurred.

We can use data x_ω, $\omega \in \mathcal{S}$ to compute empirical estimates for the expected value:

$$\hat{\mathbb{E}}[X] = \frac{1}{S}\sum_{\omega\in\mathcal{S}} x_\omega. \tag{3.69}$$

Here, we note that each observation x_ω is an n-dimensional vector and thus $\hat{\mathbb{E}}[X]$ is also an n-dimensional vector. To compute an empirical approximation for the expectation function, we simply use:

$$\hat{\mathbb{E}}[\varphi(X)] = \frac{1}{S}\sum_{\omega\in\mathcal{S}} \varphi(x_\omega). \tag{3.70}$$

The empirical approximation of the conditional expectation is:

$$\hat{\mathbb{E}}[\varphi(X)|X \in \mathcal{A}] = \frac{1}{S}\sum_{\omega \in \mathcal{S}} \mathbf{1}[x_\omega \in \mathcal{A}] \cdot \varphi(x_\omega). \tag{3.71}$$

3.7.2 Covariance and Correlation

In the univariate case with RV X, we were interested in understanding the variability of the RV because this is a measure of its uncertainty. In the multivariate case with RV $X = (X_1, X_2)$, we are interested in the variability of the individual RVs X_1 and X_2, but we are also interested in understanding how X_1 and X_2 *vary together*. This gives rise to the concept of covariance and correlation. These concepts are fundamental in statistics, and their discussion will require us to revisit basic concepts of linear algebra.

The *covariance* of $X = (X_1, X_2)$ is a *matrix* $\text{Cov}[X] \in \mathbb{R}^{2\times 2}$ given by:

$$\text{Cov}[X] = \mathbb{E}\left[(X - \mathbb{E}[X])(X - \mathbb{E}[X])^T\right]. \tag{3.72}$$

We can see that this is indeed a 2×2 matrix from the fact that $(X - \mathbb{E}[X])$ is a 2×1 column vector and $(X - \mathbb{E}[X])^T$ is a 1×2 row vector. When convenient, we will also use the short-hand notation Cov_X to denote the covariance.

To simplify notation, we define the expected value $\mathbb{E}[X]$ as the vector $m = (m_1, m_2)$ with $m_1 = \mathbb{E}[X_1]$ and $m_2 = \mathbb{E}[X_2]$; with this, we can write the covariance matrix as:

$$\begin{aligned}
\text{Cov}[X] &= \mathbb{E}\left[(X - m)(X - m)^T\right] \\
&= \mathbb{E}\left[\begin{bmatrix} X_1 - m_1 \\ X_2 - m_2 \end{bmatrix} [X_1 - m_1 \;\; X_2 - m_2]\right] \\
&= \mathbb{E}\begin{bmatrix} (X_1 - m_1)^2 & (X_1 - m_1)(X_2 - m_2) \\ (X_2 - m_2)(X_1 - m_1) & (X_2 - m_2)^2 \end{bmatrix} \\
&= \begin{bmatrix} \mathbb{E}[(X_1 - m_1)^2] & \mathbb{E}[(X_1 - m_1)(X_2 - m_2)] \\ \mathbb{E}[(X_2 - m_2)(X_1 - m_1)] & \mathbb{E}[(X_2 - m_2)^2] \end{bmatrix}.
\end{aligned} \tag{3.73}$$

We define the individual entries of the matrix as follows:

$$\begin{aligned}
\text{Cov}[X]_{11} &= \mathbb{E}[(X_1 - m_1)^2] \\
\text{Cov}[X]_{12} &= \mathbb{E}[(X_1 - m_1)(X_2 - m_2)] \\
\text{Cov}[X]_{21} &= \mathbb{E}[(X_2 - m_2)(X_1 - m_1)] \\
\text{Cov}[X]_{22} &= \mathbb{E}[(X_2 - m_2)^2].
\end{aligned} \tag{3.74}$$

Note that $\text{Cov}[X]_{12} = \text{Cov}[X]_{21}$, which indicates that the covariance matrix is *symmetric*. One can also show that the matrix is positive semidefinite (has eigenvalues that are positive or zero). Note also that:

$$\begin{aligned}
\text{Cov}[X]_{11} &= \mathbb{V}[X_1] \\
\text{Cov}[X]_{22} &= \mathbb{V}[X_2].
\end{aligned} \tag{3.75}$$

In other words, the diagonal entries of the covariance matrix are the *variances* of X_1 and X_2. Note also that the variances are nonnegative quantities and thus the diagonal entries of a covariance matrix are nonnegative.

The variance $\text{Cov}[X]_{11}$ tells us how, on average, X_1 varies (around its expected value m_1). Similarly, $\text{Cov}[X]_{22}$ tells us how, on average, X_2 varies (around its expected value m_2). The covariance $\text{Cov}[X]_{12} = \text{Cov}[X]_{21}$ tells us how, *on average*, X_1 varies with X_2; specifically:

- If $\text{Cov}[X]_{12} > 0$, we say that we have *positive covariance*: The event $X_1 \geq m_1$ is, on average, associated with $X_2 \geq m_2$ and event $X_1 \leq m_1$ is, on average, associated with $X_2 \leq m_2$. In other words, X_1 and X_2 move in the same direction (on average).
- If $\text{Cov}[X]_{12} < 0$, we say that we have *negative covariance*: The event $X_1 \geq m_1$ is, on average, associated with $X_2 \leq m_2$ and event $X_1 \leq m_1$ is, on average, associated with $X_2 \geq m_2$. In other words, X_1 and X_2 move in opposite directions (on average).
- If $\text{Cov}[X]_{12} = 0$, we say that we have *no covariance*: On average, variations in $(X_1 - m_1)$ are not associated with variations in $(X_2 - m_2)$. If X_1 andX_2 are *independent*, then we can show that $\text{Cov}[X]_{12} = 0$. However, $\text{Cov}[X]_{12} = 0$ does not imply independence, because X_1, X_2 might vary together (but not on average).

The *covariance* $\text{Cov}[X]_{12}$ is often used as a measure of the strength of *interdependence* between RVs (strength of connectivity); in other words, the higher the covariance, the higher the interdependence. Moreover, it is important to emphasize that covariance between variables can arise because of *physical or structural connections* between variables. For instance, if X_1 and X_2 are connected as $X_2 = \alpha \cdot X_1$ (e.g., conversion is proportional to pressure), then we can show that $\text{Cov}[X]_{12} = \alpha \cdot \mathbb{V}[X_1]$ and thus a larger α (stronger connectivity) increases the covariance. This means that any structural connections between variables are encoded in the covariance. These types of structural relationships between variables can be discovered from data using estimation techniques for *structural models*, as we discuss in Chapter 5. Structural models will allow us to make predictions for one variable based on another variable; note that the ability to predict relies on the fact that the variables are related (that they covary). On the other hand, if X_1 and X_2 are *not connected* (they are *independent*), we have that $\text{Cov}[X]_{12} = 0$; in other words, a multivariate RV $X = (X_1, X_2)$ with independent X_1 and X_2 has a *diagonal* covariance matrix. When variables are independent, it does not make sense to try to predict one as a function of the other. Note also that the discussion of covariance is related to the notion of conditional probabilities; specifically, if variables covary, we expect that knowledge of one will reduce the uncertainty of the other.

The *autocorrelation* matrix is another summarizing statistic that is used to measure variability between RVs. This is defined as:

$$\text{ACorr}[X] = \mathbb{E}[XX^T] = \begin{bmatrix} \mathbb{E}[X_1X_1] & \mathbb{E}[X_1X_2] \\ \mathbb{E}[X_2X_1] & \mathbb{E}[X_2X_2] \end{bmatrix}. \tag{3.76}$$

Note that the definition is similar to that of the covariance, but we do not use the expected value as a reference point. The autocorrelation matrix is related to the covariance matrix as:

$$\text{Cov}[X] = \text{ACorr}[X] - \mathbb{E}[X]\mathbb{E}[X]^T. \tag{3.77}$$

The *Pearson* correlation matrix (named after biostatistician Karl Pearson) is another important statistic. This is defined as:

$$\text{PCorr}[X] = D^{-1}\text{Cov}[X]D^{-1}, \tag{3.78}$$

where $D = \sqrt{\text{diag}(\text{Cov}[X])}$ is a 2×2 matrix that contains the square roots of the diagonal elements of matrix Cov[X]. Specifically, we have that:

$$D = \begin{bmatrix} \sqrt{\text{Cov}[X]_{11}} & \\ & \sqrt{\text{Cov}[X]_{22}} \end{bmatrix}. \tag{3.79}$$

We thus see that D contains the *standard deviations* $\mathbb{SD}[X_1]$ and $\mathbb{SD}[X_2]$. From this, we can observe that the Pearson matrix is given by:

$$\begin{aligned} \text{PCorr}[X] &= D^{-1}\text{Cov}[X]D^{-1} \\ &= \begin{bmatrix} \dfrac{\mathbb{E}[(X_1 - m_1)^2]}{\mathbb{SD}[X_1]\mathbb{SD}[X_1]} & \dfrac{\mathbb{E}[(X_1 - m_1)(X_2 - m_2)]}{\mathbb{SD}[X_1]\mathbb{SD}[X_2]} \\ \dfrac{\mathbb{E}[(X_2 - m_2)(X_1 - m_1)]}{\mathbb{SD}[X_2]\mathbb{SD}[X_1]} & \dfrac{\mathbb{E}[(X_2 - m_2)^2]}{\mathbb{SD}[X_2]\mathbb{SD}[X_2]} \end{bmatrix}. \end{aligned} \tag{3.80}$$

We define the individual entries of the matrix as follows:

$$\begin{aligned} \text{PCorr}[X]_{11} &= \frac{\mathbb{E}[(X_1 - m_1)^2]}{\mathbb{SD}[X_1]\mathbb{SD}[X_1]} \\ \text{PCorr}[X]_{12} &= \frac{\mathbb{E}[(X_1 - m_1)(X_2 - m_2)]}{\mathbb{SD}[X_1]\mathbb{SD}[X_2]} \\ \text{PCorr}[X]_{21} &= \frac{\mathbb{E}[(X_2 - m_2)(X_1 - m_1)]}{\mathbb{SD}[X_2]\mathbb{SD}[X_1]} \\ \text{PCorr}[X]_{22} &= \frac{\mathbb{E}[(X_2 - m_2)^2]}{\mathbb{SD}[X_2]\mathbb{SD}[X_2]}. \end{aligned} \tag{3.81}$$

Note that $\text{PCorr}[X]_{12} = \text{PCorr}[X]_{21}$ and thus this matrix is symmetric, and one can show that this matrix is also positive semidefinite. Note that $\mathbb{E}[(X_1 - m_1)^2] = \mathbb{V}[X_1] = \mathbb{SD}[X_1]\mathbb{SD}[X_1]$ and $\mathbb{E}[(X_2 - m_2)^2] = \mathbb{V}[X_2] = \mathbb{SD}[X_2]\mathbb{SD}[X_2]$. Consequently, we have that $\text{PCorr}[X]_{11} = 1$ and $\text{PCorr}[X]_{22} = 1$. Importantly, one can also show that $\text{PCorr}[X]_{12} \in [-1, 1]$. An extreme value of -1 indicates that there is strong negative covariance, while an extreme value of $+1$ indicates that there is a strong positive covariance. The quantity $\text{PCorr}[X]_{12}$ (equal to $\text{PCorr}[X]_{21}$) is known as the Pearson *correlation coefficient*.

The properties of the Pearson matrix facilitate analysis and interpretability. Specifically, the entries of the Pearson matrix have the same significance as those of the covariance matrix (i.e., positive correlation implies positive covariance, negative correlation implies negative covariance, and zero correlation implies zero covariance). However, the Pearson matrix is a scaled matrix that sets bounds to the strength of correlation (while covariance does not, it can take any arbitrary values). Moreover, the Pearson matrix has *no units*, while the covariance and autocorrelation matrix can have complicated units. For example, imagine that X_1 is temperature (in K) and X_2 is pressure (in atm). The correlation $\mathrm{PCorr}[X]_{12}$ has no units, while $\mathrm{Cov}[X]_{12}$ has units of K · atm, which are difficult to interpret. Scaling variables is also beneficial in trying to compare how strongly some variables are related to others; such comparisons are difficult when variables have values of different orders of magnitude.

Example 3.14 Covariance of Lifetimes for Thermostat

Consider the thermostat example and address the following:

- Determine the covariance of the component lifetimes.
- Determine the Pearson correlation of the component lifetimes.
- Use these statistics to determine how lifetimes depend on one another.

We have already established that $\mathbb{E}[X] = (\mu_1, \mu_2)$. The entries of the covariance matrix are:

$$\begin{aligned}
&\mathrm{Cov}[X]_{11}\\
&= \mathbb{E}[(X_1 - \mu_1)(X_1 - \mu_1)]\\
&= \int_{(x_1,x_2)\in\mathcal{D}} (x_1 - \mu_1)(x_1 - \mu_1)\left(\frac{1}{\mu_1}e^{-x_1/\mu_1}\right)\left(\frac{1}{\mu_2}e^{-x_2/\mu_2}\right) dx_1 dx_2\\
&= \int_{x_1\in\mathcal{D}_1} (x_1 - \mu_1)(x_1 - \mu_1)\left(\frac{1}{\mu_1}e^{-x_1/\mu_1}\right) dx_1 \int_{x_2\in\mathcal{D}_2}\left(\frac{1}{\mu_2}e^{-x_2/\mu_2}\right) dx_2\\
&= -\,(x_1^2 + \mu_1^2)e^{-x_1/\mu_1}\Big|_0^\infty\\
&= (0 + \mu_1^2)e^{-0/\mu_1} - (\infty + \mu_1^2)e^{-\infty/\mu_1} = \mu_1^2.
\end{aligned}$$

We can use a similar logic to establish $\mathrm{Cov}[X]_{22} = \mu_2^2$. Now we note that:

$$\begin{aligned}
&\mathrm{Cov}[X]_{12}\\
&= \mathbb{E}[(X_1 - \mu_1)(X_2 - \mu_2)]\\
&= \int_{(x_1,x_2)\in\mathcal{D}} (x_1 - \mu_1)(x_2 - \mu_2)\left(\frac{1}{\mu_1}e^{-x_1/\mu_1}\right)\left(\frac{1}{\mu_2}e^{-x_2/\mu_2}\right) dx_1 dx_2\\
&= \int_{x_1\in\mathcal{D}_1} (x_1 - \mu_1)\left(\frac{1}{\mu_1}e^{-x_1/\mu_1}\right) dx_1 \int_{x_2\in\mathcal{D}_2} (x_2 - \mu_2)\left(\frac{1}{\mu_2}e^{-x_2/\mu_2}\right) dx_2\\
&= \left(x_1 \cdot e^{-x_1/\mu_1}\Big|_0^\infty\right)\left(x_2 \cdot e^{-x_2/\mu_2}\Big|_0^\infty\right) = 0.
\end{aligned}$$

We thus have that the covariance matrix is:

$$\mathrm{Cov}[X] = \begin{bmatrix} \mu_1^2 & 0 \\ 0 & \mu_2^2. \end{bmatrix}. \tag{3.82}$$

This makes sense because the variances of the individual component lifetimes are μ_1^2, μ_2^2 (since the marginals are exponential RVs). Also, the component lifetimes are independent and thus the off-diagonal terms should be zero.

We see that $\mathbb{SD}[X_1] = \mu_1$ and $\mathbb{SD}[X_2] = \mu_2$, and thus, we can see that Pearson correlation matrix is:

$$\mathrm{PCorr}[X] = \begin{bmatrix} 1 & 0 \\ 0 & 1 \end{bmatrix}. \tag{3.83}$$

The correlation matrix confirms that lifetimes are independent. We note, however, that we cannot conclude from correlation alone that the lifetimes are independent. Independence needs to be established by showing that $f(x_1,x_2) = f_1(x_1) \cdot f_2(x_2)$ (which we established in Example 3.10). □

Generalizing the covariance and correlation to the n-dimensional case is straightforward. The covariance matrix of $X = (X_1, X_2, \ldots, X_n)$ is a matrix $\mathrm{Cov}[X] \in \mathbb{R}^{n \times n}$ of the form:

$$\mathrm{Cov}[X] = \begin{bmatrix} \mathrm{Cov}[X]_{11} & \mathrm{Cov}[X]_{12} & \cdots & \mathrm{Cov}[X]_{1n} \\ \mathrm{Cov}[X]_{21} & \mathrm{Cov}[X]_{22} & \cdots & \mathrm{Cov}[X]_{2n} \\ \vdots & \vdots & \ddots & \vdots \\ \mathrm{Cov}[X]_{n1} & \mathrm{Cov}[X]_{n2} & \cdots & \mathrm{Cov}[X]_{nn} \end{bmatrix}, \tag{3.84}$$

where:

$$\mathrm{Cov}[X]_{ij} = \mathbb{E}[(X_i - \mathbb{E}[X_i])(X_j - \mathbb{E}[X_j])],\ i,j = 1,\ldots,n. \tag{3.85}$$

The covariance matrix for arbitrary n can also be computed from Eq. (3.72).

The autocorrelation matrix $\mathrm{ACorr}[X] \in \mathbb{R}^{n \times n}$ has the same structure as that of the covariance matrix, but we replace entries $\mathrm{Cov}[X]_{ij}$ with $\mathrm{ACorr}[X]_{ij} = \mathbb{E}[X_i X_j]$. The Pearson correlation matrix of $X = (X_1, X_2, \ldots, X_n)$ is a matrix $\mathrm{PCorr}[X] \in \mathbb{R}^{n \times n}$ of the form:

$$\mathrm{PCorr}[X] = \begin{bmatrix} 1 & \mathrm{PCorr}[X]_{12} & \cdots & \mathrm{PCorr}[X]_{1n} \\ \mathrm{PCorr}[X]_{21} & 1 & \cdots & \mathrm{PCorr}[X]_{2n} \\ \vdots & \vdots & \ddots & \vdots \\ \mathrm{Pcorr}[X]_{n1} & \mathrm{PCorr}[X]_{n2} & \cdots & 1 \end{bmatrix}, \tag{3.86}$$

where:

$$\begin{aligned} \mathrm{PCorr}[X]_{ij} &= \frac{\mathbb{E}[(X_i - \mathbb{E}[X_i])(X_j - \mathbb{E}[X_j])]}{\mathbb{SD}[X_i]\mathbb{SD}[X_j]} \\ &= \frac{\mathrm{Cov}[X]_{ij}}{\mathbb{SD}[X_i]\mathbb{SD}[X_j]},\ i = 1,\ldots,n,\ j = 1,\ldots,n. \end{aligned} \tag{3.87}$$

The entries of this matrix are the Pearson correlation coefficients, and these satisfy $\text{PCorr}[X]_{ij} \in [-1, 1]$. The Pearson matrix can also be computed directly from Eq. (3.78).

As with any summarizing statistic, we can use data to compute empirical estimates for the covariance and correlation matrices. Imagine that we have observations $x_\omega = (x_{\omega,1}, x_{\omega,2}, \ldots, x_{\omega,n})$, $\omega \in \mathcal{S}$ of the RV $X = (X_1, X_2, \ldots, X_n)$. The empirical covariance matrix (also known as the *sample covariance matrix*) is a matrix $\hat{\text{Cov}}[X] \in \mathbb{R}^{n \times n}$ with entries:

$$\hat{\text{Cov}}[X]_{ij} = \frac{1}{S} \sum_{\omega \in \mathcal{S}} (x_{\omega,i} - \hat{m}_i)(x_{\omega,j} - \hat{m}_j), \; i, j = 1, \ldots, n, \tag{3.88}$$

where:

$$\hat{m}_i = \hat{\mathbb{E}}[X_i] = \frac{1}{S} \sum_{\omega \in \mathcal{S}} x_{\omega,i}, \; i = 1, \ldots, n. \tag{3.89}$$

Here, $\hat{m}_i$ are the empirical expectations of X_i, $i = 1, \ldots, n$. Similar logic can be used to compute $\hat{\text{PCorr}}[X]$ and $\hat{\text{ACorr}}[X]$.

The sample covariance and correlation matrix encode important information about how variables are connected. These connections also point out *redundancies* that are present in our data. For example, if variables are strongly correlated, it means that there is a lot of redundancy in our matrices (as we can infer some variables from others). These redundancies manifest in the presence of linearly dependent columns or rows and in small eigenvalues. In Chapter 5, we will see that these redundancies can introduce complications when seeking to build predictive models from data (as the data is not particularly informative). In Chapter 6, we will discuss techniques (principal component analysis and singular value decomposition) that can help us systematically identify and eliminate redundancies from our data.

Example 3.15 Expected Value, Covariance, and Correlation for Gibbs Reactor

We revisit the reactor under which the reaction $CO + 2H_2 \leftrightarrow CH_3OH$ takes place. Recall that the reaction is favored by high pressure. We have data available for $n = 5$ RVs given by pressure, conversion, and output flow for reactants (CO, H_2) and product CH_3OH.

Consider the following questions:

- Do you expect a positive or negative covariance (correlation) between conversion and pressure?
- Do you expect a positive or negative covariance (correlation) between pressure and output flow of CO?
- How are the output flows for CO, H_2, and CH_3OH related?

Without even looking at the data, our physical understanding of the reactor gives us some intuition as to how the variables are connected. We know that the reaction is favored by high pressure and thus we expect that the conversion is positively correlated with pressure (i.e., higher pressures tend to give higher conversions). From basic conservation laws, we also know that the output flows of reactants decrease as we increase the output flow of product (with higher conversion). Consequently, we expect to see that the output flow of CO is negatively correlated with pressure.

We also expect that the output flow of CH_3OH is negatively correlated with the output flows of CO and H_2. Also, we expect that the output flow of the reactant H_2 increases with that of reactant CO, and thus, they are positively correlated. These variable interdependencies are confirmed from the data shown in Figure 3.15. Here, we show plots between different pairs of variables (these are known as parity plots). We can see that, in general, conversion increases with pressure and that the output flows of reactants decrease as the output flow of product increases. Note that, when the variables are positively correlated, the cloud of data points is tilted toward the right; on the other hand, when the variables are negatively correlated, the cloud is tilted toward the left.

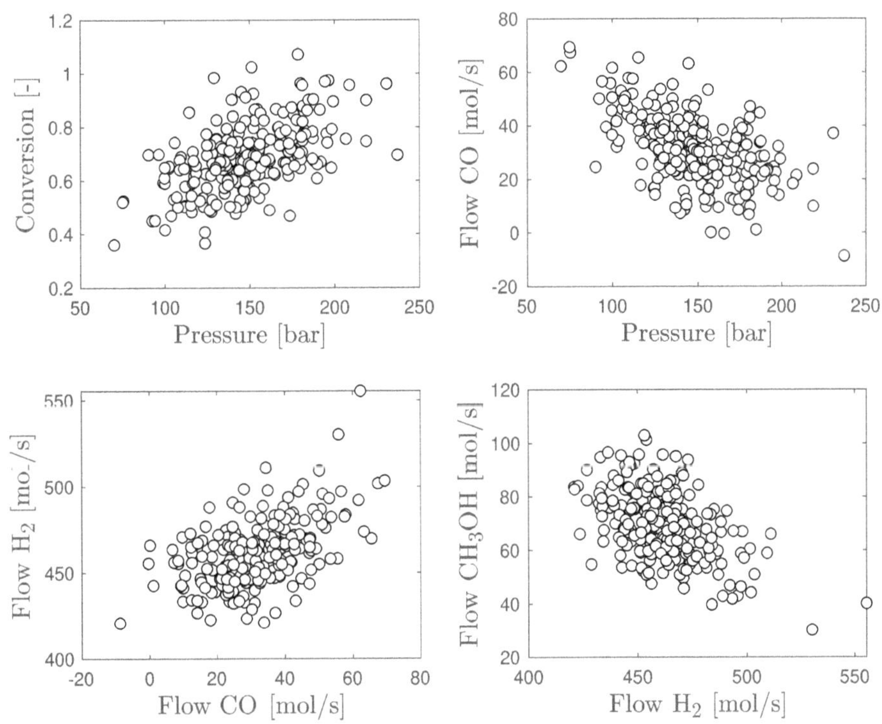

Figure 3.15 Realizations for different random variables in Gibbs reactor.

Note that the visual trends are not as clear as one might have hoped for, and we note that it is difficult to appreciate the strength of the interactions visually. This highlights the importance of using summarizing statistics such as the covariance and correlation.

We are thus interested in using the available data for computing the sample covariance and Pearson correlation matrices. The covariance matrix is:

$$\hat{\text{Cov}}[X] = \begin{bmatrix} 776.94 & 1.83 & -195.91 & -396.57 & 181.45 \\ 1.83 & 0.02 & -0.54 & -1.11 & 0.40 \\ -195.91 & -0.54 & 163.81 & 117.59 & -58.17 \\ -396.57 & -1.11 & 117.59 & 338.35 & -116.12 \\ 181.45 & 0.40 & -58.17 & -116.12 & 161.70 \end{bmatrix}.$$

The rows and columns are arranged as: (1) pressure, (2) conversion, and (3–5) output flows of CO, H_2, and CH_3OH. The sample mean is:

$$\hat{\mathbb{E}}[X] = \begin{bmatrix} 147.87 \\ 0.69 \\ 31.39 \\ 461.95 \\ 69.24 \end{bmatrix}.$$

The sample correlation matrix is:

$$\hat{\text{PCorr}}[X] = \begin{bmatrix} 1.00 & 0.53 & -0.55 & -0.77 & 0.51 \\ 0.53 & 1.00 & -0.34 & -0.49 & 0.25 \\ -0.55 & -0.34 & 1.00 & 0.50 & -0.36 \\ -0.77 & -0.49 & 0.50 & 1.00 & -0.50 \\ 0.51 & 0.25 & -0.36 & -0.50 & 1.00 \end{bmatrix}.$$

The sample covariance and correlation matrix confirm our observations. From the first row, we have that pressure is positively correlated with conversion (with correlation coefficient $\hat{\text{PCorr}}[X]_{1,2} = 0.53$), pressure is negatively correlated with output flow of CO and H_2 (with coefficients $\hat{\text{PCorr}}[X]_{1,3} = -0.55$ and $\hat{\text{PCorr}}[X]_{1,4} = -0.77$), and pressure is positively correlated with output flow of CH_3OH (with coefficient $\hat{\text{PCorr}}[X]_{1,5} = 0.51$). The entries of the correlation matrix give us an entire picture on how all variables are related to each other. For instance, the output flow of reactant CO is negatively correlated with the output flow of product CH_3OH (with coefficient $\hat{\text{PCorr}}[X]_{3,5} = -0.36$). We can confirm that the covariance and correlation matrices are symmetric.

We note that the signs of the entries of the covariance and correlation matrix coincide (both reflect positive and negative interactions), but the correlation matrix is more intuitive to analyze because its values are bounded between -1 and 1. This gives us a quick mental picture of how strong different interactions are. The covariances, on the other hand, have values that cover a wide range of magnitudes, and it is thus difficult to make conclusions on how strong the interactions are from these values. Recall that an issue with the covariance matrix is that variables have different magnitudes and units (e.g., pressure is in bar, while conversion is adimensional). This can be seen from the empirical expectation. This is why we often prefer to analyze variable dependencies based on the correlation matrix. On the other hand, note that the diagonal entries of the covariance matrix tell us the variance of the individual RVs, and this information can be used to quickly assess their level of variability and uncertainty. Consequently, both of these statistics are relevant.

From the plots, we note that the data clouds appear to concentrate around a center point given by the sample means and that the clouds "diffuse" quickly as we move away from the center point. In other words, the density of points quickly decreases as we move away from the center point. For example, the data cloud for conversion and pressure seems to concentrate around the point (150,0.7), which is close to the sample means. In Section 3.7.3, we will see why this is the case.

Script to reproduce results: `ch3_gibbs_covariance` □

As we have discussed, one can obtain misleading insights with covariance calculations when variables exhibit different orders of magnitude. A strategy to alleviate these issues involves *re-scaling* variables. This can be done by simply redefining RVs in units that equilibrate the magnitude of the variables (e.g., from kilograms to tons). A more general strategy is to scale RVs based on their mean and standard deviation as:

$$X_i^s = \frac{X_i - \mathbb{E}[X_i]}{\mathbb{SD}[X_i]}, \; i = 1, \ldots, n. \tag{3.90}$$

This has the useful property that the rescaled RVs X^s have a covariance matrix that satisfies $\text{Cov}[X^s] = \text{PCorr}[X]$. This can be seen from:

$$\begin{aligned} \mathbb{E}[X_i^s] &= \mathbb{E}\left[\frac{X_i - \mathbb{E}[X_i]}{\mathbb{SD}[X_i]}\right] \\ &= \frac{\mathbb{E}[X_i] - \mathbb{E}[X_i]}{\mathbb{SD}[X_i]} \\ &= 0. \end{aligned} \tag{3.91}$$

From this, we have that:

$$\begin{aligned} \text{Cov}[X^s]_{ij} &= \mathbb{E}[(X_i^s - \mathbb{E}[X_i^s])(X_j^s - \mathbb{E}[X_j^s])] \\ &= \mathbb{E}[X_i^s X_j^s] \\ &= \mathbb{E}\left[\frac{(X_i - \mathbb{E}[X_i])(X_j - \mathbb{E}[X_j])}{\mathbb{SD}[X_i]\mathbb{SD}[X_j]}\right] \\ &= \text{PCorr}[X]_{ij}. \end{aligned} \tag{3.92}$$

This reveals that the Pearson correlation is the covariance matrix under the scalarization of the RVs. This also reveals that $\mathbb{E}[X_i^s] = 0$ and $\mathbb{V}[X_i^s] = 1$ for all $i = 1, \ldots, n$ (the scaled RVs have zero mean and variance of one).

3.7.3 Event Probabilities

Summarizing statistics such as the expectation and the covariance matrix are useful tools because, as the name suggests, they can help summarize (reduce) large amounts of data. This reduction process facilitates comparisons between RVs and decision-making. Event probabilities can also be used as summarizing statistics;

these probabilities are effective data reduction tools because they can filter out data by using logical conditions (constraints). In Chapter 7, we will see that event probabilities can be used as measures of risk.

As we have seen, computing the event probability $\mathbb{P}(X \in \mathcal{A})$ requires the joint pdf of X, and it can require complex operations with integrals (for the continuous case). As such, it is much more practical to use available data to obtain the empirical approximation:

$$\hat{\mathbb{P}}(X \in \mathcal{A}) = \frac{1}{S} \sum_{\omega \in \mathcal{S}} \mathbf{1}[x_\omega \in \mathcal{A}]. \tag{3.93}$$

This formula is simple but powerful because it can be combined with properties of probabilities to compute diverse types of event probabilities. For instance, consider the following empirical approximations for some of the event probabilities that we have discussed:

$$\hat{\mathbb{P}}(X \in \mathcal{A}_1 \cap \mathcal{A}_2) = \frac{1}{S} \sum_{\omega \in \mathcal{S}} \mathbf{1}[x_\omega \in \mathcal{A}_1 \ \& \ x_\omega \in \mathcal{A}_2] \tag{3.94a}$$

$$\hat{\mathbb{P}}(X \in \mathcal{A}_1 \cup \mathcal{A}_2) = \frac{1}{S} \sum_{\omega \in \mathcal{S}} \mathbf{1}[x_\omega \in \mathcal{A}_1 \text{ or } x_\omega \in \mathcal{A}_2] \tag{3.94b}$$

$$\hat{\mathbb{P}}(X \in \mathcal{A}_1 | X \in \mathcal{A}_2) = \frac{\hat{\mathbb{P}}(X \in \mathcal{A}_1 \ \& \ X \in \mathcal{A}_2)}{\hat{\mathbb{P}}(X \in \mathcal{A}_2)}. \tag{3.94c}$$

The key idea behind these approximations is simply to use the indicator function as a counting/logic operator that determines if a specific outcome satisfies the conditions of a specific event. In Chapter 4, we will see that these approximations become exact as we accumulate data.

3.8 Multivariate Gaussian Model

Our previous discussion has focused on general properties for multivariate RVs. As with univariate RVs, our goal is to use data to infer the properties of the RV with the goal of developing an RV model. Such an RV model is key to predicting behavior that might not be readily apparent from the data. In the univariate RV case, we saw that there is a wide variety of RV models that can help us model behavior (e.g., exponential, Gaussian, and Poisson). Unfortunately, there are few established multivariate RV models. The most used multivariate RV model is the *Gaussian* RV model; this model has many useful properties and is used in a wide variety of applications, such as estimation, control, uncertainty quantification, and machine learning. As such, it is important to understand its properties.

3.8.1 Domain and Joint Density Functions

An n-dimensional Gaussian X has a *joint pdf* of the form:

$$\begin{aligned} f(x) &= f(x_1, x_2, \ldots, x_n) \\ &= \frac{1}{(2\pi)^{n/2}|\Sigma|^{1/2}} \exp\left(-\frac{1}{2}(x-\mu)^T \Sigma^{-1} (x-\mu)\right), \quad x \in \mathcal{D}. \end{aligned} \tag{3.95}$$

We state the fact that X follows a Gaussian model by using $X \sim \mathcal{N}(\mu, \Sigma)$, where $\mu \in \mathbb{R}^n$ (a vector) and $\Sigma \in \mathbb{R}^{n \times n}$ (a matrix) are the model parameters (these fully define the Gaussian RV model).

In the definition of the pdf, we recall that $|\Sigma|$ is the determinant of matrix Σ (the product of its eigenvalues). We thus have that the Gaussian model is well defined only if Σ is a positive definite matrix (i.e., its determinant is positive and thus all eigenvalues are positive). Note that if Σ is positive semidefinite, then at least one of its eigenvalues is zero and the determinant is zero; as a result, the pdf as written is not defined. This situation is the generalization of the special case for a univariate Gaussian $\mathcal{N}(\mu, \sigma^2)$ with $\sigma^2 = 0$. Dealing with a positive semidefinite matrix Σ is possible, but such a case is significantly more complicated to analyze and will not be explored here [30].

The *domain* of the Gaussian RV is $\mathcal{D} = (-\infty, \infty)^n \subseteq \mathbb{R}^n$. This indicates that observations x_ω of X satisfy $-\infty \leq x_{\omega,i} \leq \infty$ for all $i = 1, \ldots, n$. The cdf of a multivariate Gaussian RV is a complicated mathematical object and is often evaluated *numerically*.

The Gaussian RV has the key and useful property that $\mu = \mathbb{E}[X]$ and $\Sigma = \text{Cov}[X]$ (this is a generalization of the univariate Gaussian case). This result is important because, as in the univariate case, it indicates that the parameters can be estimated using empirical approximations of the expected value and covariance matrix.

By differentiating the joint pdf $f(x)$ with respect to x, one can find that its maximum value (the mode $\mathbb{MO}_X$) occurs at μ (the mode and expected value coincide). This insight is also rather obvious from the exponential form of the joint pdf. Recall that the mode is the point of maximum density (where observations of X concentrate the most) and thus μ is the point at which the observations of X concentrate the most. The maximum density is given by $f(\mu) = 1/(2\pi)^{n/2}|\Sigma|^{1/2}$.

In Figure 3.16, we plot the pdf and cdf of a bivariate Gaussian with $\mu = (0, 1)$ and $\Sigma = \begin{bmatrix} 1 & -1/2 \\ -1/2 & 1 \end{bmatrix}$. Note that the pdf has a peak at μ and note how the pdf defines a bell-like surface and that the density decays exponentially in both x_1 and x_2 directions. We can also see that the cdf is a surface that increases monotonically with x_1 and x_2 and converges to a value of 1. Here, we also visualize the pdf and cdf projections on a 2D plane (these are known as pdf and cdf fields). From the pdf projection, we can see that the pdf becomes more concentrated (dense) in the center (indicating a higher pdf value). We also see that the concentration is maximum in the center (corresponding to the mean) and diffuses progressively as we move away from the center. We also see that the concentration forms *ellipsoidal* patterns; we will explain soon why this is the case. From the cdf projection, we see that the cdf becomes progressively more concentrated as the threshold values x_1 and x_2 increase (the increase in concentration is monotonic).

For the case with $n = 2$, the joint pdf can be written as:

$$f(x_1, x_2) = c \cdot \exp\left(-\frac{1}{2}\begin{bmatrix} x_1 - \mu_1 \\ x_2 - \mu_2 \end{bmatrix}^T \begin{bmatrix} \Sigma_{11} & \Sigma_{12} \\ \Sigma_{21} & \Sigma_{22} \end{bmatrix}^{-1} \begin{bmatrix} x_1 - \mu_1 \\ x_2 - \mu_2 \end{bmatrix}\right), \tag{3.96}$$

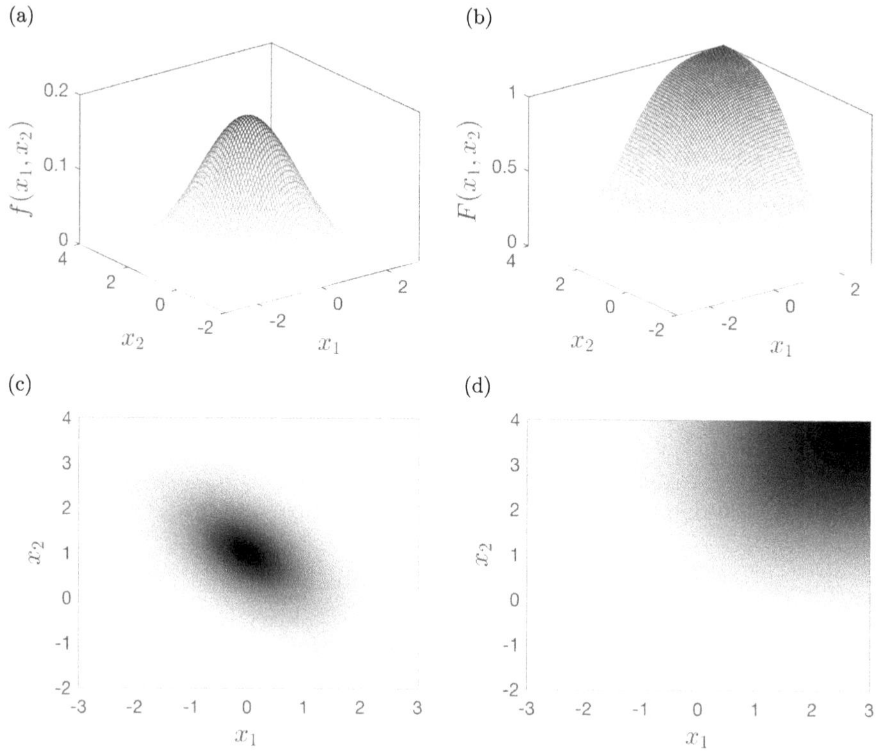

Figure 3.16 Joint pdf (a) and joint cdf (b) for multivariate Gaussian and corresponding 2D projections (c and d).

where we define:

$$c = \frac{1}{2\pi \begin{vmatrix} \Sigma_{11} & \Sigma_{12} \\ \Sigma_{21} & \Sigma_{22} \end{vmatrix}^{1/2}}. \tag{3.97}$$

Note also that:

$$\begin{aligned}
\Sigma_{11} &= \text{Cov}[X]_{11} = \mathbb{E}[(X_1 - \mu_1)^2] \\
\Sigma_{12} &= \text{Cov}[X]_{12} = \mathbb{E}[(X_1 - \mu_1)(X_2 - \mu_2)] \\
\Sigma_{21} &= \text{Cov}[X]_{21} = \mathbb{E}[(X_2 - \mu_2)(X_1 - \mu_1)] \\
\Sigma_{22} &= \text{Cov}[X]_{22} = \mathbb{E}[(X_2 - \mu_2)^2].
\end{aligned} \tag{3.98}$$

Recall now that, if (X_1, X_2) are independent, we have that $\Sigma_{12} = \Sigma_{21} = 0$ and thus the covariance matrix becomes diagonal. One can use this property to show that:

$$f(x_1, x_2) = \frac{1}{\sqrt{2\pi\Sigma_{11}}} \exp\left(\frac{-(x_1 - \mu_1)^2}{2\Sigma_{11}}\right) \cdot \frac{1}{\sqrt{2\pi\Sigma_{22}}} \exp\left(\frac{-(x_2 - \mu_2)^2}{2\Sigma_{22}}\right). \tag{3.99}$$

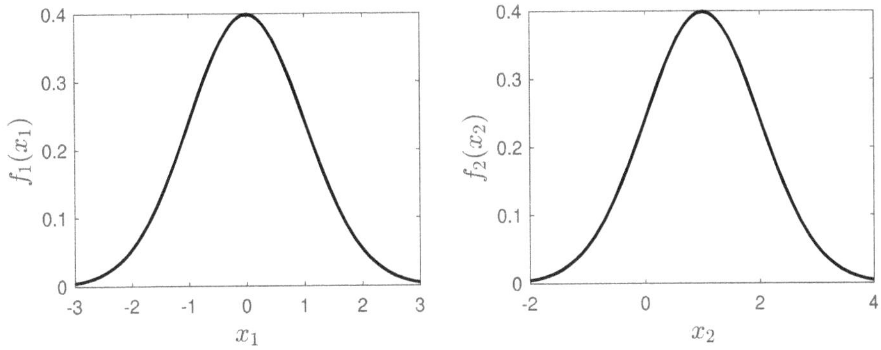

Figure 3.17 Marginal pdfs for multivariate Gaussian.

3.8.2 Marginal Density Functions

Recall that the marginals of a multivariate RV $X = (X_1, X_2)$ are given by:

$$\begin{aligned} f_1(x_1) &= \int_{x_2 \in \mathcal{D}_2} f(x_1, x_2) dx_2, \; x_1 \in \mathcal{D}_1 \\ f_2(x_2) &= \int_{x_1 \in \mathcal{D}_1} f(x_1, x_2) dx_1, \; x_2 \in \mathcal{D}_2. \end{aligned} \tag{3.100}$$

By plugging Eq. (3.95) in these expressions, we can establish that:

$$\begin{aligned} f_1(x_1) &= \frac{1}{\sqrt{2\pi\Sigma_{11}}} \exp\left(\frac{-(x_1 - \mu_1)^2}{2\Sigma_{11}}\right) \\ f_2(x_2) &= \frac{1}{\sqrt{2\pi\Sigma_{22}}} \exp\left(\frac{-(x_2 - \mu_2)^2}{2\Sigma_{22}}\right). \end{aligned} \tag{3.101}$$

Note that these correspond to the pdfs of the univariate Gaussians $X_1 \sim \mathcal{N}(\mu_1, \Sigma_{11})$ and $X_2 \sim \mathcal{N}(\mu_2, \Sigma_{22})$. By comparing Eq. (3.99) and Eq. (3.101), we can confirm that, if $X = (X_1, X_2)$ are independent, we have that $f(x_1, x_2) = f_1(x_1) \cdot f_2(x_2)$. Moreover, we conclude that a multivariate Gaussian has the key property that its marginal pdfs are univariate Gaussians.

In Figure 3.17, we plot the marginal pdfs for a Gaussian with parameters:

$$\mu = (0, 1) \qquad \Sigma = \begin{bmatrix} 1 & -1/2 \\ -1/2 & 1 \end{bmatrix}.$$

The marginals correspond to the pdfs of $X_1 \sim \mathcal{N}(0, 1)$ and $X_2 \sim \mathcal{N}(1, 1)$.

3.8.3 Conditional Density Functions

Recall that the conditional pdfs of an RV $X = (X_1, X_2)$ are given by:

$$\begin{aligned} f(x_1|x_2) &= \frac{f(x_1, x_2)}{f_2(x_2)}, \; x_1 \in \mathcal{D}_1 \\ f(x_2|x_1) &= \frac{f(x_1, x_2)}{f_1(x_1)}, \; x_2 \in \mathcal{D}_2. \end{aligned} \tag{3.102}$$

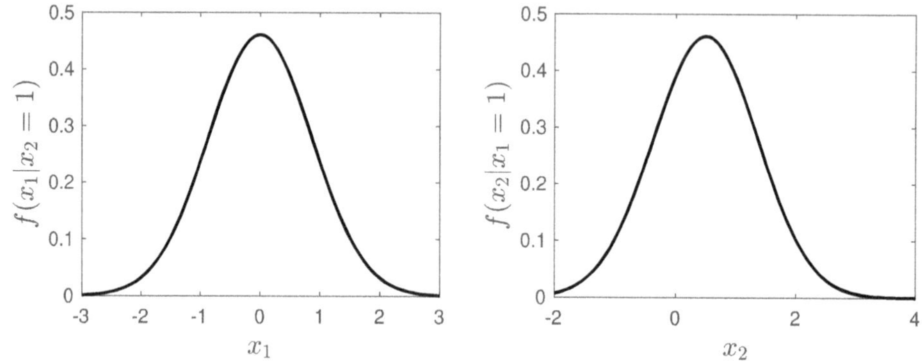

Figure 3.18 Conditional pdfs for multivariate Gaussian.

By inserting the joint pdf (3.95) and the marginals (3.101) into the above definitions, one can show that the conditionals have the form:

$$\begin{aligned} f(x_1|x_2) &= \frac{1}{\sqrt{2\pi\Sigma_{1|2}}}\exp\left(\frac{-(x_1-\mu_{1|2})^2}{2\Sigma_{1|2}}\right) \\ f(x_2|x_1) &= \frac{1}{\sqrt{2\pi\Sigma_{2|1}}}\exp\left(\frac{-(x_2-\mu_{2|1})^2}{2\Sigma_{2|1}}\right), \end{aligned} \tag{3.103}$$

where:

$$\begin{aligned} \mu_{1|2} &= \mu_1 + \Sigma_{12}\Sigma_{22}^{-1}(x_2-\mu_2) \\ \Sigma_{1|2} &= \Sigma_{11} - \Sigma_{12}\Sigma_{22}^{-1}\Sigma_{21} \\ \mu_{2|1} &= \mu_2 + \Sigma_{21}\Sigma_{11}^{-1}(x_1-\mu_1) \\ \Sigma_{2|1} &= \Sigma_{22} - \Sigma_{21}\Sigma_{11}^{-1}\Sigma_{12}. \end{aligned} \tag{3.104}$$

This indicates that the conditional pdfs correspond to those of the Gaussians $X_1|X_2 \sim \mathcal{N}(\mu_{1|2},\Sigma_{1|2})$ and $X_2|X_1 \sim \mathcal{N}(\mu_{2|1},\Sigma_{2|1})$. In other words, the conditional pdfs of a multivariate Gaussian are also Gaussian. To illustrate this, in Figure 3.18, we plot the conditional pdfs $f(x_1|x_2 = 1)$ and $f(x_2|x_1 = 1)$ for a Gaussian with

$$\mu = (0,1) \qquad \Sigma = \begin{bmatrix} 1 & -1/2 \\ -1/2 & 1 \end{bmatrix}.$$

For this RV, the conditionals correspond to the pdfs of $X_1|X_2 \sim \mathcal{N}(0,3/4)$ and $X_2|X_1 \sim \mathcal{N}(1/2,3/4)$.

The expressions (3.104) provide interesting insights into how the covariance between X_1 and X_2 helps resolve the uncertainty of one variable when we have knowledge of the other variable. Recall that the variance $\Sigma_{11} = \mathbb{V}[X_1]$ reflects the uncertainty of X_1 if we ignore X_2 (variance of marginal $\mathcal{N}(\mu_1,\Sigma_{11})$) and that $\Sigma_{22} = \mathbb{V}[X_2]$ reflects the uncertainty of X_2 if we ignore X_1 (variance of marginal $\mathcal{N}(\mu_2,\Sigma_{22})$). Recall also that $\Sigma_{12} = \Sigma_{21}$ and that Σ_{11} and Σ_{22} are positive quantities. We thus have that the terms $\Sigma_{12}\Sigma_{22}^{-1}\Sigma_{21}$ and $\Sigma_{21}\Sigma_{11}^{-1}\Sigma_{12}$ are non-negative quantities. This indicates that the variance (uncertainty) of X_1 *always* "shrinks" (or remains the same) when we have knowledge of X_2 and that the uncertainty of X_2 shrinks (or remains the same) when we have knowledge of X_1. In other words, knowledge of one RV cannot increase the uncertainty of the other.

Consider now the following extreme cases:

- If X_1 and X_2 are *independent*, we have $\Sigma_{12} = \Sigma_{21} = 0$, $\Sigma_{12}\Sigma_{22}^{-1}\Sigma_{21} = 0$, and $\Sigma_{1|2} = \Sigma_{11}$. We thus have that knowledge of X_2 does not shrink the uncertainty of X_1. In other words, no uncertainty in X_1 is resolved by knowing the fact that $X_2 = x_2$. The same can be concluded by setting $\Sigma_{21} = 0$ (no uncertainty in X_2 is resolved by knowing that $X_1 = x_1$).
- If X_1 and X_2 are *strongly correlated*, we have that $\Sigma_{12} = \Sigma_{21} \gg 0$ or $\Sigma_{12} = \Sigma_{21} \ll 0$. We thus have that the term $\Sigma_{12}\Sigma_{22}^{-1}\Sigma_{21} \gg 0$. Consequently, we have that $\Sigma_{12} \ll \Sigma_{11}$; as such, when there is a strong correlation between X_1 and X_2, knowledge that $X_2 = x_2$ resolves a significant amount of uncertainty on X_1. The same logic can be applied when analyzing how knowledge in X_1 resolves uncertainty in X_2.
- Now, note that uncertainty reduction in X_1 is affected by its covariance with X_2 (Σ_{12}) and by the variance of X_2 (Σ_{22}). This indicates that, even if the correlation between X_1 and X_2 is strong, a high variance of X_2 will limit the amount of reduction achieved. This makes sense because, if X_2 contains a lot of variability, then knowing its value does not reveal much information about its overall behavior. On the other hand, if the variability of X_2 is small, then knowing its value provides a lot of information about its behavior. This tells us that variability and information are closely related concepts. We will discuss this connection in Chapter 6, as part of our discussion of principal component analysis.

Note that knowledge of X_2 has the effect of *shifting* the mean of X_1; the shift is affected by the known value x_1, the covariance Σ_{12}, and the variance Σ_{22}. If X_1 and X_2 are independent, we have that the shift is zero.

Example 3.16 Uncertainty Reduction in Gibbs Reactor (Part I)

The pressure and conversion are given by the RVs X_1 and X_2. Assume that $X = (X_1, X_2)$ is Gaussian with covariance and correlation matrix:

$$\text{Cov}[X] = \begin{bmatrix} 776.94 & 1.83 \\ 1.83 & 0.02 \end{bmatrix}$$

$$\text{PCorr}[X] = \begin{bmatrix} 1 & 0.53 \\ 0.53 & 1 \end{bmatrix}.$$

We are interested in addressing the following questions:

- How does having knowledge of pressure affect uncertainty in conversion?
- How does having knowledge of conversion affect uncertainty in pressure?

From the covariance matrix, we see that the marginal variance for conversion is $\Sigma_{22} = 0.02$; this is a measure of its uncertainty, given no knowledge of pressure. From the covariance matrix, we see that pressure and conversion are positively correlated (the off-diagonal entry is positive), and thus we expect that having knowledge of pressure decreases uncertainty in conversion. To quantify how much uncertainty in conversion is reduced by having knowledge of pressure, we compute the variance of the conditional pdf $f(x_2|x_1)$. We know that the conditional is Gaussian and has variance given by:

$$\begin{aligned}\Sigma_{2|1} &= \Sigma_{22} - \Sigma_{21}\Sigma_{11}^{-1}\Sigma_{12}\\ &= 0.02 - 1.83 \cdot (776.94)^{-1} \cdot 1.83\\ &= 0.015.\end{aligned}$$

Consequently, uncertainty in conversion is reduced from 0.02 (no knowledge of pressure) to 0.015 (with knowledge of pressure). This is a relative reduction of *about 25%*. Note that Σ_{22} has no units (conversion is adimensional), the units of Σ_{12} and Σ_{21} are atm, and the units of Σ_{11} are atm^2. We thus have that $\Sigma_{2|1}$ has no units.

Now consider that we want to quantify how knowledge of the conversion affects uncertainty in pressure. The variance of $f(x_1|x_2)$ is:

$$\begin{aligned}\Sigma_{1|2} &= \Sigma_{11} - \Sigma_{12}\Sigma_{22}^{-1}\Sigma_{21}\\ &= 776.94 - 1.83 \cdot (0.02)^{-1} \cdot 1.83\\ &= 609.49.\end{aligned}$$

This is a relative reduction of 21%; we can thus see that knowledge of conversion has a similar effect on pressure than the other way around (in relative terms). We can also confirm that the units of $\Sigma_{1|2}$ are atm^2. □

The conditional pdf for n-dimensional Gaussians $X = (X_1,\ldots,X_n) \sim \mathcal{N}(\mu,\Sigma)$ can be easily obtained by using variable grouping. Imagine that we partition our RV vector into two subvectors $X_I = (X_1,\ldots,X_m)$ and $X_{II} = (X_{m+1},\ldots,X_n)$. One can show that the joint pdfs are given by:

$$\begin{aligned} f(x_I|x_{II}) &= \frac{1}{(2\pi)^{m/2}|\Sigma_{I|II}|^{1/2}}\exp\left(-\frac{1}{2}(x_I-\mu_{I|II})^T\Sigma_{I|II}^{-1}(x_I-\mu_{I|II})\right)\\ f(x_{II}|x_I) &= \frac{1}{(2\pi)^{(n-m)/2}|\Sigma_{II|I}|^{1/2}}\exp\left(-\frac{1}{2}(x_{II}-\mu_{II|I})^T\Sigma_{II|I}^{-1}(x_{II}-\mu_{II|I})\right),\end{aligned} \tag{3.105}$$

where:

$$\begin{aligned}\mu_{I|II} &= \mu_I + \Sigma_{I,II}\Sigma_{II,II}^{-1}(x_{II}-\mu_{II})\\ \Sigma_{I|II} &= \Sigma_{I,I} - \Sigma_{I,II}\Sigma_{II,II}^{-1}\Sigma_{II,I}\\ \mu_{II|I} &= \mu_{II} + \Sigma_{II,I}\Sigma_{I,I}^{-1}(x_I-\mu_I)\\ \Sigma_{II|I} &= \Sigma_{II,II} - \Sigma_{II,I}\Sigma_{I,I}^{-1}\Sigma_{I,II}.\end{aligned} \tag{3.106}$$

Here, $\mu_I \in \mathbb{R}^m$ and $\Sigma_{I,I} \in \mathbb{R}^{m\times m}$ are the mean vector and covariance matrix of X_I, while $\mu_{II} \in \mathbb{R}^{n-m}$ and $\Sigma_{II,II} \in \mathbb{R}^{(n-m)\times(n-m)}$ are the mean vector and covariance matrix of X_{II}. The matrix $\Sigma_{I,II} \in \mathbb{R}^{m\times(n-m)}$ is the covariance between X_I and X_{II}. The matrix $\Sigma_{II,I} \in \mathbb{R}^{(n-m)\times m}$ is the covariance between X_{II} and X_1, and we note that $\Sigma_{II,I} = \Sigma_{I,II}^T$. The vectors x_I and x_{II} are the known values for X_I and

X_{II}, respectively. The conditional pdfs correspond to those of the Gaussians $X_I \sim \mathcal{N}(\mu_{I|II}, \Sigma_{I|II})$ and $X_{II} \sim \mathcal{N}(\mu_{II|I}, \Sigma_{II|I})$.

Example 3.17 Uncertainty Reduction in Gibbs Reactor (Part II)

The covariance matrix for all the variables in the Gibbs reactor is given by:

$$\text{Cov}[X] = \begin{bmatrix} 776.94 & 1.83 & -195.91 & -396.57 & 181.45 \\ 1.83 & 0.02 & -0.54 & -1.11 & 0.40 \\ -195.91 & -0.54 & 163.81 & 117.59 & -58.17 \\ -396.57 & -1.11 & 117.59 & 338.35 & -116.12 \\ 181.45 & 0.40 & -58.17 & -116.12 & 161.70 \end{bmatrix}.$$

The rows and columns are arranged as: (1) pressure, (2) conversion, and (3–5) output flows of CO, H_2, and CH_3OH. The expected value is:

$$\hat{\mathbb{E}}[X] = \begin{bmatrix} 147.87 \\ 0.69 \\ 31.39 \\ 461.95 \\ 69.24 \end{bmatrix}.$$

Assume that pressure and output flows are known and given by their expected values. Address the following questions:

- What is the conditional density of conversion?
- By how much is conversion uncertainty reduced (relative to its marginal) by having knowledge of pressure and flows?

For convenience, we reorder the variables so that X_1 is conversion, X_2 is pressure, and X_3, X_4, and X_5 are the output flows. This variable reordering rearranges the covariance matrix and expected value as:

$$\text{Cov}[X] = \begin{bmatrix} 0.02 & 1.83 & -0.54 & -1.11 & 0.40 \\ 1.83 & 776.94 & -195.91 & -396.57 & 181.45 \\ -0.54 & -195.91 & 163.81 & 117.59 & -58.17 \\ -1.11 & -396.57 & 117.59 & 338.35 & -116.12 \\ 0.40 & 181.45 & -58.17 & -116.12 & 161.70 \end{bmatrix}, \quad \hat{\mathbb{E}}[X] = \begin{bmatrix} 0.69 \\ 147.87 \\ 31.39 \\ 461.95 \\ 69.24 \end{bmatrix}.$$

We want the conditional $X_I|X_{II}$ with $X_I = X_1$ and $X_{II} = (X_2, X_3, X_4, X_5)$. Note that $n = 5$, $m = 1$, and $(n - m) = 4$. The covariance matrices are:

$$\Sigma_{I,I} = 0.02$$

$$\Sigma_{II,I} = \begin{bmatrix} 1.83 \\ -0.54 \\ -1.11 \\ 0.40 \end{bmatrix}$$

$$\Sigma_{II,II} = \begin{bmatrix} 776.94 & -195.91 & -396.57 & 181.45 \\ -195.91 & 163.81 & 117.59 & -58.17 \\ -396.57 & 117.59 & 338.35 & -116.12 \\ 181.45 & -58.17 & -116.12 & 161.70 \end{bmatrix}.$$

The mean vectors are $\mu_I = 0.69$ and $\mu_{II} = (147.87, 31.39, 461.95, 69.24)$. By using Eq. (3.106) and by performing matrix operations, we obtain that $\mu_{I|II} = 147.87$ and $\Sigma_{I|II} = 0.01$. The conditional density is that of the Gaussian $X_{I|II} \sim \mathcal{N}(147.87, 0.01)$. From the covariance matrix, we have that variance of the marginal of X_I is $\text{Cov}[X]_{11} = 0.02$, while the conditional variance is 0.01. We thus have that the uncertainty is reduced by 100%. Note that the expected value of the marginal is 0.69, and the mean of the conditional is 0.69 (the mean is not shifted). This is because the known values of X_{II} are assumed to be their expected values.

Script to reproduce results: `ch3_gibbs_covariance` □

3.8.4 Uncertainty Propagation

As in the univariate case, multivariate Gaussians have the nice property that their propagation through a linear model (a linear transformation) yields a Gaussian RV. Specifically, consider the n-dimensional Gaussian $X \sim \mathcal{N}(\mu, \Sigma)$ and consider the linear model:

$$Y = \varphi(X) = AX + b, \tag{3.107}$$

with matrix $A \in \mathbb{R}^{p \times n}$ and vector $b \in \mathbb{R}^p$. We thus have that the output Y is a p-dimensional RV. The matrix A captures structural connections between X and Y (e.g., that might result from physics); for instance, this might capture how the demand for electricity affects the cost of a manufacturing facility.

One can show that Y is a p-dimensional Gaussian RV of the form:

$$Y \sim \mathcal{N}(A\mu + b, A\Sigma A^T). \tag{3.108}$$

This implies that $\mathbb{E}[Y] = A\mathbb{E}[X] + b$ and $\text{Cov}[Y] = A\text{Cov}[X]A^T$. This result is important because it highlights how the variability of Y captures the structural connection with X (given by matrix A). This highlights again how covariance matrices encode structural dependencies between variables (e.g., as one variable increases, the other tends to increase). Unraveling such dependencies from data is key in developing predictive modes, as discussed in Chapter 5.

Example 3.18 Linear Transformation of a Multivariate Gaussian

Consider the Gaussian *mixture*:

$$Y = X_1 + X_2,$$

with independent $X_1 \sim \mathcal{N}(\mu_1, \sigma_1^2)$ and $X_2 \sim \mathcal{N}(\mu_2, \sigma_2^2)$. Show that Y is Gaussian and determine its parameters.

The Gaussian mixture can be expressed as a linear transformation $Y = AX + b$ with $X = (X_1, X_2)$, $\mu = (\mu_1, \mu_2)$, diagonal matrix Σ (with entries $\Sigma_{11} = \sigma_1^2$ and $\Sigma_{22} = \sigma_2^2$), $A = (1, 1)$, and $b = 0$. Consequently, we have that $Y \sim \mathcal{N}(A\mu, A\Sigma A^T)$ with:

$$A\mu = [1 \;\; 1]\begin{bmatrix} \mu_1 \\ \mu_2 \end{bmatrix} = \mu_1 + \mu_2,$$

and

$$A\Sigma A^T = [1 \;\; 1]\begin{bmatrix} \sigma_1^2 & \\ & \sigma_2^2 \end{bmatrix}\begin{bmatrix} 1 \\ 1 \end{bmatrix} = \sigma_1^2 + \sigma_2^2.$$

We thus have that $Y \sim \mathcal{N}(\mu_1 + \mu_2, \sigma_1^2 + \sigma_2^2)$. □

Example 3.19 Mixture of Revenue Streams

Imagine that you have two investments with associated revenues F_1 and F_2 and the total revenue collected is $F_3 = \kappa_1 \cdot F_1 + \kappa_2 \cdot F_2$. The revenues are uncertain and are modeled as Gaussian RVs $F_1 \sim \mathcal{N}(\mu_1, \sigma_1^2)$ and $F_2 \sim \mathcal{N}(\mu_2, \sigma_2^2)$ with $\mu_1 = 10$ million USD per year, $\mu_2 = 5$ million USD per year, $\sigma_1 = 10$ million USD per year, and $\sigma_2 = 1$ million USD per year. Note that the first investment has a high mean yield but also high uncertainty (risk), while the second investment has a low mean yield but also low uncertainty (risk). We also know that the revenues F_1 and F_2 are independent (involve investments in different markets). Since F_1 and F_2 are uncertain, the total revenue F_3 is also uncertain; however, we can manipulate the total mean revenue and its risk by adjusting the *proportions* $0 \leq \kappa_1 \leq 1$ and $0 \leq \kappa_2 \leq 1$. The system is sketched in Figure 3.19; note that this problem is mathematically analogous to a flow mixture problem (the proportions can be interpreted as "valves" that are strategically opened/closed to control the output).

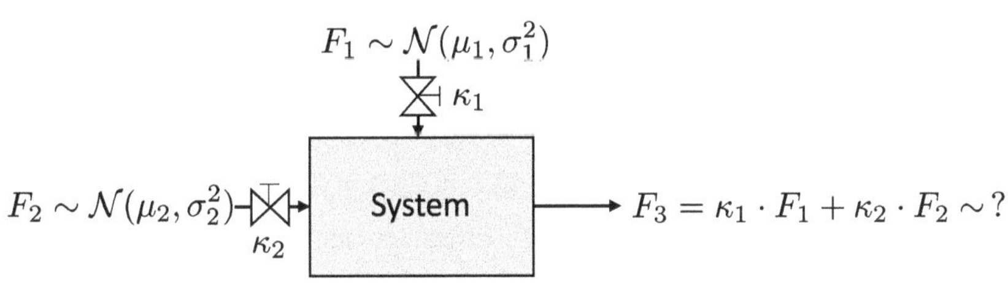

Figure 3.19 Sketch of revenue mixing problem.

We are interested in addressing the following questions:

- What is the expected value and risk (variance) of the total revenue F_3?
- How are these quantities affected by the investment strategy (selection of κ_1 and κ_2)?

- What strategy would you use to maximize the mean revenue $\mathbb{E}[F_3]$? What strategy would you use to minimize the risk $\mathbb{V}[F_3]$?

To address these questions, we write F_3 as a linear combination of F_1 and F_2 of the form:

$$F_3 = \begin{bmatrix} \kappa_1 & \kappa_2 \end{bmatrix} \begin{bmatrix} F_1 \\ F_2 \end{bmatrix} = \kappa_1 F_1 + \kappa_2 F_2.$$

We thus have that $F_3 \sim \mathcal{N}(\mu_3, \sigma_3^3)$ with:

$$\mu_3 = \begin{bmatrix} \kappa_1 & \kappa_2 \end{bmatrix} \begin{bmatrix} \mu_1 \\ \mu_2 \end{bmatrix} = \kappa_1 \mu_1 + \kappa_2 \mu_2$$

$$\sigma_3^2 = \begin{bmatrix} \kappa_1 & \kappa_2 \end{bmatrix} \begin{bmatrix} \sigma_1^2 & 0 \\ 0 & \sigma_2^2 \end{bmatrix} \begin{bmatrix} \kappa_1 \\ \kappa_2 \end{bmatrix} = \kappa_1^2 \sigma_1^2 + \kappa_2^2 \sigma_2^2.$$

We can manipulate the amount of revenue that we collect from our individual investments (using κ_1 and κ_2) in a way that we maximize the expected revenue μ_3 and that we minimize our uncertainty σ_3^2. Clearly, if the objective is to maximize mean revenue, we should set $\kappa_1 = \kappa_2 = 1$. If the objective is to minimize risk, we should set $\kappa_1 = \kappa_2 = 0$ (do not invest). □

Another useful application of the linear transformation property is that of standardization. Specifically, consider the multivariate standard normal $Z \sim \mathcal{N}(\mu_Z, \Sigma_Z)$, with $\Sigma_Z = \mathbf{I}$ and $\mu_Z = 0$. Here, $\mathbf{I}$ denotes the identity matrix. We have that $X = \sqrt{\Sigma} \cdot Z + \mu$ is a linear transformation of Z with $A = \sqrt{\Sigma}$ and $b = \mu$. We thus have that:

$$A \Sigma_Z A^T = \sqrt{\Sigma} \cdot \mathbf{I} \cdot \sqrt{\Sigma} = \Sigma, \tag{3.109}$$

and

$$A \mu_Z + b = \sqrt{\Sigma} \cdot 0 + \mu = \mu. \tag{3.110}$$

Consequently, $X \sim \mathcal{N}(\mu, \Sigma)$. Here, we use the fact that Σ is symmetric and thus $\Sigma^T = \Sigma$. This provides a procedure to obtain properties of X from properties of Z. This is important because Z is a much simpler RV; for instance, the covariance of Z is diagonal (given by identity matrix $\mathbf{I}$). This approach is a generalization of the standardization procedure for univariate RVs that we discussed in Chapter 2.

3.8.5 Geometry

Understanding the geometry of multivariate Gaussian RVs facilitates *data visualization*. We will see that the exponential form of the Gaussian pdf generates confidence regions that form ellipses. Moreover, we will see that the size and orientation of those ellipses are directly related to the covariance and correlation between variables.

We explain these concepts by using the $n = 2$ case with joint pdf defined in Eq. (3.96). First note that we can write the covariance between X_1 and X_2 as:

$$\Sigma_{12} = \rho\sqrt{\Sigma_{11}\Sigma_{22}}, \tag{3.111}$$

where $\rho = \text{PCorr}[X]_{12}$ is the Pearson correlation coefficient. With this, we can write the covariance matrix as:

$$\text{Cov}[X] = \begin{bmatrix} \Sigma_{11} & \rho\sqrt{\Sigma_{11}\Sigma_{22}} \\ \rho\sqrt{\Sigma_{11}\Sigma_{22}} & \Sigma_{22} \end{bmatrix}. \tag{3.112}$$

The determinant of this matrix is:

$$\begin{aligned} \begin{vmatrix} \Sigma_{11} & \rho\sqrt{\Sigma_{11}\Sigma_{22}} \\ \rho\sqrt{\Sigma_{11}\Sigma_{22}} & \Sigma_{22} \end{vmatrix} &= \Sigma_{11}\Sigma_{22} - \rho^2\Sigma_{11}\Sigma_{22} \\ &= (1-\rho^2)\Sigma_{11}\Sigma_{22}. \end{aligned} \tag{3.113}$$

Inserting these forms of the determinant and the covariance matrix in the joint pdf Eq. (3.96), we can express this as:

$$f(x_1,x_2) = \frac{1}{2\pi\sqrt{\Sigma_{11}\Sigma_{22}}}\exp\left(-E(x_1,x_2)\right), \tag{3.114}$$

where:

$$E(x_1,x_2) = \frac{1}{2(1-\rho^2)}\left[\frac{(x_1-\mu_1)^2}{\Sigma_{11}} + \frac{(x_2-\mu_2)^2}{\Sigma_{22}} - 2\rho\frac{(x_1-\mu_1)(x_2-\mu_2)}{\sqrt{\Sigma_{11}\Sigma_{22}}}\right]. \tag{3.115}$$

If we fix $f(x_1,x_2)$ to a given probability value α, we can write Eq. (3.114) as:

$$\ell = E(x_1,x_2), \tag{3.116}$$

for some constant ℓ. Equation (3.116) defines an ellipse whose size scales with ℓ. Specifically, for a fixed value of ℓ, all pairs (x_1,x_2) that satisfy $E(x_1,x_2) = \ell$ lie on the boundary of the ellipse. The set of pairs (x_1,x_2) satisfying $\ell = E(x_1,x_2)$ is known as the level set of the function $E(x_1,x_2)$.

We thus have that the level sets of the joint pdf $f(x_1,x_2)$ define ellipses; these ellipses have the following interesting geometric properties:

- The ellipse $\ell = E(x_1,x_2)$ is centered around (μ_1,μ_2); specifically, note that $E(\mu_1,\mu_2) = 0$.
- The ellipse length of the axis in the x_1 direction is defined by the variance Σ_{11} (the larger the variance, the longer the axis). The length of the axis in the x_2 direction is defined by the variance Σ_{22}. As such, the size of the ellipse in each direction is a function of the uncertainty/variability of the RVs X_1 and X_2.
- The correlation between X_1 and X_2 defines the *orientation of the ellipse*; specifically:
  - If $\rho > 0$ (positive correlation), the ellipse is tilted to the right.
  - If $\rho < 0$ (negative correlation), the ellipse is tilted to the left.
  - If $\rho = 0$ (no correlation), the ellipse has no tilt.

Correlation does not affect the size of the ellipsoid (only affects its orientation).

We are interested in building *confidence regions*, which are regions over which we expect to find X with a given probability. We recall that, for the univariate Gaussian, we defined confidence regions as 1D intervals (along a line). In the multivariate case, confidence regions are n-dimensional and can take more complicated forms (e.g., boxes, ellipses, or polytopes).

Imagine that we want to determine a rectangular (box) region $\mathcal{B} \subseteq \mathcal{D}$ that contains X with probability $1 - \alpha$. This confidence region satisfies:

$$\mathbb{P}(X \in \mathcal{B}) = 1 - \alpha. \tag{3.117}$$

For $n = 1$ with $X \sim \mathcal{N}(\mu,\sigma^2)$, one can show that the box is a line interval given by:

$$\mathcal{B} = \{ x \in [\mu \pm \sqrt{\mathbb{Q}(1-\alpha)}\sigma]\}, \tag{3.118}$$

where $\mathbb{Q}(1 - \alpha)$ is the $(1 - \alpha)$-quantile of $\chi^2(1)$ (a chi-squared distribution with $r = 1$ degree of freedom). Note that this is exactly the confidence interval computed in Section 2.2.2.

The confidence region tells us that a realization of $\mathcal{N}(\mu,\sigma^2)$ will land in $\mathcal{B}$ with a probability of $1 - \alpha$. For $n = 2$ with $X = (X_1,X_2)$ and $X_1 \sim \mathcal{N}(\mu_1,\sigma_1^2)$ and $X_2 \sim \mathcal{N}(\mu_2,\sigma_2^2)$, the box is:

$$\mathcal{B} = \{ x_1 \in [\mu_1 \pm \sqrt{\mathbb{Q}(1-\alpha)}\sigma_1] \,\&\, x_2 \in [\mu_2 \pm \sqrt{\mathbb{Q}(1-\alpha)}\sigma_2]\}. \tag{3.119}$$

This box is known as the *marginal box* and does not capture correlations in X_1 and X_2. The marginal box is just the intersection of 1D confidence intervals defined for the corresponding univariate RVs.

For a multivariate Gaussian $X \sim \mathcal{N}(\mu,\Sigma)$, we have seen that the pdf has level sets that form ellipses. This means that realizations of X concentrate in ellipses. It thus makes sense to try to find an *ellipsoidal* confidence region $\mathcal{E} \subseteq \mathcal{D}$ that contains X with some probability. In other words, we want to find an ellipse $\mathcal{E}$ such that:

$$\mathbb{P}(X \in \mathcal{E}) = 1 - \alpha. \tag{3.120}$$

One can show that the ellipsoidal region is given by:

$$\mathcal{E} = \{(x-\mu)^T\Sigma^{-1}(x-\mu) \le \mathbb{Q}(1-\alpha)\}, \tag{3.121}$$

where $\mathbb{Q}(1 - \alpha)$ is the $(1 - \alpha)$-quantile of $\chi^2(n)$ (a chi-squared RV with n degrees of freedom). The quantile defines the size of the ellipsoid. As expected, the size of the ellipse increases as the probability $1 - \alpha$ increases. The confidence region $\mathcal{E}$ is the region under which we can find $\mathcal{N}(\mu,\Sigma)$ with probability $1 - \alpha$. The larger the probability $1 - \alpha$, the larger the ellipsoid (the more likely it is to find $\mathcal{N}(\mu,\Sigma)$ in $\mathcal{E}$).

One can show that the *tightest box* that encloses ellipsoid $\mathcal{E}$ is given by:

$$\mathcal{B} = \{ x_1 \in [\mu_1 \pm \sqrt{\mathbb{Q}(1-\alpha)}\sigma_1] \,\&\, x_2 \in [\mu_2 \pm \sqrt{\mathbb{Q}(1-\alpha)}\sigma_2] \}, \tag{3.122}$$

where $\mathbb{Q}(1-\alpha)$ is $(1-\alpha)$-quantile of $\chi^2(n)$. Note that the quantile used here is different from that of the marginal box Eq. (3.119) (which uses $n = 1$). The ellipsoidal and the enclosing box regions capture correlations in X_1 and X_2, while the marginal does not.

Example 3.20 Geometry of 2D Gaussian Ellipsoid

In Figure 3.20, we visualize the joint pdf and marginal pdfs for a Gaussian with:

$$\mu = \begin{bmatrix} 0 \\ 0 \end{bmatrix}, \qquad \Sigma = \begin{bmatrix} 1 & -1/2 \\ -1/2 & 1 \end{bmatrix}.$$

Here, we also present the 95% ellipsoidal confidence region and compare this against its bounding box and the marginal box. In Figure 3.21, we present the 95% ellipsoidal region and 1,000 realizations from the Gaussian. From these figures, we can make several important observations:

- From the marginal pdfs, note that their 95% confidence intervals align with the sides of the marginal box.

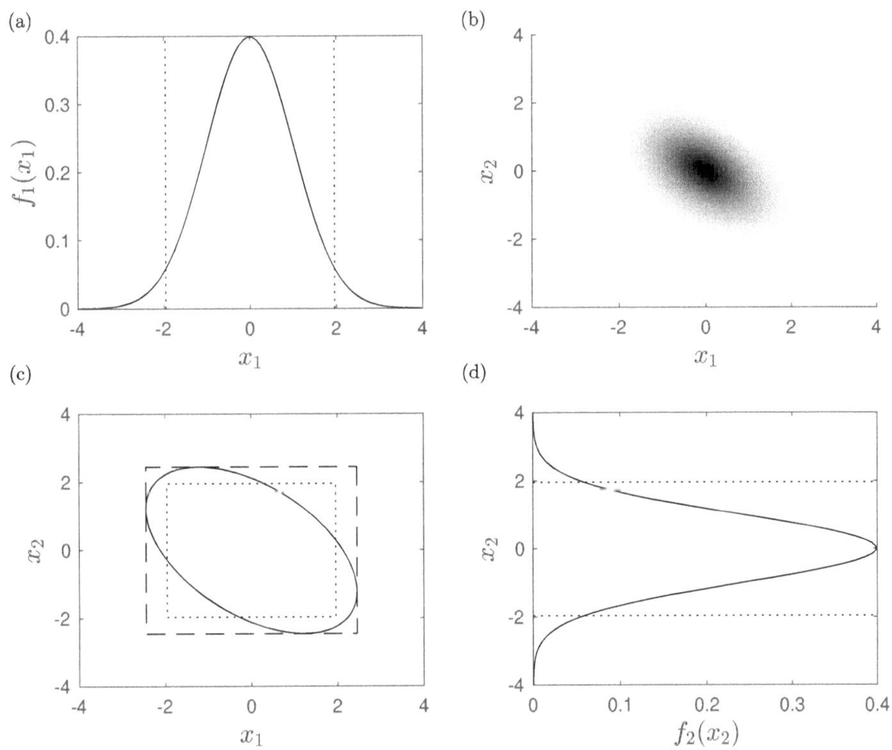

Figure 3.20 Joint pdf visualized as a heat map (b), marginal pdfs (a and d), and 95% confidence regions for multivariate Gaussian (c). In (c), note how the enclosing box encapsulates the confidence ellipsoid (while the marginal box does not). Note also how the borders of the marginal box coincide with the confidence intervals of the marginal pdfs.

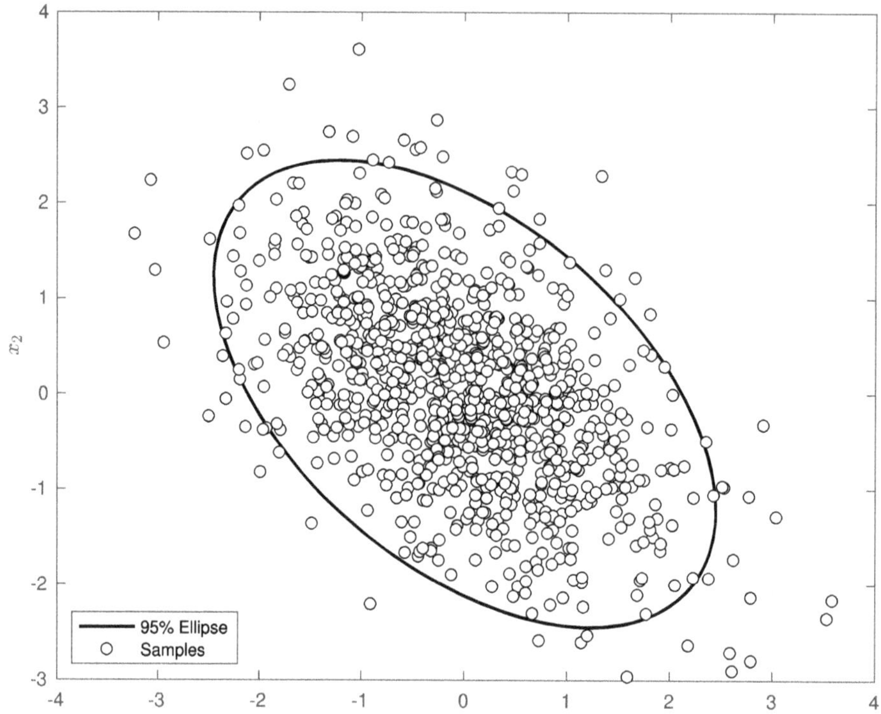

Figure 3.21 95% ellipsoidal confidence region and 1,000 realizations.

- The ellipsoidal confidence region is tilted to the left, indicating that X_1 and X_2 are negatively correlated; this can be confirmed from the covariance matrix since we have $\mathrm{Cov}[X]_{12} = -1/2$.
- The axes along the x_1 and x_2 directions have the same length. This is because the variance of X_1 and X_2 is the same ($\Sigma_{11} = \Sigma_{22}$).
- The marginal box does not capture its shape (its tilt), while the enclosing box does capture its shape.
- We have that 942 out of $S = 1{,}000$ realizations (94.2%) lie inside the 95% ellipsoid. Note that the fraction of realizations that are inside the ellipsoid is not exactly 95%. This is because the fraction is an empirical approximation of the probability $\mathbb{P}(X \in \mathcal{E})$. Specifically, the fraction is the estimate $\hat{\mathbb{P}}(X \in \mathcal{E}) = \frac{1}{S}\sum_{\omega \in \mathcal{S}} \mathbf{1}[x_\omega \in \mathcal{E}]$. The fraction will converge to 95% as we increase the number of realizations.

Script to reproduce results: `ch3_gaussian_geometry` and `ch3_gaussian_samples_ellipse`. □

Establishing the origin of the ellipsoidal region Eq. (3.121) is important because it merges several key concepts that we have discussed and introduces new ones that will become relevant later. We begin by noticing the rather "suspicious" appearance of the quantile of $\chi^2(n)$ in the definition of the ellipsoidal region. This indicates that, somehow, there is a relationship between an n-dimensional Gaussian $X \in \mathcal{N}(\mu, \Sigma)$ and a chi-squared RV $\chi^2(n)$. This relationship, in fact, exists and is established as follows; consider an eigenvalue decomposition of the covariance of the form:

$$\Sigma = Q\Lambda Q^T = \sum_{i=1}^{n} \lambda_i q_i q_i^T, \tag{3.123}$$

Here, $Q \in \mathbb{R}^{n\times n}$ is an eigenvector matrix; the i-th column of this matrix is the eigenvector $q_i \in \mathbb{R}^n$. The matrix $\Lambda \in \mathbb{R}^{n\times n}$ is an eigenvalue matrix; this is a diagonal matrix whose i,i-th entry is the eigenvalue $\lambda_i \in \mathbb{R}$.

The eigenvector matrix has the property that it is orthogonal and thus $Q^{-1} = Q^T$ and $Q^T Q = QQ^T = \mathbf{I}$. Now recall that we are interested in finding an ellipsoid of the form $\mathcal{E} = \{(x-\mu)^T\Sigma^{-1}(x-\mu) \leq \ell\}$ such that $\mathbb{P}(X \in \mathcal{E}) = 1-\alpha$. We rewrite the ellipsoid by using the eigenvalue decomposition as:

$$\begin{aligned}\mathcal{E} &= \{(x-\mu)^T\Sigma^{-1}(x-\mu) \leq \ell\}\\ &= \{(x-\mu)^T Q\Lambda^{-1}Q^T(x-\mu) \leq \ell\}\\ &= \{z^T z \leq r\}\\ &= \left\{\sum_{i=1}^{n} z_i^2 \leq \ell\right\},\end{aligned} \tag{3.124}$$

where we have defined $z = \sqrt{\Lambda^{-1}}Q^T(x-\mu)$. Note that this is a linear transformation (from the space of x to a space of z). The ellipsoid defined in the z space is $\mathcal{E} = \left\{\sum_{i=1}^{n} z_i^2 \leq \ell\right\}$; this defines a perfect ball, which has no tilt and has axes of the same length.

Now note that z is the realization of the RV $Z = \sqrt{\Lambda^{-1}}Q^T X - \sqrt{\Lambda^{-1}}Q^T\mu$ and note that Z is a linear transformation of our Gaussian RV $X \sim \mathcal{N}(\mu,\Sigma)$ of the form $Z = AX + b$ with $A = \sqrt{\Lambda^{-1}}Q^T$ and $b = -\sqrt{\Lambda^{-1}}Q^T\mu$. This implies that $Z \sim \mathcal{N}(A\mu + b, A\Sigma A^T)$ with:

$$\begin{aligned}A\mu + b &= \sqrt{\Lambda^{-1}}Q^T\mu - \sqrt{\Lambda^{-1}}Q^T\mu\\ &= 0,\end{aligned} \tag{3.125}$$

and

$$\begin{aligned}A\Sigma A^T &= \sqrt{\Lambda^{-1}}Q^T\Sigma Q\sqrt{\Lambda^{-1}}\\ &= \sqrt{\Lambda^{-1}}Q^T Q\Lambda Q^T Q\sqrt{\Lambda^{-1}}\\ &= \mathbf{I}.\end{aligned} \tag{3.126}$$

We thus have that $Z \sim \mathcal{N}(0,\mathbf{I})$, and this implies that $Z_i \sim \mathcal{N}(0,1)$ for all $i = 1,\ldots,n$. Now recall that this implies that $Y = \sum_{i=1}^{n} Z_i^2 \sim \chi^2(n)$. With this, we conclude that Z is a transformation of the Gaussian X. The final observation is that:

$$\mathbb{P}(X \subset \mathcal{E}) = \mathbb{P}(Y \leq \ell). \tag{3.127}$$

This property arises from the fact that variable transformations *preserve volume* of their corresponding domains. This is key because we know that the value of r

satisfying $\mathbb{P}(Y \leq \ell) = 1 - \alpha$ is the $(1-\alpha)$-quantile of $\chi^2(n)$. From this, we conclude that, if we define the ellipsoid $\mathcal{E}$ with $\ell = \mathbb{Q}(1-\alpha)$, then $\mathbb{P}(X \in \mathcal{E}) = 1 - \alpha$, as desired.

The previous derivations reveal some interesting properties of Gaussian RVs. Note that one can always transform a Gaussian RV $X \in \mathcal{N}(\mu, \Sigma)$ into a standard normal RV $Z \sim \mathcal{N}(0, \mathbf{I})$ by using a transformation of the form $Z = \sqrt{\Lambda^{-1}}Q^T X - \sqrt{\Lambda^{-1}}Q^T \mu$, with Q and Λ being eigenvectors and eigenvalues of Σ. The volume preservation property Eq. (3.127) indicates that transforming the ellipsoid from the original space to a new space has the effect of reshaping the ellipsoid (to obtain a perfect ball), but it does not affect its volume.

Eigenvalues and eigenvectors are powerful mathematical tools that play an important role in statistics and data analysis. Specifically, they allow us to transform the original space of a given RV X into a different space that is easier to work with. This is particularly important when analyzing and visualizing high-dimensional spaces (large n). For instance, in Chapter 6, we will see that the eigenvalue decomposition of the covariance matrix:

$$\Sigma = \sum_{i=1}^{n} \lambda_i q_i q_i^T, \tag{3.128}$$

will allow us to identify dominant variables that explain most of the variability in the data. These concepts will also be used to derive strategies to compress and visualize high-dimensional datasets (with the goal of identifying "peculiar" data points).

3.9 Summary

This chapter provides a discussion on multivariate RVs; the following are important take-home messages:

- Multivariate RVs are collections of RVs and have the key characteristic that their realizations are vectors.
- The domains of multivariate RVs are high-dimensional and can be difficult to visualize. The subdomains of their domain define events; these events can be constructed by using conditions or constraints that the realizations need to satisfy. The construction of these events allows decision-makers to model complex logic.
- Multivariate RVs are defined by their joint pdf; this describes how densely realizations concentrate in different parts of the domain. The joint pdf is used to compute the probability of events.
- There are several important probability density functions that are useful in the context of multivariate RVs. Specifically, the marginal pdfs determine the probabilities of individual RVs, while the conditional pdf determines probabilities of individual RVs given knowledge that other RVs have taken specific values.
- Important summarizing statistics of multivariate RVs are the expected value (which is a vector) and the covariance (which is a matrix). These quantities are useful for reducing data and to enable analysis and comparisons of

high-dimensional RVs. The covariance matrix captures how different individual RVs vary together. Related to the covariance matrix is the correlation matrix, which is a useful summarizing statistic that scales values and enables more systematic comparisons. Probabilities of events are also effective summarizing statistics that allow one to navigate large datasets, as they help filter out data using logical conditions and constraints.

- Computing summarizing statistics requires the joint pdf, which is rarely available. Moreover, computing these quantities can require complex integral operations. In practice, it is much more convenient to compute empirical approximations for summarizing statistics directly from data; the computation of such approximations requires simple counting operations.
- There are few theoretical models for multivariate RVs, with the Gaussian model being by far the most used one. The Gaussian model has many important properties and applications. For instance, linear transformations of a Gaussian RV lead to another Gaussian RV.

The thermostat examples provided in this chapter are extensions of an example provided in [8]. Detailed derivations for multivariate Gaussian RVs are provided in [30].

3.10 Exercises

Exercise 3.1. Multivariate Domains and Probabilities.
Your company has invested in a couple of wind farms in Wisconsin (these are located 50 miles apart). The weekly revenues of these wind farms (denoted as R_1 and R_2) are highly variable because the extracted power varies with wind speed conditions at the farm location. Use the historical revenue data for the farms (in MUSD) provided in `wind_revenues.dat` to determine the following:

(a) Create a scatter plot that shows your observations for R_1 and R_2. Determine the individual domains for R_1 and R_2 and their joint domain. Show the domains in the scatter plot.
(b) Determine the probability that R_1 is profitable (revenue is nonnegative) and show the domain associated with this event in the scatter plot.
(c) Determine the probability that R_2 is profitable and show the domain associated with this event in the scatter plot.
(d) Determine the probability that both R_1 and R_2 are profitable and show the domain associated with this event in the scatter plot.
(e) Determine the probability that R_1 is higher than R_2 and show the domain associated with this event together in the scatter plot.
(f) Determine the probability that $R_1 \geq 1$ MUSD and the probability that $R_2 \geq 1$ MUSD and determine whether these events are independent.
(g) Determine the marginal and joint pdfs. Provide a visualization of the joint pdf in 3D and as a 2D projection.

(h) Determine the mean, covariance, and Pearson correlation matrix for the revenues. Which farm has a higher mean revenue? How are the revenues correlated? Why could the revenues be correlated in a positive or negative direction?

Exercise 3.2. Discrete Events.

You are provided large amounts of data for the real-time electricity prices of Texas (see `RTM_ERCOT.csv`), which is operated by the Electric Reliability Council of Texas (ERCOT) [9]. Your data contains prices for different locations; the prices are recorded every 15 min (in USD/MWh). You would like to do a quick exploration of the prices at three specific locations (Austin, San Antonio, and the Panhandle) to explore potential connections between them. To do so, you categorize the prices as "low" (price is $\leq$5 USD/MWh), "medium" (price is >5 USD/MWh and $\leq$50 USD/MWh), and "high" (price is >50 USD/MWh).

(a) Determine the probability that the prices at Austin, San Antonio, and the Panhandle are simultaneously high. Determine the probability that the prices at Austin, San Antonio, and the Panhandle are simultaneously medium. Determine the probability that the prices at Austin, San Antonio, and Panhandle are simultaneously low. Comment on your observations.
(b) Determine the probability that the prices at Austin are high. Determine the probability that the prices at Austin are high, given that you know that the prices at San Antonio are high. Determine the probability that the prices at Austin are high, given that you know that the prices at San Antonio and the Panhandle are high. Does information about San Antonio and the Panhandle prices convey information about prices in Austin?
(c) Determine the probability that prices in Austin and San Antonio have medium values simultaneously. Determine the probability that the prices in Austin are medium or that the probability of prices in San Antonio are medium. Compute also the marginal probabilities of the events. Confirm that the "or" probability is given by the sum of the marginal probabilities minus the "and" (joint) probability.
(d) Determine the probability that the prices in Austin are medium, given that you know that the prices at San Antonio are medium. Compute the probability that the prices at San Antonio are medium, given that you know that the prices at Austin are medium. Confirm that the Bayes' theorem holds.

Exercise 3.3. Physical Connections and Correlations.

Correlations often arise because of physical connections between variables. Consider the chemical process with recycle shown below:

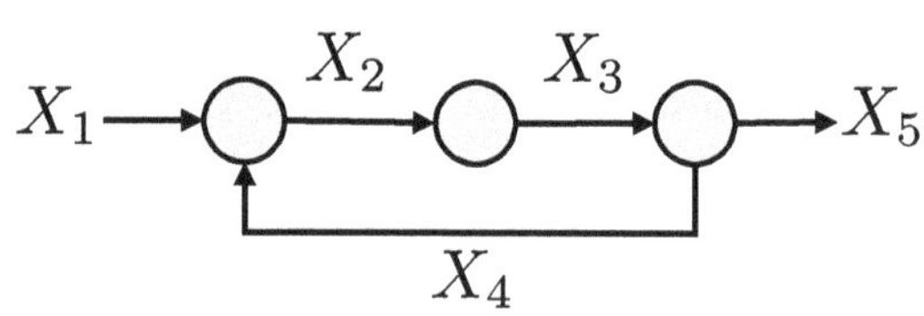

The input flow of this system is $X_1 \sim \mathcal{N}(\mu,\sigma^2)$. The rest of the flows in the process can be obtained from the following material balances:

$$
\begin{aligned}
X_2 &= X_1 + X_5 \\
X_3 &= \alpha \cdot X_2 \\
X_5 &= (1-r) \cdot X_3 \\
X_4 &= r \cdot X_3.
\end{aligned}
$$

Here, X_5 is the output flow of the process, $\alpha \in [0,1]$ is a reactor yield factor, and $r \in [0,1]$ is the recycle fraction.

(a) Reformulate the material balances to obtain an expression that relates the output process flow X_5 to the input flow X_1. Show that $X_5 \sim \mathcal{N}(\mu_5,\sigma_5^2)$ and determine μ_5 and σ_5^2 as a function of μ, σ, α, and r.
(b) Determine the covariance between the input flow X_1 and the output flow X_5 as a function of μ, σ, α, and r. Comment on the effect of the recycle fraction on the degree of correlation between X_1 and X_5; specifically, comment on the extreme cases in which $r = 0$ (no recycle) and $r = 1$ (full recycle).
(c) What is the covariance matrix of (X_1, X_5)?

Exercise 3.4. Portfolio Optimization.
You have a collection of investments with varying degrees of cash flow and uncertainty. The cash flows (revenues) of these investments can be modeled as $X_k \sim \mathcal{N}(\mu_k,\sigma_k^2)$ for $k = 1,\ldots,n$.

Mixing investments is a portfolio optimization strategy that can be used to mitigate/dilute uncertainty (analogous to mixing liquid streams with different concentrations to achieve a target concentration). Specifically, you want to determine the mix proportions $w_k \in [0,1]$, $k = 1,\ldots,n$ (note that the proportions satisfy $\sum_{k=1}^{n} w_k = 1$) such that the total cash flow:

$$
Y = \sum_{k=1}^{n} w_k \cdot X_k
$$

has an acceptable level of risk. The risk here is defined as the probability that the total cash flow is less than 4×10^5 USD.

(a) Show that $Y \sim \mathcal{N}(\mu_Y,\sigma_Y^2)$ and determine the parameters.
(b) Consider $n = 3$ investments with the following statistics: Investment 1 ($\mu_1 = 5 \times 10^5$ USD, $\sigma_1 = 1 \times 10^5$ USD), Investment 2 ($\mu_2 = 4 \times 10^5$ USD, $\sigma_2 = 1 \times 10^2$ USD), and Investment 3 ($\mu_3 = 1 \times 10^6$ USD, $\sigma_1 = 5 \times 10^5$ USD). Determine the risk of individual investments (mixtures with proportions (1,0,0), (0,1,0), (0,0,1)). Provide some intuition as to why some investments lead to higher/lower risk than others.
(c) Play around with mixtures of different proportions and determine whether mixing portfolios can indeed help reduce the risk of the individual investments.

Exercise 3.5. Electricity Markets.
Consider the data for the real-time electricity market of Texas (see `RTM_ERCOT.csv`). The Panhandle region is of particular interest because there are large amounts of wind power in this region that may lead to low electricity prices. The Austin region is also of interest because there is a growing tech sector in this region that will require large amounts of power.

(a) Determine the mean and Pearson correlation matrix for the prices at both locations. Based on this, comment on which location has the most favorable market conditions (on average) and comment on the degree of correlation that you observe between the different locations.
(b) Construct the joint pdf for the prices in Austin and the Panhandle; construct the joint pdf over individual domains consisting of the range of −30 to 100 USD/MWh and discretize this range using 10 bins. Visualize the prices using a scatter plot and visualize the joint pdf as a heat map. Comment on how the degree of correlation explains the nature of joint pdf and of the scatter plot.
(c) Determine the bin of the joint pdf in which you see the highest density of prices. Compute the probability that the prices land in this bin.
(d) From the scatter plot, comment on how likely it is that the prices in Austin are higher than those in the Panhandle. Compute the probability of this event.
(e) From the scatter plot, comment on how often Austin has negative prices (the consumer gets paid for using electricity) and comment on how often the Panhandle has negative prices. Compute the corresponding probabilities to confirm your observations.

Exercise 3.6. Uncertainty Propagation in Dynamical Systems.
Understanding how uncertainty propagates in dynamical systems is important in the field of process control. Consider a system with an uncertain initial state $X_0 \sim \mathcal{N}(0, P_0)$, where P_0 is its covariance matrix. This state propagates in time via the dynamical system:

$$X_{k+1} = AX_k,\ k = 0, \ldots, N-1,$$

where A is a matrix that captures the time evolution and N is the number of time steps. Our goal is to understand how does the uncertainty of state X_k (its covariance matrix) evolve in time. For a dynamical system with two states, the evolution can be written as:

$$\begin{bmatrix} X_{k+1,1} \\ X_{k+1,2} \end{bmatrix} = \begin{bmatrix} A_{11} & A_{12} \\ A_{21} & A_{22} \end{bmatrix} \begin{bmatrix} X_{k,1} \\ X_{k,2} \end{bmatrix}.$$

Here, A_{11} and A_{22} can be interpreted as individual decay rates for the first and second states. The terms A_{12} and A_{22} can be interpreted as cross-decay rates that capture the coupling between the first and second states.

(a) Determine a formula for the covariance matrix of the state X_k; express this formula as a recursion that relates the covariance matrix of X_{k+1} with the covariance matrix of X_k.

(b) Consider a dynamical system with:

$$A = \begin{bmatrix} 0.9 & 0.1 \\ 0.1 & 0.5 \end{bmatrix} \qquad P_0 = \begin{bmatrix} 10 & 0 \\ 0 & 10 \end{bmatrix}.$$

Visualize the 95% confidence ellipsoids for the state X_k at $k = 0, \ldots, 10$. Comment on how the uncertainty shrinks for the two states; why does the uncertainty of $X_{k,1}$ shrinks faster than that of $X_{k,2}$?

(c) Consider a dynamical system with:

$$A = \begin{bmatrix} 0.9 & 0 \\ 0 & 0.9 \end{bmatrix} \qquad P_0 = \begin{bmatrix} 10 & 0 \\ 0 & 10 \end{bmatrix}.$$

Visualize the 95% confidence ellipsoids for the state X_k at times $k = 0, 1, \ldots, 10$. Compare how the uncertainty shrinks for the dynamical system of the previous question. Comment on how the structure of the matrix of the dynamical system influences this.

(d) For the dynamical system of (b), obtain samples of the initial state X_0 and propagate these through the dynamical system over $N = 3$ steps to obtain samples of the final state X_3. Visualize the time trajectories for each sample (on the same plot as the 95% confidence ellipse) and confirm that the final state of the trajectories lands in the ellipse.

Exercise 3.7. Brain Analysis.

Neuroscientists use functional magnetic resonance imaging (FMRI) data to monitor brain activity in human subjects when they are exposed to specific stimuli (e.g., watching a movie or conducting a task) [10]. This is done by collecting time series data at different locations of their brain and by detecting locations that exhibit strong/weak correlations. Strong correlations indicate brain locations that are being simultaneously activated by a stimulus (or that communicate in the presence of such stimulus). You have FMRI data for a child and an adult participant (see `child_data.csv` and `adult_data.csv`); these datasets consist of time series data from 39 different brain locations (visualized below).

(a) Construct the Pearson correlation matrix for the child and adult dataset and visualize these as heat maps.

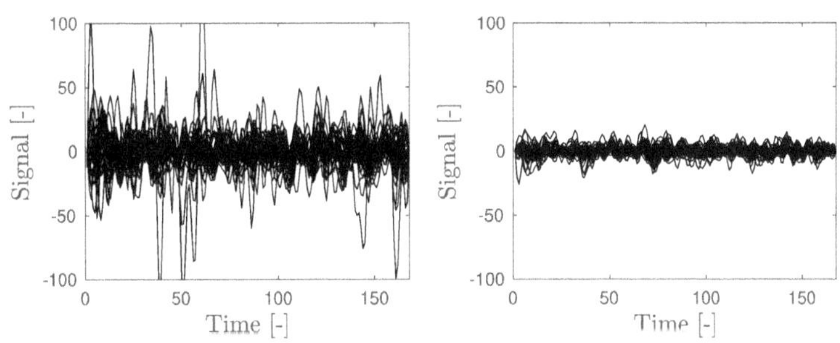

(b) Remove entries from the correlation matrices that have an absolute value below a threshold of 0.6 (set the corresponding values to zero). Visualize the filtered matrices as heat maps and comment on whether you observe differences in the correlation structure of the child and adult brains. Comment also on whether you see more areas of the brain activating simultaneously for the child or the adult.

Exercise 3.8. Convergence of Sample Covariance.
Data is commonly used to estimate the covariance matrix of an RV (this approximation is known as the sample covariance matrix). We would like to convince ourselves that the sample covariance converges to the true covariance matrix as we increase the amount of data used in the approximation. Consider a Gaussian RV with zero mean and covariance matrix:

$$\mathrm{Cov}[X] = \begin{bmatrix} 2 & 1 \\ 1 & 1 \end{bmatrix}.$$

Use samples to obtain sample covariances with $S = 10, 100, 1000$. Compare the 95% confidence ellipsoid of the true RV with the ellipsoid defined by the sample covariances. Comment on how many samples are sufficient to approximate the true ellipsoid.

Exercise 3.9. Scenario Trees.
Scenario/event trees provide a useful approach to visualizing conditional probabilities and are widely used in decision-making under uncertainty. Consider, for instance, the tree shown below constructed for RVs X_1 and X_2; here, each level represents an RV and a branch represents a possible value that the RV can take. In the example tree below, we assume that each RV can take three possible values. Each branch has an associated conditional probability (conditional to previous events leading to that branch).

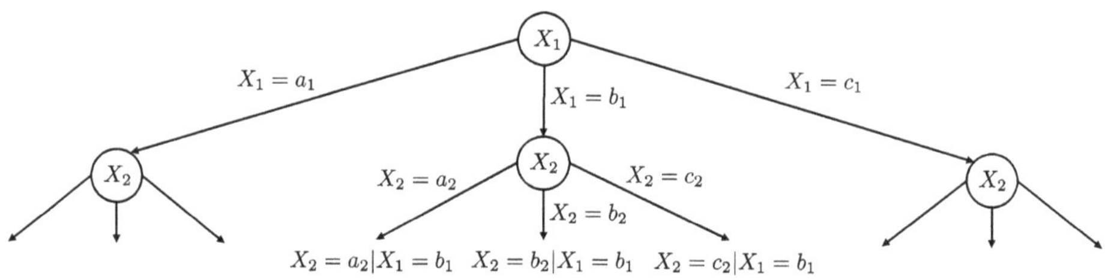

Use the electricity market data (see `RTM_ERCOT.csv`) and categorize the prices as "low" (price is ≤5 USD/MWh), "medium" (price is >5 USD/MWh and ≤50 USD/MWh), or "high" (price is >50 USD/MWh). Define the prices for Austin as X_1 and the prices for the Panhandle as X_2.

Address the following:

(a) Populate all the probabilities of the scenario tree. What are the most and least likely events in the last level of the tree?
(b) Do a "balance" of probabilities around each node of the tree (each circle) and confirm the balance holds. Confirm also that, at each level of the tree, the probabilities of all branches add up to one.

Exercise 3.10. Laplace Approximation.

The Laplace approximation is a technique that is used to approximate the distribution of an output variable Y that is obtained from a nonlinear transformation of an input Gaussian variable X. Specifically, consider $X \sim \mathcal{N}(\mu, \Sigma)$ and consider a nonlinear transformation of the form $Y = g(X)$. Because the function $g(X)$ is nonlinear, the resulting output variable is non-Gaussian and thus obtaining statistics for this requires sampling (e.g., Monte Carlo). To avoid using Monte Carlo, we can also approximate the nonlinear function using a Taylor expansion of the form:

$$Y \approx g(\mu) + \nabla g(\mu)(X - \mu),$$

where $\nabla g(\mu)$ is the matrix of first derivatives of the function $g(X)$ (the gradient) evaluated at μ. Given that the Taylor expansion is a linear function of X, we can obtain an *analytical* Gaussian approximation of Y (this is known as the Laplace approximation). The motivation behind this approximation is that, if the function $g(X)$ is "mildly" nonlinear, we can expect that the output behaves close to Gaussian (when the function is linear, we know that this is indeed the case).

(a) Consider the nonlinear function:

$$\begin{bmatrix} Y_1 \\ Y_2 \end{bmatrix} = \begin{bmatrix} g_1(X) \\ g_2(X) \end{bmatrix} = \begin{bmatrix} \exp(-X_1) + X_2 + X_3 \\ X_2 \end{bmatrix}.$$

Obtain the Laplace approximation of Y, given $X \sim \mathcal{N}(\mu, \Sigma)$ with:

$$\mu = \begin{bmatrix} 1 \\ 1 \\ 1 \end{bmatrix} \qquad \Sigma = \begin{bmatrix} 1 & 0 & 0 \\ 0 & 1 & 0 \\ 0 & 0 & 1 \end{bmatrix}.$$

Visualize the mean and confidence ellipsoids for the Laplace approximation. Obtain 1,000 samples of the nonlinear transformation $Y = g(X)$ using Monte Carlo and visualize the samples and the empirical mean in the same plot of the Laplace approximation. Comment on the ability of the Laplace approximation to approximate the true mean and the density of samples near and far from the mean.

(b) Consider the nonlinear function:

$$\begin{bmatrix} Y_1 \\ Y_2 \end{bmatrix} = \begin{bmatrix} g_1(X) \\ g_2(X) \end{bmatrix} = \begin{bmatrix} X_1^2 + X_2^2 + X_3 \\ X_2^2 \end{bmatrix}.$$

Visualize the mean and confidence ellipsoids for the Laplace approximation. Visualize also the Monte Carlo approximation of the true nonlinear transformation $Y = g(X)$ and visualize the samples and the empirical mean in the same plot of the Laplace approximation. By comparing these results to those of the previous question, comment on how the nature of the nonlinear function can make the Laplace approximation less/more effective.

Part II

Intermediate Concepts

4 Estimation for Random Variables

We now have a general idea of the types of theoretical random variable (RV) models that are available to explain different random phenomena. We have seen that these models have *parameters* that can be tuned to fit the model to available data. We have also discussed how we can use realizations of an RV (obtained from data or from samples of an RV model) to obtain *empirical estimates* that approximate theoretical quantities of interest, such as summarizing statistics, event probabilities, and the cdf/pdf of an RV X or of an RV Y that results from the propagation of X through a system $Y = \varphi(X, u)$ (using Monte Carlo [MC] simulation).

We have leveraged the properties of RV models to relate its parameters to empirical estimates of summarizing statistics or of the cdf/pdf. Moreover, we have seen that one can easily approximate theoretical quantities using empirical estimates and MC (e.g., by using simple counting operations). However, at this point, we still do not have a good idea if estimates obtained with such procedures can provide any guarantees. Specifically, we would like to answer the following questions:

- How do we know that estimated quantities obtained from data are the *best* ones?
- Moreover, how do we know if the data available is sufficient to obtain estimates that are reliable?

We now proceed to develop procedures to determine if an RV model adequately fits a particular dataset available. To do so, it is necessary to find parameters for a proposed model that provide the *best* fit to the data. In other words, if we want to determine if a given RV model is adequate or not, we need a guarantee that there is no better set of parameters that can improve its fit. If an RV model cannot explain the data well, even with the best set of parameters found, then this indicates that an alternative RV model needs to be proposed.

Our first step will be to *analyze the data* so that we can identify some general features/trends of the random phenomenon. In doing so, we will use data to compute *empirical estimates* for the pdf/cdf and for summarizing statistics. We will use these quantities and our knowledge of RV models (and their properties) to propose/postulate a suitable model. Our postulated model will be defined by its theoretical pdf/cdf (which we will denote as $f(x|\theta)$ and $F(x|\theta)$) and of its associated summarizing statistics. Here, θ is a vector containing all the RV model parameters; for instance, for an exponential RV model, we have $\theta = \beta$, and for a Gaussian RV model, we have $\theta = (\mu,\sigma)$. Our next step will be to find the θ that gives the best fit of the RV model to our data and empirical approximations of the pdf/cdf and summarizing statistics. If the fit is not adequate, we will postulate another RV model. To understand the notion of what we mean by "best," we will introduce concepts of *optimization*. Optimization will be an extremely useful tool, as it will provide guarantees that our parameters are the best given data; moreover, we will see that optimization provides interesting insights into what empirical quantities (e.g., mean, variance, or quantiles) we should use to estimate the parameters.

A key issue that arises when estimating model parameters and when estimating theoretical quantities using empirical quantities is to determine whether the available data is sufficient to obtain estimates that are *reliable*. Specifically, we want to make sure that our estimates will not vary much if we obtain them with other sets of data (i.e., we want to ensure that our results are *reproducible*). This is important because this would indicate that our RV model *generalizes* well to situations that have not been observed yet or because this would indicate that our empirical estimates are accurate representations of theoretical quantities. Our goal will be to understand whether the quality of empirical approximations improves as we increase the amount of data available, and we will see that data collection procedures play a key role here.

In Chapters 5–7, we will see that many of the concepts discussed here are also applicable to estimating parameters for structural models and for making decisions under uncertainty.

4.1 Point Estimation

We will begin our discussion by introducing estimation methods to determine parameters for RV models. These methods can be classified in terms of a couple of general paradigms:

- Point estimation (method of moments and least-squares).
- Maximum likelihood estimation (MLE).

Recall that the pdf/cdf provides a full characterization of an RV model; consequently, if the cdf/pdf of our RV model matches the empirical estimate of the cdf/pdf, we would have a good indication that the model is adequate. As such, it makes sense to try to tune parameters in a way that the theoretical cdf/pdf matches the empirical cdf/pdf; this motivates the method of least-squares (LS). In some situations, we have that the model parameters are directly related to their summarizing statistics (e.g., the parameters of the Gaussian are directly related to the mean and variance). As such, it makes sense to estimate the parameters by *matching* the empirical estimates of summarizing statistics; this motivates the method of moments. The point estimation paradigm aims thus to find parameters that *best match* the data. MLE is a completely different paradigm to estimate parameters; specifically, this method aims to find the *most likely* set of parameters given the data. Note how the notion of *best* is different in the MLE and point estimation paradigms; this is important and will have important implications in the insights that you can obtain with the different paradigms.

4.1.1 Method of Moments

The moments of an RV model are related to a wide range of summarizing statistics (e.g., mean, variance, kurtosis, and skewness), and thus, we can use them as general quantities to estimate parameters. Recall that the kth *central moments* of an RV model X (with pdf $f(x|\theta)$ and parameters θ) are given by:

$$\mathbb{CMO}_k(\theta) := \mathbb{E}[(X - \mathbb{E}_X)^k], \qquad k = 1,2,\ldots,N. \tag{4.1}$$

Here, we use the notation $\mathbb{CMO}_k(\theta)$ to highlight the dependence of the model moment values on the parameters θ.

The method of moments uses data x_ω, $\omega \in \mathcal{S}$ to obtain the empirical (sample) approximations:

$$\hat{\mathbb{CMO}}_k = \frac{1}{S}\sum_{\omega\in\mathcal{S}}(x_\omega - \hat{\mathbb{E}}_X)^k, \; k = 1,2,\ldots,N, \tag{4.2}$$

where $\hat{\mathbb{E}}_X = \frac{1}{S}\sum_{\omega\in\mathcal{S}} x_\omega$ is the empirical mean.

Our goal is to find θ that solves the moment-matching *equations*:

$$\mathbb{CMO}_k(\theta) = \hat{\mathbb{CMO}}_k, \; k = 1,2,\ldots,N. \tag{4.3}$$

In other words, we want to find θ that matches the model moments to the empirical moments. An issue that can arise here is that the matching equations might not have a solution; this can happen, for instance, if our postulated RV model is wrong or if the data is corrupted. We will see shortly how to overcome this issue.

Example 4.1 Estimation for Gaussian RV Using Moment-Matching

We illustrate how to estimate $\theta = (\mu,\sigma^2)$ using moment-matching for the RV model $\mathcal{N}(\mu,\sigma^2)$ by using data x_ω, $\omega\in\mathcal{S}$.

We recall that a Gaussian model has the property that $\mu = \mathbb{E}_X$ and $\sigma^2 = \mathbb{V}_X$. We thus use the observations to obtain the sample mean and sample variance:

$$\hat{\mathbb{E}}_X = \frac{1}{S}\sum_{\omega\in\mathcal{S}} x_\omega$$

$$\hat{\mathbb{V}}_X = \frac{1}{S} \sum_{\omega \in \mathcal{S}} (x_\omega - \hat{\mathbb{E}}_X)^2.$$

Intuition tells us that we should simply set the parameters μ, σ to the sample mean and variance; we now show that the method of moments justifies this intuition.

From the definition of the first central moment, we have:

$$\mathbb{CMO}_1(\theta) = \mathbb{E}[(X - \mathbb{E}_X)] = \mathbb{E}_X - \mu = 0.$$

This tells us that we should set $\hat{\mu} = \hat{\mathbb{E}}_X$. From the definition of the second central moment, we have that $\mathbb{CMO}_2(\theta) = \mathbb{V}_X = \sigma^2$. This tells us that we should set $\hat{\sigma}^2 = \hat{\mathbb{V}}_X$.

It is important to note that $\hat{\mu} = \hat{\mathbb{E}}_X$ and $\hat{\sigma}^2 = \hat{\mathbb{V}}_X$ are *estimates* of the actual/theoretical parameters $\mu = \mathbb{E}_X$ and $\sigma^2 = \mathbb{V}_X$ (which we do not know). Moreover, note that the estimates depend on the data available. For instance, we can expect that the estimates will be different if we use fewer or more data points.

We can thus ask ourselves: How far are our estimates from the actual values? How much data do I need? To get some insights into these questions, we use moment-matching with $S = 10$ and $S = 100$ data points to estimate parameters. If we use $S = 10$, the estimates are $\hat{\mu} = 0.45$ and $\hat{\sigma} = 1.43$; if we use $S = 100$, the estimates are $\hat{\mu} = 0.9$ and $\hat{\sigma} = 2$. The true parameters are $\mu = 1$ and $\sigma = 2$; we can thus see that we get closer as we increase S.

In Figure 4.1, we present the empirical pdf/cdf obtained with the available data and the theoretical pdf/cdf predictions obtained with the estimated parameters $\hat{\mu}, \hat{\sigma}$.

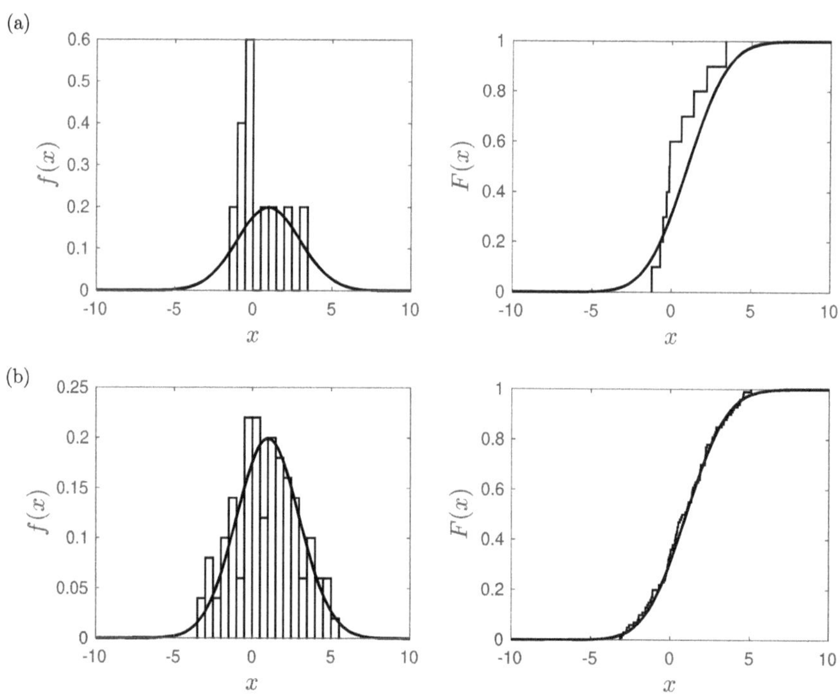

Figure 4.1 Empirical pdf/cdf for Gaussian RV obtained from data and model predictions obtained with estimated parameters. Results for $S = 10$ (a) and for $S = 100$ (b).

Note that the pdf/cdf cannot be matched well for $S = 10$, but they are matched well with $S = 100$. Also note that, for $S = 100$, the empirical pdf/cdf better reveals the nature of the RV and suggests that the phenomenon indeed follows a Gaussian model.

Script to reproduce results: `ch4_gaussian_moments_example` □

Example 4.2 Estimation for Weibull RV Using Moment-Matching

Now consider that we want to estimate $\theta = (\xi, \beta)$ by using moment-matching for Weibull(ξ, β). As in the previous example, we use the data to obtain the empirical mean $\hat{\mathbb{E}}_X$ and variance $\hat{\mathbb{V}}_X$. For the Weibull RV, the parameters ξ, β are not the expected value and variance of the RV (compared to the Gaussian RV case). Instead, we need to find β, ξ that match the model moments to their empirical counterparts. This is done by solving the following equations for β and ξ:

$$\hat{\mathbb{E}}_X = \beta\Gamma(1 + 1/\xi)$$
$$\hat{\mathbb{V}}_X = \beta^2 \left(\Gamma(1 + 2/\xi) - \Gamma(1 + 1/\xi)^2\right).$$

Note that this is a coupled set of equations (the parameters appear on both equations). We also recall that the Gamma function is a highly complex nonlinear function; as a result, solving this set of equations cannot be done analytically (it requires numerical techniques).

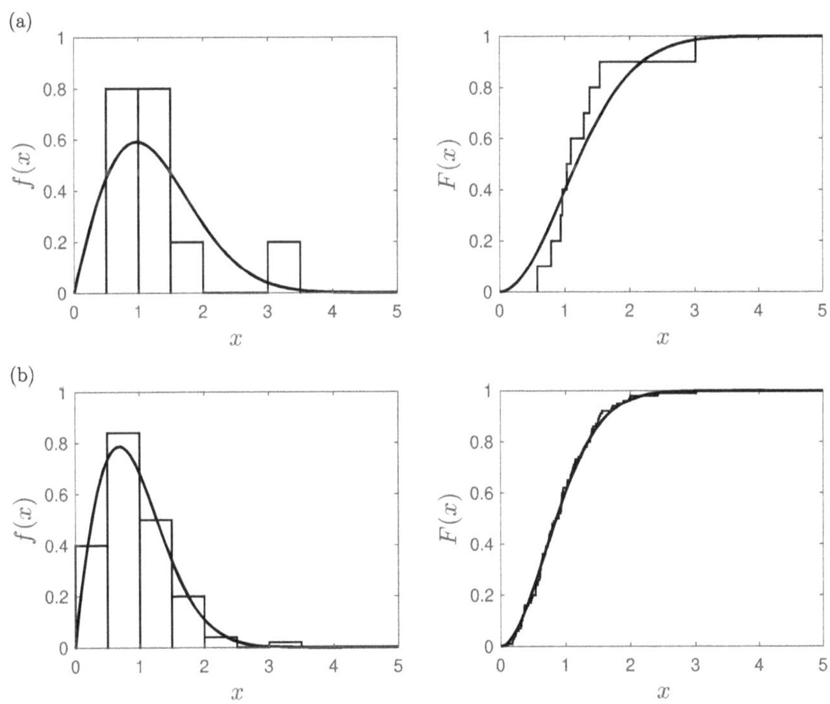

Figure 4.2 Empirical pdf/cdf for Weibull RV obtained from data and model predictions obtained with estimated parameters. Results for $S = 10$ (a) and for $S = 100$ (b).

We illustrate the effectiveness of the moment-matching approach by estimating parameters using $S = 10$ and $S = 100$ data points. We first use the data to obtain the empirical estimates; for $S = 10$, we have $\hat{\mathbb{E}}_X = 1.26$ and $\hat{\mathbb{V}}_X = 0.46$, and for $S = 100$, we have $\hat{\mathbb{E}}_X = 0.92$ and $\hat{\mathbb{V}}_X = 0.26$. We then use the estimates to solve the nonlinear set of equations and with this determine the corresponding parameters. The values obtained for $S = 10$ are $\hat{\beta} = 1.42$ and $\hat{\xi} = 1.93$ and for $S = 100$ are $\hat{\beta} = 1.04$ and $\hat{\xi} = 1.85$. The true parameter values are $\beta = 1$ and $\xi = 2$. The corresponding empirical and predicted pdf/cdfs are shown in Figure 4.2. The estimates and corresponding pdf/cdf vary significantly but improve as we increase S (again highlighting the importance of having sufficient data).

Script to reproduce results: `ch4_weibull_moments_example` □

By comparing the Weibull and the Gaussian examples, we can see that the moment-matching approach is not always straightforward to implement (e.g., the moments of some RV models are complex functions of the parameters). We also highlight that, even if the parameters match the empirical moments, there is no guarantee that the resulting RV model will be an adequate representation of the pdf/cdf. Specifically, recall that moments do not provide a full characterization of an RV (the full pdf/cdf are needed for this). Consequently, it is always important to verify that the parameters obtained with moment matching actually provide a good fit of the pdf/cdf. This also raises the possibility of estimating parameters by directly matching the empirical pdf/cdf. We proceed to discuss how to do this.

4.1.2 Least-Squares

The LS method is flexible and can be used to fit empirical approximations for the pdf/cdf and for any summarizing statistic beyond moments (e.g., median, quantiles, and mode).

For instance, assume that we want to find the parameters that match the RV model cdf $F(t|\theta)$ to its empirical approximation $\hat{F}(t)$. To do so, we use available data x_ω, $\omega \in \mathcal{S}$ for X and propose a set of threshold values $t_k, k = 1,2,\ldots,N$ to compute the empirical cdf:

$$\hat{F}(t_k) = \frac{1}{S}\sum_{\omega\in\mathcal{S}} \mathbf{1}[x_\omega \leq t_k],\ k = 1,2,\ldots,N. \tag{4.4}$$

To simplify notation, we define $\hat{F}_k = \hat{F}(t_k)$. We find the parameters by solving the optimization problem:

$$\min_\theta \frac{1}{2}\sum_{k=1}^{N}(F(t_k|\theta) - \hat{F}_k)^2. \tag{4.5}$$

Note that this problem aims to find the parameters θ that minimize the squared discrepancy (squared error) between the model cdf and the empirical cdf. As such, the parameters obtained have the guarantee that they are the *best* ones (there are no better ones). The ability to obtain optimality guarantees is a key benefit of using optimization techniques. However, note that the parameters obtained with LS are still a function of the amount of data used. We thus need to ensure that we have an adequate amount of data. We also highlight that, in many cases, the LS optimization problem cannot be solved analytically (by hand), and we thus need numerical techniques.

A similar procedure can be used to match other values; for instance, we can aim to find the parameters that best match the model quantiles $\mathbb{Q}(\alpha_k)$, $k = 1,\ldots,N$ to the empirical counterparts $\hat{\mathbb{Q}}(\alpha_k)$, $k = 1,\ldots,N$.

Example 4.3 Estimation for Weibull Using Least-Squares

We want to estimate $\theta = (\xi,\beta)$ for the RV model $X \sim \text{Weibull}(\xi,\beta)$. We are given data x_ω, $\omega \in \mathcal{S}$, and we use these to obtain empirical values for the cdf $\hat{F}(t_k)$ for t_k, $k = 1,\ldots,N$. Recall that the cdf of the Weibull model has the form

$$F(t|\theta) = 1 - e^{-(t/\beta)^\xi}.$$

We thus obtain the parameters by solving the LS problem:

$$\min_{\beta,\xi} \; \frac{1}{2}\sum_{k=1}^{N}\left((1 - e^{-(t_k/\beta)^\xi}) - \hat{F}_k\right)^2.$$

This is a nonlinear optimization problem that cannot be solved by hand and requires numerical techniques to be solved. We apply this methodology to the dataset of Example 4.2 ($S = 100$) and find that the parameters that best match the data are $\hat{\beta} = 1.04$ and $\hat{\xi} = 1.95$; note that these estimates are different from those obtained with moment-matching ($\hat{\beta} = 1.04$ and $\hat{\xi} = 1.85$). Specifically, the LS estimates are closer to the true values ($\beta = 1$ and $\xi = 2$); this highlights how the use of optimization techniques can help find the best set of parameters.

In Figure 4.3, we compare the empirical cdf $\hat{F}(t_k)$ and the cdf predicted by the model $F(t_k|\hat{\theta})$; we can see that the Weibull model matches the data well. We further validate this by comparing the model and empirical pdfs.

When cdfs have exponential forms, it is convenient to simplify these using logarithmic transformations. For instance, by applying a logarithmic transformation to the cdf, we find that:

$$\log(1 - F(t|\theta)) = -(t/\beta)^\xi.$$

Furthermore, note that:

$$\log(-\log(1 - F(t|\theta))) = \xi \cdot \log t - \xi \cdot \log \beta.$$

As such, we can also estimate the parameters by solving the LS problem:

$$\min_{\beta,\xi} \; \frac{1}{2}\sum_{k=1}^{N}\left(-\xi \cdot \log \beta + \xi \cdot \log t_k - \log(-\log(1 - \hat{F}_k))\right).$$

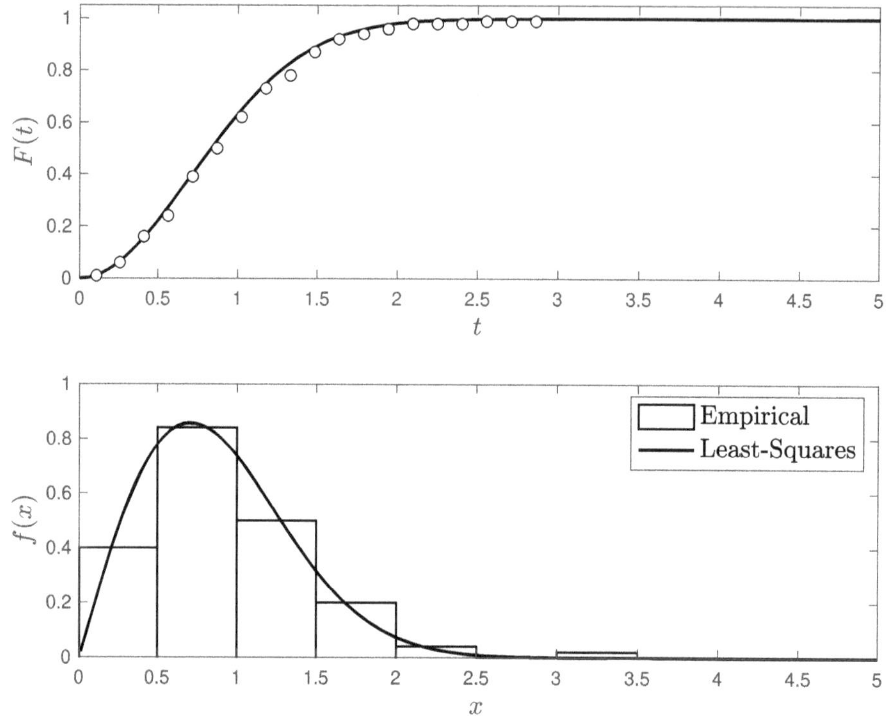

Figure 4.3 Estimation of Weibull parameters using least-squares method.

This LS problem looks quite complicated but can be simplified by making the following definitions:

$$
\begin{aligned}
\theta_0 &= -\xi \cdot \log \beta \\
\theta_1 &= \xi \\
x_k &= \log t_k \\
y_k &= \log(-\log(1 - \hat{F}_k)),
\end{aligned}
$$

to obtain the alternative LS problem:

$$\min_{\theta_0, \theta_1} \frac{1}{2} \sum_{k=1}^{N} (\theta_0 + \theta_1 \cdot x_k - y_k)^2.$$

Note that this problem is much simpler; in particular, the cdf model has been replaced by a simple linear model $\theta_0 + \theta_1 \cdot x_k$. The solution of this problem can be obtained analytically (by hand) as opposed to using numerical techniques; we will show how to do this in Chapter 5. The solution of the problem gives the estimates $\hat{\theta}_0 = -0.16$ and $\hat{\theta}_1 = 1.89$ from which we can back-calculate our parameter estimates as $\hat{\beta} = 1.08$ and $\hat{\xi} = 1.89$. Note that the estimates are close to the true values and that there are small discrepancies with the estimates previously obtained (this is because of differences between analytical and numerical calculations).

Script to reproduce results: `ch4_leastsquares_example` □

4.2 Maximum Likelihood Estimation

Maximum likelihood estimation is a method that is conceptually similar to the LS method in that it tries to find the best possible set of parameters that match the data. However, in MLE, by "best" we mean parameters that are most probable (likely) given the data; in LS, by "best" we mean parameters that best fit the data. We thus note that *MLE has a natural statistical interpretation*, while LS does not.

Assume that the observations x_ω, $\omega \in \mathcal{S}$ are collected at *random*. In other words, the collection of one observation does not influence the collection of another one. As we will discuss in Section 4.3, this implies that the observations are independent RVs.

We postulate the model $f(x|\theta)$ for the RV X and recall that $f(x_\omega|\theta)$ can be interpreted as the probability/likelihood that X takes observation x_ω. We seek to find θ that maximizes the *joint* probability that X takes observations x_ω, $\omega \in \mathcal{S}$. Because the observations are independent RVs, their joint probability is given by $\prod_{\omega \in \mathcal{S}} f(x_\omega|\theta)$. We thus find θ by solving the optimization problem:

$$\max_\theta \; L(\theta) = \prod_{\omega \in \mathcal{S}} f(x_\omega|\theta). \tag{4.7}$$

Here, $L(\theta)$ is known as the *likelihood function*.

The pdf $f(x_\omega|\theta)$ often has small values (close to zero) and thus their product can lead to a likelihood function that has extremely small values. As such, it is often convenient to apply a logarithmic transformation to solve the equivalent problem:

$$\max_\theta \; \log L(\theta) = \sum_{\omega \in \mathcal{S}} \log f(x_\omega|\theta). \tag{4.8}$$

The function $\log L(\theta)$ is known as the *log-likelihood function*. The logarithmic transformation transforms the product into a summation and scales the values of the pdfs. The maximization problem (4.8) is equivalent to the minimization problem:

$$\min_\theta \; -\log L(\theta) = -\sum_{\omega \in \mathcal{S}} \log f(x_\omega|\theta). \tag{4.9}$$

MLE problems can be solved analytically (by hand) when the pdf of the RV is simple but require numerical techniques when the pdf is complex. The solution of the MLE problem yields the estimate $\hat{\theta}$; this is interpreted as the *most likely parameter* given the available data.

Example 4.4 Estimation for Exponential Using MLE

Assume that we want to use MLE to estimate β for the exponential RV model $\text{Exp}(\beta)$ from the observations x_ω, $\omega \in \mathcal{S}$ shown in Figure 4.4.

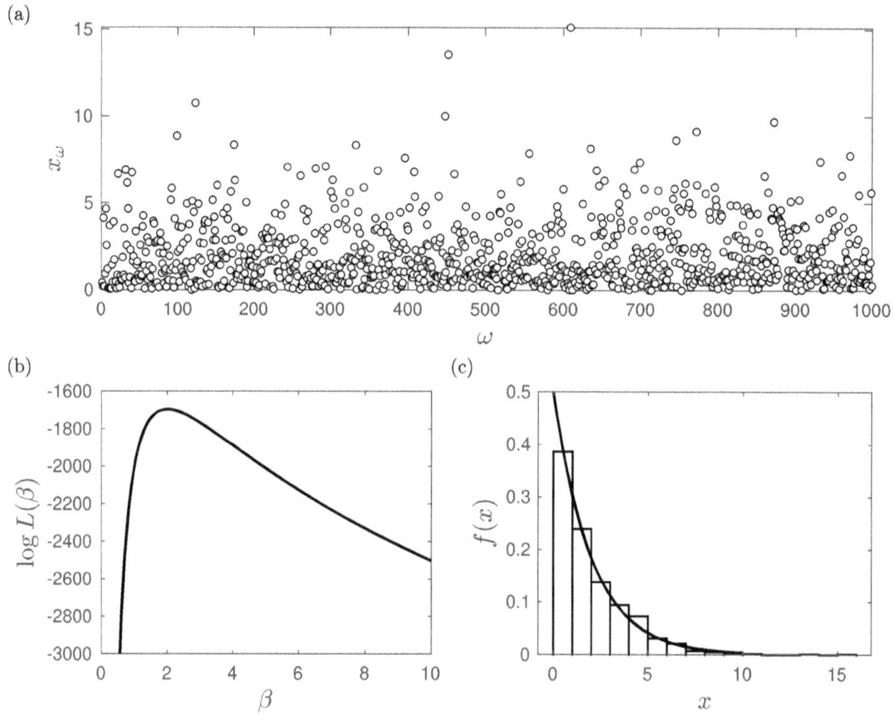

Figure 4.4 Observations (a), log-likelihood function (b), and fit of pdf of Exp(2) to empirical pdf (c).

Recall that the pdf of the exponential RV is $f(x|\beta) = \frac{1}{\beta}e^{-x/\beta}$ and thus the likelihood function is:

$$L(\beta) = \left(\frac{1}{\beta}e^{-x_1/\beta}\right)\left(\frac{1}{\beta}e^{-x_2/\beta}\right)\cdots\left(\frac{1}{\beta}e^{-x_S/\beta}\right)$$
$$= \frac{1}{\beta^S}\exp\left(-\frac{1}{\beta}\sum_{\omega=1}^{S}x_\omega\right).$$

We now find β that maximizes the log-likelihood $\log L(\beta)$:

$$\max_{\beta}\ \log L(\beta) = -S\log\beta - \frac{1}{\beta}\sum_{\omega=1}^{S}x_\omega.$$

This is equivalent to the minimization of $-\log L(\beta)$:

$$\min_{\beta}\ -\log L(\beta) = S\log\beta + \frac{1}{\beta}\sum_{\omega=1}^{S}x_\omega.$$

The value of β that solves this problem must satisfy:

$$\frac{\partial(-\log L(\beta))}{\partial\beta} = 0.$$

The first derivative can be obtained analytically:

$$\frac{\partial\left(-\log L(\beta)\right)}{\partial \beta}=\frac{S}{\beta}-\frac{1}{\beta^2}\sum_{\omega=1}^{S}x_\omega.$$

We thus find that the solution is:

$$\hat{\beta}=\frac{1}{S}\sum_{\omega=1}^{S}x_\omega.$$

This reveals an interesting and important insight; MLE tells us that the most likely parameter value $\hat{\beta}$ is the empirical mean of the observations $\hat{\mathbb{E}}_X=\frac{1}{S}\sum_{\omega=1}^{S}x_\omega$. This provides justification that, for the exponential model, it is appropriate to use the empirical mean as the estimate of β (as this is its most likely value). Note that this also makes sense because the exponential model has the property that $\beta=\mathbb{E}_X$.

From the data provided, we find that the value of the estimated parameter $\hat{\beta}=\hat{\mathbb{E}}_X=2$. From Figure 4.4, we note that this is precisely the location where $\log L(\beta)$ is maximized (where $-\log L(\beta)$ is minimized). From Figure 4.4, we also note that the predicted pdf $\text{Exp}(\hat{\beta})$ matches the empirical pdf well (the exponential model satisfactorily represents the data).

It is important to highlight that the values of the likelihood function $L(\theta)$ are extremely small. Specifically, note that the maximum value of the log likelihood function is $\log L(2)=-1700$, which corresponds to a likelihood value $L(2)=\exp(-1700)\approx 0$. This is because the likelihood function is the product of S pdf values that are *all* quite small in magnitude. For example, note that $f(1|2)=\frac{1}{2}e^{-1/2}=0.3$ and the product of $S=1{,}000$ such values is $0.3^{1{,}000}\approx 0$. This is why making a logarithmic transformation of the likelihood function is necessary.

Script to reproduce results: `ch4_example_mle_exp` □

MLE is a general paradigm for estimating models from data. This approach also provides interesting insights into how parameters of RV models are naturally associated with certain summarizing statistics of models. For example, in the previous example, we showed that the MLE estimate of β for the model $\text{Exp}(\beta)$ is $\hat{\beta}=\hat{\mathbb{E}}_X$. Similarly, one can show that the MLE estimates of μ,σ^2 for the model $\mathcal{N}(\mu,\sigma^2)$ are $\hat{\mu}=\hat{\mathbb{E}}_X$ and $\hat{\sigma}^2=\hat{\mathbb{V}}_X$. Interestingly, one can also show that the MLE estimate for μ,Σ for the multivariate $\mathcal{N}(\mu,\Sigma)$ are $\mu=\hat{\mathbb{E}}_X$ and $\Sigma=\widehat{\text{Cov}}[X]$. This provides some justification behind approaches used to determine parameters for RV models from empirical estimates.

4.3 Sampling and Asymptotic Properties

We have so far assumed that we have data $x_\omega,\ \omega\in\mathcal{S}$ available of a given random phenomenon (characterized by the RV X). However, we have not yet discussed how this data is being collected (how it is being *sampled*).

It turns out that data collection procedures have *key* implications on what we can conclusively say about the underlying random phenomenon of interest. Specifically, we might want to know if the available data captures the inherent randomness of the system and/or if the data is sufficient to obtain accurate empirical approximations and parameter estimates. Moreover, we might want to know how these quantities behave as we accumulate data. For example, if data is collected in a way that the domain of the RV is not properly covered, then the data might not be particularly useful to infer properties of the RV over its entire domain.

We begin our discussion by making the following important definitions:

- The data samples x_ω, $\omega \in \mathcal{S}$ are observations of an underlying random phenomenon X. These observations are collected from the total population Ω of the underlying RV by a defined procedure; in other words, sampling is a data collection *procedure*. Think that you have a population of pinballs of different sizes; you can sample pinballs by using a random procedure or by using some selection logic (e.g., sample only the large ones).
- The data samples x_ω, $\omega \in \mathcal{S}$ are called a *random sequence* if each sample x_ω is drawn from the *same* underlying phenomenon X (with associated pdf f_X) and if it is selected *independently* from the others. If this is the case, the samples are said to be *independent and identically distributed* (i.i.d). For instance, if you select a sequence of pinballs from your population randomly, then your sequence is random.
- If a sample x_ω is selected at random, then the *sample itself is an RV*. This is because, if we repeat the random selection process, we might get a different outcome. To highlight this fact, one typically writes the sample as X_ω, $\omega \in \mathcal{S}$; here, each X_ω has the same underlying pdf f_X (because it is sampled from the same system).
- A random sample X_ω has the *same probability* $1/S$ of being selected as any other sample. This guarantees that the sample is *unbiased*. The lack of bias indicates that $\mathbb{E}[X_\omega] = \mathbb{E}[X]$; in other words, drawing a sample several times and averaging the results gives the same expected value of the underlying RV X. A biased sample is one in which there is a higher probability of selecting some elements of the population over others (there is favoring/biasing toward some elements). Note, therefore, that random sampling has a flavor of *fairness*. For instance, if you select a sequence of pinballs randomly, you can anticipate that the mean size of your sequence is a reflection of the mean size of the entire population. On the other hand, if you only sample the pinballs of large sizes, the mean size of your sequence is not a reflection

of the mean of the population. Moreover, by selecting only the large pinballs, you are introducing a bias (you are not giving equal chance to all pinballs of being selected); your selection process can thus be deemed unfair.

Sample *randomness* is essential to obtain *empirical approximations* and *parameter estimates* with desirable asymptotic properties; specifically, we want that these quantities become increasingly accurate as we accumulate data and converge to the true values of the underlying RV X. Sample randomness is the basis of MC (used to obtain empirical approximations of the cdf/pdf and of summarizing statistics) and of MLE (used to obtain parameter estimates).

Collecting samples (data) at random is not as easy as it sounds. Specifically, one must ensure that there is no bias in selecting a sample (i.e., there is no *hidden mechanism or logic* when selecting a sample). For instance, there are advanced algorithms available in numerical packages that aim to generate random samples from an RV model; the development of such algorithms is an important area of statistics.

Humans are notoriously bad at collecting data at random; specifically, humans have the tendency to select samples in an "ordered" or "logical" manner, and this naturally introduces biases. On the other hand, there is also an element of *efficiency* that arises here because collecting data in an ordered/logical manner (as opposed that at random) tends to decrease the amount of data needed. For example, we can populate a domain $\mathcal{D}_X$ more effectively by placing samples *strategically* in a way that we cover the domain (than by placing the samples at random). Efficient data collection procedures aim to achieve a small variance of empirical estimates, while unbiased procedures aim to obtain estimates that are close to the actual mean. Designing data collection procedures that are both efficient and unbiased is challenging because these are conflicting goals.

The topics of data collection and sampling are also the basis of an important area of statistics known as *experimental design*. The goal of experimental design is to systematically collect data in a way that model parameters are unbiased and have a small variance (uncertainty). We will discuss this in Chapters 5–7.

Example 4.5 Random and Nonrandom Sample Sequences

In Figure 4.5, we show a random and nonrandom sample sequence and their corresponding averages. The random sequence $X_\omega, \omega \in \mathcal{S}$ is generated by using an algorithm that generates random samples for the model $X \sim \mathcal{N}(0,1)$. We confirm that the samples generated by the algorithm distribute evenly around the mean of X ($\mathbb{E}[X] = 0$). For this sequence, we find that the bias is $\frac{1}{S}\sum_{\omega \in \mathcal{S}} X_\omega - \mathbb{E}[X] \approx 0$.

The nonrandom sequence $X_\omega, \omega \in \mathcal{S}$ is generated by drawing a random sample X_ω from $\mathcal{N}(0,1)$ and by discarding this sample if $X_\omega \geq 1$. Note that this logic introduces a

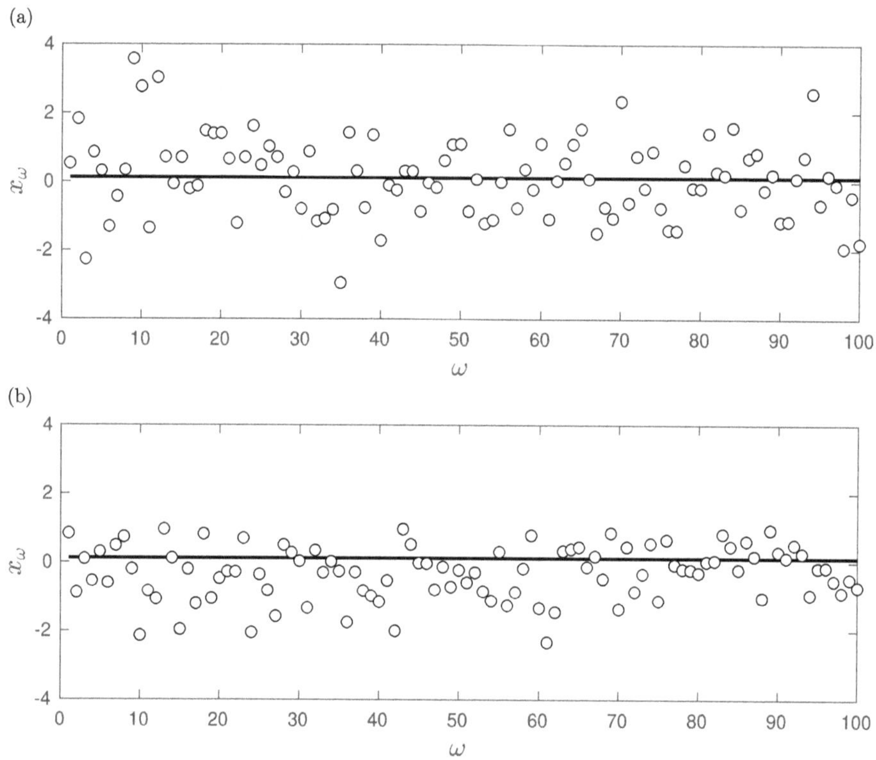

Figure 4.5 Random (a) and nonrandom (b) sequences for $\mathcal{N}(0,1)$.

selection/discrimination mechanism that biases the sample. This can be confirmed by visualizing the sequence; the selected samples do not distribute evenly around $\mathbb{E}[X] = 0$. For this sample sequence, we find that the bias is $\frac{1}{S}\sum_{\omega\in\mathcal{S}} X_\omega - \mathbb{E}[X] = -0.43$ (shifted toward negative values).

Script to reproduce results: `ch4_random_sample` □

4.3.1 Monte Carlo Approximations

Monte Carlo simulation is a computational technique that uses *random samples* to estimate summarizing statistics of an RV X and derived quantities (e.g., functions of X). As discussed in Chapter 1, MC is widely used to analyze how uncertainty propagates through systems. In Chapter 7, we will see that MC also plays a key role in decision-making.

Consider a random sample x_ω, $\omega \in \mathcal{S}$ of X. MC uses the samples to compute a sample (empirical) approximation of an *expectation* function. This approximation can be stated in general terms as:

$$\hat{\mathbb{E}}^S[\phi(X)] = \frac{1}{S}\sum_{\omega\in\mathcal{S}} \phi(x_\omega) \approx \mathbb{E}[\phi(X)], \tag{4.10}$$

where $\phi(X)$ is a vector-valued function of X.

We recall that the true (theoretical) expectation of a function of a continuous RV is given by:

$$\mathbb{E}[\phi(X)] = \int_{x \in \mathcal{D}_X} \phi(x) f_X(x) dx, \tag{4.11}$$

where f_X is the joint pdf of X. From this, we see that computing the expectation requires that we know f_X and involves computing an integral (a multidimensional integral for multivariate X). Integration cannot be carried out analytically in most applications. MC provides a numerical technique to approximate such an integral by using random samples of X that cover the domain $\mathcal{D}_X$. This approach differs from other numerical techniques to compute integrals (e.g., quadrature rules). Specifically, such approaches systematically place samples (called support points) in $\mathcal{D}_X$ to minimize the approximation error. The quadrature approach is efficient but uses biased samples; the presence of bias makes it difficult to infer properties of the approximation error. This is important because we will not know the actual (theoretical) value of the actual expectation. As we will see shortly, the use of random samples in MC allows us to infer properties of the approximation error; specifically, we can guarantee that the error decreases as we increase the sample size.

Another key benefit of MC is that it can be used *with or without having knowledge of the pdf* f_X. Specifically, if we do not have a pdf f_X (we do not have a model for X), we can directly use data $x_\omega, \omega \in \mathcal{S}$ collected from the system to compute the MC approximation. If we know the pdf f_X (e.g., we know that X is Gaussian), we can draw random samples $x_\omega, \omega \in \mathcal{S}$ from the RV model to compute the MC approximation.

We recall that the expectation of a discrete RV is given by:

$$\mathbb{E}[\phi(X)] = \sum_{x \in \mathcal{D}_X} \phi(x) f_X(x). \tag{4.12}$$

Note that this operation is easy to carry out if we know the pdf f_X (we have a model). In the absence of a pdf, however, we can still use data $x_\omega, \omega \in \mathcal{S}$ to approximate the expectation via MC.

Many quantities of interest can be expressed as expectation operations and thus can be computed via MC. Some examples include:

- Expectation of X: $\hat{\mathbb{E}}_X^S = \frac{1}{S} \sum_{\omega \in \mathcal{S}} x_\omega \approx \mathbb{E}[X]$,
- Variance of X: $\hat{\mathbb{V}}_X^S = \frac{1}{S} \sum_{\omega \in \mathcal{S}} (x_\omega - \hat{\mathbb{E}}_X^S)^2 \approx \mathbb{V}[X]$,
- CDF of X: $\hat{F}_X^S(x) = \frac{1}{S} \sum_{\omega \in \mathcal{S}} \mathbf{1}[x_\omega \le x] \approx F_X(x)$,
- Expectation of function $\varphi(X,u)$: $\hat{\mathbb{E}}_\varphi^S = \frac{1}{S} \sum_{\omega \in \mathcal{S}} \varphi(x_\omega, u) \approx \mathbb{E}[\varphi(X,u)]$.

Here, we make the dependence of the empirical approximations on the number of samples S explicit in the notation. Empirical approximations are also often called *data-driven approximations* (because we use data to compute them) and *sample average approximations* (because the expectation is approximated using a sample average).

The empirical approximation of the CDF uses the key property:

$$\mathbb{P}(X \leq x) = \mathbb{E}[\mathbf{1}[X \leq x]]. \tag{4.13}$$

In fact, more generally, one can show that:

$$\mathbb{P}(X \in \mathcal{A}) = \mathbb{E}[\mathbf{1}[X \in \mathcal{A}]], \tag{4.14}$$

where $\mathcal{A} \subseteq \mathcal{D}_X$ is an event. For a continuous RV, this property arises from the observation that:

$$\begin{aligned} \mathbb{P}(X \in \mathcal{A}) &= \int_{x \in \mathcal{A}} x \cdot f_X(x)dx \\ &= \int_{x \in \mathcal{D}_X} \mathbf{1}[x \in \mathcal{A}] \cdot f_X(x)dx = \mathbb{E}[\mathbf{1}[X \in \mathcal{A}]]. \end{aligned} \tag{4.15}$$

For a discrete RV, we simply replace the integral with summation. We thus see that we can use MC to compute approximations for probabilities of diverse events of interest (see Chapter 3 for a detailed discussion of event probabilities).

Certain summarizing statistics such as the median cannot be expressed directly as an expectation operation; however, one can compute them indirectly via an optimization problem that is cast in terms of an expectation operator. For example, one can show that the median $\mathbb{M}[X]$ is the solution of the optimization problem:

$$\mathbb{M}[X] \in \underset{t}{\operatorname{argmin}}\ \mathbb{E}[|X - t|], \tag{4.16}$$

where $|X - t|$ is the absolute value of $X - t$. We can thus use MC to compute an empirical approximation as:

$$\hat{\mathbb{M}}_X^S \in \underset{t}{\operatorname{argmin}}\ \frac{1}{S} \sum_{\omega \in \mathcal{S}} [|x_\omega - t|]. \tag{4.17}$$

Similarly, the expectation $\mathbb{E}[X]$ is the solution of the optimization problem:

$$\mathbb{E}[X] \in \underset{t}{\operatorname{argmin}}\ \mathbb{E}[(X - t)^2]. \tag{4.18}$$

This property is interesting but not necessarily of practical relevance because we can compute the expectation directly using MC.

The quantile $\mathbb{Q}_X(\alpha)$ is the solution of the optimization problem:

$$\begin{aligned} \mathbb{Q}_X(\alpha) \in \underset{t}{\operatorname{argmin}}\ & t \\ \text{s.t. } & \mathbb{P}(X \leq t) \geq \alpha. \end{aligned} \tag{4.19}$$

Because we have that $\mathbb{P}(X \leq t) = \mathbb{E}[\mathbf{1}[X \leq t]]$, the problem can also be solved using MC approximations.

The previous formulations reveal that we can use MC to solve optimization problems that involve expectation operations. For example, imagine that we want to find a decision u that minimizes the expected cost of a system $\mathbb{E}[\varphi(X,u)]$. This can be done by solving the problem:

$$\min_{u} \mathbb{E}[\varphi(X,u)]. \tag{4.20}$$

Because event probabilities can be written as expectations, we can also formulate a problem that minimizes $\mathbb{P}(\varphi(X,u) > t)$ (probability that the cost is higher than a value t). These types of problems will be the basis of decision-making strategies that seek to minimize risk and will be covered in Chapter 7. All the previous observations highlight the fact that MC is a flexible and important statistical technique.

Now that we have convinced ourselves that MC is a versatile technique, we proceed to ask the following questions related to MC approximations:

- Do MC approximations become exact as we accumulate data?
- How accurate are the approximations when we have limited amounts of data?
- How much data do we need to obtain a desired accuracy?
- How stable/reliable are our approximations when we change our data?

We will now explore important tools to answer these questions. The use of these tools is enabled by the fact that samples are collected at random. Because, as we have seen, many quantities of interest use an expectation operation, we restrict our discussion to the behavior of the estimate $\hat{\mathbb{E}}_X^S \approx \mathbb{E}[X]$. We will see that $\hat{\mathbb{E}}_X^S$ indeed converges to $\mathbb{E}[X]$ as S increases. We will also see that the estimate $\hat{\mathbb{E}}_X^S$ exhibits Gaussian-like behavior for *all RVs* as we accumulate data (i.e., as $S \to \infty$); this will provide the basis to quantify the quality of the estimate and to understand how much data we actually need to yield stable estimates. To simplify the discussion, we focus our attention on the behavior of univariate X (these concepts generalize to the multivariate case).

4.3.2 Law of Large Numbers

We first assess how the estimate $\hat{\mathbb{E}}_X^S$ behaves as $S \to \infty$. To do so, we consider an i.i.d. random sample sequence $X_1, X_2, \ldots, X_S$ for the univariate RV X. Because the samples are i.i.d, they have same underlying pdf f_X with expected value $\mathbb{E}[X]$. Moreover, because the samples are random, they are unbiased and thus $\mathbb{E}[X_1] = \mathbb{E}[X_2] = \cdots = \mathbb{E}[X_S] = \mathbb{E}[X]$.

Consider now the MC approximation (empirical estimate) of $\mathbb{E}[X]$:

$$\hat{\mathbb{E}}_X^S = \frac{1}{S} \sum_{\omega \in \mathcal{S}} X_\omega. \tag{4.21}$$

Because the samples X_ω, $\omega \in \mathcal{S}$ are random, the approximation $\hat{\mathbb{E}}_X^S$ is also random (i.e., different samples give different approximations). This implies that there is inherent uncertainty/variability in the estimate $\hat{\mathbb{E}}_X^S$. The uncertainty captures the *accuracy* of the approximation. The next property (known as the law of large numbers [LLN]) indicates that uncertainty vanishes as the size of our sample S becomes large.

The law of large numbers (LLN) states that:

$$\lim_{S \to \infty} \hat{\mathbb{E}}_X^S = \mathbb{E}[X]. \tag{4.22}$$

The LLN is a fundamental result that guarantees *stable long-run* behavior of sample averages. Intuitively, if the sampling process is random, samples will fluctuate around $\mathbb{E}[X]$, and any discrepancies in our samples will "average out." Conversely, if there is a bias in the way we collected samples, discrepancies will not average out (they will accumulate). Importantly, the LLN indicates that MC approximations become *exact* as $S \to \infty$. This result holds for any RV (e.g., discrete, continuous, univariate, and multivariate). Note, however, that the LLN does not quantify the accuracy of the approximation (just guarantees that the error eventually vanishes).

Example 4.6 Monte Carlo Approximations

We generate random samples X_ω, $\omega \in \mathcal{S}$ from $X \sim$ Weibull(2, 1) and compute MC approximations for $\mathbb{E}[X], \mathbb{E}[\log(X)]$, $\mathbb{V}[\exp(-X) + X^2]$, and $\mathbb{P}(X \leq 1)$ for different values of S.

The results are shown in Figure 4.6. We can see that the approximations exhibit significant variability (they are unstable) when the number of samples is small

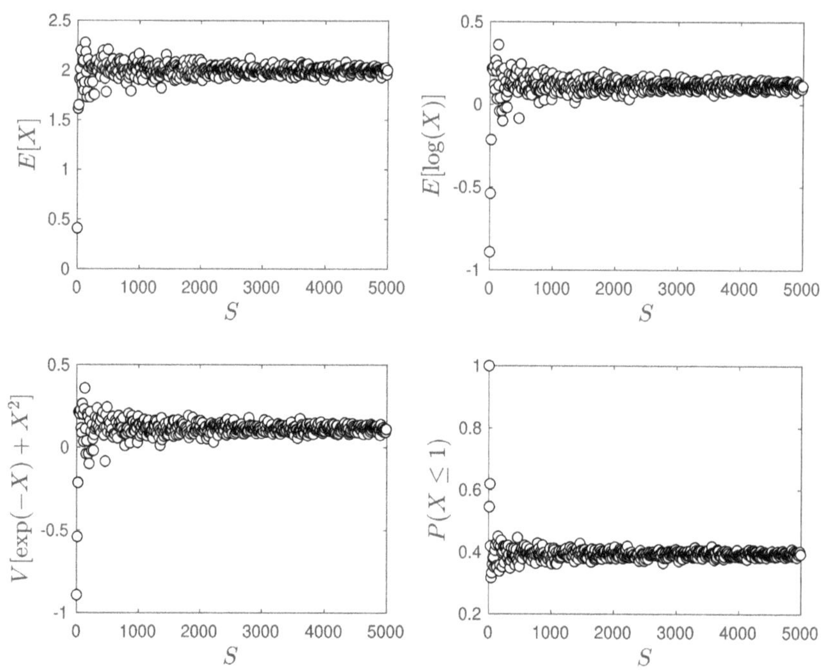

Figure 4.6 Convergence of Monte Carlo approximations as S increases.

(e.g., $S \leq 100$), but the approximations become increasingly accurate as we increase S. This illustrates that MC provides consistent approximations for many quantities of interest. This desirable behavior is enabled by the LLN.

Script to reproduce results: `ch4_montecarlo_approx` □

The results of the previous example highlight that it can be dangerous to use a small number of samples to compute empirical approximations (there is high variability in the approximation). We must ensure that we have enough data to mitigate such variability because our estimates are not reliable/reproducible (they vary from sample to sample). We also notice that, even if we have lots of samples, there is still some variability present. This highlights the fact that the approximation is an RV and thus, we cannot be absolutely certain that the approximation is exact. In other words, we can only provide statistical guarantees on the quality of the approximation. We discuss this in more detail next.

The results of the example can also be interpreted from the perspective of *experiment reproducibility*. Imagine that you run an experiment in a laboratory reactor to determine the yield of a reaction. Every time you run the reactor (even if you run it at the same conditions), you will notice fluctuations in the observed yield. The fluctuations can be the result of an inherent random phenomenon occurring in the reactor that is difficult to control (e.g., equipment vibrations) or due to inherent randomness in the set up of the reactor. In other words, *experiments are not perfectly reproducible*. To mitigate this issue, imagine that you conduct the experiment 100 times to get the mean yield and then repeat the experiment another 100 times to get the mean yield; you will notice discrepancies in the mean yields obtained with the two batches of experiments. Now think what would happen if you conduct the experiment only three times to get the mean yield and then conduct the experiment three more times and get the mean yield; you will likely see larger discrepancies in the mean yields of the two batches of experiments. This is because small sample sizes do not provide a good picture of the actual randomness of your system (while large samples do). The LLN tells you that, when the number of experiments in the batch (the sample size) is sufficiently large, your mean yield will be stable across batches and converge to the true mean.

4.3.3 Central Limit Theorem

Motivated by our previous discussion, we now turn our attention to assessing the accuracy of the sample mean (estimate) $\hat{\mathbb{E}}_X^S$. Because the estimate is an RV, accuracy is measured in terms of the variability/uncertainty of the sample mean. Now, if we want to measure the uncertainty of an RV, we need to know its underlying pdf. Obtaining the pdf of the sample mean $\hat{\mathbb{E}}_X^S$ can achieved by using a powerful result known as the central limit theorem (CLT).

Consider a random sample sequence $X_1, X_2, \ldots, X_S$ that is i.i.d. with *true* expected value $e = \mathbb{E}[X]$ and variance $v^2 = \mathbb{V}[X]$. The samples X_ω are RVs and thus the sample mean $\hat{\mathbb{E}}_X^S = \frac{1}{S}\sum_{\omega \in \mathcal{S}} X_\omega$ is an RV.

The CLT states that:

$$\lim_{S\to\infty} \hat{\mathbb{E}}_X^S \sim \mathcal{N}(e, v^2/S). \tag{4.23}$$

This result is one of the most surprising and useful results in statistics. We highlight the following implications:

- The CLT states that, *regardless* of the underlying nature of X (e.g., Weibull, Poisson, and exponential), the empirical approximation of its expected value (the sample mean $\hat{\mathbb{E}}_X^S$) will behave as a Gaussian RV as S increases. Note that this holds provided that a random sample is used for computing the approximation.
- The CLT states that the variance of the estimator $\mathbb{V}[\hat{\mathbb{E}}_X^S]$ is v^2/S. The standard deviation is $v/\sqrt{S}$. As such, the variance of the sample mean shrinks as we increase the sample size S. In other words, $\hat{\mathbb{E}}_X^S$ becomes more certain (more accurate and more stable) as S increases.
- Since we know that $\hat{\mathbb{E}}_X^S \sim \mathcal{N}(e, v^2/S)$, we have the pdf for $\hat{\mathbb{E}}_X^S$, and we can thus compute quantities of interest for it. For example, we can compute *confidence regions* for the sample mean:

$$\mathbb{P}\left(e - z_{\alpha/2}v/\sqrt{S} \leq \hat{\mathbb{E}}_X^S \leq e + z_{\alpha/2}v/\sqrt{S}\right) = 1 - \alpha. \tag{4.24}$$

Consequently, for a given sample size S, we can determine how confident we are that $\hat{\mathbb{E}}_X^S$ is in a specific region. It is convenient to express the confidence interval as $\hat{\mathbb{E}}_X^S \in [\mathbb{E}_X \pm z_{\alpha/2}v/\sqrt{S}]$, and we note that this is equivalent to:

$$\mathbb{E}_X \in \left[\hat{\mathbb{E}}_X^S \pm z_{\alpha/2}v/\sqrt{S}\right]. \tag{4.25}$$

In words, we can infer that the *true* expected value $\mathbb{E}_X$ is somewhere in the interval $\hat{\mathbb{E}}_X^S \pm z_{\alpha/2}v/\sqrt{S}$ with probability $1 - \alpha$. This interval shrinks as S increases (because the variance shrinks); as such, we become more certain of where $\mathbb{E}_X$ is.

The CLT assumes that we know the variance of X (v^2); specifically, we need to know this to obtain the confidence interval. In practice, one often replaces v^2 with the sample variance:

$$\hat{v}^2 = \frac{1}{S}\sum_{\omega\in\mathcal{S}}(x_\omega - \hat{\mathbb{E}}_X^S)^2. \tag{4.26}$$

The sample variance can also be computed as:

$$\hat{v}^2 = \frac{1}{S-1}\sum_{\omega\in\mathcal{S}}(x_\omega - \hat{\mathbb{E}}_X^S)^2. \tag{4.27}$$

One can show that this is an unbiased estimate of the true variance, while the former is not when S is small. However, these estimates become close when S is large. We will elaborate on this issue shortly.

It is important to note that v^2 is the variance of the RV X, while v^2/S is the variance of the sample mean $\hat{\mathbb{E}}_X^S$. The variance of the sample mean can be estimated by using a *batching* procedure. Specifically, we define a set of $j = 1, \ldots, B$ batches of data (each of size S). For a given batch j, we obtain the sample mean $m_j = \hat{\mathbb{E}}_X^S$. We then use the collected values to estimate the variance

$$v^2/S \approx \frac{1}{B-1}\sum_{j=1}^{B}(m_j - \bar{m})^2, \tag{4.28}$$

where $\bar{m} = \frac{1}{B}\sum_{j=1}^{B} m_j$. From the estimate of v^2/S, we can then back-calculate an estimate of the variance v^2 (because we know S).

Example 4.7 Simulating the Central Limit Theorem

We conduct computer simulations to confirm that the CLT indeed holds. To do this, we draw a batch of S random samples $X_1, X_2, \ldots, X_S$ from $X \sim$ Weibull(2, 1) and compute the sample mean for the batch $\bar{X} = \hat{\mathbb{E}}_X^S = \frac{1}{S}\sum_{\omega=1} X_\omega$. We repeat this for 1,000 different batches of samples (to obtain 1,000 realizations of $\hat{\mathbb{E}}_X^S$) and visualize the empirical pdf of the sample mean $\hat{\mathbb{E}}_X^S$. We repeat this procedure for different sample sizes S to verify that the pdf follows that of a Gaussian as the sample size S increases.

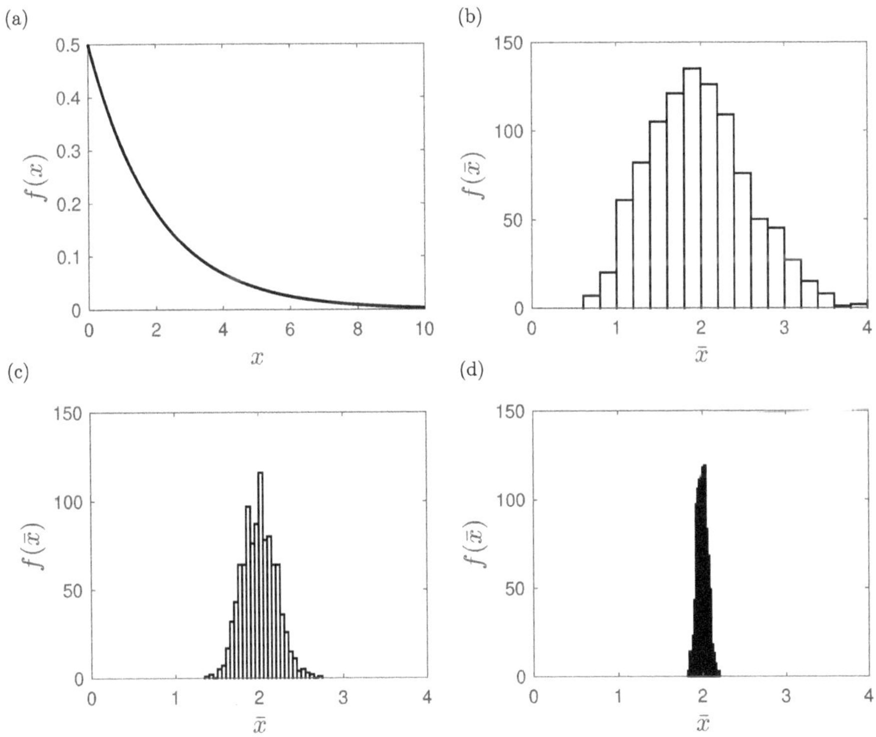

Figure 4.7 Pdf for $X \sim$ Weibull(2, 1) (a) and sample average $\bar{X}$ with $S = 10$ (b), $S = 100$ (c), and $S = 1{,}000$ (d).

In Figure 4.7, we show the empirical pdf of Weibull(2, 1) and of $\bar{X}$ for $S = 10, 100, 1{,}000$ (we express the pdf in terms of frequency). It is clear that the pdf of the sample average $\bar{X}$ becomes increasingly closer to Gaussian and that the uncertainty narrows down. To verify this, we fit the empirical pdf of $\hat{\mathbb{E}}_X$ for $S = 1{,}000$ to a Gaussian $\mathcal{N}(\mu, \sigma^2/S)$ and find that $\mu = 1.99$, $\sigma^2/S = 0.0038$, and $\sigma^2 = 3.83$. For a Weibull(2, 1), we know that $\mathbb{E}[X] = e = 2$ and $\mathbb{V}[X] = v^2 = 4$.

We can thus confirm that the sample average and the variance coincide with that of the RV Weibull(2, 1). The results also highlight that the estimate becomes more certain as we increase the sample size; specifically, for $S = 1{,}000$, the variance of the sample average is small (0.0038). We also used a batching process to estimate the variance of the sample average and found $v^2/S \approx 0.0038$ from which we backcalculate an estimate of the variance $v^2 \approx 3.83$; the estimate coincides with that obtained from the Gaussian fit.

Finally, we highlight that the CLT holds even if the underlying RV X is not a Gaussian. We will see in Chapter 6 that this can be explained using a mathematical tool known as convolution.

Script to reproduce results: `ch4_clt_example` □

4.3.4 Extreme Value Theorem

In the CLT, we are interested in the limiting behavior of the *sample mean* $\hat{\mathbb{E}}_X^S = \frac{1}{S}\sum_{\omega \in \mathcal{S}} X_\omega$. But, what if we are interested in a different summarizing statistic? Specifically, consider the *sample maximum*:

$$X_{max}^S = \max\{X_1, X_2, \cdots, X_S\}. \tag{4.29}$$

There is another important result in statistics that characterizes the pdf of X_{max}^S as $S \to \infty$. The result is known as the *extreme value theorem* (EVT).

Consider, as before, an i.i.d. sequence $X_1, X_2, \ldots, X_S$ for RV X. The EVT states that:

$$\lim_{S\to\infty} X_{max}^S \sim \text{GEV}(a, b, c), \tag{4.30}$$

where $GEV(a, b, c)$ is known as the generalized extreme value (GEV) RV with parameters a, b, and c. The GEV RV has a pdf of the form:

$$f_{X_{max}}(x) = \begin{cases} (1 + cs)^{(-1/c)-1}\exp(-(1 + cs)^{-1/c}) & c \neq 0 \\ \exp(-s)\exp(-\exp(-s)) & c = 0, \end{cases} \tag{4.31}$$

and a cdf of the form:

$$F_{X_{max}}(x) = \begin{cases} \exp(-(1 + cs)^{-1/c}) & c \neq 0 \\ \exp(-\exp(-s)) & c = 0, \end{cases} \tag{4.32}$$

where $s = (x - a)/b$ is a standardized variable. The GEV is a general RV that includes the reverse Weibull model (for $c < 0$), the Fréchet model (for $c > 0$), and the Gumbel model (for $c = 0$). The reverse Weibull RV is a variant of the Weibull model that we explored in Chapter 2.

The GEV RV is widely used in failure and safety analysis because the sample maximum statistic characterizes *peak (extreme) events*. For example, imagine that you want to characterize the peak/largest demand of electricity of a city. This is relevant because such events often have small probabilities (are *rare* events) but can have catastrophic effects (e.g., blackouts). The GEV thus gives a mechanism to estimate probabilities for rare events.

As with the CLT, the EVT does not depend on the underlying nature of X (this is a general result). The CLT and EVT are powerful results that are used in many branches of science and engineering because they allow us to infer key properties of an RV X (e.g., means and peaks) without needing a full characterization of X (without knowing its pdf).

As we did with the sample average and the CLT, we can also infer properties of the sample max using the EVT. For example, if we know the cdf $F_{X_{max}}(x)$, we can determine the probability that the peak value in a sequence (X^S_{max}) is above or below a certain threshold value.

Example 4.8 Simulating the Extreme Value Theorem

Our goal now is to use computer simulations to verify that the EVT holds. We collect random samples $X_1, X_2 \ldots, X_S$ for $X \sim \mathcal{N}(2,1)$ and compute the peak (largest value) of the batch $X^S_{max} = \max\{X_1, X_2, \cdots, X_S\}$. As X^S_{max} is an RV, we repeat this procedure for different batches of samples to obtain realizations of X^S_{max} and visualize its empirical pdf. We do this for different sample sizes S to verify that the pdf follows that of a GEV as we increase S.

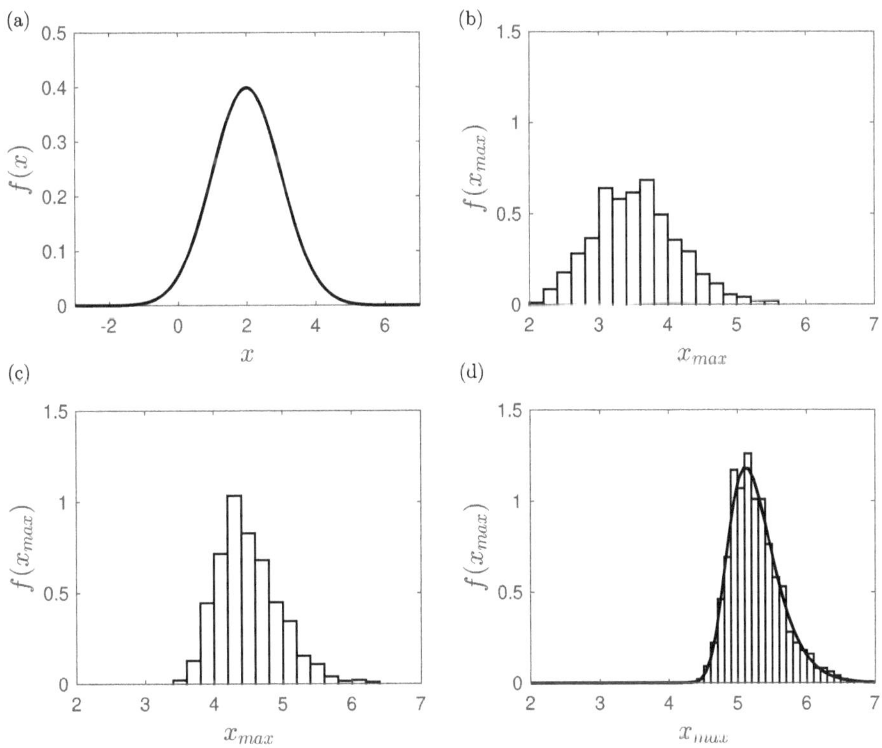

Figure 4.8 Pdf for $X \sim \mathcal{N}(2,1)$ (a) and sample max X^S_{max} with $S = 10$ (b), $S = 100$ (c), and $S = 1{,}000$ (d).

In Figure 4.8, we present the pdf of $X \sim \mathcal{N}(2,1)$ and the empirical pdf for X^S_{max} for $S = 10, 100, 1{,}000$. We can see that the pdf of the sample maximum converges to that of a GEV. To confirm this, for $S = 1{,}000$, we fit the empirical pdf to a GEV and find that the parameters are $a = 5.1, b = 0.31$, and $c = -0.68$. We can see that the GEV matches the empirical pdf quite well. Because $c < 0$, we have that the GEV is a reverse Weibull RV.

The GEV obtained also tells that the extreme (peak) values of $X \sim \mathcal{N}(2,1)$ are in the range of $[4,7]$ (this is consistent with the right tail of the Gaussian). Specifically, we can use the cdf to compute the probability that the peak in a sequence (X^S_{max}) has a value above a given threshold. For example, we have that $\mathbb{P}(X^S_{max} > 5) = 0.7444$ and $\mathbb{P}(X^S_{max} > 7) = 3.6 \times 10^{-4}$. Consequently, given a random sequence of $S = 1{,}000$ elements, it is quite likely that the peak value in such a sequence is above 5, but it is quite unlikely that the peak value is above 7. Note that we can infer these quantities without knowing the underlying nature of X.

Script to reproduce results: `ch4_evt_example` □

4.3.5 Bias and Efficiency

The MC approximation (sample mean) $\hat{\mathbb{E}}^S_X$ is an *estimator* of the statistic $\mathbb{E}_X$. Importantly, because the samples $X_\omega, \omega \in \mathcal{S}$ used by the MC estimator are random (they have the same probability $1/S$ and therefore are unbiased), we can show that the sample $\hat{\mathbb{E}}_X$ is an *unbiased estimator* of $\mathbb{E}_X$.

An unbiased estimator is desirable because it guarantees that, as we change our samples, the approximation error $\hat{\mathbb{E}}^S_X - \mathbb{E}_X$ will gravitate around zero and average out (regardless of the number of samples S). In other words, the estimator does not contain a systematic error.

To establish that the MC estimator is indeed unbiased, recall that $\hat{\mathbb{E}}^S_X$ is an RV and that $\mathbb{E}_X$ is a number (we define $e = \mathbb{E}_X$ to make this clear). The estimator bias is defined as:

$$\text{Bias}(\hat{\mathbb{E}}^S_X) = \mathbb{E}[\hat{\mathbb{E}}^S_X - e]. \tag{4.33}$$

The estimator is thus unbiased if $\text{Bias}(\hat{\mathbb{E}}^S_X) = 0$. This can also be written as $\mathbb{E}[\hat{\mathbb{E}}^S_X - e] = 0$ or as $\mathbb{E}[\hat{\mathbb{E}}_X] = e$. In other words, the expected value of the sample mean is the true expected value (the sample mean gravitates around the true expected value).

We can prove that the MC estimator is unbiased by noticing that:

$$\begin{aligned} \mathbb{E}[\hat{\mathbb{E}}^S_X] &= \mathbb{E}\left[\frac{1}{S}\sum_{\omega\in\mathcal{S}} X_\omega\right] \\ &= \frac{1}{S}\sum_{\omega\in\mathcal{S}} \mathbb{E}\left[X_\omega\right]. \end{aligned} \tag{4.34}$$

Because the samples are unbiased, we have $\mathbb{E}[X_\omega] = \mathbb{E}_X = e$ and thus:

$$\mathbb{E}[\hat{\mathbb{E}}^S_X] = \frac{1}{S}\sum_{\omega\in\mathcal{S}} e = e. \tag{4.35}$$

We thus conclude that $\hat{\mathbb{E}}_X$ is an unbiased estimator of $\mathbb{E}_X$; the estimator is unbiased *regardless* of the sample size S.

The MC approximation $\hat{\mathbb{E}}^S[\phi(X)]$ is an unbiased estimator of $\mathbb{E}[\phi(X)]$ for any function $\phi(X)$ and for any sample size S. Note, however, that the variance involves a nested expectation. As a result, the sample variance $\hat{\mathbb{V}}_X^S$ is not an unbiased estimator of $\mathbb{V}_X$ for any sample size S. To ensure unbiasedness, the estimator needs to be corrected as done in Eq. (4.27), where we replace S for $S-1$ in the denominator. The effect of this correction disappears as S becomes large but highlights that certain MC estimators might be biased.

Another important issue related to estimators is that of *efficiency*. Efficiency is concerned with the number of samples needed to obtain an estimate with a given accuracy (precision). Estimator accuracy is measured in terms of the variance of the estimator

$$\mathrm{Eff}(\hat{\mathbb{E}}_X^S) = \mathbb{V}[\hat{\mathbb{E}}_X^S], \tag{4.36}$$

while bias is measured in terms of the mean. The variance of an estimator is also known as the *precision*.

For an MC estimator, one can show that the estimator variance is:

$$\mathbb{V}[\hat{\mathbb{E}}_X^S] = \mathbb{V}\left[\frac{1}{S}\sum_{\omega\in\mathcal{S}} X_\omega\right] = \frac{\mathbb{V}_X}{S} = \frac{v^2}{S}. \tag{4.37}$$

This reveals that the estimator variance is proportional to the inherent variance of X and that it decreases with the sample size S (as predicted also by the CLT). Specifically, the estimator precision increases with S. This desirable behavior also results from the fact that the MC estimator uses random samples. However, one might be interested in decreasing variance at a faster rate to reduce the amount of data needed. To do so, one might consider *variance-reduction sampling* techniques such as Latin hypercube sampling and importance sampling. These techniques use samples that are strategically chosen to improve precision and give more efficient estimators than those obtained with random samples. We say that an estimator $\hat{\mathbb{E}}_X^S$ is more efficient than another estimator $\tilde{\mathbb{E}}_X^S$ if $\mathbb{V}[\hat{\mathbb{E}}_X^S] < \mathbb{V}[\tilde{\mathbb{E}}_X^S]$ (for the same number of samples S). Alternatively, we say that an estimator is more efficient if it requires less samples to achieve the same precision (same variance).

Unfortunately, variance-reduction techniques tend to give biased estimators (while random samples give zero bias). Consequently, estimators often have a fundamental *efficiency-bias conflict or trade-off* (mean-variance conflict). A measure that captures this inherent conflict is the *mean squared error* (MSE), which is given by:

$$\mathrm{MSE}(\hat{\mathbb{E}}_X^S) = \mathbb{E}[(\hat{\mathbb{E}}_X^S - \mathbb{E}_X)^2]. \tag{4.38}$$

Specifically, one can show that:

$$\mathrm{MSE}(\hat{\mathbb{E}}_X^S) = \mathrm{Bias}(\hat{\mathbb{E}}_X^S)^2 + \mathrm{Eff}(\hat{\mathbb{E}}_X^S). \tag{4.39}$$

We can thus use MSE to simultaneously compare the bias and efficiency of estimators. Specifically, we select an estimator if $\mathrm{MSE}(\hat{\mathbb{E}}_X^S) \leq \mathrm{MSE}(\tilde{\mathbb{E}}_X^S)$.

4.4 Statistical Inference

Statistical inference is a set of techniques that aims to draw conclusions about a population based on sample data. Hypothesis testing is a specific set of techniques that aims to determine if there is enough evidence in our sample data to draw conclusions about the population. This set of techniques are key to determine, for instance, if a postulated (hypothesized) model fits available data or if different sets of data come from the same model. So far, we have been conducting informal comparisons between our data and empirical estimates to theoretical counterparts; our goal is now to discuss techniques that we can use to conclusively make these comparisons.

The area of hypothesis testing involves many techniques; at the core, however, all of these techniques have the same elements. Specifically, to use a technique we first need to postulate what is known as a *null hypothesis*, which we will aim to reject or not based on available data (evidence). Another element of these tests is an *alternative hypothesis* (mutually exclusive with the null hypothesis). Every test also uses a so-called *statistic* (a measure of deviation) and a *p-value* (a measure of statistical significance of such deviation). The p-value is compared against a specified *significance level* to obtain a final outcome of the test. A hypothesis test has a couple of possible outcomes: (i) fails to reject the null hypothesis in favor of the alternative and (ii) rejects the null hypothesis in favor of the alternative.

To illustrate some uses of hypothesis tests, we focus on a couple of important ones: the z-test and the Kolmogorov–Smirnov (KS) test.

In the z-test, we determine if a given sample $x_\omega,\ \omega \in \mathcal{S}$ comes from a postulated model given by a normal RV with given mean and variance μ, σ. The test statistic for the z-test is the z-score:

$$z = \frac{\bar{x} - \mu}{\sigma / \sqrt{S}}, \tag{4.40}$$

where $\bar{x}$ is the sample mean and S is the sample size. The null hypothesis of the z-test is that the sample comes from the postulated normal RV. This is done by determining how far the sample mean is to the true mean of the postulated model. If the difference is large, then we can anticipate that the null hypothesis will be rejected (the sample data does not come from the postulated model); on the other hand, if the difference is small, we can anticipate that we cannot reject the null hypothesis. Note that the z-score uses the variance of the sample mean $\bar{x}$, which is given by σ^2/S (from the CLT).

The alternative hypothesis of the z-test can take different forms; for instance, it can be that the sample mean is not the true mean, that the sample mean is greater than the true mean, and that the sample mean is lower than the true mean. These alternative hypotheses are useful when conducting different types of comparisons (i.e., we compare if there is enough evidence to reject or not the null against the alternative). It is important to note that a "successful" test does not mean that we accept the null hypothesis; a successful test only fails to reject the null hypothesis. In other words, our data does not provide enough evidence to reject the null hypothesis (but this does not necessarily imply that the null hypothesis is true). We also highlight that the z-test requires us to know the variance σ^2 of the model; when this is not available, one can use the so-called t-test.

In the KS test, we determine if a given sample $x_\omega,\ \omega \in \mathcal{S}$ follows a postulated model. More specifically, we aim to determine if the empirical cdf of our data follows the theoretical cdf of a specified model. As such, the KS test uses a test statistic that captures the maximum distance (error) between the empirical and theoretical cdf:

$$d = \max_{x} |\hat{F}(x) - F(x)|. \tag{4.41}$$

The use of this distance as a measure of error provides a justification for the LS formulation discussed in Section 4.1.2 (which aims to minimize the distance between the empirical and model cdf).

The null hypothesis of the KS test is that the sample comes from the postulated model. As in the z-test, there are several alternative hypothesis that the null can be benchmarked against. For instance, we can have the alternative hypothesis that the sample cdf is larger (or smaller) than the postulated cdf. The KS test can be used to determine if our data fits a specific model well enough (the KS test can be used as a "goodness-of-fit" test). For instance, imagine that we use data $x_\omega,\ \omega \in \mathcal{S}$ to determine the parameters $\hat{\theta}$ to obtain a cdf model $F(x|\hat{\theta})$. To determine if the fit is sufficiently good, we can conduct a KS test using $F(x|\hat{\theta})$ as the postulated model; the KS test will aim to determine if the data is generated from the model. It is important to note that the KS test only applies to continuous models.

There are several variants of the KS test; for instance, the two-sample KS test determines whether two samples $x_\omega,\ \omega \in \mathcal{S}$ and $x'_\omega,\ \omega \in \mathcal{S}$ come from the same RV model. This test will aim to measure the distance between the empirical cdfs of the two samples. The Lilliefors test is a special type of KS test that determines whether a sample $x_\omega,\ \omega \in \mathcal{S}$ comes from a normal model; this test first fits the data to a normal model (by estimating the parameters).

Example 4.9 Hypothesis Testing

We obtain a random sample from $\mathcal{N}(2,1)$ with $S = 100$ and use the z-test and the KS test to determine if the data comes from a wrong postulated model $\mathcal{N}(1,1)$. We also use the Lilliefors test to determine if the random sample comes from a normal distribution.

Both the z-test and KS test reject the null hypothesis (that the data comes from $\mathcal{N}(1,1)$). As such, these tests seem to identify that the data comes from a different model than the postulated one. The Lilliefors test fails to reject the null hypothesis;

this is because this test first fits the data to a normal model and then conducts the test. From this, we can infer that the data likely originates from a normal model.

Script to reproduce results: `ch4_example_hypothesis` □

4.5 Summary

In this chapter we have discussed techniques that help us estimate parameters and summarizing statistics for RVs from data. Some of the important take-home messages are the following:

- Parameters for RV models can be estimated from data by using different techniques. Point estimation techniques aim to match statistics of RV models to empirical counterparts obtained from data. Moment-matching techniques are easy-to-use and intuitive but do not provide optimality guarantees. The LS method is a flexible optimization-based method that can be used to find the best set of parameters that fit the data. The use of optimization provides a guarantee that there is no better set of parameters; however, the LS approach does not have a natural statistical interpretation.
- Maximum likelihood estimation provides a general paradigm to find the most likely set of parameters from data. This approach provides optimality guarantees and has a statistical interpretation. In addition, this approach provides important insights into what are the best summarizing statistics to use to obtain estimates of parameters.
- Both point estimation and maximum likelihood ultimately relate model parameters to empirical quantities of an RV that are obtained from data; as such, it is important to understand how to collect data and to determine how much data is needed to obtain reliable parameter estimates.
- One can obtain empirical estimates of summarizing statistics and of cdf/pdf by using MC. At the core, this approach uses realizations (either data or random samples obtained from a model) to obtain a sample-based estimate of an expectation operation (known as the sample average or sample mean). The MC approach is powerful because it can use data directly, without having an RV model, to obtain estimates. Moreover, when a model is available, the approach can be used to bypass the need to compute complex integrals. We showed that different quantities of interest (e.g., cdf, probabilities of events, and the median) can be expressed in terms of an expectation operation, making MC a highly flexible approach.
- MC relies on the key property that realizations are collected at random; this guarantees that the MC estimate converges to the theoretical counterpart as we accumulate data (as established by the LLN). In addition, the use of random samples guarantees that the sample mean follows a Gaussian model (as established by the CLT); this result allows us to quantify the precision of the estimate and to determine how much data is needed to obtain desired levels of precision.

- The use of random samples also guarantees that the MC estimate is unbiased (gravitates around the true value). In addition to unbiasedness, we have discussed the need for estimators to be efficient (to obtain good approximations with small amounts of data). We have seen that obtaining estimators that are both unbiased and efficient is challenging due to inherent trade-offs that appear between these goals.
- The generation of random samples for MC is not as easy as it sounds; there are advanced algorithms developed that try to obtain these types of samples. Humans are notoriously bad at collecting data at random, as we typically follow hidden procedures/logic to collect data; such procedures introduce biases (a form of discrimination or unfairness) and can prevent us to obtain statistical guarantees for our estimates.
- We discussed hypothesis testing, which is an important statistical inference technique that can help us determine if available data is sufficient to conclude if a phenomenon follows a specific type of RV model. The KS test, for instance, can be used to determine if a dataset was generated from a specified RV model; to do so, the KS approach measures the distance between the empirical cdf and the model cdf. There are a broad range of hypothesis tests available for different types of models and for different types of metrics.

4.6 Exercises

Exercise 4.1. Adhesive Failure.
You run 100 experiments for a new adhesive material to determine its failure time. The outcomes of the experiments are recorded in the data file `failure_data_mle.dat` (in months). It is known that the failure time follows an exponential distribution.

(a) Determine the maximum likelihood estimate (MLE) for the parameter of the exponential model. Confirm the fit by comparing the empirical and theoretical pdf and cdf.
(b) Use the CLT to determine the 95% confidence interval for the MLE.
(c) Determine the precision of the estimate (its standard deviation). Comment whether such a precision is satisfactory (compare precision to estimate value).
(d) Determine the number of experiments needed to obtain a precision that is 1% of the value of your estimate.

Exercise 4.2. Building Confidence in Your Confidence Intervals.
You are skeptical (as you should be) that the formulas for confidence intervals derived by your statistics instructor are correct. You compute the confidence intervals for the RV $\mathcal{N}(1,2)$ at probability levels $1-\alpha = 0.9$ and $1-\alpha = 0.5$ using the formulas provided. To convince yourself that the intervals are correct, you obtain 100 random samples for your RV and count how many samples lie on the confidence intervals.

(a) Determine the confidence intervals and determine the fractions of your samples that land on the intervals.

(b) Compare these fractions to the probability levels of the confidence intervals. Are they the same or at least similar? If not the same, explain why.
(c) Repeat the procedure with 1,000 and 10,000 samples; compare the fraction of samples in the intervals against the probability levels. What do you observe?
(d) Provide a justification (based on the LLN) of why you should expect that the fraction of samples should match the probability levels of the confidence intervals as $S \to \infty$.

Exercise 4.3. Buffering Power Grid Variability.
A highly volatile power signal from the power grid will be used to run an electrochemical manufacturing facility. The facility will be paid by the grid to absorb this signal because this can help the grid mitigate volatility of wind/solar power. The peaks of the electrical signal, however, damage the electrochemical cells; as such, you would like to install a battery that will act as a buffer and absorb these peaks. The power grid has provided historical data for the signal with a resolution of 2 seconds over a period of 100 hr (units in MW), see `gev_data_signal.dat` and the figure below.

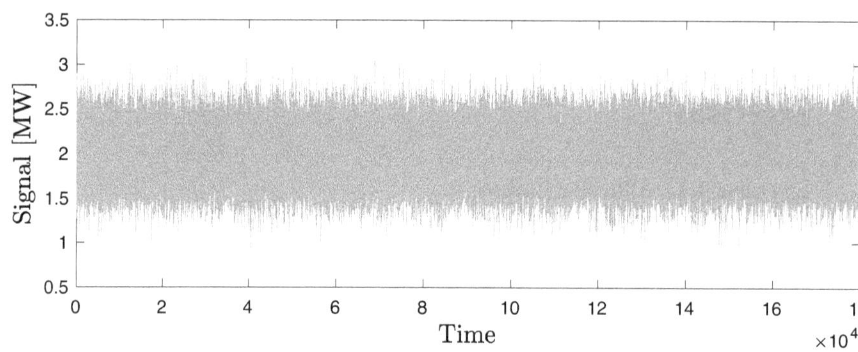

(a) Split the signal in periods of 1 hr and determine the peak signal over each period. Generate a plot of the peak values at each hour. Comment on how much variability you observe on the peaks and if you perceive any trends.
(b) Determine whether the 1-hr peak follows a GEV model (conduct a KS test to confirm this) and report the parameters of the model.
(c) Use the model to determine the capacity of the battery (in MW) that can tolerate the 1-hr peak signal with a probability of 99.99%.
(d) Use the model to determine the capacity of the battery (in MW) that can *always* tolerate the 1-hr peak signal. Compare this against the largest 1-hr peak observed in the data. Why are they different?

Exercise 4.4. MLE for Gaussian Using Analytical Calculations.
Assume that you have a set of random observations x_ω, $\omega \in \mathcal{S}$. Show (using analytical calculations) that the MLE estimates of the parameters of a Gaussian model $\mathcal{N}(\mu, \sigma^2)$ are given by the empirical mean $\mu = \hat{\mathbb{E}}_X$ and empirical variance $\sigma^2 = \hat{\mathbb{V}}_X$.

Exercise 4.5. MLE for Gaussian using Numerical Optimization.
One can apply MLE to analytically obtain general expressions/formulas for the parameters of simple distributions. However, when the distributions are complex, it is also possible to solve the MLE problem numerically.

(a) Use MLE and a numerical optimization solver to find the parameters of $\mathcal{N}(\mu, \sigma^2)$ that best fit the data in `data_mlegauss.dat`.
(b) Verify that the solver crashes if you provide an initial guess for the parameters of zero. Why does it crash?
(c) Make a 3D plot of the log-likelihood function against its parameters μ, σ^2 and verify that its maximum is located at the estimates obtained by the solver. Comment on the shape of the likelihood and on whether there might be multiple parameter values that maximize the log-likelihood.
(d) Confirm that the values obtained by the solver are the empirical mean and the empirical variance (as shown via the formulas obtained in Exercise 4.4). If you encounter differences, provide an explanation on potential sources of discrepancy.

Exercise 4.6. Particle Size Distribution.
Laser diffraction was used to determine the particle size distribution (PSD) of a drug product, and this reports the data shown below.

Particle size (μm)	90	200	400
Fraction of particles below size (–)	0.1	0.5	0.9

(a) Determine whether the data follows a log normal distribution and, if so, report the parameters that best fit the data.
(b) If you have a sample of 200 g, predict the amount of material that has a particle size below 500 μm.
(c) In the pharma industry, the spread of a PSD is typically measured using the distance $(\mathbb{Q}(0.9) - \mathbb{Q}(0.1))/\mathbb{Q}(0.5)$. Determine the spread of the PSD.

Exercise 4.7. Monte Carlo to the Rescue (Again).
You are interested in quantifying how the uncertainty of an initial investment x evolves over time. You have an initial investment x that has a lognormal random rate of return $R \sim \log \mathcal{N}(\mu, \sigma^2)$. The investment value after n years is:

$$Y = (1 + R)^n \cdot x.$$

Unfortunately, because the rate of return R enters nonlinearly in this expression, there is no direct analytical approach to obtain the distribution of Y. Monte Carlo simulation provides an approach to overcome this limitation.

(a) Assume that $x = 10{,}000$ USD, $\mu = -2.5$, and $\sigma = 0.1$; use MC with $S = 1{,}000$ random samples to estimate the probability that the investment reaches (or is above) a target of 100,000 USD in 30 years.

(b) You are interested in determining if $S = 1{,}000$ provides an estimate of adequate quality/precision. To do so, compute the estimate using 100 different batches of samples (each batch has $S = 1{,}000$ samples) and determine the empirical pdf of the estimate. Fit the empirical pdf to a Gaussian model and use the model to obtain the 95% confidence interval of your estimate. Use your confidence interval to justify whether $S = 1{,}000$ is an adequate number of samples.
(c) Compute the confidence interval for $S = 10{,}000$ and comment on the effect that this has on precision.

Exercise 4.8. Accelerated Life-Testing and the Arrhenius–Weibull Model. Testing the lifetime of materials can consume significant experimental resources; for instance, if a material has in general a long lifetime (e.g., months), then the associated testing experiments can take a long time. As such, practitioners typically use accelerated life testing techniques; one of such techniques involves reducing lifetime by using thermal stress.

A widely used model in accelerated life testing is the Arrhenius–Weibull (AW) model [11, 42]; we recall that the Weibull model is:

$$F(x) = 1 - \exp(-(x/\beta)^{\xi}).$$

In this context, the RV X is the time to failure and β is the characteristic life. The AW model assumes that this characteristic life β is a function of temperature t, and this dependence follows the Arrhenius law:

$$\beta(t) = A \cdot \exp\left(E_a/t\right),$$

where A is a positive parameter and E_a is the activation energy, which can be negative (temperature accelerates degradation) or positive (temperature slows down degradation). These parameters depend on the material being tested and thus need to be estimated from data. Specifically, you can conduct lifetime tests at different temperatures, determine the effect of temperature on lifetime, and then use the AW model to make predictions of lifetime and probability of failures at different temperatures.

(a) You have data (see `arrhenius_weibull_data.mat`) for the characteristic lifetime of a material (in hours) as a function of temperature (in K). Determine the parameters A, E_a and assess the ability of the Arrhenius model to explain the impact of temperature on lifetime.
(b) Use the Weibull model to show the lifetime pdf at 430 K and 500 K. Here, use the shape parameter $\xi = 3$. Comment on the impact of temperature.
(c) Use the Weibull model to compute the probability that the material fails before 3,000 hr at 430 K and 450 K.
(d) Practitioners like to compute the "acceleration factor," which is defined as the ratio of lifetime at a given temperature and lifetime at a reference temperature. This allows them to assess how much benefit they can get from accelerated tests. What is the acceleration factor for 450 K and assuming a reference of 430 K?

Exercise 4.9. Membrane Pore Sizes.
The log-normal distribution has been used extensively in characterizing the pore sizes of membranes; such characterization is important, as the distribution affects transport properties and helps guide membrane fabrication. It has been reported that the pore size X of an ultrafiltration membrane has the following statistics (all in Å):

$$\begin{aligned} \mathbb{M}[X] &= 42.9 \\ \mathbb{MO}[X] &= 31.35 \\ \mathbb{E}[X] &= 50.18 \\ \mathbb{SD}[X] &= 30.45. \end{aligned}$$

(a) Assuming that $X \sim \log\mathcal{N}(\mu, \sigma^2)$, use the available information to determine the parameters μ and σ.
(b) Show the pdf of the pore size and the location of the mode, mean, and median. Is this a positively or negatively skewed distribution?

5 Estimation for Structural Models

Our discussion so far has centered on using data to characterize and model random phenomena (see Chapters 2–4). Our approach has been to obtain a random variable (RV) model that best explains the data, and we have used estimation techniques to determine the parameters for such models.

From our discussion of multivariate RVs in Chapter 3, we have also observed that correlations between RVs encode dependencies between variables (e.g., reactor conversion depends on pressure). We now turn our attention to developing techniques that use data to develop models that capture these *structural* dependencies between variables.

Structural models embed *knowledge* that allows us to make predictions for a set of variables from another set of variables. Structural models might originate from fundamental laws (mechanistic models), from empirical observations (empirical models), or from combinations of both (hybrid models). As we did with RV models, our goal here will be to postulate a model with associated parameters and to use estimation techniques to determine the parameters that best fit the data and that have desirable behavior (e.g., lead to model predictions that generalize well). We will also develop techniques to *quantify the uncertainty* of our model parameters and of our model predictions. In Chapter 7, we will see that quantifying uncertainty is key for making decisions and for guiding data collection.

In the following discussion, we will make extensive use of optimization and linear algebra; we will see that such tools play a key role in helping us build and analyze complex models that might contain many parameters, variables, and nonlinearities (as those encountered in applications). Concepts of optimization and linear algebra are also essential for developing and using more sophisticated methods from data science and machine learning, as we will see in Chapter 6.

5.1 Scalar Linear Models

We begin our discussion by considering scalar models that establish linear relationships between univariate RVs Y and X. This technique is known as linear estimation or *linear regression*. Consider a set of experiments $\omega \in \mathcal{S}$ and consider that each experiment generates an observation/data pair (y_ω, x_ω). For instance, x_ω can be

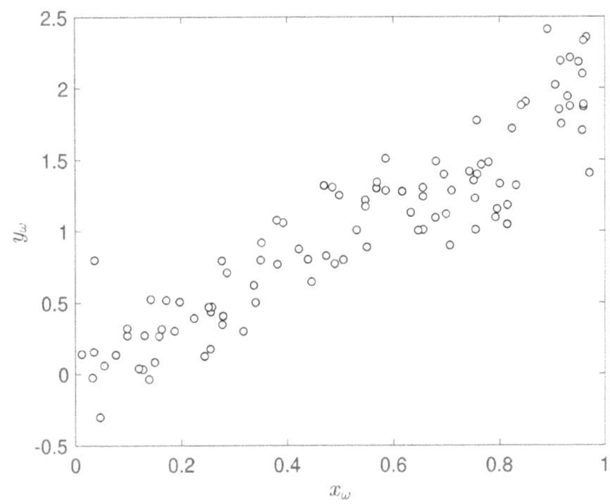

Figure 5.1 Observations showing apparent dependence of Y on X.

the reactor pressure and y_ω can be the resulting conversion observed in an experiment. The input variable X is typically (but not always) a variable that we can control, while the output variable Y is the outcome of the experiment.

To motivate the discussion, consider the observations shown in Figure 5.1. It is clear that the pairs (y_ω, x_ω) follow a *trend* (as X increases, we have that Y increases). If we use the data to compute the correlation coefficient between Y and X, we have that this is 0.91 (they are strongly positively correlated). In our discussion of correlation in Chapter 3, we highlighted that structural (e.g., physical) relationships between variables often manifest as high levels of covariance/correlation.

From Figure 5.1, we also note that the linear trend is not perfect (it contains variability or "noise"). For example, for a given value of X (say 0.9), there are multiple values of Y. This seems to indicate that there is some *inherent randomness* in the experiment generating the data; for example, if we run multiple experiments under the exact same conditions, we might likely observe different outcomes. The variability observed is often the result of uncontrollable effects such as sample preparation (which is difficult to reproduce perfectly), measurement error (sensors have limited accuracy), or a random phenomenon occurring inside the system (e.g., equipment vibrations, material failures, or molecular fluctuations).

The main goal of the estimation techniques that we will learn is to *extract* the trend from the data. Based on the linear trend that appears to be embedded in the data, we postulate that any observed behavior in Y is due to a systematic/structural dependence on X of the form:

$$Y = \theta X + \epsilon. \tag{5.1}$$

Here, θ is a parameter that captures a *linear* trend between the input X and the output Y. For now, we will simplify our discussion and assume that X and θ are deterministic variables and, therefore, any variability observed in Y (for a fixed value of X and θ) is the result of the random effect ϵ. In Section 5.5, we explore the case in which X and θ are also treated as RVs.

The postulated model (5.1) thus contains a *deterministic component* θX. This component will be called the *structural model* and captures the mechanism by which Y depends on X. The structure here is assumed to be linear, but we do not know the associated parameter θ (the slope of the trend). The structural model θX (the trend) captures our knowledge about Y and allows us to make predictions as a function of X. Specifically, if we know X and θ, then we can predict that the output is θX.

The postulated model also obtains a *random component*, which is modeled as an RV ϵ. Here, we will assume that the RV model is of the form $\epsilon \sim \mathcal{N}(0,\sigma^2)$. The random component will be called the *random model* (or noise model), and we note that this contains the parameter σ. The random model captures aspects that cannot be explained by the structural model $X\theta$. Specifically, our prediction θX might not perfectly match Y because of some unknown behavior that our structural model cannot capture.

Now, if the structural model $X\theta$ is to be used for *predicting* Y, our hope is that the model (the knowledge) dominates the random component ϵ (the unknown). Otherwise, if the random component dominates, the model will not be particularly useful (i.e., the system is not predictable). For example, if the variability of the preparation of an experiment is too strong, it will be difficult to observe a trend in the experiments. Similarly, if we have a highly inaccurate sensor, it will be difficult to observe a trend from the experiments.

Having postulated a model, our goal is to find the parameter θ (of the deterministic model component) and the parameter σ (of the random model component) that best explain the observed data. To do so, we collect data (given by observation pairs (y_ω, x_ω)) and assume that these are related as:

$$y_\omega = \theta x_\omega + \epsilon_\omega,\ \omega \in \mathcal{S}. \tag{5.2}$$

Here, we assume that ϵ_ω is a random sample of the RV ϵ. As such, ϵ_ω is an RV and y_ω is an RV. Because X is assumed to be deterministic, we assume that its value in observation ω is known and given by x_ω. The parameter θ is assumed to be deterministic, but we do not know its actual value.

The expression (5.2) can also be written as:

$$\epsilon_\omega = y_\omega - \theta x_\omega. \tag{5.3}$$

This highlights the fact that the noise captures any discrepancy (prediction error) between the observed output y_ω and the model prediction θx_ω. One often refers to the prediction error ϵ_ω as the *residual*. If our trend is indeed linear, any prediction error of the structural model θx_ω should be the result of a *pure* random effect ϵ_ω (as opposed to some trend). Here, ϵ_ω has the same underlying statistics of ϵ, and we can thus expect that the model predictions distribute as $\epsilon_\omega \sim \mathcal{N}(0,\sigma^2)$.

Our goal will be to maximize the knowledge extracted (θx_ω) from the data x_ω, y_ω. The aspects of the data that cannot be explained with the model (ϵ_ω) represent the unknown; consequently, our goal is equivalent to minimize the residual (minimize what the model cannot predict). This *knowledge extraction* process is illustrated in Figure 5.2. An interesting analogy of this process is that of a separation system in

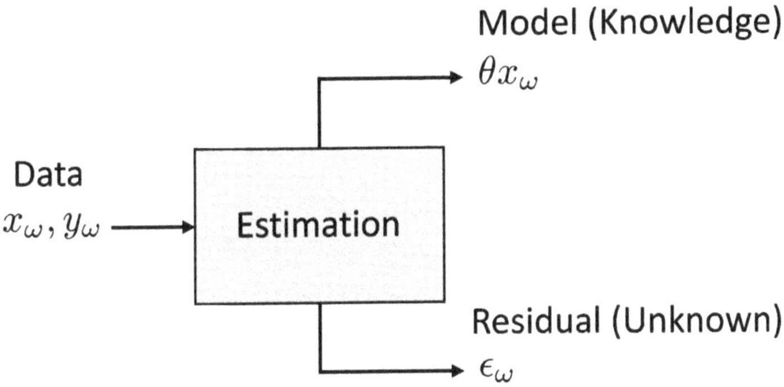

Figure 5.2 Extraction of knowledge from raw data. The structural model represents our knowledge (what can be predicted), while the random model represents the unknown (what cannot be predicted with the structural model).

which raw data is an input stream that contains a valuable product (the trend) and a contaminant (the noise).

Now recall that y_ω is an RV because ϵ_ω is an RV and recall that θx_ω is a deterministic variable. As such, we have that y_ω is a linear transformation of the Gaussian ϵ_ω and thus $y_\omega \sim \mathcal{N}(\theta x_\omega, \sigma^2)$ with expectation θx_ω and variance σ^2. In other words, the output is Gaussian, its mean is the deterministic model, and its variance (its uncertainty) is the variance of the noise.

We have made several assumptions in the postulation of our model; in other words, we are making a *hypothesis*. Specifically, we are hypothesizing that Y is indeed related to X and that the relationship is linear. In other words, we are assuming that there is a nonzero value of θ that explains the data. Now, how do we know that this hypothesis is correct? Here, it is useful to think about what would happen if any of our assumptions do not hold. For instance, imagine that Y and X are not related, or imagine that the actual relationship between Y and X is nonlinear (as opposed to our linear assumption), or imagine that Y depends on another input variable that we are missing in our model. In such cases, the residual $\epsilon_\omega = y_\omega - \theta x_\omega$ might contain a trend and will violate our assumption that it is purely random $\epsilon_\omega \sim \mathcal{N}(0, \sigma^2)$.

Our strategy will thus be to *carefully monitor the residuals* to make sure that the unknown component is indeed purely random and does not capture any systematic trend. Now, in order to make a conclusive statement about our hypothesis, we also need to make sure that we have *sufficient data*. Specifically, we need to ensure that the data is sufficient to obtain *precise* (stable) estimates of θ. We will see that this is analogous to our discussion in Chapter 4 on the precision of Monte Carlo estimates. Specifically, estimating parameters with few data points will lead to estimates that vary significantly with the data. Ultimately, these "safety" checks need to ensure that our model has suitable prediction capabilities in situations that are not captured by the available data (it is generalizable). This is important because, if we can use the model to make predictions, then we can also use it to make decisions (e.g., find optimal conditions to run our system in order to maximize performance).

5.1.1 Maximum Likelihood

Maximum likelihood estimation (MLE) is a widely used technique for parameter estimation. We have previously used this technique to estimate parameters for RV models. Here, we will show that this technique can also be used to estimate parameters for structural models.

Recall that $f(y_\omega|x_\omega,\theta,\sigma)$ can be interpreted as the probability (likelihood) of y_ω given values of x_ω, θ, and σ. Because our observations y_ω are *random* samples, we have that these are i.i.d. (see Chapter 4). Because the observations are independent, their *joint likelihood* is the product of their individual likelihoods and thus:

$$f(\mathbf{y}|\mathbf{x},\theta,\sigma) = \prod_{\omega\in\mathcal{S}} f(y_\omega|x_\omega,\theta,\sigma). \tag{5.4}$$

We thus aim to find the parameters θ that maximize the likelihood (equivalently, the log-likelihood) by solving the optimization problem:

$$\max_\theta \sum_{\omega\in\mathcal{S}} \log f(y_\omega|x_\omega,\theta,\sigma). \tag{5.5}$$

In the context of optimization, the function that is minimized (or maximized) is known as the *objective function*. For instance, in the above problem, the objective function is the log-likelihood.

Because $y_\omega \sim \mathcal{N}(\theta x_\omega,\sigma^2)$, the logarithm of $f(y_\omega|x_\omega,\theta,\sigma)$ is:

$$\log f(y_\omega|x_\omega,\theta,\sigma) = -\log\sqrt{2\pi\sigma^2} - \frac{(y_\omega-\theta x_\omega)^2}{2\sigma^2}. \tag{5.6}$$

The MLE problem can thus be written as:

$$\max_\theta -\frac{S}{2}\log 2\pi - \frac{S}{2}\log\sigma^2 - \sum_{\omega\in\mathcal{S}} \frac{(y_\omega-\theta x_\omega)^2}{2\sigma^2}. \tag{5.7}$$

It is often convenient to write this problem as a minimization problem:

$$\min_\theta \frac{S}{2}\log 2\pi + \frac{S}{2}\log\sigma^2 + \sum_{\omega\in\mathcal{S}} \frac{(y_\omega-\theta x_\omega)^2}{2\sigma^2}. \tag{5.8}$$

We assume for now that σ is a known value (a constant); in such a case, the first and second terms are fixed constants that do not affect the solution of the optimization problem. Similarly, σ in the third term is a constant and does not affect the solution. We can thus express the problem in the simple form:

$$\min_\theta \frac{1}{2}\sum_{\omega\in\mathcal{S}} (y_\omega-\theta x_\omega)^2. \tag{5.9}$$

This is a *least-squares (LS) problem* that estimates θ by minimizing the discrepancy between the observed outputs y_ω and the deterministic model predictions θx_ω. The objective function to be minimized in this problem is known as the *sum of squared errors (SSE)* and will be denoted as $SSE(\theta)$. The SSE function technically does not include the 1/2 factor, but this is often included for facilitating analytical derivations (e.g., taking the derivative of the quadratic terms will cancel out the 1/2 factor). The value of θ that solves problem (5.9) is known as the *maximum likelihood estimate* and is denoted as $\hat{\theta}$. It is important to note that, because the observations used to estimate $\hat{\theta}$ are random, the estimate is an RV (the estimate will change if a different set of data is used).

The value of $\hat{\theta}$ that minimizes $SSE(\theta)$ must satisfy the condition:

$$\frac{\partial SSE(\hat{\theta})}{\partial \theta} = -\sum_{\omega \in \mathcal{S}} x_\omega (y_\omega - \hat{\theta} x_\omega) = 0. \tag{5.10}$$

Satisfying this condition guarantees that $\hat{\theta}$ is a stationary point for $SSE(\theta)$ but does not guarantee that this is a minimum point for $SSE(\theta)$ (i.e., a maximum point also satisfies this condition). We thus say that the condition (5.10) is *a necessary but not a sufficient* condition.

The sufficient condition that guarantees that $SSE(\hat{\theta})$ is a minimum is:

$$\frac{\partial^2 SSE(\hat{\theta})}{\partial \theta^2} > 0. \tag{5.11}$$

This condition ensures that the SSE function (the objective function) has positive curvature at the minimum $\hat{\theta}$ and that the estimate is unique (there is no other value of the parameters that minimizes the SSE). A maximum point has negative curvature ($\frac{\partial^2 SSE(\hat{\theta})}{\partial \theta^2} < 0$), which is clearly not useful because we aim to minimize the SSE function. It is also possible that the minimum point $\hat{\theta}$ has no curvature ($\frac{\partial^2 SSE(\hat{\theta})}{\partial \theta^2} = 0$); this would indicate that there are multiple values of $\hat{\theta}$ that minimize the SSE function. This is illustrated in Figure 5.3.

The necessary condition tells us that the estimate is given by:

$$\hat{\theta} = \frac{\sum_{\omega \in \mathcal{S}} x_\omega y_\omega}{\sum_{\omega \in \mathcal{S}} x_\omega^2}. \tag{5.12}$$

Note that the estimate can be obtained analytically, as computing derivatives for the SSE function is straightforward in this case. We will see later that this is not always the case (when the model is nonlinear), and we will need to resort to

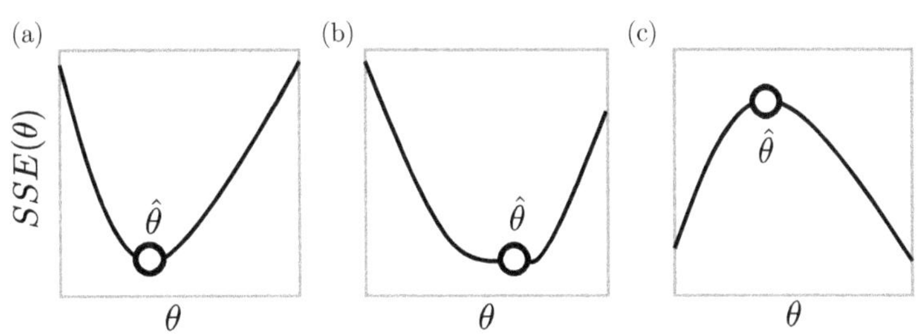

Figure 5.3 Illustration of an estimate $\hat{\theta}$ that sits at a minimum with positive curvature (a), minimum with zero curvature (b), and maximum with negative curvature (c).

numerical techniques. The sufficient condition (5.11) is satisfied because:

$$\frac{\partial^2 SSE(\hat{\theta})}{\partial \theta^2} = \sum_{\omega \in \mathcal{S}} x_\omega^2 > 0. \tag{5.13}$$

The only way that this condition would not hold is if all our input data x_ω is zero (which is not practical). Consequently, the estimate $\hat{\theta}$ minimizes the SSE and the estimate is unique. Note also that it is not possible for our problem to have a maximum point; this is a property of our problem at hand (which involves a linear model). We will see later that maxima and flat minima can appear in estimation problems that involve multiple input variables and nonlinear models.

We now make the following observations about the estimate $\hat{\theta}$:

- Equation (5.12) tells us that the estimate $\hat{\theta}$ captures *dependencies* in the input–output pairs x_ω, y_ω.
- Equation (5.13) tells us that the estimate $\hat{\theta}$ is *unique* and becomes better defined as we add more data (observations). Because the solution is unique, we have a guarantee that there is no other value $\hat{\theta}$ that minimizes the SSE.
- Recall that $y_\omega \sim \mathcal{N}(\hat{\theta} x_\omega, \sigma^2)$ and thus the model prediction (with the estimated parameters) $\hat{\theta} x_\omega$ is the *most likely outcome* (the mode of y_ω).
- The estimate $\hat{\theta}$ gives the residuals $\hat{\epsilon}_\omega = y_\omega - \hat{\theta} x_\omega$. If these estimates follow our assumption that $\hat{\epsilon}_\omega \sim \mathcal{N}(0, \sigma^2)$, then the available data and postulated model are satisfactory. If this is not satisfied, another type of model might be needed (e.g., a nonlinear model).

Example 5.1 Estimation with Correct Postulated Model

We consider that X, Y have a structural relationship of the form $Y = \theta X$ and $\theta = 2$. We assume that the observations y_ω of Y are corrupted by the random noise $\epsilon_\omega \sim \mathcal{N}(0, 0.0625)$. We use this model to generate $S = 100$ observations (x_ω, y_ω), shown in Figure 5.4(a). Note how we can perceive a linear trend, but this trend is hidden (blurred) by noise.

We postulate a model of the form $Y = \theta X + \epsilon$ with structural component θX and random component $\epsilon \in \mathcal{N}(0, \sigma^2)$. We proceed to estimate θ and σ using MLE and check if these are consistent with the known quantities. We recall that σ does not affect the solution of the MLE problem, so we would need to obtain this value from the residuals. Of course, in a real application, we will not know the real value of θ and σ; but our goal in this example is to validate that the MLE methodology actually works.

The predictions obtained with the estimated model $\hat{y}_\omega = \hat{\theta} x_\omega$ and the corresponding residuals $\epsilon_\omega = y_\omega - \hat{y}_\omega$ are shown in Figure 5.4. From these results, we make the following observations:

- From Figure 5.4(a), we see that the linear trend defined by the estimated model $(x_\omega, \hat{y}_\omega)$ follows the general trend of the data (x_ω, y_ω).

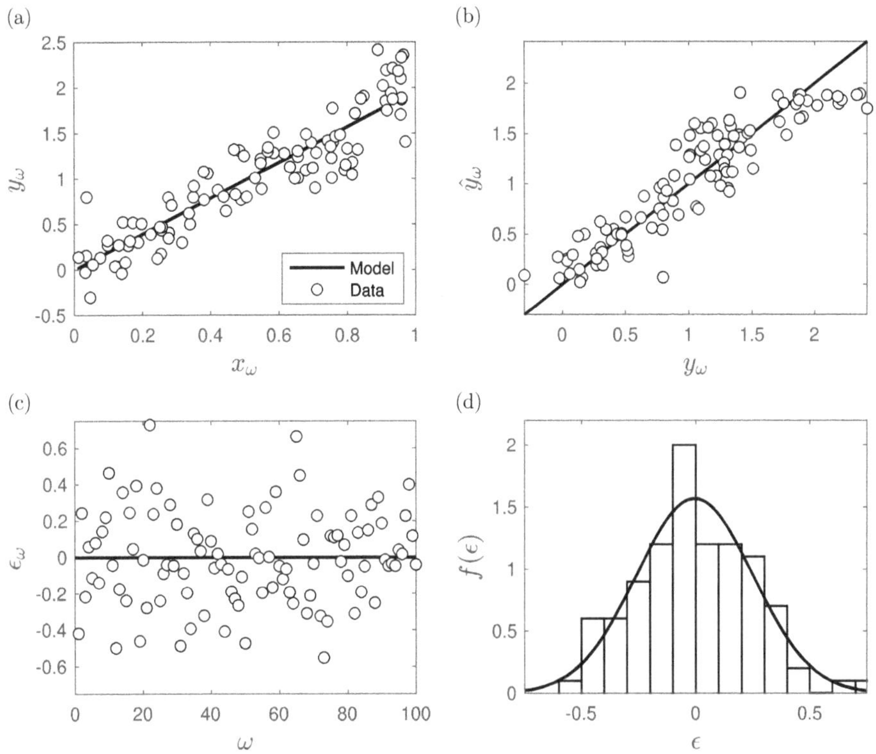

Figure 5.4 Model predictions using estimate $\hat{\theta}$ and observed data (a). Parity plot between predicted and observed outputs (b). Residuals for observations (c) and corresponding pdf (d).

- Figure 5.4(b) is a *parity plot* that compares the predicted outputs $\hat{y}_\omega$ and the observed outputs y_ω. If the model $\hat{\theta}X$ is perfect, the points would lie on the diagonal line. Here, we see deviations around the diagonal, and thus, we conclude that the model is not perfect; this is because we have noise that cannot be explained by the structural model. Note, however, that the pairs seem to distribute evenly along the diagonal line (indicating that the residuals are random).
- Figure 5.4(c) and (d) show the residuals for the different observations and the empirical pdf for such residuals. We can see that the residuals seem to follow a Gaussian distribution, and thus, it seems that the postulated noise model is correct. To validate this, we fitted the empirical pdf of the residuals to a Gaussian $\mathcal{N}(\mu,\sigma^2)$ and found that the estimated mean is $\mu = -0.004$ and the estimated standard deviation is $\sigma^2 = 0.0647$. We note that the estimated variance of the residuals matches that of the true noise used to corrupt the data. This confirms that the MLE procedure is consistent.

In Figure 5.5, we plot the function $SSE(\theta)$ for different amounts of data $S = 10, 100, 1{,}000$. We note that the surface becomes better defined as we add data and reveals the true value of the parameter $\theta = 2$. If we use $S = 10$ observations, the $SSE(\theta)$ is minimized at $\hat{\theta} = 2.2$. If we use $S = 100$ or $S = 1{,}000$ observations, the

$SSE(\theta)$ is minimized at a value that is close to $\hat{\theta} = 2$. This highlights that adding more data helps identify the true parameter.

Script to reproduce results: `ch5_linear_est_example` □

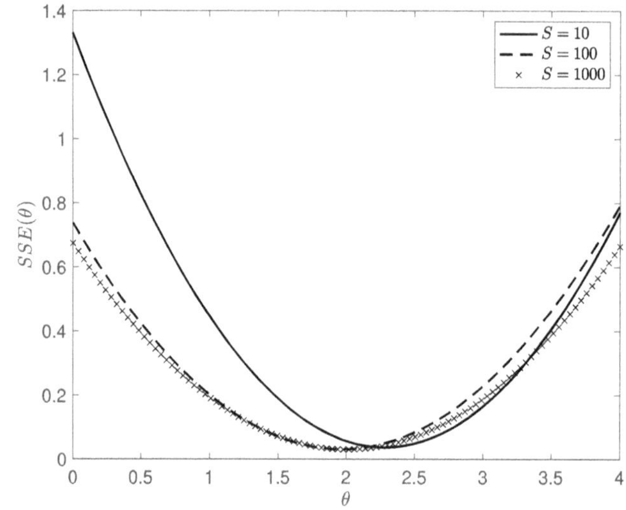

Figure 5.5 Sum-of-squared errors $SSE(\theta)$ with different amounts of data.

Example 5.2 Estimation with Incorrect Postulated Model

Assume now that the true relationship between X and Y is nonlinear and of the form $Y = \theta X^2$ with $\theta = 2$. As before, we assume that the observations y_ω of Y are corrupted by random noise $\epsilon_\omega \sim \mathcal{N}(0, 0.0625)$. We use this model to generate $S = 100$ observations (x_ω, y_ω), shown in Figure 5.6(a). Note that there is a nonlinear trend in the data.

Assume now that we postulate the wrong (linear) model $Y = \theta X + \epsilon$ to match the observations. The results are shown in Figure 5.6. From Figure 5.6(a), it is clear that the linear trend of the model does not follow the quadratic trend of the data (particularly for large values of x_ω because the nonlinearity becomes strong).

From the parity plot in Figure 5.6(b), it is clear that the pairs no longer distribute evenly along the diagonal, and we see a clear deviation on the right (this corresponds to the large values of x_ω). This indicates that there is a systematic error in the model. This is confirmed in Figure 5.6(d); here, we see that the pdf of the residuals is no longer Gaussian. This indicates that the noise model contains a trend that cannot be explained by the linear structural component that we postulated (the noise is not purely random). This example illustrates how residual analysis plays a key role in determining the appropriateness of the postulated model.

Script to reproduce results: `ch5_linear_est_example_wrong` □

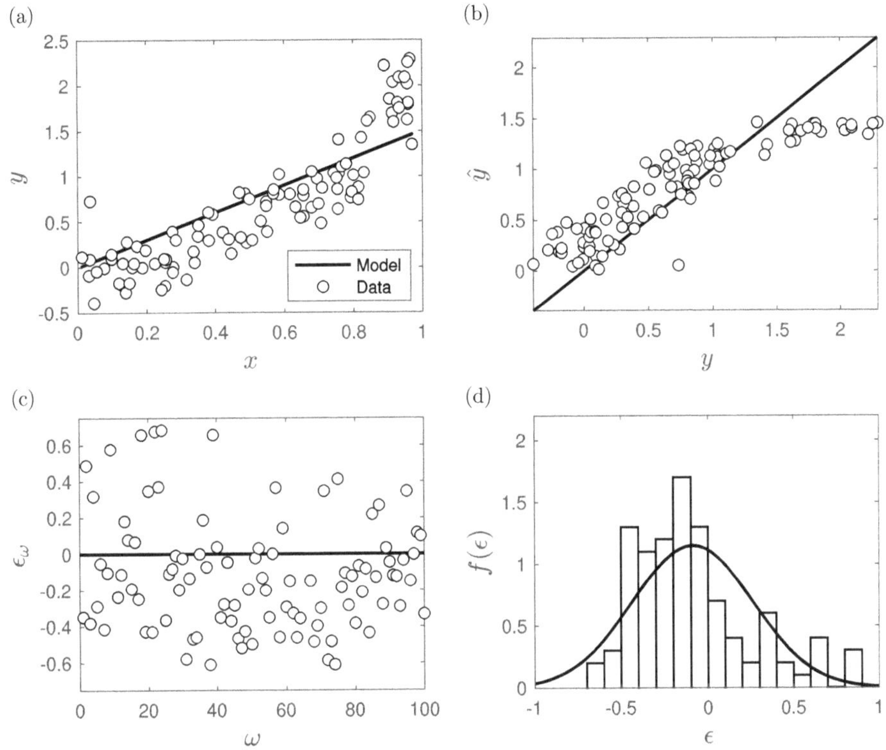

Figure 5.6 Model predictions using estimate $\hat{\theta}$ and observed data (a). Parity plot between predicted/observed outputs (b). Residuals for the observations (c) and pdf for residuals (d).

5.1.2 Bias, Precision, and Information

The MLE methodology has the key property that it delivers an *unbiased* estimate $\hat{\theta}$ of the true parameters θ. As we discussed in Chapter 4, this is important because the estimate will gravitate around the true parameter. Moreover, the *variance (precision)* of the estimate $\hat{\theta}$ is tightly related to the *sharpness* of the log-likelihood. Specifically, the sharpness of the likelihood function will tell us how easy (or difficult) it is to uncover the estimate from the noisy data. In other words, we will see that a sharp likelihood indicates that the data contains significant *information* about θ and the estimate can be easily uncovered (observed) from the data. Conversely, if the data is not informative, the likelihood will be flat and the estimates will be difficult to uncover from the data.

To gain additional insight into these issues, consider the MLE problem (5.5) written in the following form:

$$\min_{\theta} \; -\log L(\theta) = -\log f(\mathbf{y}|\theta) = -\sum_{\omega \in \mathcal{S}} \log f(y_\omega|\theta). \tag{5.14}$$

Here, note that we replaced the maximization for minimization; the function $-\log L(\theta)$ is known as the *negative log-likelihood*. Because we are only interested in estimating θ, we have eliminated $\mathbf{x}$ and σ from the notation. Moreover, we capture all the observations y_ω, $\omega \in \mathcal{S}$ in the vector $\mathbf{y}$, and thus we write the joint likelihood as $f(\mathbf{y}|\theta)$.

Any solution of the MLE problem (our estimate $\hat{\theta}$) must satisfy the necessary condition:

$$-\frac{\partial \log f(\mathbf{y}|\hat{\theta})}{\partial \theta} = 0. \tag{5.15}$$

In other words, the first derivative of the negative log-likelihood must vanish at $\hat{\theta}$. The derivative $-\frac{\partial \log f(\mathbf{y}|\theta)}{\partial \theta}$ is known as the *score function*.

Now, recall that the observations $\mathbf{y}$ are RVs, and thus, the score function $-\frac{\partial \log f(\mathbf{y}|\theta)}{\partial \theta}$ is also an RV. In addition, because the estimate $\hat{\theta}$ depends on the score function, then the estimate is also an RV. In other words, as we change our observations $\mathbf{y}$, the estimate $\hat{\theta}$ will change.

Imagine now that we use a set of observations to obtain an estimate $\hat{\theta}$ satisfying $-\frac{\partial \log f(\mathbf{y}|\theta)}{\partial \theta} = 0$, and we repeat this process multiple times to obtain different estimates $\hat{\theta}$. Can we guarantee that this process delivers estimates satisfying $\mathbb{E}[\hat{\theta}] = \theta$? In other words, can we show that $\hat{\theta}$ delivered by MLE is an unbiased estimator of the true θ?

If the estimate delivered by MLE is unbiased, we need that any changes in the score function (due to changes in the observations $\mathbf{y}$) average out. In other words, we need that:

$$\mathbb{E}\left[-\frac{\partial \log f(\mathbf{y}|\theta)}{\partial \theta}\right] = 0. \tag{5.16}$$

Here, the expectation is taken with respect to $\mathbf{y}$ (with pdf given by the joint likelihood $f(\mathbf{y}|\theta)$). We can see that this is indeed the case from:

$$\begin{aligned}
\mathbb{E}\left[-\frac{\partial \log f(\mathbf{y}|\theta)}{\partial \theta}\right] &= \int -\frac{\partial \log f(\mathbf{y}|\theta)}{\partial \theta} f(\mathbf{y}|\theta) d\mathbf{y} \\
&= \int -\frac{\frac{\partial f(\mathbf{y}|\theta)}{\partial \theta}}{f(\mathbf{y}|\theta)} f(\mathbf{y}|\theta) d\mathbf{y} \\
&= \int -\frac{\partial f(\mathbf{y}|\theta)}{\partial \theta} d\mathbf{y} \\
&= -\frac{\partial}{\partial \theta} \int f(\mathbf{y}|\theta) d\mathbf{y} \\
&= -\frac{\partial}{\partial \theta} 1 = 0.
\end{aligned} \tag{5.17}$$

We thus have that the estimate $\hat{\theta}$ delivered by MLE is unbiased.

We now explore how does our estimate $\hat{\theta}$ vary as we vary the observations $\mathbf{y}$. The variance will tell us how certain we are that our estimate $\hat{\theta}$ is the true value θ and recall also that the level of uncertainty is a measure of the estimate *precision* (analogous to our discussion of Monte Carlo estimates). To assess how much the estimate $\hat{\theta}$ varies, we need to determine how much does the score function vary. The variance of the score function is:

$$\begin{aligned}\mathcal{I}(\theta) &= \mathbb{V}\left[-\frac{\partial \log f(\mathbf{y}|\theta)}{\partial \theta}\right] \\ &= \mathbb{E}\left[\left(\frac{\partial \log f(\mathbf{y}|\theta)}{\partial \theta}\right)^2\right].\end{aligned} \tag{5.18}$$

Here, we used the fact that the expected value of the score function is zero, as shown in Eq. (5.17).

The variance of the score function is known as the *Fisher information* (FI), and this is denoted as $\mathcal{I}(\theta)$. One can show (not an easy derivation) that the FI can also be written as:

$$\mathcal{I}(\theta) = \mathbb{E}\left[-\frac{\partial^2 \log f(\mathbf{y}|\theta)}{\partial \theta^2}\right]. \tag{5.19}$$

This indicates that the FI is related to the *second derivative* of the log-likelihood. This observation is important because we recall that the estimate $\hat{\theta}$ minimizes the negative log-likelihood function if the second derivative evaluated at $\hat{\theta}$ satisfies:

$$-\frac{\partial^2 \log f(\mathbf{y}|\hat{\theta})}{\partial \theta^2} \geq 0. \tag{5.20}$$

Also, as we discussed before, a sharp (unique) minimum has a second derivative that is large and positive, while a flat minimum has a second derivative that is zero. A flat minimum tells us that the estimate is nonunique (different values give the same minimum value of the negative log-likelihood).

The FI is a fundamental quantity because it gives us a limit of the best possible estimator variance (precision) that we can achieve. Specifically, one can show that:

$$\mathbb{V}[\hat{\theta}] \geq \frac{1}{\mathcal{I}(\theta)}. \tag{5.21}$$

This inequality is known as the *Cramer–Rao bound* and is one of the most important results in statistics. The bound tells us that the variance of the MLE estimate *can never* be below $\frac{1}{\mathcal{I}(\theta)}$. Note, however, that the FI is a function of the true parameter θ (which we do not know). Consequently, it is difficult to verify this bound in practice (unless we somehow know the true parameter). However, the Cramer–Rao bound gives us some key insights into how the variance of MLE estimates behave.

The FI is the expected value of the second derivative of the negative log-likelihood function. The Cramer–Rao bound thus tells us that if, on average, the second derivative is large (we have a sharp minimum), then the FI is large and $1/\mathcal{I}(\theta)$ is small; consequently, the variance of the estimate $\mathbb{V}[\hat{\theta}]$ can reach low values. This indicates that the data contains significant information to determine the estimate to high precision (there is low uncertainty). If, on average, the second derivative is small (close to zero), then the FI is small and $1/\mathcal{I}(\theta)$ is large; consequently, the variance of the estimate will be high. This indicates that the data is not sufficient to determine the estimate to high precision (there is high uncertainty on the estimate).

This discussion highlights the intricate connections between *information, uncertainty, and flatness.* When we have lots of information, our estimates will have low uncertainty and they will sit at a shaper minimum of the negative likelihood. Conversely, when we have little information, our estimates will have high uncertainty and they will sit at a flat minimum of the negative likelihood. In Section 5.2, we will provide more details on this subject.

We have mentioned that the FI, in general, cannot be computed because the true parameter θ is not known. Moreover, note that the FI requires an expectation operation (multidimensional integral) with respect to the output data $\mathbf{y}$. In Section 5.3, we will derive procedures to approximate the FI. In some special situations, however, the FI does not depend on the parameters and the data $\mathbf{y}$. This is the case for the linear structural models discussed in this section; to see this, note that the negative log-likelihood of the linear estimation problem is given by:

$$-\log f(\mathbf{y}|\theta) = \frac{S}{2}\log 2\pi + \frac{S}{2}\log\sigma^2 + \sum_{\omega\in\mathcal{S}} \frac{(y_\omega - \theta x_\omega)^2}{2\sigma^2}. \tag{5.22}$$

If we take the second derivative, we obtain that:

$$-\frac{\partial^2 \log f(\mathbf{y}|\theta)}{\partial\theta^2} = \frac{1}{\sigma^2}\sum_{\omega\in\mathcal{S}} x_\omega^2. \tag{5.23}$$

The second derivative does not depend on θ and $\mathbf{y}$; we thus have that the FI is:

$$\mathbb{E}\left[-\frac{\partial^2 \log f(\mathbf{y}|\theta)}{\partial\theta^2}\right] = \frac{1}{\sigma^2}\sum_{\omega\in\mathcal{S}} x_\omega^2. \tag{5.24}$$

This tells us that, as we increase our input data (as S becomes large), the FI increases (the likelihood becomes sharper). This also tells us that, as the uncertainty of our output data decreases (as σ^2 becomes small), the FI increases. This confirms our intuition that we can trust our parameters more as we add more information (data) to the problem and as our noise becomes less strong.

5.1.3 Estimation of Structural and Random Model Parameters

In our previous discussion, we assumed that the noise variance σ is a deterministic variable with a known value. This variance captures the noise associated with the outputs Y. We saw that the value of σ has no direct impact on the parameter estimate because the MLE problem reduces to minimizing the SSE. We also saw that we can estimate the value of σ from the statistics of the residual (once we have obtained the parameter estimate $\hat{\theta}$). Note, however, that σ is also a parameter of our postulated model (of the random component) and thus can also be estimated from the data (as we did for the parameter θ of the structural component). This is done by solving the MLE problem:

$$\min_{\theta,\sigma}\; -\log L(\theta,\sigma) = -\sum_{\omega\in\mathcal{S}} \log f(y_\omega|x_\omega,\theta,\sigma). \tag{5.25}$$

By writing down the density explicitly and rearranging, we obtain:

$$\min_{\theta,\sigma} \frac{S}{2}\log 2\pi + \frac{S}{2}\log \sigma^2 + \frac{1}{2\sigma^2}\sum_{\omega\in\mathcal{S}}(y_\omega - \theta x_\omega)^2. \tag{5.26}$$

The estimates obtained from the MLE problem must satisfy:

$$\begin{aligned} -\frac{\partial \log L(\theta,\sigma)}{\partial \theta} &= 0 \\ -\frac{\partial \log L(\theta,\sigma)}{\partial \sigma^2} &= 0. \end{aligned} \tag{5.27}$$

Note that we take the derivative with respect to σ^2 (and not σ) because σ always appears in a squared form. By writing down the log-likelihood explicitly, we obtain that the first condition is:

$$\frac{1}{\sigma^2}\sum_{\omega\in\mathcal{S}} x_\omega(y_\omega - \theta x_\omega) = 0. \tag{5.28}$$

By multiplying through by σ^2, we obtain:

$$\sum_{\omega\in\mathcal{S}} x_\omega(y_\omega - \theta x_\omega) = 0. \tag{5.29}$$

From this condition, we can recover the estimate $\hat{\theta}$ as:

$$\hat{\theta} = \frac{\sum_{\omega\in\mathcal{S}} x_\omega y_\omega}{\sum_{\omega\in\mathcal{S}} x_\omega^2}. \tag{5.30}$$

Note that this is the exact same estimate that we obtained before in Eq. (5.12). The second condition is:

$$\frac{S}{2}\frac{1}{\sigma^2} - \frac{1}{2(\sigma^2)^2}\sum_{\omega\in\mathcal{S}}(y_\omega - \theta x_\omega)^2 = 0. \tag{5.31}$$

Using the estimated $\hat{\theta}$ and rearranging, we obtain:

$$\hat{\sigma}^2 = \frac{1}{S}\sum_{\omega\in\mathcal{S}}(y_\omega - \hat{\theta} x_\omega)^2. \tag{5.32}$$

We thus have that the estimate of the noise variance is the sample variance of the model residuals. This is consistent with our assumption that the noise model is Gaussian $\epsilon \sim \mathcal{N}(0,\sigma^2)$. Consequently, if the noise variance σ^2 is not known, we can estimate it using the variance of our model residuals.

In our previous analysis, we treated the output observations y_ω as RVs (with uncertain values), and we characterized the output uncertainty by using the random noise model. On the other hand, we treated the input observations x_ω as deterministic variables (with known values). But what if the input observations are also uncertain (e.g., also due to sensor uncertainty)? In such a case, we will need to postulate a random model associated with the input observations, and we will need to estimate its parameters from the data. We will address this more general setting in Section 5.5.

5.2 Multidimensional Linear Models

We now generalize the linear model to account for connections between the output Y and multiple inputs $(X_1,X_2,\ldots,X_n)$ (e.g., reactor conversion depends on temperature, pressure, and concentration of reactants). Specifically, we postulate the following model:

$$Y = \theta_0 + \sum_{i=1}^{n} \theta_i X_i + \epsilon. \tag{5.33}$$

We can think about this model as a *mixing system* in which multiple input effects are combined to obtain an output. Specifically, the parameters $\theta_1,\ldots,\theta_n$ act as weights for the corresponding inputs. The parameter θ_0 does not act as a weight but is a constant term that adds flexibility to the model. The mixing analogy of the model is useful, as it provides some perspective on what inputs have the most impact on the output and/or if the impact of some inputs cancels out (e.g., the inputs are redundant/not relevant and thus some can be eliminated). This is important because some models can contain hundreds or thousands of inputs. We will discuss these issues in more detail in Chapter 6.

The observation pairs (y_ω,x_ω) satisfy the equations:

$$y_\omega = \theta_0 + \sum_{i=1}^{n} \theta_i x_{\omega,i} + \epsilon_\omega,\ \omega \in \mathcal{S}. \tag{5.34}$$

This set of equations can be expressed compactly using matrix notation:

$$\mathbf{y} = \mathbf{X}\theta + \mathbf{e}, \tag{5.35}$$

where:

$$\mathbf{y} = \begin{bmatrix} y_1 \\ y_2 \\ \vdots \\ y_S \end{bmatrix} \quad \mathbf{X} = \begin{bmatrix} 1 & x_{1,1} & x_{1,2} & \ldots & x_{1,n} \\ 1 & x_{2,1} & x_{2,2} & \ldots & x_{2,n} \\ \vdots & & & \vdots & \\ 1 & x_{S,1} & x_{S,2} & \ldots & x_{S,n} \end{bmatrix} \tag{5.36}$$

$$\theta = \begin{bmatrix} \theta_0 \\ \theta_1 \\ \vdots \\ \theta_n \end{bmatrix} \quad \mathbf{e} = \begin{bmatrix} \epsilon_1 \\ \epsilon_2 \\ \vdots \\ \epsilon_S \end{bmatrix}. \tag{5.37}$$

As we begin working more with matrix notation, it is important to keep track of the dimensions of the different quantities of interest. The output vector $\mathbf{y} \in \mathbb{R}^S$ is a vector of dimension S (the number of observations). The parameter vector $\theta \in \mathbb{R}^{n+1}$ has dimension $n + 1$ (the number of parameters, which equals the number of input variables plus one). The input data matrix $\mathbf{X} \in \mathbb{R}^{S\times(n+1)}$ is a matrix with S rows and $n + 1$ columns; the rows contain observations and the columns contain input variables. We denote the ω-th row of the data matrix as the vector:

$$\mathbf{x}_\omega = (1, x_{\omega,1}, x_{\omega,2}, \ldots, x_{\omega,n}) \in \mathbb{R}^{n+1}. \tag{5.38}$$

The random noise is an S-dimensional RV of the form $\mathbf{e} \sim \mathcal{N}(0, \Sigma)$. Here, $\Sigma \in \mathbb{R}^{S \times S}$ is the noise covariance matrix. Typically, it is assumed that $\epsilon_\omega, \omega \in \mathcal{S}$ are independent RVs and thus the covariance is a diagonal matrix with entries $\Sigma_{\omega,\omega} = \sigma_\omega^2$ for $\omega \in \mathcal{S}$. This captures situations in which the experiment is conducted using the same devices (e.g., the same reactor running under different conditions). It is also possible, however, to assume that the noise variables are not independent and thus that the covariance matrix is not diagonal (there are correlations). This captures situations in which measurement errors for different variables might be correlated.

We first assume that the input data $\mathbf{X}$, the model parameters θ, and the noise parameters Σ are deterministic variables. Under these assumptions, the output $\mathbf{y}$ is a linear transformation of the Gaussian noise $\mathbf{e}$, and it is thus a Gaussian $\mathbf{y} \sim \mathcal{N}(\mathbf{X}\theta, \Sigma)$, with pdf $f(\mathbf{y}|\mathbf{X}, \theta, \Sigma)$ (the likelihood). The best estimates for the model parameters $\hat{\theta}$ and the random model parameters $\hat{\Sigma}$ are found by maximizing the log joint likelihood $\log f(\mathbf{y}|\mathbf{X}, \theta, \Sigma)$. This problem is given by:

$$\max_{\theta, \Sigma} \; \log f(\mathbf{y}|\mathbf{X}, \theta, \Sigma), \tag{5.39}$$

where the log-likelihood function is:

$$\log f(\mathbf{y}|\mathbf{X}, \theta, \Sigma) = -\log\left((2\pi)^{S/2}|\Sigma|^{1/2}\right) - \frac{1}{2}(\mathbf{y} - \mathbf{X}\theta)^T \Sigma^{-1} (\mathbf{y} - \mathbf{X}\theta). \tag{5.40}$$

This is the most general form of the MLE problem that simultaneously estimates θ and Σ. We now make different assumptions to obtain formulations of different complexity. This will reveal that, under certain assumptions, the MLE problem again reduces to a simple least-squares (LS) problem.

Analogous to the scalar case, we assume that the noise covariance Σ is given. Under this assumption, the MLE problem can be expressed as:

$$\min_{\theta} \; -\log f(\mathbf{y}|\mathbf{X}, \theta, \Sigma). \tag{5.41}$$

Because Σ is treated as a constant in the previous minimization problem, the first term in the log-likelihood function can be treated as a constant and can be eliminated. The problem can thus be reduced to:

$$\min_{\theta} \; \frac{1}{2}(\mathbf{y} - \mathbf{X}\theta)^T \Sigma^{-1} (\mathbf{y} - \mathbf{X}\theta). \tag{5.42}$$

This is an LS minimization problem and notice the appearance of the covariance matrix as a *weighting* matrix. If we assume that the covariance matrix is diagonal with entries $\Sigma_{\omega,\omega} = \sigma_\omega^2$ for $\omega \in \mathcal{S}$, the previous problem can be reduced to:

$$\min_{\theta} \; \frac{1}{2} \sum_{\omega \in \mathcal{S}} \frac{1}{\sigma_\omega^2} \left(y_\omega - \mathbf{x}_\omega^T \theta\right)^2. \tag{5.43}$$

From this form, we can clearly see that the reciprocal of the variance $1/\sigma_\omega^2$ acts as *weight* for observation ω. If the variance σ_ω^2 is small (we trust this observation), the weight is large; on the other hand, if the variance is large (we do not trust the observation), the weight is small. Consequently, the above formulation naturally places more emphasis on observations that are more certain.

If we further assume that all the noise variances are equal ($\sigma_\omega^2 = \sigma^2$ for all $\omega \in \mathcal{S}$), the problem further reduces to:

$$\min_{\theta} \; \frac{1}{2} \sum_{\omega \in \mathcal{S}} \left(y_\omega - \mathbf{x}_\omega^T \theta \right)^2 . \tag{5.44}$$

This is the simplest form of the MLE problem; here, we place equal emphasis on all observations. We will explore this formulation in detail to make some important observations. In Section 5.3, we will generalize these observations by returning to the MLE formulation (5.41).

The SSE function can be written compactly using matrix notation as:

$$\begin{aligned} SSE(\theta) &= \frac{1}{2} \sum_{\omega \in \mathcal{S}} (y_\omega - \mathbf{x}_\omega^T \theta)^2 \\ &= \frac{1}{2} (\mathbf{y} - \mathbf{X}\theta)^T (\mathbf{y} - \mathbf{X}\theta) \\ &= \frac{1}{2} \|\mathbf{y} - \mathbf{X}\theta\|^2 . \end{aligned} \tag{5.45}$$

Here, $\|\mathbf{y}-\mathbf{X}\theta\|^2$ is the squared Euclidean norm of the residual vector $\mathbf{e} = \mathbf{y}-\mathbf{X}\theta$. This quantity is simply the sum of the individual squared residuals $\epsilon_\omega^2 = (y_\omega - \mathbf{x}_\omega^T \theta)^2$, $\omega \in \mathcal{S}$. Here, we again notice that the SSE function includes a 1/2 factor; this is kept for facilitating analytical derivations (the factor will cancel out when taking derivatives of the function).

The solution of the LS problem (5.44) (our estimate $\hat{\theta}$) must satisfy the necessary conditions:

$$\nabla_\theta SSE(\theta) = 0. \tag{5.46}$$

Here, $\nabla_\theta SSE(\theta) \in \mathbb{R}^{n+1}$ is known as the gradient vector, whose entries are the partial derivatives of the SSE function with respect to each parameter. Specifically, we have that:

$$\nabla_\theta SSE(\theta) = \begin{bmatrix} \frac{\partial SSE(\theta)}{\partial \theta_0} \\ \frac{\partial SSE(\theta)}{\partial \theta_1} \\ \vdots \\ \frac{\partial SSE(\theta)}{\partial \theta_n} \end{bmatrix} . \tag{5.47}$$

For the SSE function (5.45), we have that the gradient vector is given by:

$$\nabla_\theta SSE(\theta) = -\mathbf{X}^T (\mathbf{y} - \mathbf{X}\theta). \tag{5.48}$$

The conditions (5.46) are thus given by:

$$-\mathbf{X}^T (\mathbf{y} - \mathbf{X}\theta) = 0. \tag{5.49}$$

These conditions are a set of $n + 1$ linear equations that must be solved *simultaneously* to obtain the estimate $\hat{\theta}$. These equations are known as the *score equations* and indicate that the estimate $\hat{\theta}$ must be such that the partial derivatives of the SSE with respect to all parameters are zero.

The solution of the score equations can be found *analytically* by using matrix manipulations. Specifically, we rearrange Eq. (5.49) to obtain:

$$-\mathbf{X}^T\mathbf{y} + \mathbf{X}^T\mathbf{X}\theta = 0. \tag{5.50}$$

By moving $\mathbf{X}^T\mathbf{y}$ to the right-hand side and by inverting the matrix $\mathbf{X}^T\mathbf{X}$ to isolate θ, we obtain that the solution of the set of equations is:

$$\hat{\theta} = (\mathbf{X}^T\mathbf{X})^{-1}\mathbf{X}^T\mathbf{y}. \tag{5.51}$$

Recall that this estimate is only guaranteed to be a stationarity point for the SSE function. Specifically, a stationary point can be a minimum or a maximum; moreover, such minimum and maximum might be sharp or flat. A flat minimum indicates that multiple values of θ reach the same minimum value for the SSE function $SSE(\theta)$; in other words, the estimate is *nonunique*.

We would like to guarantee that the estimate $\hat{\theta}$ is a unique minimum of the SSE function. The estimate $\hat{\theta}$ obtained from Eq. (5.51) is guaranteed to be a *unique minimum* of $SSE(\theta)$ if the following sufficient condition holds:

$$\nabla^2_{\theta} SSE(\theta) \succ 0. \tag{5.52}$$

We now explore carefully what this condition means. The quantity $\nabla^2_{\theta} SSE(\theta) = \nabla_{\theta}(\nabla_{\theta} SSE(\theta))$ is the *Hessian matrix* (matrix of second partial derivatives of the SSE function):

$$\nabla^2_{\theta} SSE(\theta) = \begin{bmatrix} \frac{\partial^2 SSE(\theta)}{\partial\theta_0\partial\theta_0} & \frac{\partial^2 SSE(\theta)}{\partial\theta_1\partial\theta_0} & \cdots & \frac{\partial^2 SSE(\theta)}{\partial\theta_n\partial\theta_0} \\ \frac{\partial^2 SSE(\theta)}{\partial\theta_0\partial\theta_1} & \frac{\partial^2 SSE(\theta)}{\partial\theta_1\partial\theta_1} & \cdots & \frac{\partial^2 SSE(\theta)}{\partial\theta_n\partial\theta_1} \\ \vdots & \vdots & \ddots & \vdots \\ \frac{\partial^2 SSE(\theta)}{\partial\theta_0\partial\theta_n} & \frac{\partial^2 SSE(\theta)}{\partial\theta_1\partial\theta_n} & \cdots & \frac{\partial^2 SSE(\theta)}{\partial\theta_n\partial\theta_n} \end{bmatrix}. \tag{5.53}$$

The sufficient condition (5.52) is a generalization of the scalar linear case (in which it is required that the second derivative with respect to the parameter is positive). The generalization when we have multiple parameters is the requirement that *all the eigenvalues of the Hessian must be strictly positive*. In other words, the Hessian must be *positive definite*, which is expressed mathematically as $\nabla^2_{\theta} SSE(\theta) \succ 0$.

For the LS problem of interest (5.44), we have that the Hessian of the SSE function is given by:

$$\nabla^2_{\theta} SSE(\theta) = \mathbf{X}^T\mathbf{X}. \tag{5.54}$$

We can thus see that the Hessian of this problem is fully defined by the input data matrix $\mathbf{X}$.

Eigenvalues are important quantities that characterize the Hessian matrix; we will be discussing these quantities in more detail in Chapter 6. In our MLE context,

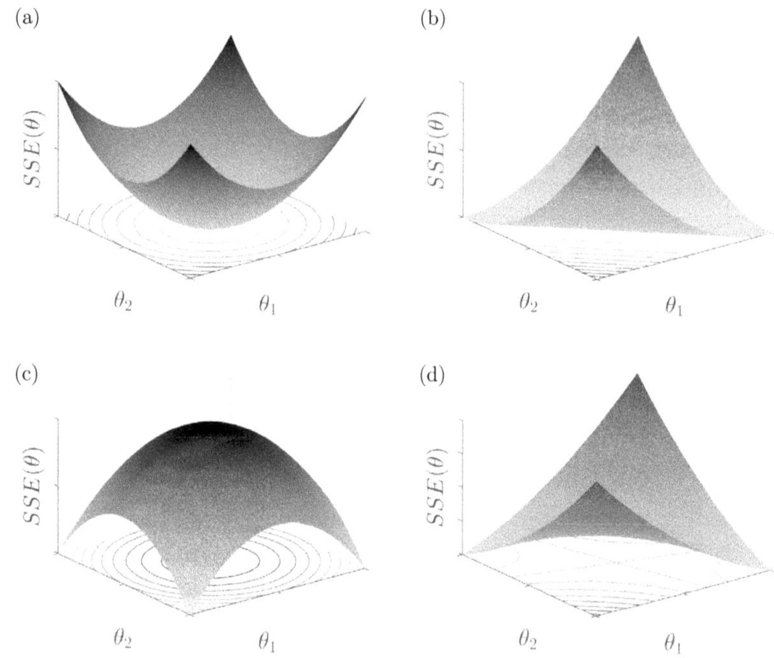

Figure 5.7 Example of $SSE(\theta)$ function for different Hessian matrices. Hessian with positive eigenvalues defines a unique minimum (a). Hessian with one positive and one zero eigenvalue defines a flat minimum (b). Hessian with negative eigenvalues (c). Hessian with one positive and one negative eigenvalue (d).

eigenvalues are important because they characterize the shape of the SSE function (e.g., has a sharp or flat minimum). As shown in Figure 5.7, a Hessian with positive eigenvalues defines a quadratic (parabolic) function with a unique minimum. A Hessian that contains zero eigenvalues defines a quadratic function with a flat minimum (we cannot estimate some of the parameters uniquely from the data). Hessians can have negative eigenvalues (the optimum sits at a maximum) or a combination of negative and positive eigenvalues (the optimum sits at a "saddle point"). Negative eigenvalues are undesirable, as they indicate that our estimates maximize the SSE (as opposed to minimize it). Fortunately, negative eigenvalues do not appear in estimation problems with linear models, but they can appear in nonlinear models (as we will discuss in Section 5.3).

Example 5.3 Estimating Heat Capacity Parameters

You are interested in conducting calorimetry experiments to determine how does the heat capacity of a gas varies with temperature. You hypothesize that there exists a structural relationship of the following form:

$$Y = \theta_0 + \theta_1 X + \theta_2 X^2 + \epsilon,$$

where Y is the heat capacity (in cal/mol-K) and X is the temperature (in K). This model has three parameters and the relationship established between Y and X is nonlinear (quadratic). The model, however, is linear in the parameters. To see this more clearly, we write the model in the more convenient form (5.33):

$$Y = \theta_0 + \theta_1 X_1 + \theta_2 X_2 + \epsilon,$$

by defining $X_1 = X$ as the temperature and $X_2 = X^2$ as the squared temperature.

To determine the parameters of this structural model, you conduct a total of $S = 30$ experiments. Specifically, you conduct 10 experiments by setting the temperature at 300 K, 10 experiments at 600 K, and 10 experiments at 900 K. You repeat experiments because you want to make sure that the experiments are performed right (there is reproducibility) and because you do not fully trust the heat capacity measured (due to inherent variations in the experiment). At the same time, your experiments cover a wide range of temperatures to make sure that the trend is persistent.

The data collected from your experiments is shown in Figure 5.8. From here, we can see that, indeed, the measured heat capacity varies widely over the experiment repetitions; as a result, it is not directly obvious what is the actual value of the heat capacity at each temperature. Here, we also show the mean value for the output and use this as a representative value. From the mean values, we see that the trend appears to be linear; as a result, we can anticipate that the quadratic component of the model is not that strong in the temperature range explored.

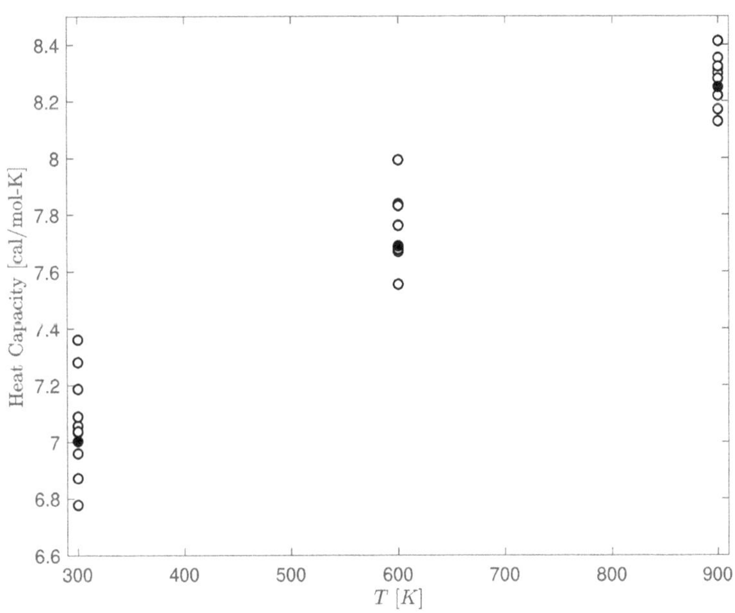

Figure 5.8 Data for calorimetry experiments showing dependence between heat capacity and temperature. Black dots represent the mean values.

As is often the case, we do not have information about the sensor accuracy and thus we do not have a basis to determine the noise model. Here, we assume that the noise covariance is a diagonal matrix with entries $\Sigma_{\omega,\omega} = \sigma^2$ for $\omega \in \mathcal{S}$. Under this assumption, the MLE problem reduces to minimization of the SSE and has a solution given by (5.51). The solution of this problem tells us that the estimates are given by:

$$\hat{\theta} = \begin{bmatrix} 6.19 \times 10^{+0} \\ 3.17 \times 10^{-3} \\ -9.60 \times 10^{-7} \end{bmatrix}. \tag{5.55}$$

From here, we confirm that the parameter corresponding to the quadratic term is fairly small. Upon computing the eigenvalues of the Hessian $\nabla^2_\theta SSE(\theta) = \mathbf{X}^T\mathbf{X}$, we confirm that all of them are strictly positive and thus the estimate is unique. Using the estimated parameters, we compute the predictions and compare them against the observed measurements. From this, we also compute the residuals and check their distribution. The results are shown in Figure 5.9; here, we can see that the model seems to capture the trend well (relative to the mean observed outputs) and that the residuals appear to be normal. This indicates that the variability observed in the data is the result of a purely random effect.

Script to reproduce results: `ch5_heat_capacity_lin_est` □

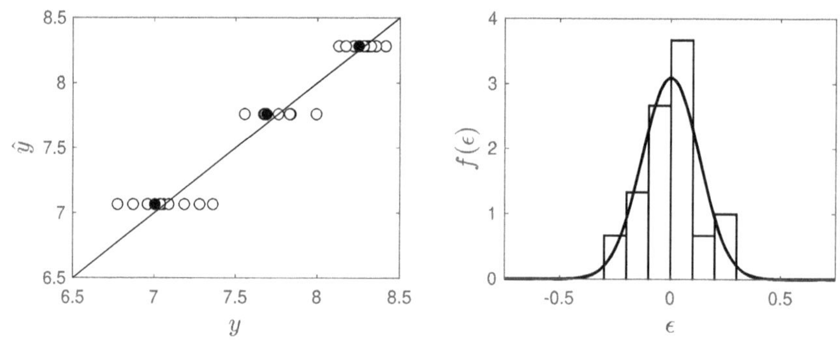

Figure 5.9 Parity plot showing model fit and residuals.

Recall now that our postulated model indicates that the observations are random and given by $\mathbf{y} = \mathbf{X}\theta + \mathbf{e}$, where θ denotes the true parameter. Consequently, we can express the estimate $\hat{\theta}$ as:

$$\begin{aligned} \hat{\theta} &= (\mathbf{X}^T\mathbf{X})^{-1}\mathbf{X}^T(\mathbf{X}\theta + \mathbf{e}) \\ &= (\mathbf{X}^T\mathbf{X})^{-1}(\mathbf{X}^T\mathbf{X})\theta + (\mathbf{X}^T\mathbf{X})^{-1}\mathbf{X}^T\mathbf{e} \\ &= \theta + (\mathbf{X}^T\mathbf{X})^{-1}\mathbf{X}^T\mathbf{e}. \end{aligned} \tag{5.56}$$

This reveals that the estimate $\hat{\theta}$ is an RV that results from a *linear transformation* of the Gaussian noise $\mathbf{e} \sim \mathcal{N}(0, \Sigma)$ (with $\Sigma_{\omega,\omega} = \sigma^2$ for $\omega \in \mathcal{S}$). Consequently, one can use the properties of Gaussian RVs to show that:

$$\hat{\theta} \sim \mathcal{N}(\theta, (\mathbf{X}^T\mathbf{X})^{-1}\sigma^2). \tag{5.57}$$

From this result, we can make the following important observations:

- The expected value of the estimate is the true parameter $\mathbb{E}[\hat{\theta}] = \theta$. This indicates that $\hat{\theta}$ is an *unbiased* estimator of θ.
- The covariance of the estimate is $\text{Cov}[\hat{\theta}] = (\mathbf{X}^T\mathbf{X})^{-1}\sigma^2$. From this expression, we see that, as the noise variance σ^2 increases, so does the covariance of $\hat{\theta}$. This indicates that the parameter covariance can be interpreted as the *sensitivity* of the estimates to noise.
- The Hessian matrix $\mathbf{X}^T\mathbf{X}$ plays a key role in determining the covariance of the estimate. Specifically, this matrix can dampen or magnify sensitivity to noise (we will analyze this in more precise terms later).

We now consider an MLE formulation with a general covariance matrix Σ. Here, we want to estimate $\hat{\theta}$ by solving the MLE problem (5.41). The estimate $\hat{\theta}$ that solves Eq. (5.41) must satisfy the necessary conditions:

$$-\nabla_\theta \log f(\mathbf{y}|\theta) = 0. \tag{5.58}$$

Here, we have eliminated $\mathbf{X}$ and Σ from the notation in order to simplify the discussion. The vector function $-\nabla_\theta \log f(\mathbf{y}|\theta)$ are the score functions. The score functions are:

$$-\nabla_\theta \log f(\mathbf{y}|\theta) = -\mathbf{X}^T\Sigma^{-1}(\mathbf{y} - \mathbf{X}\theta). \tag{5.59}$$

From this expression, we can see that the estimate is given by:

$$\hat{\theta} = (\mathbf{X}^T\Sigma^{-1}\mathbf{X})^{-1}\mathbf{X}^T\Sigma^{-1}\mathbf{y}. \tag{5.60}$$

The estimate is a unique minimum of the negative log-likelihood if:

$$-\nabla_\theta^2 \log f(\mathbf{y}|\theta) \succ 0, \tag{5.61}$$

where $-\nabla_\theta^2 \log f(\mathbf{y}|\theta)$ is the Hessian matrix of the negative log-likelihood. For the linear estimation problem (5.41), we have that the Hessian is given by:

$$-\nabla_\theta^2 \log f(\mathbf{y}|\theta) = \mathbf{X}^T\Sigma^{-1}\mathbf{X}. \tag{5.62}$$

As in the scalar linear case, note that this matrix does not depend on the parameters and output data; it only depends on the data matrix and the covariance matrix of the noise.

It is easy to verify that the above expressions reduce to the previous expressions derived for the special case in which the noise covariance Σ is diagonal with entries $\Sigma_{\omega,\omega} = \sigma^2$; in such a case, the covariance matrix can be written as $\Sigma = \sigma^2\mathbf{I}$.

We can express the estimate $\hat{\theta}$ as a function of the noise as:

$$\begin{aligned}\hat{\theta} &= (\mathbf{X}^T\Sigma^{-1}\mathbf{X})^{-1}\mathbf{X}^T\Sigma^{-1}(\mathbf{X}\theta + \mathbf{e})\\ &= \theta + (\mathbf{X}^T\Sigma^{-1}\mathbf{X})^{-1}\mathbf{X}^T\Sigma^{-1}\mathbf{e}.\end{aligned} \tag{5.63}$$

We again have that the estimate is a linear transformation of the Gaussian noise $\mathbf{e} \sim \mathcal{N}(0,\Sigma)$, with general covariance Σ. From this observation, we can conclude that the estimate is Gaussian:

$$\hat{\theta} \sim \mathcal{N}(\theta, (\mathbf{X}^T \Sigma^{-1} \mathbf{X})^{-1}). \tag{5.64}$$

This indicates that the estimate is unbiased and that the covariance of the estimate is $\text{Cov}[\hat{\theta}] = (\mathbf{X}^T \Sigma^{-1} \mathbf{X})^{-1}$. If we assume that the covariance is diagonal with entries given by σ^2, the covariance matrix reduces to $\text{Cov}[\hat{\theta}] = (\mathbf{X}^T \mathbf{X})^{-1} \sigma^2$.

Example 5.4 Mean and Covariance for Heat Capacity Parameters

We return to our previous example of estimating parameters for the heat capacity from calorimetry data. We recall that we assumed that the covariance matrix Σ is diagonal with entries σ^2. From the fit of the residuals to a Gaussian RV, we find $\epsilon \sim \mathcal{N}(\hat{\mu}, \hat{\sigma}^2)$ with $\hat{\mu} = 5.02 \times 10^{-14}$ and $\hat{\sigma} = 0.13$. From this, we conclude that the noise has a mean value of zero and thus the estimate $\hat{\theta}$ is unbiased; in other words, if we were to conduct multiple rounds of experiments and corresponding estimates, the mean of such estimates will be the true parameter value. From this, we also find that the covariance matrix is given by:

$$\begin{aligned} \text{Cov}[\hat{\theta}] &= (\mathbf{X}^T \hat{\Sigma}^{-1} \mathbf{X})^{-1} = (\mathbf{X}^T \mathbf{X})^{-1} \hat{\sigma}^2 \\ &= \begin{bmatrix} 3.15 \times 10^{-2} & -1.16 \times 10^{-4} & 9.23 \times 10^{-8} \\ -1.16 \times 10^{-4} & 4.52 \times 10^{-7} & -3.69 \times 10^{-10} \\ 9.23 \times 10^{-8} & -3.69 \times 10^{-10} & 3.07 \times 10^{-13} \end{bmatrix}, \end{aligned}$$

where $\hat{\sigma}^2$ is the empirical estimate of the true output noise σ^2 (which is not known). The estimate of the noise is obtained from the residuals of our model. Note also that, in this problem, the noise covariance matrix is assumed to be $\Sigma = \sigma^2 \mathbf{I}$ and thus its estimate is $\hat{\Sigma} = \hat{\sigma}^2 \mathbf{I}$.

We see that the parameter variances (diagonal entries of the covariance matrix) are small, indicating that the estimated parameters have a small uncertainty. In other words, we are quite certain that the estimates are close to the true values.

Script to reproduce results: `ch5_heat_capacity_lin_est` □

5.2.1 Bias, Precision, and Information

We can use a similar procedure as that used for the scalar case to show that the MLE estimate of the multivariate model is unbiased, to characterize the precision of the estimate, and to quantify the information contained in our data.

We begin by noticing that the score functions in the multivariate case satisfy:

$$\mathbb{E}[-\nabla_\theta \log f(\mathbf{y}|\theta)] = 0. \tag{5.65}$$

This indicates that perturbations in the output data $\mathbf{y}$ cancel out (on average). From this observation, we can also conclude that the MLE estimate $\hat{\theta}$ is an unbiased estimator of the true parameter θ. In other words, we have that $\mathbb{E}[\hat{\theta}] = \theta$. This validates the result (5.64), from which we can also conclude that the MLE estimate of the multivariate linear model is unbiased. However, we note that Eq. (5.64) was obtained by expressing the estimate as an explicit function of the noise. The result established here is more general and applies to any type of structural model.

The FI in the multivariate case is a matrix of the same dimension as the number of parameters and of the form:

$$\mathcal{I}(\theta) = \mathbb{E}[-\nabla_\theta^2 \log f(\mathbf{y}|\theta)]. \tag{5.66}$$

To analyze parameter precision, we are interested in analyzing the covariance matrix $\mathrm{Cov}[\hat{\theta}]$. We can use the FI matrix to generalize the Cramer–Rao bound as:

$$\mathrm{Cov}[\hat{\theta}] \geq \mathcal{I}(\theta)^{-1}. \tag{5.67}$$

Here, the symbol $\geq$ indicates that the covariance matrix is larger than the inverse of the FI matrix. Here, by larger we mean that the matrix $\mathrm{Cov}[\hat{\theta}] - \mathcal{I}(\theta)^{-1}$ is positive definite. This form of the Cramer–Rao bound is not particularly useful in practice, but one can show that we can express the bound for the diagonal entries of the covariance as:

$$\mathrm{Cov}[\hat{\theta}]_{jj} \geq \mathcal{I}(\theta)^{-1}_{jj}, \quad j = 0, \ldots, n. \tag{5.68}$$

This indicates that the variances of the individual parameters are greater than the diagonal entries of the inverse FI matrix. High variances of the parameters indicate that they are more uncertain (less precise); we can thus see that less information leads to less precision. We will also see shortly that we can use the parameter covariance matrix to quantify parameter precision in the form of confidence intervals.

As in the single-parameter case, the Cramer–Rao bound tells us that the precision of the parameters is dictated by the *sharpness* of the minimum of the negative log-likelihood function. In a multivariate setting, the sharpness of the minimum can be measured in different ways. For instance, one can use the determinant of the FI matrix (the product of its eigenvalues) or one can use the trace of the FI matrix (the sum of its eigenvalues). In general, what we want is that the eigenvalues of the FI are positive and as large as possible; such a case will indicate that our data encodes sufficient information to uncover the parameters from the data in a reliable manner. FI matrices that have eigenvalues that are close to zero indicate that our data is not as informative and it is thus more difficult to uncover the parameters from our data; this case manifests as a minimum of the negative log-likelihood function that is flat.

The Cramer–Rao bound is a theoretical bound that gives us an indication of the best possible precision (covariance) that we can achieve for the MLE estimate, but it does not give us an expression for the covariance matrix of the MLE estimate. Moreover, computing the FI matrix is often not possible because it depends on the true parameter and involves an expectation operator. In some special settings, however, one can compute the covariance and FI matrix explicitly. For the multivariate linear model under study, we have seen that the covariance matrix for an MLE estimate with noise covariance Σ is given by:

$$\mathrm{Cov}[\hat{\theta}] = (\mathbf{X}^T \Sigma^{-1} \mathbf{X})^{-1}. \tag{5.69}$$

Moreover, we recall that the negative log-likelihood is $-\log f(\mathbf{y}|\theta) = \log\left((2\pi)^{S/2}|\Sigma|^{1/2}\right) + \frac{1}{2}(\mathbf{y} - \mathbf{X}\theta)^T \Sigma^{-1}(\mathbf{y} - \mathbf{X}\theta)$. Consequently, the second derivative matrix (the Hessian matrix) is given by:

$$-\nabla_\theta^2 \log f(\mathbf{y}|\theta) = \mathbf{X}^T \Sigma^{-1} \mathbf{X}. \tag{5.70}$$

This does not depend on the parameter or on the output data; as such, the FI matrix is:

$$\begin{aligned} \mathcal{I}(\theta) &= \mathbb{E}\left[-\nabla_\theta^2 \log f(\mathbf{y}|\theta)\right] \\ &= \mathbf{X}^T \Sigma^{-1} \mathbf{X}. \end{aligned} \tag{5.71}$$

We thus have that, for the multivariate linear case:

$$\mathrm{Cov}[\hat{\theta}] = \mathcal{I}(\theta)^{-1}. \tag{5.72}$$

In other words, the Cramer–Rao bound holds with equality. This indicates that, in the linear case, the covariance obtained with MLE is optimal (it is the best we can possibly get). Moreover, we note that the FI matrix is the inverse of the covariance matrix and is the Hessian of the negative log-likelihood. This again reveals interesting *connections between information, uncertainty, precision, and minimum sharpness* (e.g., more information leads to less parameter uncertainty, higher parameter precision, and sharper minima).

Computing the exact parameter covariance and FI matrices for general nonlinear models will be difficult or impossible. However, our discussion so far already points to potential approaches to approximate these matrices. Specifically, we will see that one can approximate the FI matrix by using the Hessian matrix of the negative likelihood function and that this can also be used to approximate the covariance matrix. The *Hessian matrix* thus plays a central role in MLE methods. We will also use the observation that the estimate $\hat{\theta}$ is a transformation of the random noise $\mathbf{e}$ (as observed in Eq. (5.63)) to derive a Monte Carlo procedure to obtain an empirical estimate of the covariance matrix.

The *eigenvalues of the FI matrix* play a key role in quantifying the total *information content* of our data. Imagine that we compute the eigenvalues λ_j, $j = 0, \ldots, n$ of the FI matrix; note that the FI matrix contains $n + 1$ eigenvalues (same as the dimension of the matrix and the number of parameters). We have that the total information content can be expressed as the sum of the eigenvalues $\sum_{j=0}^n \lambda_j$ (which is the trace of the FI matrix) or as the product of the eigenvalues $\prod_{j=0}^n \lambda_j$ (which is the determinant of the matrix). In general, the larger the eigenvalues, the more information content of our data (the larger the trace and the determinant).

Example 5.5 Fisher Information Matrix for Heat Capacity Parameters

Because our model is linear, the FI matrix for the parameters is simply the inverse of the covariance matrix and is given by:

$$\mathcal{I}(\theta) = \mathbf{X}^T \hat{\Sigma}^{-1} \mathbf{X} = \begin{bmatrix} 1.80 \times 10^3 & 1.08 \times 10^6 & 7.57 \times 10^8 \\ 1.08 \times 10^6 & 7.57 \times 10^8 & 5.84 \times 10^{11} \\ 7.57 \times 10^8 & 5.84 \times 10^{11} & 4.77 \times 10^{14} \end{bmatrix}.$$

The matrix has large diagonal entries, and its eigenvalues are large and given by $(4.77 \times 10^{14}, 4.19 \times 10^7, 3.16 \times 10^1)$. From this, we can conclude that the input data contains significant information to estimate the parameters. Specifically, this indicates that the estimates lie on a sharp minimum of the log-likelihood function. Moreover, because the FI matrix is the inverse of the covariance, we can expect that the parameters have low uncertainty (high precision); this confirms the observations made in Example 5.4.

Script to reproduce results: `ch5_heat_capacity_lin_est` □

5.2.2 Residual Analysis

We now proceed to develop strategies to *quantify* the quality of an MLE estimate $\hat{\theta}$. We will be interested in analyzing the statistics of the residuals and in determining the confidence of the estimates.

Consider the following question:

> Are the residuals obtained with the estimate $\hat{\theta}$ following the postulated random noise model?

Recall that our postulated model has made the assumption that the random noise model is Gaussian. Moreover, we have established that the noise is precisely the model residuals $\epsilon_\omega = y_\omega - \hat{y}_\omega$ for $\omega \in \mathcal{S}$ (the model prediction errors). After we have obtained the estimate $\hat{\theta}$, it is important to verify that the residuals are indeed Gaussian; otherwise, the estimated model might contain systematic errors that are not explained by the postulated linear model.

A useful technique to visually confirm that the residuals are Gaussian consists of comparing the empirical quantiles $\hat{\mathbb{Q}}_\epsilon(\alpha)$ (obtained with our estimate $\hat{\theta}$) against the theoretical quantiles $\mathbb{Q}_\epsilon(\alpha)$ of the Gaussian model $\mathcal{N}(0, \Sigma)$. This is equivalent to comparing the empirical and theoretical cdfs. One can create a parity plot for these quantiles to see if the empirical and the theoretical quantiles coincide. This parity plot is known as a *Q–Q plot* and is used to check for the "Gaussianity" of the residuals.

Consider also the following question:

> How much of the variability in the output data $\mathbf{y}$ can be explained by our knowledge (model predictions $\hat{\mathbf{y}}$) and how much of it cannot be explained (model error ϵ)?

Answering this question is of relevance because it will give us an idea of how strong is our extracted trend (our knowledge) relative to noise. This question can be addressed by using a technique called *analysis of variance* (ANOVA). The total variability of the data is given by:

$$S_y = \sum_{\omega \in \mathcal{S}} (y_\omega - \bar{y})^2, \tag{5.73}$$

where:

$$\bar{y} = \frac{1}{S} \sum_{\omega \in \mathcal{S}} y_\omega, \tag{5.74}$$

is the sample average of the output data. The total variability can be decomposed into contributions as:

$$S_y = \underbrace{\sum_{\omega \in \mathcal{S}} (\hat{y}_\omega - \bar{y})^2}_{S_m} + \underbrace{\sum_{\omega \in \mathcal{S}} (y_\omega - \hat{y}_\omega)^2}_{S_e}. \tag{5.75}$$

Here, S_m is known as the *model sum-of-squares* and S_e is the *sum-of-squared errors* (SSE). Based on these quantities, we define an index, known as the *coefficient of determination*:

$$R^2 = \frac{S_m}{S_y} = 1 - \frac{S_e}{S_y}. \tag{5.76}$$

This index is the fraction of variability captured by model. Note that, by definition, we have that $R^2 \in [0, 1]$. The fraction of the variance that is left unexplained by the model is $1 - R^2 = S_e/S_y$; consequently, $R^2 = 1$ ($S_e = 0$) indicates that the model fully explains the variability. Ideally, we want that our estimate has an R^2 value that is equal to one. One can also show that the coefficient of determination R^2 is the square of the Pearson correlation coefficient ρ between the observed outputs y_ω and the model-predicted outputs $\hat{y}_\omega$. When the observed and model outputs are close, they become strongly correlated; this means that they lie on a diagonal line in a parity plot (the slope of such line is the Pearson correlation). Note, however, that the coefficient of determination does not tell us much about the nature of the statistics of the residuals; as such, it should be used with care.

The sum-of-squared errors S_e (SSE) is also a widely used measure of model quality because this is the measure that most estimation procedures are aiming to minimize and because it is intuitive. One often uses the mean squared error (MSE), which is a scaled version of the SSE:

$$MSE = \frac{1}{S}\sum_{\omega \in \mathcal{S}}(y_\omega - \hat{y}_\omega)^2 = \frac{1}{S}\sum_{\omega \in \mathcal{S}} \epsilon_\omega^2. \tag{5.77}$$

The MSE is particularly useful when comparing alternative models (e.g., models with different input variables or different structures). For instance, we can claim that a model is better than another if it has a lower MSE. One also often considers the *root-mean-squared error (RMSE)* $= \sqrt{MSE}$. Now, because the squared residual $\epsilon_\omega^2 = (y_\omega - \hat{y}_\omega)^2$ is an RV, one may also consider comparing models and their associated errors using other summarizing statistics. Specifically, the cdf of the squared error can be used to benchmark models. For instance, as we will see in Chapter 7, we can say that a model is better than another if the cdf of its errors dominates the cdf of the errors of the alternative model. This concept is known as *stochastic dominance* in decision-making and provides a powerful approach to compare RVs.

Example 5.6 Residual Analysis for Heat Capacity Parameters

From the observations y_ω and the model predictions $\hat{y}_\omega = \mathbf{X}\hat{\theta}$, we compute the following quantities:

$$\begin{aligned} \bar{y} &= 7.70 \\ S_y &= 7.92 \\ S_m &= 7.44 \\ S_e &= 0.48. \end{aligned}$$

From this, we confirm that $S_y = S_m + S_e$, and we can compute the index $R^2 = S_m/S_y = 0.94$. This indicates that 94% of the variability in the data can be explained by the model. The fraction of the variability that is left unexplained is $S_e/S_y = 0.06$ (6%). The Pearson correlation is $\rho = \sqrt{R^2} = 0.9695$; this indicates that the observed and model outputs are strongly correlated and lie on a diagonal line (as observed in the parity plot of Figure 5.9).

In Figure 5.10, we present a Q–Q plot of the residuals $\epsilon_\omega = y_\omega - \hat{y}_\omega$; note that the empirical estimates match the theoretical values, indicating that the residuals are Gaussian. Here, we also present the cdf for the squared residuals $\epsilon_\omega^2 = (y_\omega - \hat{y}_\omega)^2$ for a couple of competing models; Model I is our base model, while Model II ignores the contribution of X_2 (squared term for temperature). We can see that cdf of Model I is above that of Model II for most error levels. This indicates that Model I has a higher probability of having smaller errors.

Scripts to reproduce results: `ch5_heat_capacity_lin_est` and `ch5_heat_capacity_benchmark` □

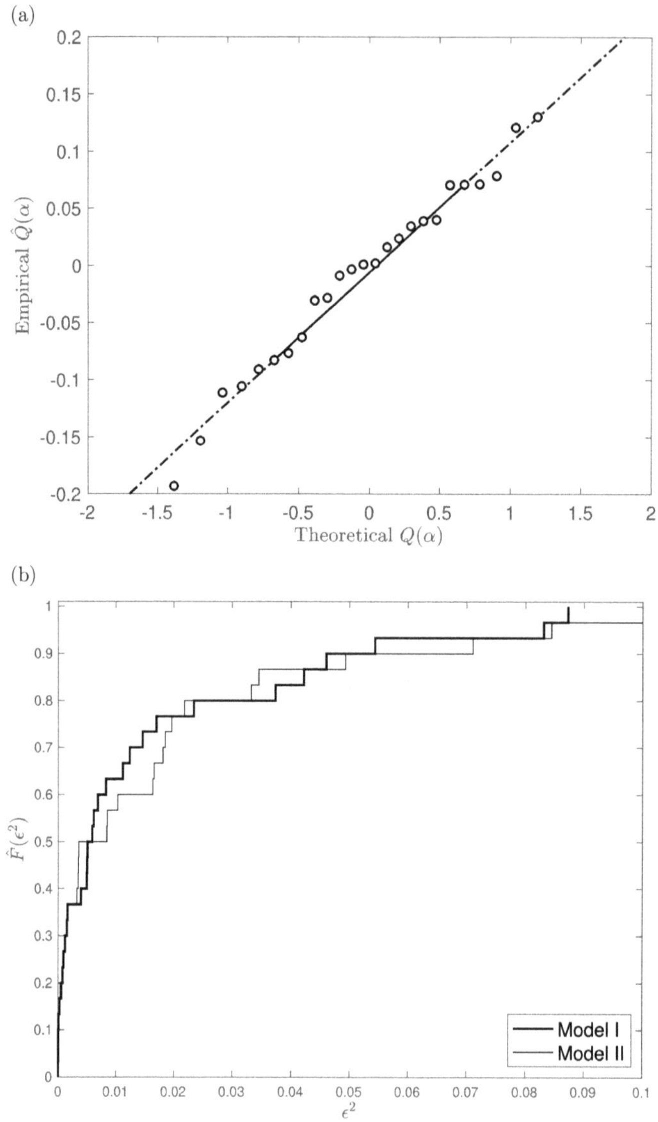

Figure 5.10 Q–Q plot showing that residuals ϵ_ω follow a Gaussian distribution (a). Comparison of cdf for squared errors (residuals) ϵ_ω^2 for a couple of competing models (b).

5.2.3 Uncertainty Quantification

Now consider the following question:

> How much can we trust the parameter estimates and the model predictions?

This question is at the core of a broad topic known as *uncertainty quantification*. Our goal here will be to develop *uncertainty models* for our parameter estimates and model predictions that will allow us to quantify our confidence in them. As we will discuss in Section 5.6 and Chapter 7, uncertainty models are essential in guiding data collection efforts, quantifying risk, and ultimately making decisions that anticipate and mitigate risk.

For our linear model, we have established that $\hat{\theta} \sim \mathcal{N}(\theta, \mathrm{Cov}[\hat{\theta}])$, with $\mathrm{Cov}[\hat{\theta}] = (\mathbf{X}^T \Sigma^{-1} \mathbf{X})^{-1}$. We also recall that the covariance is the inverse of the Hessian $-\nabla^2_\theta \log f(\mathbf{y}|\theta)$ and is the inverse of the FI matrix.

Because $\hat{\theta}$ is Gaussian, we have that this lies in the confidence ellipsoid:

$$\mathcal{E} = \left\{\theta \mid (\hat{\theta} - \theta)^T \mathrm{Cov}[\hat{\theta}]^{-1} (\hat{\theta} - \theta) \le q_{n+1}\right\}, \tag{5.78}$$

with $1-\alpha$ probability; here, q_{n+1} is the $(1-\alpha)$-quantile of $\chi^2(n+1)$. See Chapter 3 for more details on how to arrive at this result by using the properties of multivariate Gaussian RVs.

Importantly, the confidence ellipsoid (Eq. 5.78) is centered at the true parameter θ (because the estimate is unbiased). Note also that the shape of the ellipsoid (size and orientation) is dictated by the parameter covariance and by the number of parameters $n + 1$. Specifically, the region becomes larger as the variance of the parameters increases and as the quantile q_{n+1} increases with the number of parameters. A larger region indicates that we have more uncertainty in the estimate.

We also note that the covariance matrix is in turn directly related to the input data matrix $\mathbf{X}$ and to the noise covariance Σ. We thus have that, the more noise contained in the output data, the more uncertainty we will have in the estimate. In the next section, we will discuss the role of the input data and further analyze the impact of having many parameters in the model.

As we have seen, the rotation of the ellipsoid (the tilt) tells us if there are correlations between parameters. Ideally, one would want that our parameter estimates are not correlated; strong correlations between parameters indicate that they contain redundant information (e.g., and we can eliminate some of them). We will return to this topic in Chapter 6 (as part of our discussion on principal component estimation).

One often does not have the value of the true parameters θ, and it is thus not possible to define the ellipsoid. In such a case, one can also define the ellipsoid:

$$\mathcal{E} = \left\{\theta \mid \Delta\theta^T \mathrm{Cov}[\hat{\theta}]^{-1} \Delta\theta \le q_{n+1}\right\}. \tag{5.79}$$

This ellipsoid is the confidence region over which $\Delta\theta = \hat{\theta} - \theta$ (difference between estimated and true parameter) lies with probability $1-\alpha$. This ellipsoid corresponds to the multivariate Gaussian $\mathcal{N}(0, \mathrm{Cov}[\hat{\theta}])$, and we thus have that this has the same shape (size and orientation) as the standard ellipsoid, but it has been recentered around zero.

Confidence ellipsoids are difficult to visualize in more than two dimensions and are thus often analyzed based on pairwise plots (pairs of parameters). For example, for the pair $\hat{\theta} = (\hat{\theta}_i, \hat{\theta}_j)$, one constructs ellipsoids of the form:

$$\mathcal{E}_{ij} = \left\{\theta \mid (\hat{\theta} - \theta)^T C^{-1} (\hat{\theta} - \theta) \le q_2\right\}, \tag{5.80}$$

where $\theta = (\theta_i, \theta_j)$ are the true parameters and C is a 2×2 matrix with entries:

$$\begin{aligned} C_{11} &= \mathrm{Cov}[\hat{\theta}]_{ii} \\ C_{12} &= \mathrm{Cov}[\hat{\theta}]_{ij} \\ C_{21} &= \mathrm{Cov}[\hat{\theta}]_{ji} \\ C_{22} &= \mathrm{Cov}[\hat{\theta}]_{jj}, \end{aligned} \tag{5.81}$$

and we also note that q_2 is the $(1 - \alpha)$-quantile of $\chi^2(2)$. This 2D ellipsoid corresponds to that of a bivariate Gaussian RV $\mathcal{N}(\theta, C)$; in forming such an ellipsoid, we are ignoring information about the rest of the parameters. The 2D approximations thus miss important features of the true ellipsoid, but they are useful in getting a general idea of how uncertain are the parameters and if there are correlations between them.

We have also seen that the enclosing/bounding box of the full ellipsoid is given by the confidence intervals:

$$\theta_i = \hat{\theta}_i \pm m_i, \; i = 0, \ldots, n, \tag{5.82}$$

with:

$$m_i = \sqrt{q_{n+1} \mathrm{Cov}[\hat{\theta}]_{ii}}. \tag{5.83}$$

The ellipsoid and the enclosing box capture correlations between parameters, but the enclosing box is much easier to analyze. The confidence intervals defined by the bunding box give us ranges under which the true parameters lie. The size of the confidence intervals (m_i) is a measure of *parameter precision*. The size of the confidence intervals is affected by the number of parameters $n + 1$ (q_{n+1} increases with n) and by the variances of the estimated parameters $\mathbb{V}[\hat{\theta}_i] = \mathrm{Cov}[\hat{\theta}]_{ii}$ (diagonal entries of the covariance matrix). As noticed for the ellipsoid, our confidence in the parameters increases as we decrease the number of parameters and as the variance of the estimates decreases. Consequently, we deem an estimate to be precise if it has a small m_i. The confidence intervals thus provide a clear measure of the uncertainty of the parameters.

Confidence in estimates is also often expressed in terms of the *marginal* intervals:

$$\hat{\theta}_i \sim \mathcal{N}(\theta_i, \mathrm{Cov}[\hat{\theta}]_{ii}), \; i = 0, \ldots, n. \tag{5.84}$$

The marginal intervals are given by $\theta_i = \hat{\theta}_i \pm m_i$ with:

$$m_i = \sqrt{q_1 \mathrm{Cov}[\hat{\theta}]_{ii}}, \tag{5.85}$$

and where q_1 is the $(1 - \alpha)$-quantile of $\chi^2(1)$. These intervals capture the effect of the parameter variances; note, however, that these intervals *do not* capture the effect of the number of parameters. We also recall (as discussed in Chapter 3) that the marginal intervals form a confidence box that ignores correlations.

We can also quantify the uncertainty of model predictions by deriving confidence intervals for the outputs. To do so, we recall that $\hat{\mathbf{y}} = \mathbf{X}\hat{\theta}$ and $\hat{\theta} \sim \mathcal{N}(\theta, \mathrm{Cov}[\hat{\theta}])$. As such, we can propagate the uncertainty of the parameter estimate (which is a

Gaussian RV) through the linear model to obtain that the model prediction is a Gaussian of the form:

$$\hat{\mathbf{y}} \sim \mathcal{N}(\mathbf{y}, \mathbf{X}\mathrm{Cov}[\hat{\theta}]\mathbf{X}^T), \tag{5.86}$$

with $\mathbf{y} = \mathbf{X}\mathbb{E}[\hat{\theta}] = \mathbf{X}\theta$ (because the estimate is unbiased). We thus have that the expected value of the model predictions $\mathbb{E}[\hat{\mathbf{y}}]$ is the true (noise-free) model prediction $\mathbf{y} = \mathbf{X}\theta$. We also have that:

$$\begin{aligned}\mathrm{Cov}[\hat{\mathbf{y}}] &= \mathbf{X}\mathrm{Cov}[\hat{\theta}]\mathbf{X}^T \\ &= \mathbf{X}(\mathbf{X}^T\Sigma^{-1}\mathbf{X})^{-1}\mathbf{X}^T.\end{aligned} \tag{5.87}$$

The confidence intervals for the true outputs are:

$$y_\omega = \hat{y}_\omega \pm m_\omega, \; \omega \in \mathcal{S}, \tag{5.88}$$

with

$$m_\omega = \sqrt{q_S \,\mathrm{Cov}[\hat{\mathbf{y}}]_{\omega\omega}}. \tag{5.89}$$

The marginal output intervals are obtained by replacing q_S with q_1.

Confidence intervals play a central role in *hypothesis testing*; in a nutshell, hypothesis testing provides systematic tests that indicate whether our available data $x_\omega, \omega \in \mathcal{S}$ provides sufficient evidence that there is a relationship between the output Y and the input variables X. In other words, we are interested in determining if we have enough data to "make a call." Hypothesis testing can help reveal which input variables are not relevant (and therefore can be eliminated from the model). There are excellent and detailed resources on this topic, and we will thus not cover this topic here.

Another strategy that can be used to quantify the uncertainty of model predictions is by using Monte Carlo simulation. For instance, we can generate possible realizations of the estimate $\hat{\theta}$ by sampling $\mathcal{N}(\theta, (\mathbf{X}^T\Sigma^{-1}\mathbf{X})^{-1})$ and propagate the samples through the model $\hat{\mathbf{y}} = \mathbf{X}\hat{\theta}$ to obtain possible realizations of the model prediction. The realizations of the model prediction can then be used to obtain the sample covariance of $\hat{\mathbf{y}}$. The Monte Carlo approach does not make much sense in the linear case because we have a way to obtain analytical expressions, but it becomes very useful in nonlinear models (see Section 5.3 for more details). Overall, it is important to understand when it is possible to derive analytical expressions and when it is necessary to resort to Monte Carlo.

Example 5.7 Uncertainty Quantification for Heat Capacity Parameters

From our previous calculations, we have determined that the estimate is $\hat{\theta} \sim \mathcal{N}(\theta, (\mathbf{X}^T\Sigma^{-1}\mathbf{X})^{-1})$. This indicates that the $(1-\alpha)$ confidence ellipsoid:

$$\mathcal{E} = \left\{\theta \mid (\hat{\theta} - \theta)^T \mathrm{Cov}[\hat{\theta}]^{-1}(\hat{\theta} - \theta) \le q_{n+1}\right\},$$

is a 3D ellipsoid (we have a total of $n+1 = 3$ parameters). We recall that the parameter estimate is given by:

$$\hat{\theta} = \begin{bmatrix} 6.19 \times 10^{+0} \\ 3.17 \times 10^{-3} \\ -9.60 \times 10^{-7} \end{bmatrix},$$

and the covariance matrix is given by:

$$\begin{aligned} \text{Cov}[\hat{\theta}] &= (\mathbf{X}^T \hat{\Sigma}^{-1} \mathbf{X})^{-1} \\ &= \begin{bmatrix} 3.15 \times 10^{-2} & -1.16 \times 10^{-4} & 9.23 \times 10^{-8} \\ -1.16 \times 10^{-4} & 4.52 \times 10^{-7} & -3.69 \times 10^{-10} \\ 9.23 \times 10^{-8} & -3.69 \times 10^{-10} & 3.07 \times 10^{-13} \end{bmatrix}. \end{aligned}$$

Since visualizing such an ellipsoid is difficult in 3D (and we do not know the true parameters), we proceed to construct 2D ellipsoids for the parameter differences. For instance, for the parameter pair $\Delta\theta = (\Delta\theta_0, \Delta\theta_1)$, we have the ellipsoid:

$$\mathcal{E}_{01} = \left\{ \theta \mid \Delta\theta^T C^{-1} \Delta\theta \le q_2 \right\},$$

with:

$$C = \begin{bmatrix} 3.15 \times 10^{-2} & -1.16 \times 10^{-4} \\ -1.16 \times 10^{-4} & 4.52 \times 10^{-7} \end{bmatrix}.$$

To form the ellipsoids, we determine that $q_2 = 5.99$ (this is the 0.95-quantile of $\chi^2(2)$). In Figure 5.11, we visualize the 95% confidence ellipsoids (with $\alpha = 0.05$) for the pairs $(\Delta\theta_0, \Delta\theta_1)$, $(\Delta\theta_1, \Delta\theta_2)$, and $(\Delta\theta_0, \Delta\theta_2)$. We observe that the ellipsoids are centered at zero and are quite small along both axes, indicating that there is a small amount of uncertainty in the parameter differences (we are certain that the difference between the estimated and true parameters is small). From here, we observe that there are negative correlations between $\hat{\theta}_0$ and $\hat{\theta}_1$ and between $\hat{\theta}_1$ and $\hat{\theta}_2$, as well as a positive correlation between $\hat{\theta}_0$ and $\hat{\theta}_2$.

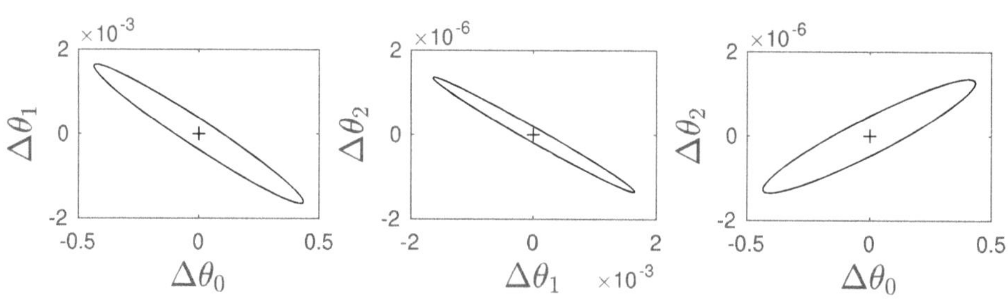

Figure 5.11 2D ellipsoids for heat capacity model.

We now compute the confidence region corresponding to the bounding box of the ellipsoid; for this, we determine $q_{n+1} = q_3 = 7.81$ (this is the 0.95 quantile of $\chi^2(3)$) and thus:

$$m_0 = \sqrt{7.81 \cdot 3.15 \times 10^{-2}} = 4.96 \times 10^{-1}$$

$$m_1 = \sqrt{7.81 \cdot 4.52 \times 10^{-7}} = 1.88 \times 10^{-3}$$
$$m_2 = \sqrt{7.81 \cdot 3.07 \times 10^{-13}} = 1.55 \times 10^{-6}.$$

From this, we can see that the sides of the enclosing box are indeed fairly small. We use these quantities and the estimated parameters to establish that the 95% confidence intervals for the true parameters are:

$$\theta_0 = 6.19 \times 10^{+0} \pm 4.96 \times 10^{-1}$$
$$\theta_1 = 3.17 \times 10^{-3} \pm 1.88 \times 10^{-3}$$
$$\theta_2 = -9.60 \times 10^{-7} \pm 1.55 \times 10^{-6}.$$

Another way to express this is that we are 95% certain that the true parameters lie in the intervals:

$$\theta_0 \in [5.70 \times 10^{+0}, 6.69 \times 10^{+0}]$$
$$\theta_1 \in [1.29 \times 10^{-3}, 5.06 \times 10^{-3}]$$
$$\theta_2 \in [-2.51 \times 10^{-6}, 5.91 \times 10^{-7}].$$

We can thus see that the range for all of them is narrow and thus we are fairly certain that the estimate is close to the true values.

For comparison, we also compute the marginal confidence intervals by first determining $q_1 = 3.84$ and:

$$m_0 = \sqrt{3.84 \cdot 3.15 \times 10^{-2}} = 3.48 \times 10^{-1}$$
$$m_1 = \sqrt{3.84 \cdot 4.52 \times 10^{-7}} = 1.31 \times 10^{-3}$$
$$m_2 = \sqrt{3.84 \cdot 3.07 \times 10^{-13}} = 1.08 \times 10^{-6}.$$

Consequently, the 95% marginal intervals are given by:

$$\theta_0 \in [5.85 \times 10^{+0}, 6.54 \times 10^{+0}]$$
$$\theta_1 \in [1.86 \times 10^{-3}, 4.49 \times 10^{-3}]$$
$$\theta_2 \in [-2.04 \times 10^{-6}, 1.27 \times 10^{-7}].$$

These intervals are smaller than those of the bounding box, but their qualitative features are similar (e.g., they have the same order of magnitude).

Finally, we compute the confidence intervals for the predicted heat capacities. To do so, we first determine that $q_S = 43.77$, which is the 0.95 quantile of $\chi^2(S)$ with $S = 30$ (the number of observations). We then compute the covariance matrix $\text{Cov}[\hat{\mathbf{y}}] = \mathbf{X}\text{Cov}[\hat{\theta}]\mathbf{X}^T$ (this is a 30×30 matrix that we do not show here), and from this, we compute the confidence intervals using (5.89). The results are summarized in Figure 5.12; here we present the model prediction (black dots), the observed outputs (white dots), and the lower and upper 95% confidence intervals. We can see that most observations lie inside the confidence intervals, and we note that the model prediction lies in the center. This tells us that the most likely value of the true output is the model prediction and that the true output lies somewhere in the interval with 95% certainty.

We can see that the intervals for the outputs are rather wide (even if the intervals for the parameters are small); this is because the noise variance is quite large (e.g., we have a lot of uncertainty on the sensor measurement).

Script to reproduce results: `ch5_heat_capacity_uncertainty` □

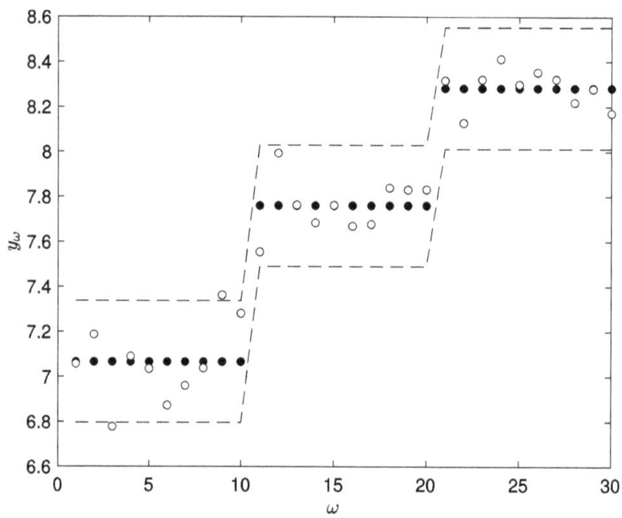

Figure 5.12 Confidence intervals for heat capacity predictions.

5.2.4 Data Sufficiency and Overfitting

Consider now the question:

Do we have enough data to estimate the model parameters?

The Hessian of the negative log-likelihood $\mathcal{H}(\hat{\theta}) = -\nabla^2_\theta \log f(\mathbf{y}|\hat{\theta})$ plays a fundamental role in answering this question. Here, we will analyze the Hessian in the context of a *linear model*, while in the next section we discuss issues that arise in nonlinear models. Analyzing the linear case is important because this is often a suitable approximation for the nonlinear case.

For the estimation of a linear model, we have seen that the Hessian of the negative log-likelihood function is $\mathcal{H}(\hat{\theta}) = \mathbf{X}^T\Sigma^{-1}\mathbf{X}$. This matrix is also often called the *kernel matrix*, and we observe that it contains all input data and that it does not depend on the estimate $\hat{\theta}$ and output data $\mathbf{y}$ (for nonlinear models, the Hessian does depend on the estimate and output data). We also recall that the Hessian is the FI matrix, and it is the inverse of the parameter covariance matrix. In summary, for a linear model, we have that:

$$\mathcal{H}(\hat{\theta}) = \text{Cov}[\hat{\theta}]^{-1} = \mathcal{I}(\hat{\theta}) = \mathbf{X}^T\Sigma^{-1}\mathbf{X}. \tag{5.90}$$

We also recall that the covariance matrix $\text{Cov}[\hat{\theta}]$ captures the sensitivity of the estimate to noise and its precision (uncertainty).

The Hessian matrix $\mathcal{H}(\hat{\theta})$ determines the sharpness of the minimum of the MLE problem and the variance (sensitivity) of the estimate $\hat{\theta}$. The eigenvalues of the Hessian matrix are quantities that can help us quantify this behavior. Specifically:

- For the estimate $\hat{\theta}$ to be a minimizer of the negative log-likelihood function $-\log f(\mathbf{y}|\theta)$, we need that all the eigenvalues of $\mathcal{H}(\hat{\theta})$ are nonnegative (either zero or positive). In the linear model case, this is guaranteed from the special structure of the Hessian $\mathcal{H}(\hat{\theta}) = \mathbf{X}^T\Sigma^{-1}\mathbf{X}$.
- If all the eigenvalues of $\mathcal{H}(\hat{\theta})$ are large and positive, the estimate $\hat{\theta}$ is a sharp minimum of the likelihood. This indicates that the estimate $\hat{\theta}$ is well defined by the data. Moreover, when the eigenvalues of the Hessian are large, we have that the eigenvalues of the covariance matrix are small; this indicates that the estimate exhibits low variance and sensitivity to noise. This also indicates that the estimate will have high precision (low uncertainty).
- If one (or more) eigenvalues of $\mathcal{H}(\hat{\theta})$ are close to zero, the estimate $\hat{\theta}$ is sitting at a minimum that is not sharp. This indicates that the estimate $\hat{\theta}$ is ill-defined by the data (cannot be easily observed/inferred from the data) and that it exhibits high variance (the covariance matrix has large eigenvalues) and sensitivity to noise. This also indicates that the estimate will have low precision (high uncertainty).
- If one eigenvalue of $\mathcal{H}(\hat{\theta})$ is exactly zero, the estimate $\hat{\theta}$ cannot be observed/inferred uniquely from the data. This indicates that the estimate is sitting at a minimum that is flat. This indicates that multiple parameter values give the same minimum value for the likelihood. This also indicates that the estimate has infinite variability (infinite uncertainty).

We have the ability to improve the quality of the estimate (e.g., increase precision) by controlling the input data $\mathbf{X}$ and, with this control, the Hessian/kernel matrix $\mathbf{X}^T\Sigma^{-1}\mathbf{X}$. This can be done by increasing the amount of data provided (add more observations). It is important to note, however, that the *amount* of data alone might not be sufficient to obtain a good estimate; we also require *quality* of data. Specifically:

- Data points that are *redundant* do not provide new information. Data redundancy results in a data matrix $\mathbf{X}$ that has dependent rows and an FI (Hessian) matrix $\mathcal{I}(\theta) = \mathbf{X}^T\Sigma^{-1}\mathbf{X}$ that has zero eigenvalues or eigenvalues that are very small. This will also result in a covariance matrix $\text{Cov}[\hat{\theta}] = \mathbf{X}^T\Sigma^{-1}\mathbf{X}$ with large eigenvalues and variances. Similarly, using input variables that are redundant leads to a data matrix that has dependent columns, and this leads to low information content and high estimate variability. Data redundancy is a common issue that appears in practice, but it is often difficult to spot and eliminate such redundancy (particularly in high-dimensional models). In Chapter 6, we will explore strategies to do this (e.g., singular value decomposition and principal component regression).

- If the selected input variables $X = (X_1, X_2, \ldots X_n)$ are not related to the output variable Y (or are weakly related to it), then the parameter estimates $\hat{\theta}$ will exhibit high variability (regardless of the number of observations used). For example, imagine that you want to build a model that relates reactor conversion to the concentration of an inert; if the inert does not impact conversion, any data provided on this variable will not be informative. Using knowledge of the application at hand to select appropriate input variables that best explain an output is extremely important. The inputs X are also known as *regressor variables*, *explanatory variables*, *features*, or *descriptors*.
- One can strategically *design experiments* to collect informative data that improves estimate precision. Specifically, we want to design the input data matrix $\mathbf{X}$ in such a way that it maximizes the Fisher information matrix $\mathcal{I}(\theta) = \mathbf{X}^T\Sigma^{-1}\mathbf{X}$. In other words, we want to design the data in a way that the estimate becomes sharp. The systematic selection of input variables and input data is part of a topic called *design of experiments*; we will discuss this topic in Chapters 6 and 7 (see also Exercises 5.10 and 5.11).

A common problem that arises in practice is that we might be unsure about which input variables best explain our output and, therefore, we might be tempted to select a large number of them to build our model. In other words, we would be tempted to use too many input variables X to explain output Y. This results in a model that has a large number of parameters, and this can lead to a phenomenon called *overfitting*. This issue can be diagnosed and mitigated as follows:

- Overfitting often results when we have more parameters than observations available to estimate them (the ratio n/S is greater than one), but this is not always the case. Ultimately, what is important is that the estimated parameters give model predictions that are accurate (that we have a good predictive model). As such, it is necessary to check that the estimate $\hat{\theta}$ obtained with the observations (y_ω, x_ω), $\omega \in \mathcal{S}$ gives good model predictions for a different set of observations. This procedure is called *cross-validation* or *out-of-sample testing*. Here, the observation set $\mathcal{S}$ is often called the training set, and the new observation set is often called the testing set. Cross-validation will ensure that the estimated model is *generalizable* (it predicts well outside the training data). Models that are overfitted do not generalize well (they follow the training set too closely). See Section 5.2.5 for more details.
- In linear models, an adjusted R^2 index is used to account for the number of parameters:

$$R^2_{adj} = 1 - \frac{S_e}{S_y}\frac{(S-1)}{(S-n)}. \tag{5.91}$$

  Here, as the number of parameters n decreases, we have that R^2_{adj} increases.
- Note also that the size of the confidence ellipsoid and of the enclosing box for the estimates $\hat{\theta}$ depends on the number of parameters $n+1$. Consequently, as we

increase the number of parameters, our confidence in them decreases. This will ultimately manifest in model predictions with high levels of uncertainty.

- If issues with generalizability or uncertainty are encountered, one would need to reduce the number of input variables (number of parameters used). This will create a *trade-off* between fitting (in-sample training) and generalizability (out-of-sample testing). You want to have enough parameters such that the model fit is adequate, but you also want the model to generalize well outside the training set. Overall, our goal should be to have the simplest model possible (with the least number of parameters).
- Another strategy used to deal with data scarcity and overfitting issues is *regularization*. Regularization seeks to embed *prior* knowledge in the estimation problem and with this eliminate parameters. See Section 5.4 for more details.

Example 5.8 Estimating Heat Capacity with Insufficient Data

To illustrate how overfitting issues can arise, imagine that we want to create a structural model to predict the heat capacity C of a chemical component as a function of temperature T:

$$C = \theta_0 + \theta_1 T.$$

Assume that you have one experimental data point available, denoted as (t_1, c_1), and consider the question:

Can θ_0, θ_1 be estimated *uniquely* from (t_1, c_1)?

The answer can be visualized in Figure 5.13(a); clearly, there are multiple (in fact, infinite) combinations of the model parameters (θ_0, θ_1) that can fit the single experimental data point given. In other words, if we were to estimate these parameters using MLE, the estimates would not be unique. This indicates that the data available is insufficient to estimate the parameters uniquely. Moreover, this indicates that the model will overfit the data (we have too many parameters for the data available).

In this simple example, we can easily visualize nonuniqueness, but in more complicated models (e.g., with many parameters), we cannot do so. We thus proceed to verify this using the Hessian matrix. Specifically, we will confirm that, when the parameters cannot be uncovered from the data, the Hessian matrix is singular (contains zero eigenvalues). Because we are not given information on the noise model, we consider the SSE minimization problem:

$$\min_{\theta_0,\theta_1} SSE(\theta_0,\theta_1) = \frac{1}{2}(c_1 - \theta_0 - \theta_1 t_1)^2.$$

The first-order derivatives of $SSE(\theta)$ are:

$$\frac{\partial SSE(0)}{\partial \theta_0} = -(c_1 - \theta_0 - \theta_1 t_1)$$
$$\frac{\partial SSE(\theta)}{\partial \theta_1} = -t_1(c_1 - \theta_0 - \theta_1 t_1).$$

The second derivatives of $SSE(\theta)$ are:

$$\frac{\partial^2 SSE(\theta)}{\partial\theta_0\partial\theta_0} = 1, \quad \frac{\partial^2 SSE(\theta)}{\partial\theta_0\partial\theta_1} = t_1$$
$$\frac{\partial^2 SSE(\theta)}{\partial\theta_1\partial\theta_0} = t_1, \quad \frac{\partial^2 SSE(\theta)}{\partial\theta_1\partial\theta_1} = t_1^2.$$

We thus have that the Hessian matrix for the SSE function is:

$$\mathcal{H}(\hat{\theta}) = \begin{bmatrix} 1 & t_1 \\ t_1 & t_1^2 \end{bmatrix}.$$

Since the model is linear, we can also obtain the Hessian by noticing that $\mathcal{H}(\hat{\theta}) = \mathbf{X}^T\mathbf{X}$ with:

$$\mathbf{X} = [1 \;\; t_1],$$

and thus

$$\mathbf{X}^T\mathbf{X} = \begin{bmatrix} 1 \\ t_1 \end{bmatrix}[1 \;\; t_1] = \begin{bmatrix} 1 & t_1 \\ t_1 & t_1^2 \end{bmatrix}.$$

Because the model is linear, we also confirm that the Hessian does not depend on the output data and on the parameters.

The determinant of the Hessian matrix is:

$$|\mathcal{H}(\hat{\theta})| = t_1^2 - t_1^2 = 0.$$

Recall that the determinant is the product of the eigenvalues of a matrix. Consequently, a determinant of zero indicates that at least one eigenvalue is zero (the Hessian is singular). Hessian analysis thus tells us that the parameters θ_0, θ_1 cannot be estimated uniquely from data for a single experiment, as expected.

Now imagine that you obtain an additional experimental data point (t_2, c_2) to estimate the model parameters. In this case, you now have two data points and two parameters. Can the parameters be estimated uniquely? Visual intuition tells us that yes, as shown in Figure 5.13(b). Specifically, two points are sufficient to fully

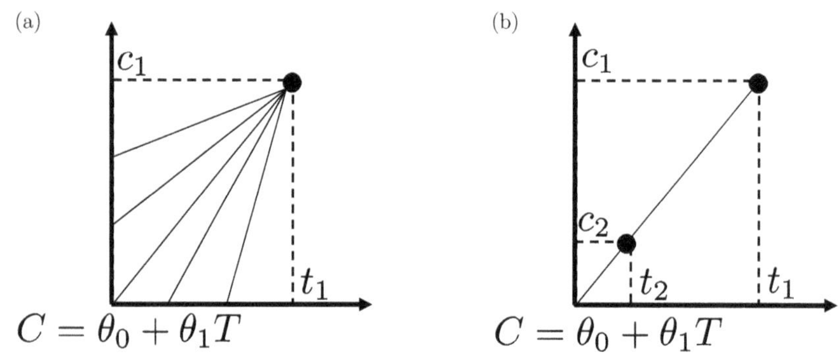

Figure 5.13 Fitting heat capacity model to a single data point (a) and to a couple of data points (b).

define the linear model. To confirm our intuition, we analyze the Hessian of the SSE minimization problem:

$$\min_{\theta_0,\theta_1} SSE(\theta_0,\theta_1) = \frac{1}{2}(c_1 - \theta_0 - \theta_1 t_1)^2 + \frac{1}{2}(c_2 - \theta_0 - \theta_1 t_2)^2.$$

The Hessian of the SSE function is:

$$\mathcal{H}(\hat{\theta}) = \mathbf{X}^T\mathbf{X} = \begin{bmatrix} 1 & 1 \\ t_1 & t_2 \end{bmatrix}\begin{bmatrix} 1 & t_1 \\ 1 & t_2 \end{bmatrix} = \begin{bmatrix} 2 & t_1 + t_2 \\ t_1 + t_2 & t_1^2 + t_2^2 \end{bmatrix}.$$

The determinant of the Hessian is:

$$\begin{aligned} |\mathcal{H}(\hat{\theta})| &= 2(t_1^2 + t_2^2) - (t_1 + t_2)^2 \\ &= t_1^2 - 2t_1t_2 + t_2^2 \\ &= (t_1 - t_2)^2. \end{aligned}$$

The determinant is positive (parameters are unique) as long as $t_1 \neq t_2$ (the experiments are different). In other words, if the second experiment is not exactly the same as the first one (the data is not redundant), the parameters will be unique. If the second experiment is redundant, then $t_1 = t_2$ and the determinant is zero (parameters are not unique). This is equivalent to the case of having a single experiment and illustrates how data redundancy can be a problem. This result also tells us that we only need two distinct experiments to ensure that the estimate is uniquely defined by the data; however, this does not mean that we should not try to obtain more experiments. Specifically, remember that our *confidence* in our parameters grows as we add data (this is particularly relevant if the data is noisy). □

5.2.5 Cross-Validation

K-fold cross-validation is an intuitive and useful technique that aims to ensure that a model is generalizable. The key idea is to *shuffle* our dataset $\mathcal{S}$ and then split the dataset into K (nonoverlapping) partitions $\mathcal{S}_k$, $k \in \mathcal{K}$, where $\mathcal{K} = \{1,2,\ldots,K\}$. The partitions are also called *folds*. Shuffling ensures that no bias is introduced in the selection of the partitions. Nonoverlapping indicates that there are no data points in one partition that are also present in other partitions. The approach is summarized in the following procedure.

For each unique group $k \in \mathcal{K}$:

- Define test dataset $\mathcal{T}_e = \mathcal{S}_k$.
- Define training set $\mathcal{T}_a = \cup_{j \in \mathcal{K}\setminus\{k\}} \mathcal{S}_j$.
- Compute $\hat{\theta}_k$ using training set $\mathcal{T}_a$ and define model $\mathbf{m}_k$.
- Test the model $\mathbf{m}_k$ in $\mathcal{T}_e$ and obtain performance metric p_k.

This procedure builds K distinct models $\mathbf{m}_k$, $k \in \mathcal{K}$ with their associated parameters and performance metrics $\hat{\theta}_k, p_k$, $k \in \mathcal{K}$. The performance metric can be any measure of quality that we deem appropriate (e.g., R^2, MSE, or the cdf of the squared errors). Note that the metrics are computed on the test set (not on the training set). From the models $\mathbf{m}_k$, $k \in \mathcal{K}$ collected, we can determine how well overall our proposed model structure behaves; this could allow us to compare a model structure with another model structure (e.g., that contains other input variables).

From the models $\mathbf{m}_k$, $k \in \mathcal{K}$, we can also select the model k that best behaves and use this as our final model; with this, we are selecting the training data that best generalizes. This procedure can also provide some insight into what is the minimum set of training data that generalizes well (this can be identified by tuning K). We also note that each data point $\omega \in \mathcal{S}$ is assigned to an individual group and stays in that group in the entire procedure. As such, each data point is given the opportunity to be used in the test set one time and is used to train the model $K - 1$ times.

Example 5.9 Cross-Validation for Heat Capacity Model

We use cross-validation to compare Model I (our base model) with Model II (which ignores squared temperature). Our dataset $\mathcal{S}$ has 30 data points, and we split this in $K = 3$ folds (partitions). This means that, for each fold, our test set $\mathcal{T}_e$ will contain 10 data points and our training set $\mathcal{T}_a$ will contain 20 data points. After running our K-fold validation procedure, we find that Model I gives a total MSE value of 0.049 (obtained by combining the MSE values obtained in the 3 folds), while Model II gives a total MSE value of 0.078, which is nearly double. This indicates that Model I generalizes better. The three MSE values obtained with Model I are 0.017, 0.013, and 0.018; we can thus see that the parameter estimates $\hat{\theta}_k$ obtained with the training set of the $k = 2$ fold lead to predictions that generalize better.

Script to reproduce results: `ch5_heat_capacity_benchmark` □

5.3 General Nonlinear Models

We now extend our discussion to general models of the form:

$$y_\omega = m(\theta, \mathbf{x}_\omega) + \epsilon_\omega, \; \omega \in \mathcal{S}, \tag{5.92}$$

where m is the model (a function of parameters and inputs). This model can be nonlinear and captures structural relationships between inputs, parameters, and outputs. For instance, this model can be a reactor model and θ are the kinetic constraints. In the linear case, define $m(\theta, \mathbf{x}_\omega) = \theta_0 + \sum_{i=1}^{n} \theta_i x_{\omega,i} + \epsilon_\omega$. The model $m(\theta, \mathbf{x}_\omega)$ can originate from physics, can also be a data-driven model (empirical model) such as a neural network, or can be a combination of physics and data-driven models (a hybrid model). As such, the discussion of this section applies to many different settings.

We express the model in the compact vector form $\mathbf{y} = \mathbf{m}(\theta) + \mathbf{e}$ and assume that $\mathbf{e} \sim \mathcal{N}(0, \Sigma)$. We will assume that θ and $\mathbf{x}_\omega$ are deterministic variables and we thus have that $\mathbf{y} \sim \mathcal{N}(\mathbf{m}(\theta), \Sigma)$.

We now proceed to use the MLE framework to estimate θ by minimizing the negative log-likelihood $-\log f(\mathbf{y}|\mathbf{m}(\theta),\Sigma)$. We assume that Σ is given and write the negative log-likelihood in the compact form $-\log f(\mathbf{y}|\theta)$. The MLE problem:

$$\min_{\theta} \; -\log f(\mathbf{y}|\theta), \tag{5.93}$$

is thus:

$$\min_{\theta} \; \log\left((2\pi)^{S/2}|\Sigma|^{1/2}\right) + \frac{1}{2}(\mathbf{y} - \mathbf{m}(\theta))^T\Sigma^{-1}(\mathbf{y} - \mathbf{m}(\theta)). \tag{5.94}$$

As in the linear model case, if assume that $\Sigma = \sigma^2\mathbf{I}$, the MLE problem reduces to the SSE minimization problem:

$$\min_{\theta} \; \frac{1}{2}\sum_{\omega \in S}(y_{\omega} - m(\theta, \mathbf{x}_{\omega}))^2. \tag{5.95}$$

For general noise covariance Σ, the solution $\hat{\theta}$ of Eq. (5.94) must satisfy the necessary conditions (score equations):

$$-\nabla_{\theta} \log f(\mathbf{y}|\theta) = -\nabla_{\theta}\mathbf{m}(\theta)^T\Sigma^{-1}(\mathbf{y} - \mathbf{m}(\theta)) = 0. \tag{5.96}$$

The score equations are now a set of n *nonlinear* equations that must be solved simultaneously to obtain $\hat{\theta}$. Here, $\nabla_{\theta}\mathbf{m}(\theta) \in \mathbb{R}^{S \times n}$ is the *Jacobian* matrix of the model and we denote this as:

$$\mathcal{J}(\theta) = \nabla_{\theta}\mathbf{m}(\theta). \tag{5.97}$$

The Jacobian contains the partial derivatives of the model for every observation with respect to each parameter:

$$\mathcal{J}(\theta)_{\omega,j} = \frac{\partial m(\theta, \mathbf{x}_{\omega})}{\partial \theta_j}, \; \omega = 1, \ldots, S, \; j = 1, \ldots, n. \tag{5.98}$$

The Jacobian is a *sensitivity matrix* that captures how the model outputs respond to changes/perturbations of the parameters. This sensitivity matrix plays an important role in parameter estimation, as it indicates if model outputs can be sufficiently "excited" by changing the parameter values (and thus helps understand which parameters are relevant or not). Weak model sensitivity often leads to parameter estimates that have high uncertainty, as the parameters cannot be easily inferred from output data.

For a linear model, we have that $\mathbf{m}(\theta) = \mathbf{X}\theta$, the Jacobian is thus $\mathcal{J}(\theta) = \mathbf{X}$, and the score equations reduce to the form previously discussed. This also reveals how the input data matrix $\mathbf{X}$ is the sensitivity matrix in a linear model. The nonlinear score equations do not have an analytic solution (as in the linear case) and require numerical techniques to be solved (e.g., Newton's method). The Jacobian is difficult to obtain analytically, and one must resort to numerical techniques to compute it (e.g., automatic differentiation or finite differences). Modern numerical software packages provide these capabilities.

To ensure that the estimate $\hat{\theta}$ minimizes $-\log f(\mathbf{y}|\theta)$, we must check that the sufficient condition:

$$\mathcal{H}(\hat{\theta}) \succ 0, \tag{5.99}$$

holds. This requires that the Hessian matrix:

$$\mathcal{H}(\hat{\theta}) = -\nabla^2_{\hat{\theta}} \log f(\mathbf{y}|\hat{\theta}) \in \mathbb{R}^{n \times n}, \tag{5.100}$$

is positive definite (all its eigenvalues are positive). This is particularly critical in the nonlinear case because such models can have minima, maxima, and saddle points (linear models only have minima). If the estimate lands on a maximum, it means that the model is maximizing the discrepancy between the model and the data (which obviously we do not want). If the estimate lands on a saddle point, we might have parameters that can be made arbitrarily large (or small). If the estimated parameters are not unique (they land on a flat surface and some eigenvalues of the Hessian are zero), it is important to conduct cross-validation to ensure that flatness is not the result of overfitting and that the model generalizes well.

The Hessian is the matrix of second derivatives of the negative log-likelihood function. Specifically, we have that:

$$\mathcal{H}(\hat{\theta})_{i,j} = -\frac{\partial^2 \log f(\mathbf{y}|\theta)}{\partial\theta_i \partial\theta_j}, \; i = 1,\ldots,n, \; j = 1,\ldots,n. \tag{5.101}$$

The Hessian matrix is given by:

$$\begin{aligned} \mathcal{H}(\hat{\theta}) &= -\nabla_\theta(\nabla_\theta \log f(\mathbf{y}|\hat{\theta})) \\ &= -\nabla_\theta(\nabla_\theta \mathbf{m}(\hat{\theta})^T \Sigma^{-1}(\mathbf{y} - \mathbf{m}(\hat{\theta}))). \end{aligned} \tag{5.102}$$

This is a complex quantity to evaluate analytically, and we thus require numerical techniques to compute it. One can show that, when the residuals $\mathbf{y} - \mathbf{m}(\hat{\theta})$ are small, the Hessian can be approximated as:

$$\mathcal{H}(\hat{\theta}) \approx \mathcal{J}(\hat{\theta})^T \Sigma^{-1} \mathcal{J}(\hat{\theta}). \tag{5.103}$$

This Hessian approximation is known as the *Gauss–Newton* approximation. For linear models, we have seen that the Hessian is given by:

$$\mathcal{H}(\hat{\theta}) = \mathbf{X}^T \Sigma^{-1} \mathbf{X}. \tag{5.104}$$

From here, we can see that the Gauss–Newton approximation is exact when the model is linear; specifically, recall that $\mathcal{J}(\hat{\theta}) = \mathbf{X}$. This indicates that the Gauss–Newton approximation is consistent, and it will be a good approximation if the model nonlinearity is moderate.

What is different about nonlinear estimation compared to the linear estimation case?

A couple of important issues are that: (1) It is more difficult to solve the MLE problem using numerical techniques and (2) it is more difficult to quantify the uncertainty of parameters and model predictions.

Regarding numerical behavior, it is important to note that numerical techniques such as gradient descent and Newton's method are often used to solve estimation problems. These algorithms need to be initialized properly (by providing a good initial guess for the parameters). In addition, it is important to note that these algorithms are mathematical tools that do not have the context of the application problem at hand; as such, they can converge to spurious parameter estimates that do not have physical meaning. As such, it is the responsibility of the modeler to ensure that the parameter estimates make practical sense (and to modify the problem formulation or the model used when they do not). We also note that, in the nonlinear case, the negative log-likelihood $-\log f(\mathbf{y}|\theta)$ might have multiple minimum points. This is an issue because numerical techniques such as gradient descent and Newton's method can only find local minima. This can create complications when obtaining empirical estimates (solution jumps between different minima depending on the initial guess and data used). One can use advanced techniques for global search, but these tend to be more computationally expensive. For a detailed discussion on the subject of numerical techniques for nonlinear parameter estimation (and for nonlinear optimization in general), you can consult [43, 44].

Example 5.10 Hougen–Watson Models

Olaf Hougen is one of the founding fathers of modern chemical engineering. Hougen, together with Kenneth Watson, proposed first-principles models to predict the behavior of heterogeneous catalytic reactions (which involve gaseous species that adsorb/desorb at a solid catalytic surface) [45]. Hougen–Watson (HW) models are widely used in reaction engineering but are highly nonlinear; the development of methods for parameter estimation for these models became an active subject of research in the 1960s and remains a challenge.

Here, we are interested in estimating parameters for the HW model:

$$Y = \frac{(\theta_0 X_2 - X_3/\theta_4)}{(1 + \theta_1 X_1 + \theta_2 X_2 + \theta_3 X_3)},$$

that best explains the available experimental data. The model captures the effect of competitive reaction and adsorption rates in a heterogeneous catalytic reaction. In this model, Y is the reaction rate, and (X_1, X_2, X_3) are species concentrations (partial pressures). The parameters $(\theta_0, \theta_1, \theta_2, \theta_3, \theta_4)$ have physical meaning (these represent adsorption, desorption, and reaction rates and thus need to be nonnegative). We have $S = 13$ observations (y_ω, x_ω) available and $n = 5$ parameters to estimate.

We are not given information about the noise, and thus, we proceed to find the parameter estimates by solving the estimation problem:

$$\min_{\theta} \frac{1}{2}\|\mathbf{y} - \mathbf{m}(\theta)\|^2.$$

It is clear that the HW model is highly nonlinear, and this can result in high nonlinearity of the sum-of-squared errors function $SSE(\theta)$. In Figure 5.14, we show the behavior of $SSE(\theta)$ for different values of $\theta_1 \in [-2, 2]$ and $\theta_2 \in [-2, 2]$ (while keeping the rest of the parameters fixed). This shows how, in a nonlinear model, we can see the emergence of maximum and multiple minimum points in $SSE(\theta)$ and complex behavior (e.g., the SSE function "blows up" at specific parameter values). This behavior differs from that observed in linear models (the SSE function is a well-behaved quadratic function) and makes nonlinear problems difficult to handle numerically. We also highlight that we are only showing the behavior of the SSE function as we span one parameter at a time, but the SSE is a function of five parameters; as such, the shape of the SSE function can be significantly more complex (but difficult to visualize).

From Figure 5.14, we note that we have *intentionally* evaluated the SSE function at negative values of the parameters. We know from physical understanding that these parameters cannot be negative. However, it is important to realize that, unless this knowledge is explicitly encoded in the estimation problem, we run the risk that the numerical solver lands at a minimum point that has nonnegative physical values. Specifically, numerical solvers are mathematical tools that do not have context of what constitutes acceptable physical behavior.

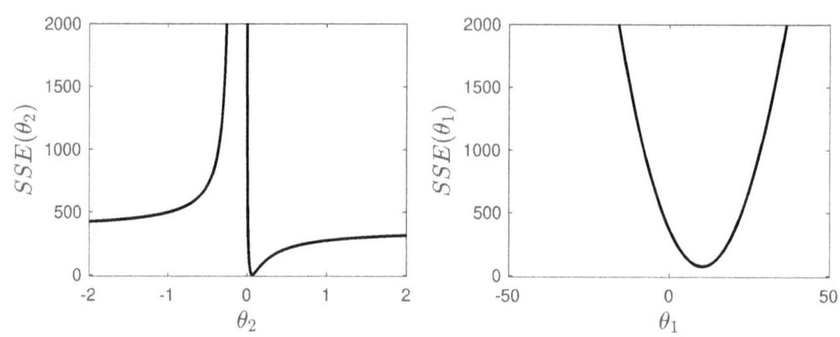

Figure 5.14 Behavior of SSE function as we span values for θ_1 and θ_2.

We solved the estimation problem by using a numerical software package; this package provides estimates and their corresponding confidence intervals. The estimates found are given by:

$$\hat{\theta}_0 = 1.25$$
$$\hat{\theta}_1 = 0.06$$
$$\hat{\theta}_2 = 0.04$$
$$\hat{\theta}_3 = 0.11$$
$$\hat{\theta}_4 = 1.19.$$

The model fit using the estimated parameters is excellent, and this is shown in Figure 5.15. The 95% confidence intervals are:

$$\theta_0 \in [-0.75, 3.25]$$
$$\theta_1 \in [-0.04, 0.16]$$
$$\theta_2 \in [-0.03, 0.11]$$
$$\theta_3 \in [-0.06, 0.29]$$
$$\theta_4 \in [-0.74, 3.12].$$

Note that such intervals do not make physical sense because they indicate that the estimates can take negative values. Moreover, the numerical package used to solve the problem does not report the Hessian matrix $\nabla^2_\theta SSE(\hat{\theta})$; because of this, it is not possible to conclude if the parameter estimates are unique.

The algorithm used by the numerical package is a Newton-like algorithm that aims to find a minimum point of the SSE function. Newton-like algorithms are sensitive to the initial guess provided; for example, if we place a parameter close to a point in which the SSE function is highly sensitive (where it blows up), the solver will diverge. To obtain the previous estimates, we guessed that the estimate was $\hat{\theta} = (1, 1, 1, 1, 1)$, which we can see is close to the estimates obtained. In this case, the Newton algorithm found the estimates that minimize the SSE function in 64 iterations. The minimum value found was $SSE(\hat{\theta}) = 0.2989$.

We now solve the estimation problem by providing the initial guess $\hat{\theta} = (-1, -1, -1, -1, -1)$; here, we intentionally ask the Newton algorithm to start in a non-physical region. We find that the Newton algorithm does not converge after 200 iterations and delivers the estimates:

$$\hat{\theta} = (-73.60, -3.70, -2.62, -6.36, -0.02),$$

which are clearly non-physical. Interestingly, however, we found that these estimated parameters yield $SSE(\hat{\theta}) = 0.3834$, which is not too far from the value found with the other set of estimates. This indicates that the *non-physical parameter estimates also provide a good fit of the data*. In Figure 5.15, we confirm that this is the case; note that the model delivers a nearly perfect fit. These numerical results highlight some of the challenges that we might encounter when dealing with nonlinear models. Specifically, numerical solvers might encounter estimates that give excellent fits but that are non-physical. Such estimates are acceptable from a purely mathematical standpoint but not from a physical standpoint. These issues can be overcome by exploiting our physical knowledge of the problem; specifically, while we might not know the actual parameter values, we might have an educated guess of their order of magnitude, and such information can be used to obtain a good initial guess. However, note that even with a good initial guess, nothing prevents the numerical solver from "landing" in an estimate with no physical meaning. To prevent this, we need to convey our physical knowledge in a different way, as we will discuss shortly.

Script to reproduce results: `ch5_hougen_watson` □

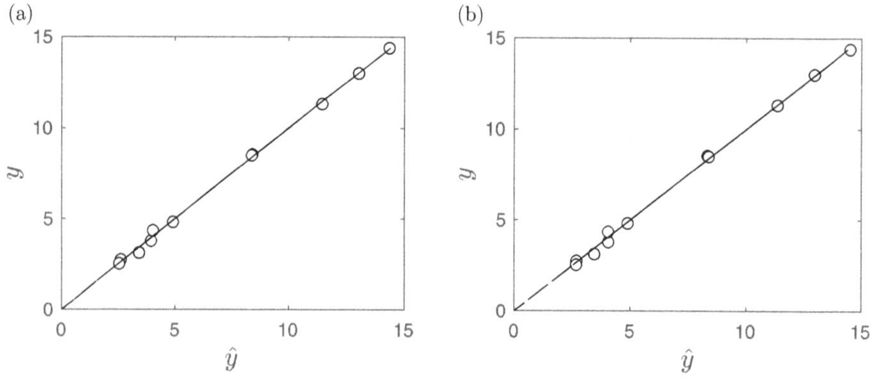

Figure 5.15 Fit of Hougen–Watson model to experimental data with physical (a) and non-physical (b) parameters.

Regarding difficulty in quantifying uncertainty in nonlinear models, we recall that, in the linear case, we concluded that the estimate $\hat{\theta}$ is Gaussian because this is obtained from a linear transformation of the noise (this is no longer true in the nonlinear case). We can quantify uncertainty for nonlinear models by using linear approximations of the model; these approximations assume close-to-linear behavior around the estimate $\hat{\theta}$. To explain this, consider the linearization of the model $\mathbf{m}(\theta)$ around the estimate $\hat{\theta}$:

$$\mathbf{y} = \mathcal{J}(\hat{\theta})(\theta - \hat{\theta}) + \mathbf{m}(\hat{\theta}), \tag{5.105}$$

where $\mathcal{J}(\hat{\theta}) = \nabla_\theta \mathbf{m}(\hat{\theta})$ is the model Jacobian evaluated at the estimate. The linearization can be written as $\mathbf{y} = \mathbf{M}\theta + \mathbf{b}$, where $\mathbf{M} = \mathcal{J}(\hat{\theta})$ and $\mathbf{b} = \mathbf{m}(\hat{\theta}) - \mathcal{J}(\hat{\theta})\hat{\theta}$. Note that this linear approximation is analogous to the linear regression model $\mathbf{y} = \mathbf{X}\theta$; as such, we can use this to derive analytical expressions that resemble those of the linear regression model. For instance, we can use the noise perturbation equation (5.63) to determine that the parameter estimate *approximately* follows the Gaussian:

$$\hat{\theta} \sim \mathcal{N}(\theta, (\mathcal{J}(\hat{\theta})^T \Sigma^{-1} \mathcal{J}(\hat{\theta}))^{-1}). \tag{5.106}$$

This approximation is known as the *Laplace approximation*. This implies that we can approximate the covariance of the parameter estimates as:

$$\text{Cov}[\hat{\theta}] \approx (\mathcal{J}(\hat{\theta})^T \Sigma^{-1} \mathcal{J}(\hat{\theta}))^{-1}. \tag{5.107}$$

Now note that $(\mathcal{J}(\hat{\theta})^T \Sigma^{-1} \mathcal{J}(\hat{\theta}))^{-1}$ is precisely the inverse of the Gauss–Newton approximation of the Hessian $\mathcal{H}(\hat{\theta})$ (Eq. 5.103). This suggests that we can use the inverse of the Hessian (when this is available) to approximate the covariance matrix. However, recall that the Hessian can be difficult to compute in complex problems; modern modeling languages can compute Hessians automatically, but some statistics packages might not provide such capabilities (in such cases, we can use the Gauss–Newton approximation).

We note that the Laplace approximation is exact when the model is linear ($\mathbf{m}(\theta) = \mathbf{X}\theta$); this is because the Jacobian reduces to $\mathcal{J}(\hat{\theta}) = \mathbf{X}$ and the covariance is $(\mathbf{X}^T\Sigma^{-1}\mathbf{X})^{-1}$, as shown in Eq. (5.64). As such, the Laplace approximation is consistent and expected to provide reasonable approximations when the model nonlinearity is mild.

Instead of using the analytical expression for the covariance $\text{Cov}[\hat{\theta}]$ using the Laplace approximation, we can also obtain an empirical estimate of the parameter covariance using Monte Carlo. Recall that $\text{Cov}[\hat{\theta}]$ tells us how the estimate $\hat{\theta}$ varies as we vary the output noise $\mathbf{e}$; in other words, the covariance captures the sensitivity of the parameter estimate to noise. This insight suggests that we can obtain the covariance by perturbing our outputs $\mathbf{y}$ with random noise (to obtain different realizations of the outputs) and by obtaining the corresponding estimates.

Specifically, consider that we have a set of observed outputs $\mathbf{y}$ (our output data); these observations are the result of a realization of a specific value of the noise $\mathbf{e} \sim \mathcal{N}(0,\Sigma)$ (which is an RV). As such, we can *simulate* alternative realizations of noise $\mathbf{e}_k$, $k = 1,\ldots,K$ by drawing random samples from $\mathcal{N}(0,\hat{\Sigma})$ (where $\hat{\Sigma}$ is an estimate of the noise covariance if this is known). With this, we obtain alternative realizations of the observations $\mathbf{y}_k = \mathbf{y} + \mathbf{e}_k$. We then use the output realizations to solve the family of MLE problems:

$$\min_{\theta} \; -\log f(\mathbf{y}_k|\theta) \qquad k = 1,\ldots,K \tag{5.108}$$

This gives realizations for the parameter estimate $\hat{\theta}_k$, $k = 1,\ldots,K$; from these, we can obtain an empirical mean:

$$\hat{\mathbb{E}}[\hat{\theta}] = \frac{1}{K}\sum_{k=1}^{K}\hat{\theta}_k, \tag{5.109}$$

and a sample covariance matrix as:

$$\hat{\text{Cov}}[\hat{\theta}] = \frac{1}{K-1}\sum_{k=1}^{K}(\hat{\theta}_k - \hat{\mathbb{E}}[\hat{\theta}])(\hat{\theta}_k - \hat{\mathbb{E}}[\hat{\theta}])^T. \tag{5.110}$$

Because the MLE estimate is unbiased, the empirical mean converges to the true parameter $\hat{\mathbb{E}}[\hat{\theta}] = \theta$ as $K \to \infty$ (from the law of large numbers). Moreover, the empirical covariance converges to the true covariance. This is a *Monte Carlo simulation* approach that is easy to implement and provides an alternative to the approximation of the covariance that uses the Hessian.

Once we have an approximation of the parameter covariance $\text{Cov}[\hat{\theta}]$ (obtained via the Laplace approximation or Monte Carlo), we can quantify the uncertainty of the model predictions by using the model linearization (5.105). This allows us to determine that the covariance of the outputs can be approximated as:

$$\text{Cov}[\hat{\mathbf{y}}] \approx \mathcal{J}(\hat{\theta})\text{Cov}[\hat{\theta}]\mathcal{J}(\hat{\theta})^T. \tag{5.111}$$

Note that this approximation reduces to Eq. (5.87) when the model is linear.

We can also quantify the uncertainty of model predictions by using Monte Carlo. Here, we draw realizations of the estimates $\hat{\theta}_k$, $k = 1,\ldots,K$ from $\mathcal{N}(\theta,\text{Cov}[\hat{\theta}])$ and propagate these realizations through the nonlinear model to obtain realizations of the output $\hat{\mathbf{y}}_k = \mathbf{m}(\hat{\theta}_k)$, $k = 1, \ldots, K$. From the output realizations, we can obtain the output sample covariance, and we can use this to obtain confidence intervals.

The above discussion highlights that, for nonlinear models, one can derive approximate analytical expressions or use sampling approaches to quantify uncertainty. These approaches have benefits and drawbacks; for instance, analytical expressions can be faster to compute (compared to conducting repeating simulations in sampling), but sampling captures nonlinear effects better than analytical expressions based on linearization. We should also emphasize that there is a broad range of sampling approaches that can be used to quantify the uncertainty of nonlinear models; e.g., importance sampling and Markov Chain Monte Carlo (MCMC) sampling.

Example 5.11 Monte Carlo Approximation of Parameter Covariance

We use the heat capacity example to illustrate how to estimate the parameter covariance matrix $\mathrm{Cov}[\hat{\theta}]$ using Monte Carlo sampling. In Example 5.4, we used analytical formulas to find that the exact covariance is:

$$\begin{aligned} \mathrm{Cov}[\hat{\theta}] &= (\mathbf{X}^T\mathbf{X})^{-1}\hat{\sigma}^2 \\ &= \begin{bmatrix} 3.15 \times 10^{-2} & -1.16 \times 10^{-4} & 9.23 \times 10^{-8} \\ -1.16 \times 10^{-4} & 4.52 \times 10^{-7} & -3.69 \times 10^{-10} \\ 9.23 \times 10^{-8} & -3.69 \times 10^{-10} & 3.07 \times 10^{-13} \end{bmatrix}, \end{aligned}$$

where $\hat{\sigma}^2$ is the empirical estimate of the true output noise σ^2. We recall that the noise covariance matrix is assumed to be $\Sigma = \sigma^2\mathbf{I}$ and thus the estimate is $\hat{\Sigma} = \hat{\sigma}^2\mathbf{I}$.

In order to determine the covariance with Monte Carlo, we obtain $K = 100$ realizations of the parameter estimate $\hat{\theta}_k$ for $k = 1, \ldots, K$. Each realization $\hat{\theta}_k$ is obtained by perturbing the output observations using the random noise as $\mathbf{y}_k = \mathbf{y} + \mathcal{N}(0, \hat{\Sigma})$ and by solving the estimation problem. We use our replicates to obtain the sample covariance (Eq. 5.110), which yields:

$$\hat{\mathrm{Cov}}[\hat{\theta}] = \begin{bmatrix} 2.79 \times 10^{-2} & -1.00 \times 10^{-4} & 7.96 \times 10^{-8} \\ -1.00 \times 10^{-4} & 3.96 \times 10^{-7} & -3.27 \times 10^{-10} \\ 7.96 \times 10^{-8} & -3.27 \times 10^{-10} & 2.77 \times 10^{-13} \end{bmatrix}.$$

We can see that the Monte Carlo estimate is close to the exact covariance. This confirms that the uncertainty of the estimate can be interpreted as a sensitivity of the estimate to perturbations in the output data.

Script to reproduce results: `ch5_heat_capacity_mc` □

5.4 Prior Knowledge

We often have an initial idea of what the actual parameter values should be or if they must respect general rules (particularly when the model has a physical origin).

For instance, our physical knowledge might tell us that a parameter cannot be negative or that it cannot be above a certain threshold (e.g., temperature, pressure, and concentration have limits). Consequently, we are interested in incorporating this *prior* knowledge into our estimation procedure. Adding prior knowledge is often needed to obtain reliable estimates and overcome numerical challenges.

Prior knowledge helps us deal with situations in which the available data is insufficient to estimate parameters uniquely (e.g., the Hessian has zero eigenvalues) or reliably (e.g., the Hessian has eigenvalues that are close to zero). Recall that a lack of sufficient data will manifest in large variances of the estimates and wide parameter confidence intervals (we have too much uncertainty on the estimates).

Prior knowledge helps us eliminate spurious estimates $\hat{\theta}$ that have no physical meaning. For instance, if the estimates are nonunique, we want to avoid those that are non-physical (e.g., negative rate constants).

Prior knowledge helps us reduce the number of parameters needed or narrow the space over which we search. Reducing the number of parameters can help us address data sufficiency issues (e.g., prevents overfitting). Narrowing the search space also helps numerical solvers to find estimates.

We can incorporate prior knowledge into our estimation problem by using several strategies:

- We can add bound constraints (limit values) on parameters:

$$\begin{aligned} \min_{\theta} \quad & -\log f(\mathbf{y}|\theta) \\ \text{s.t.} \quad & \underline{\theta} \leq \theta \leq \overline{\theta}, \end{aligned} \tag{5.112}$$

  where $\underline{\theta}, \overline{\theta} \in \mathbb{R}^n$ are allowable lower and upper bounds for the parameters (e.g., parameters cannot be negative). If we set $\underline{\theta}_i = \overline{\theta}_i$, we force the estimate to take a specific fixed value $\hat{\theta}_i = \underline{\theta}_i = \overline{\theta}_i$. Bounds are an effective way of filtering out regions that are non-physical or impractical. The region that parameters are allowed to lie on is called the *feasible region*. This is illustrated in Figure 5.16(a).
- We can add general constraints to the parameters (e.g., the sum of parameters must be equal to some value):

$$\begin{aligned} \min_{\theta} \quad & -\log f(\mathbf{y}|\theta) \\ \text{s.t.} \quad & \underline{\theta} \leq \Pi\theta \leq \overline{\theta}, \end{aligned} \tag{5.113}$$

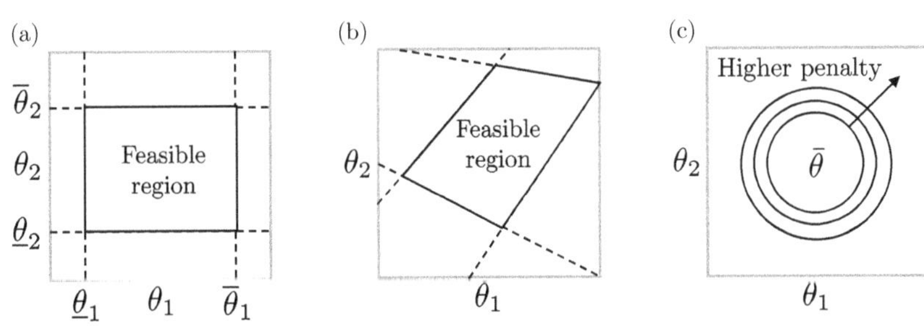

Figure 5.16 Feasible region defined by bounds (a) and general constraints (b). Penalization induced by ridge penalty (c).

where Π is a matrix that enforces some logic on the parameter values. For example, if we know that the sum of parameters has to be bounded, then we have that $\underline{\theta} \leq \sum_{i=1}^{n} \theta_i \leq \overline{\theta}$; this can be modeled as a general constraint by defining the vector $\Pi = [1, 1, \ldots, 1]$. Note how, as we discussed in Chapter 3, constraints define subdomains (in this case of the parameter space); see Figure 5.16(b). Constraints can be used to enforce complex physical knowledge and logic; e.g., mol fractions in a mixture must be equal to one or the output flow must be less than the input flow. It is also possible to enforce nonlinear constraints that capture complementary physical knowledge that is not captured in our model $\mathbf{m}(\theta)$.

- We can add a *penalty* term to the log-likelihood function to control the parameter behavior:

$$\min_{\theta} \; -\log f(\mathbf{y}|\theta) + \kappa \cdot \rho(\theta). \tag{5.114}$$

Here, $\kappa \geq 0$ is a constant that induces a trade-off between model fit (driven by the likelihood function) and allowed parameter movement (driven by the penalty term). The penalty term often seeks to prevent the movement of the parameter estimate from a given reference point. In other words, the penalty function aims to constrain the movement of the estimate by penalizing more strongly as we move away from the reference point; see Figure 5.16(c). The larger the κ value, the more weight we put on the penalty function, and this prevents too much movement of the parameter. On the other hand, the smaller the κ value, the more freedom we give the parameter estimate to move and minimize the likelihood function (fit the data). In other words, note that the optimal estimates obtained from the above formulation balance *prior knowledge* induced by the penalty $\rho(\theta)$ with *new knowledge* contained in the data. One can interpret κ as a price that we have to pay for movement/flexibility.

Common choices for the penalty function $\rho(\theta)$ are:

  - ℓ_2 norm (a.k.a ridge or Tikhonov penalty):

$$\begin{aligned} \rho(\theta) &= \frac{1}{2}\|\theta - \bar{\theta}\|_2^2 \\ &= \frac{1}{2}\sum_{i=1}^{n}(\theta_i - \bar{\theta}_i)^2 \\ &= \frac{1}{2}(\theta - \bar{\theta})^T(\theta - \bar{\theta}). \end{aligned} \tag{5.115}$$

    This penalizes squared deviations from a reference value $\hat{\theta}$.

  - ℓ_1 norm (a.k.a. lasso penalty):

$$\rho(\theta) = \|\theta - \bar{\theta}\|_1 = \sum_{i=1}^{n} |\theta_i - \bar{\theta}_i|. \tag{5.116}$$

    This penalizes absolute deviations from a reference value $\bar{\theta}$.

  - Bayes penalty function:

$$\rho(\theta) = \frac{1}{2}(\theta - \bar{\theta})^T \Sigma_{\theta}^{-1}(\theta - \bar{\theta}) \tag{5.117}$$

This is similar to the ridge penalty, but the matrix Σ_θ induces different weights on different parameter deviations (i.e., some parameters are allowed to move more than others). We will see shortly that the weight matrix can also be interpreted as a covariance matrix that captures the uncertainty that we have in our reference parameters.

The use of different norms for penalization induces different penalization regions; for instance, the circular regions shown in Figure 5.16(c) are characteristic of the ridge penalty and of the Bayes penalty (because these define ellipses). The ℓ_1 norm induces penalization regions that are rectangular.

The ridge and lasso penalties are often used to reduce the number of parameters needed in linear regression problems (to "sparsify" the parameter vector θ). This is done by setting the reference parameter $\hat{\theta} = 0$, which sets the penalty terms to $\frac{1}{2}\|\theta\|_2^2$ (for ridge) and $\|\theta\|_1$ (for lasso). Note that these penalty terms are minimized when all the parameters are equal to zero. As such, when trading off model fit with the penalty (by using different values of κ), we are trading off *model fit* vs. *model complexity* (i.e., the model is less complex when it contains fewer parameters). This is illustrated in Figure 5.17. As we have discussed, the reduction of model parameters is important to prevent overfitting.

The estimate obtained from a penalized problem depends on the value of κ chosen, and thus, we can obtain a range of solutions (for different values of κ) that we denote as $\hat{\theta}(\kappa)$. As a result, one often forms a *trade-off* curve $-\log f(\mathbf{y}|\hat{\theta}(\kappa))$ vs. $\rho(\hat{\theta}(\kappa))$ (known as a Pareto curve) and one selects the point that has the best trade-off (see Figure 5.17). Small values of κ will lead to smaller values of $-\log f(\mathbf{y}|\hat{\theta}(\kappa))$ (better fit) and to large values of $\rho(\hat{\theta}(\kappa))$ (more model complexity). Similarly, large values of κ will lead to large values of $-\log f(\mathbf{y}|\hat{\theta}(\kappa))$ (worse fit) and to small values of $\rho(\hat{\theta}(\kappa))$ (less model complexity). Every value of κ gives a different estimate $\hat{\theta}(\kappa)$ and model $\mathbf{m}(\hat{\theta}(\kappa))$; as such, one can use K-fold cross-validation on each of these models to figure out what value of κ gives the model that generalizes better. This is a strategy that is widely used in models that contain many parameters, such as machine learning models (we will discuss this in Chapter 6).

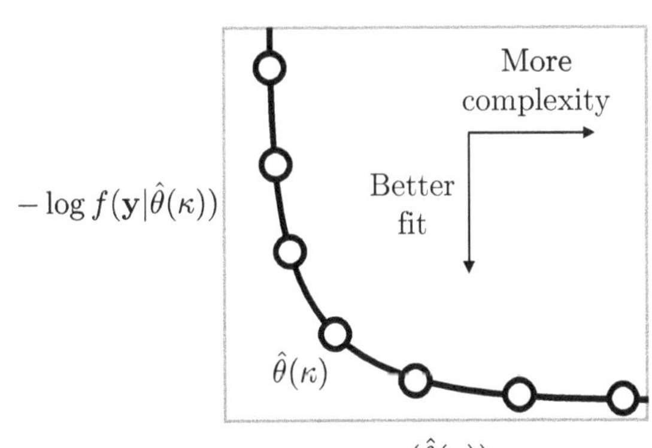

Figure 5.17 Trade-off between model complexity and model fit.

We can also obtain a trade-off curve by solving the following constrained MLE problem:

$$\begin{aligned} \min_{\theta} \ & -\log f(\mathbf{y}|\theta) \\ \text{s.t. } & \rho(\theta) \leq \kappa, \end{aligned} \tag{5.118}$$

for different values of κ. Here, κ is now interpreted as the allowed threshold for the penalty function $\rho(\theta)$. One then solves the above problem for different values of κ to form the trade-off curve. From this, we can see that one can express penalty functions implicitly as constraints. This formulation also reveals that we can think about κ as a budget of risk that we can tolerate (how much we want to "adventure" is exploring the parameter space). Specifically, as we increase our risk budget κ, we are allowing the model to be more flexible and "wild"; conversely, as we lower our risk budget, we are making the model less flexible and more conservative. We can use K-fold cross-validation to select a suitable risk budget κ.

Different penalties induce different behaviors, and some of them can be derived from statistical principles (we will see this is Section 5.5). Importantly, different constraints and penalties induce different (complementary) types of behavior and thus often one must combine multiple strategies. Numerical solvers *greatly benefit* from adding prior knowledge because sometimes the model might not be well defined in some non-physical regions and the solver might fail. For instance, by constraining the region over which a numerical solver searches on, one can avoid spurious minimum points (e.g., with no physical meaning). The incorporation of prior knowledge in MLE is also known as *regularization* because this aims to induce a more "regular" (benign) behavior of MLE problems; for instance, the incorporation of ridge penalties and Bayes penalties has the effect of adding positive curvature to the objective function of the MLE problem. This positive curvature tends to help numerical algorithms such as gradient descent and Newton's method to find the solution.

Example 5.12 Hougen–Watson Model with Ridge Penalty

In Figure 5.18, we show the behavior of the SSE function, which has been augmented using a ridge regularizer:

$$SSE(\theta) + \frac{1}{2}||\theta - \bar{\theta}||_2^2.$$

Here, we use $\kappa = 1$ and $\bar{\theta} = 0$, and we note that we use the SSE function instead of the negative likelihood function. By comparing this with the plots in Figure 5.14, we can see that the regularizer helps better define the minimum points of the function (the minimum points are sharper). However, note that this regularizer does not avoid the presence of non-physical regions. As a result, we might still encounter numerical issues if we were to apply a Newton algorithm to find the estimates.

Script to reproduce results: `ch5_hougen_watson_regularized` □

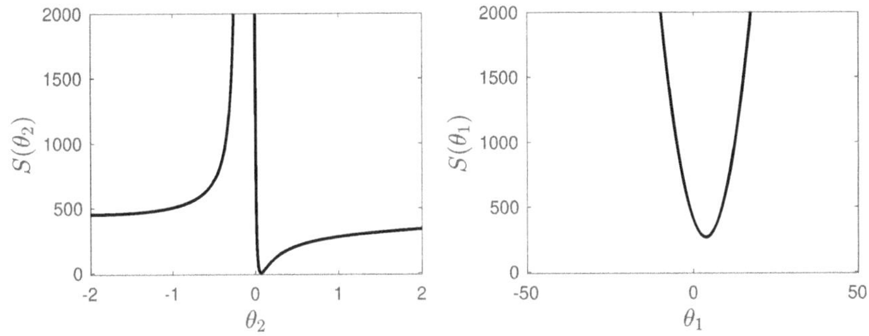

Figure 5.18 Behavior of regularized SSE function as we span θ_1 and θ_2.

Example 5.13 Hougen–Watson Model with Constraints

A highly effective way to filter out regions with non-physical meaning is to impose constraints in the MLE problem. For instance, we can estimate the parameters for the HW model by solving the constrained problem:

$$\min_{\theta} \frac{1}{2}\|\mathbf{y} - \mathbf{m}(\theta)\|^2$$
$$\text{s.t. } \underline{\theta} \leq \theta \leq \overline{\theta},$$

and we set the bounds $\underline{\theta} = (0,0,0,0,0)$ and $\overline{\theta} = (2,2,2,2,2)$. Here, the lower bounds $\underline{\theta}$ prevent the estimates from becoming negative; in other words, these bounds prevent the Newton algorithm from visiting non-physical values and from evaluating the model at such values. Similarly, the upper bounds prevent the algorithm from visiting values for the kinetic and adsorption constants that are unrealistic.

The constrained MLE problem was solved using a numerical package that uses a Newton-like algorithm to find the parameters that minimize the SSE function, but this algorithm can also handle constraints. We solve the MLE problem with an initial guess of $\hat{\theta} = (-1,-1,-1,-1,-1)$ and without using constraints; as before, we find that the algorithm does not converge and obtains estimates that are negative. Upon incorporating constraints (and using the same initial guess), we find that the algorithm identifies the estimates:

$$\hat{\theta} = (0.25, 0.06, 0.04, 0.11, 1.19),$$

in a few iterations. Note that this coincides with the estimates determined previously using a good initial guess. This clearly illustrates how the incorporation of constraints helps filter out spurious estimates.

The numerical package used does not provide Hessian information but provides the Jacobian matrix for the model $\nabla_{\theta}\mathbf{m}(\hat{\theta})$. As we have discussed, we can approximate the Hessian using the Jacobian and a Gauss-Newton approximation. This information can be used to determine if the estimated parameters are unique and to obtain a covariance approximation. The Hessian approximation obtained is given by:

$$\mathcal{H}(\hat{\theta}) = \begin{bmatrix} 619.12 & -4305.49 & -6345.59 & -940.39 & 99.98 \\ -4305.49 & 37792.30 & 44531.04 & 4721.95 & -434.18 \\ -6345.59 & 44531.04 & 75036.07 & 9325.14 & -663.07 \\ -940.39 & 4721.95 & 9325.14 & 2149.56 & -193.73 \\ 99.98 & -434.18 & -663.07 & -193.73 & 39.38 \end{bmatrix}.$$

When we obtain the eigenvalues of this Hessian approximation, we find that all of them are positive and given by:

$$\lambda = (1.06 \times 10^5, 8.35 \times 10^3, 9.52 \times 10^2, 3.53 \times 10^1, 2.55 \times 10^{-2}).$$

This provides an indication that $\hat{\theta}$ is sitting at a minimum point of the SSE function that is sharp. As such, the estimated parameters are unique in their neighborhood. Because the covariance matrix can be approximated using the inverse of the Hessian, large positive eigenvalues for the Hessian tell us that the variance of the parameters will be mild. Indeed, we can see that variances of the estimates are small and given by:

$$\mathbb{V}[\hat{\theta}] = (5.01 \times 10^{-1}, 1.26 \times 10^{-3}, 6.36 \times 10^{-4}, 3.76 \times 10^{-3}, 4.66 \times 10^{-1}).$$

Script to reproduce results: `ch5_hougen_watson_constraints` □

Example 5.14 Trade-Off Fit vs. Complexity

We would like to develop a linear model for the Gibbs reactor that predicts conversion as a function of different operating variables. We are unsure about what variables can best predict conversion so we consider data for all variables available. The model we consider has the form:

$$Y = \theta_0 + \theta_1 X_1 + \theta_2 X_2 + \theta_3 X_3 + \theta_4 X_4 + \theta_5 X_5 + \epsilon. \tag{5.119}$$

Here, Y is the conversion, X_1 is the pressure, X_2 is the output flow of CO, X_3 is the output flow of H_2, X_4 is the output flow of CH_3OH, and X_5 is the temperature. We have a total of $n + 1 = 6$ parameters and $S = 250$ observations. Because the input variables have different magnitudes, we scale them via standardization; specifically, for each variable, we subtract its mean and divide it by its standard deviation.

We begin by solving the estimation problem to obtain the estimates $\hat{\theta}$; these are obtained by using a numerical solver. The fit obtained with such parameters is shown in Figure 5.19(a); we can see that these parameters give a good fit. Because this problem is linear, we also obtain the estimates by using an analytical solution and find that the estimate is significantly different from that delivered by the numerical solver. However, both estimates provide the same fit; specifically, both of them deliver an MSE of 2.44×10^{-3}.

The fact that there are two different parameter sets that give the same fit indicates that the estimate is not unique. Upon inspection, we find that the Hessian matrix has zero eigenvalues. This tells us that there are redundancies in our input data matrix

X. As we have discussed, data redundancy can lead to overfitting because we have input variables (and associated parameters) that are not needed. A way to mitigate this issue is to reduce the number of parameters (number of input variables); to do so, we formulate an estimation problem in which we regularize the objective function as $MSE(\theta) + \kappa \, \|\theta\|_1$. In other words, we introduce a lasso penalty $\rho(\theta) = \|\theta\|_1$ that aims to force some parameters to become zero. We control how many parameters we want to keep by tuning the trade-off parameter κ.

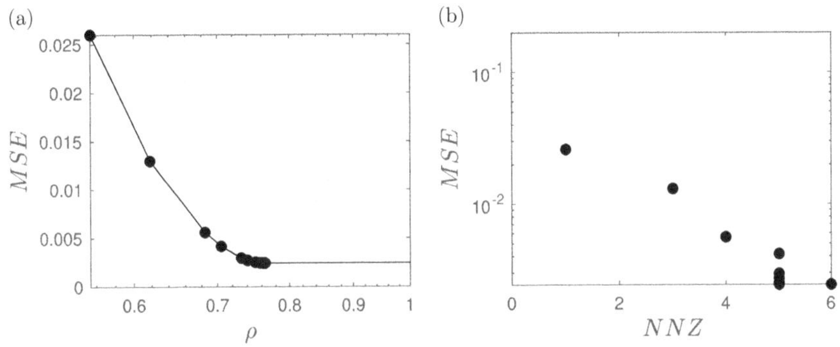

Figure 5.19 Trade-off curve of MSE vs. lasso penalty (a) trade-off curve of MSE vs. number of non-zero parameters (b).

In Figure 5.19, we can see the trade-off curve of the MSE vs. ridge penalty for different values of κ. In addition, we can see how the number of non-zero parameters (denoted as NNZ) is affected by κ; this is shown as a trade-off curve of MSE vs. NNZ. We can think of NNZ as a measure of model complexity; the higher the NNZ, the more complex the model (as it contains more parameters). Some interesting behavior and conclusions can be drawn from the trade-off curves. Note that the MSE is quite high when the penalty is small (when κ is high); this is because a small penalty restricts model flexibility. In other words, a small penalty leads to models with few parameters (less complex models). As the penalty increases, we give the model more flexibility (include more parameters and the model is more complex), and the MSE decreases sharply.

By inspecting a parameter estimate with three NNZ values, we notice that this has the following values:

$$\hat{\theta} = \begin{bmatrix} 6.14 \times 10^{-1} \\ 3.45 \times 10^{-12} \\ -2.22 \times 10^{-3} \\ -2.11 \times 10^{-3} \\ 1.39 \times 10^{-6} \\ -8.99 \times 10^{-8} \end{bmatrix}.$$

This indicates that variables X_1, X_4, and X_5 have a very small effect on the output (by at least three orders of magnitude). As such, we eliminate the data of these variables from the problem and estimate the parameters for a sparse (low-complexity) model with only three parameters. The fit of this reduced model is shown in Figure 5.20(b);

this reduced model has an MSE value of 2.45×10^{-3}. The ability to reduce model complexity (by identifying input variables that might not be relevant) is an important use of regularization approaches. In Chapter 6, we will also explore other approaches that identify data redundancies and help reduce model complexity.

Script to reproduce results: `ch5_gibbs_tradeoff` □

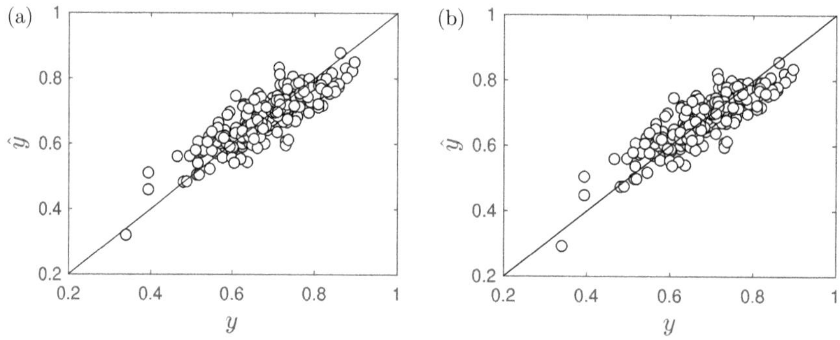

Figure 5.20 Fit of model under standard (a) and reduced model with three parameters (b).

5.5 Bayesian Estimation

So far we have used MLE as the main paradigm to determine parameters from data. In that paradigm, the idea is to find the estimate that maximizes the conditional pdf $f(\mathbf{y}|\mathbf{X}, \theta, \Sigma)$. We recall that this approach implicitly assumes that $\mathbf{X}$, θ, and Σ are deterministic variables. This is limiting because the input data $\mathbf{X}$ might also contain uncertainty (i.e., such observations might also come from sensors) or because we might want to convey prior knowledge on $\mathbf{X}, \theta$, and Σ in statistical terms.

There is a more general paradigm known as Bayesian estimation that allows us to overcome the limitations of MLE. This approach uses Bayes' theorem as a statistical basis to determine estimates. The key fundamental difference in Bayesian estimation is that $\theta, \mathbf{X}$, and Σ are treated as RVs. For simplicity in the discussion, we will assume for now that $\mathbf{X}$ and Σ are deterministic variables and focus on θ; generalizing the arguments is relatively straightforward (e.g., we will see that we can think of $\mathbf{X}$ also as parameters that need to be estimated).

In the context of estimation, Bayes' theorem states that:

$$f(\theta|\mathbf{y}) = \frac{f(\mathbf{y}|\theta)f(\theta)}{f(\mathbf{y})}. \tag{5.120}$$

The elements of the Bayes' theorem are:

- $f(\theta|\mathbf{y})$ is the probability that the parameters take value θ given knowledge that the output takes value $\mathbf{y}$. This pdf is known as the posterior pdf.
- $f(\theta)$ is the probability that the parameters take value θ given no knowledge of the output. This pdf is known as the prior pdf, and we note that this is simply the marginal pdf of the parameters.

- $f(\mathbf{y}|\theta)$ is the probability that the output takes value $\mathbf{y}$ given knowledge that parameters take value θ. This pdf is the likelihood function maximized in MLE.
- $f(\mathbf{y})$ is the marginal probability of the outputs; this pdf is often neglected, as it does not carry knowledge of the parameters and thus one often expresses Bayes' theorem as:

$$f(\theta|\mathbf{y}) \propto f(\mathbf{y}|\theta)f(\theta). \tag{5.121}$$

The goal of Bayesian estimation is to maximize the probability of the parameters $f(\theta|\mathbf{y})$. This approach makes more sense than MLE, in which one maximizes the probability of the outputs $f(\mathbf{y}|\theta)$. We thus find the Bayesian estimate $\hat{\theta}$ by solving the optimization problem:

$$\max_{\theta} \quad \log f(\mathbf{y}|\theta) + \log f(\theta). \tag{5.122}$$

This problem is similar to that of MLE but note that we incorporate the prior term $\log f(\theta)$. As a result, this approach naturally carries prior knowledge on the parameters.

Consider now that we have a model of the form $\mathbf{y} = \mathbf{m}(\theta) + \mathbf{e}$ with $\mathbf{e} \sim \mathcal{N}(0,\Sigma)$. Moreover, consider that we have the prior knowledge that $\theta \sim \mathcal{N}(\bar{\theta},\Sigma_\theta)$ and thus:

$$\begin{aligned}\log f(\theta) &= -\log\left((2\pi)^{n/2}|\Sigma_\theta|^{1/2}\right) - \frac{1}{2}(\theta-\bar{\theta})^T\Sigma_\theta^{-1}(\theta-\bar{\theta})\\ \log f(\mathbf{y}|\theta) &= -\log\left((2\pi)^{S/2}|\Sigma|^{1/2}\right) - \frac{1}{2}(\mathbf{y}-\mathbf{m}(\theta))^T\Sigma^{-1}(\mathbf{y}-\mathbf{m}(\theta)).\end{aligned}$$

After dropping constant terms and changing from maximization to minimization, problem (5.122) can be expressed as:

$$\min_{\theta} \quad \frac{1}{2}(\mathbf{y}-\mathbf{m}(\theta))^T\Sigma^{-1}(\mathbf{y}-\mathbf{m}(\theta)) + \frac{1}{2}(\theta-\bar{\theta})^T\Sigma_\theta^{-1}(\theta-\bar{\theta}). \tag{5.123}$$

From this, we can make the following observations:

- The Gaussian prior $f(\theta)$ achieves its maximum at the reference point $\bar{\theta}$, while $f(\mathbf{y}|\theta)$ achieves its maximum when θ fits perfectly the data.
- The Bayesian estimation problem thus naturally seeks to strike a *balance* (trade-off) between what we previously knew about θ and the new knowledge gained through the data $\mathbf{y}$.
- If we ignore prior knowledge, all that we know about θ is obtained through the data $\mathbf{y}$ (as in MLE). This might lead to ambiguity if the data is insufficient to estimate the parameters uniquely. Consequently, adding prior knowledge seeks to reduce *ambiguity*; in other words, the prior term used in Bayesian estimation naturally *regularizes* the estimation problem.
- The reference point $\bar{\theta}$ can be interpreted as a prior estimate in the sense that this is our best estimate prior to observing the new data $\mathbf{y}$. How much we trust our prior estimate is dictated by the covariance matrix Σ_θ. Specifically, if the variance of the prior (diagonal entries) is low, then we have high confidence in

our prior estimate. As such, the prior term will dominate the log-likelihood term. The non-diagonal entries of the covariance matrix allow us to capture knowledge of parameter correlations. This can be particularly useful when processing data sequentially, as we explain next.

Example 5.15 Bayesian Estimation for a Linear Model

In this example, we seek to illustrate how Bayesian estimation allows us to balance/trade-off prior knowledge and knowledge gained by data. We consider a simple linear estimation problem with a candidate model $Y = \theta X + \epsilon$ and assume that we have a single observation that we denote as $(x = 1, y = 2)$.

We assume that the information provided by prior knowledge is given by $\theta \sim \mathcal{N}(0, 1)$, and this defines the marginal $f(\theta)$. The information provided by the data (x, y) is given by $y \sim \mathcal{N}(\theta x, 1)$, and this defines the likelihood $f(y|\theta)$. The combined information is given by the posterior $f(\theta|y) \propto f(y|\theta)f(\theta)$.

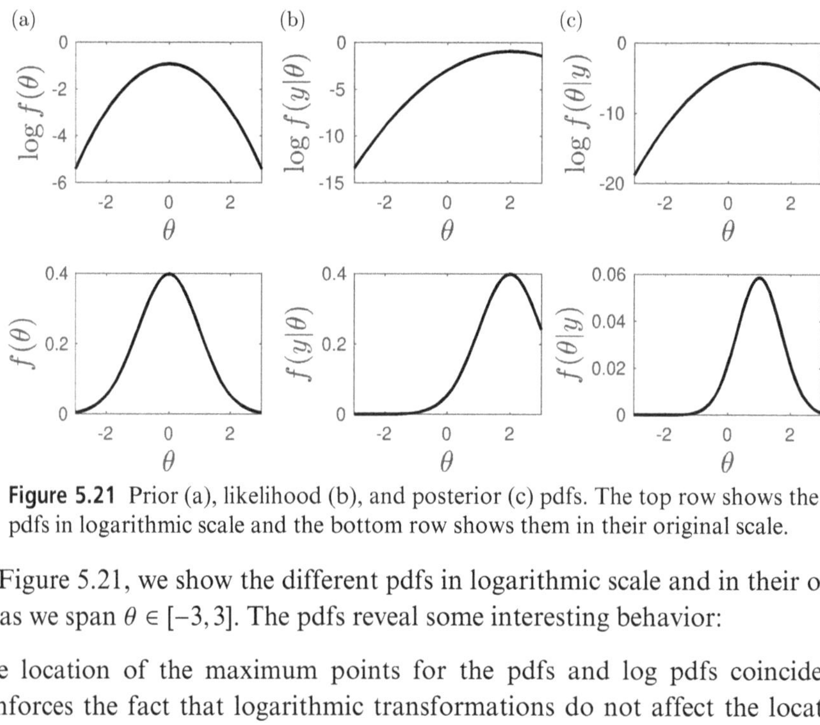

Figure 5.21 Prior (a), likelihood (b), and posterior (c) pdfs. The top row shows the pdfs in logarithmic scale and the bottom row shows them in their original scale.

In Figure 5.21, we show the different pdfs in logarithmic scale and in their original scale as we span $\theta \in [-3, 3]$. The pdfs reveal some interesting behavior:

- The location of the maximum points for the pdfs and log pdfs coincides; this reinforces the fact that logarithmic transformations do not affect the location of stationarity points.
- According to the prior information (embedded in$f(\theta)$), we have that $\theta = 0$ is the most likely value.
- According to the new information (embedded in $f(y|\theta)$), we have that $\theta = 2$ is the most likely value. This would be the estimate obtained with MLE (ignoring prior information).

- According to the combined prior and new information (embedded in $f(\theta|y)$), we have that $\theta = 1$ is the most likely value. We thus see that Bayesian estimation naturally "blends" old and new information and finds an estimate that is a trade-off point.

Script to reproduce results: `ch5_bayes_estimation` □

In order to quantify the uncertainty of the parameter estimates, we need the parameter covariance. For Bayesian estimation, the covariance of the parameter estimates is obtained using:

$$\mathrm{Cov}[\hat{\theta}] = (\mathcal{J}(\hat{\theta})^T \Sigma^{-1} \mathcal{J}(\hat{\theta}) + \Sigma_{\theta}^{-1})^{-1}. \tag{5.124}$$

Here, we recall that $\mathcal{J}(\hat{\theta}) = \nabla_{\theta}\mathbf{m}(\hat{\theta})$ is the Jacobian of the model; for linear models, we have that $\mathcal{J}(\hat{\theta}) = \mathbf{X}$. Note that this formula includes the contribution of the likelihood (as we discussed in MLE) and the prior covariance. This expression for the covariance can be obtained by deriving the optimality conditions of the Bayesian estimation problem (5.123). The expression is exact for linear models and is used as an approximation for nonlinear models. One can use the parameter covariance to quantify the uncertainty of the model predictions. For a linear model, this can be done by propagating the uncertainty of the parameters through the linear model $\mathbf{y} = \mathbf{X}\theta$.

Bayesian estimation naturally lends itself to the *sequential processing of data*. Specifically, imagine that you have an initial set of data $\mathbf{y}_1$ and you do not have any prior knowledge. We can thus compute the estimate $\hat{\theta}_1$ by solving the problem:

$$\min_{\theta} \quad -\log f(\mathbf{y}_1|\theta), \tag{5.125}$$

and, from this, we can determine the covariance $\mathrm{Cov}[\hat{\theta}_1]$. Now consider that a new set of data $\mathbf{y}_2$ becomes available. We can define the prior $\mathcal{N}(\hat{\theta}_1, \mathrm{Cov}[\hat{\theta}_1])$ and obtain the estimate $\hat{\theta}_2$ by solving:

$$\min_{\theta} \quad -\log f(\mathbf{y}_2|\theta) + \frac{1}{2}(\theta - \hat{\theta}_1)^T \mathrm{Cov}[\hat{\theta}_1]^{-1}(\theta - \hat{\theta}_1). \tag{5.126}$$

Here, we can see that the prior term *carries over* information obtained from the first dataset $\mathbf{y}_1$. Note, however, that the solution to this problem might differ from the solution to the estimation problem in which we use the data $\mathbf{y}_1, \mathbf{y}_2$ all at once. Specifically, such a problem takes the form:

$$\min_{\theta} \ -\log f(\mathbf{y}|\theta). \tag{5.127}$$

with $\mathbf{y} = (\mathbf{y}_1, \mathbf{y}_2)$. The sequential process can be repeated over a number of cycles $k = 1, \ldots, K$. At each cycle, one processes a batch of data to obtain an estimate $\hat{\theta}$ and its covariance $\mathrm{Cov}[\hat{\theta}]$; the estimate and the covariance are used as the prior for the next cycle. This procedure is useful when *large amounts of data* need to be processed (e.g., data accumulates over time) and thus solving the estimation problem all-at-once (for all data available) is impossible. This is the situation in applications such as weather forecasting (we estimate parameters that capture atmospheric conditions) and large chemical processes (we estimate parameters that capture equipment conditions). The sequential Bayes procedure is the basis of advanced estimation techniques such as the Kalman filter and moving horizon estimation [46, 47].

Example 5.16 Recursive Bayesian Estimation for Heat Capacity

We revisit our heat capacity estimation problem of Example 5.3. Here, we would like to use Bayesian estimation to estimate the parameters of the model and to obtain their covariance by processing data in a recursive manner (by batches).

We first formulate the Bayesian estimation problem using an entire dataset that contains $S = 60$ observations. From the solution of the problem, we find that the parameter estimate is:

$$\hat{\theta} = (6.41, 2.26 \times 10^{-3}, -2.43 \times 10^{-7}).$$

and the covariance of the estimate is:

$$\mathrm{Cov}[\hat{\theta}] = \begin{bmatrix} 9.41 \times 10^{-3} & -3.46 \times 10^{-5} & 2.75 \times 10^{-8} \\ -3.46 \times 10^{-5} & 1.34 \times 10^{-7} & -1.10 \times 10^{-10} \\ 2.75 \times 10^{-8} & -1.10 \times 10^{-10} & 9.18 \times 10^{-14} \end{bmatrix}.$$

We now compare this solution against a recursive Bayesian procedure; here, we process batches of $K = 10$ data points per cycle and run the procedure over six cycles. After each cycle, we obtain an estimate $\hat{\theta}$ and its covariance $\mathrm{Cov}[\hat{\theta}]$; the estimate and the covariance are used to define the prior for the next cycle (thus carrying information from cycle to cycle). The parameter estimates in the 1st, 2nd, and 6th cycle are:

$$\begin{aligned} \hat{\theta} &= (6.08, 3.00 \times 10^{-3}, -6.98 \times 10^{-7}) \\ \hat{\theta} &= (6.44, 1.86 \times 10^{-3}, \ 1.38 \times 10^{-7}) \\ \hat{\theta} &= (6.36, 2.43 \times 10^{-3}, -3.78 \times 10^{-7}). \end{aligned}$$

We can thus see that the estimate after the 6th cycle is quite close to that obtained with the full dataset. The covariance of the estimates in the 1st and 6th cycles are:

$$\mathrm{Cov}[\hat{\theta}] = \begin{bmatrix} 6.54 \times 10^{-2} & -2.21 \times 10^{-4} & 1.68 \times 10^{-7} \\ -2.21 \times 10^{-4} & 8.01 \times 10^{-7} & -6.31 \times 10^{-10} \\ 1.68 \times 10^{-7} & -6.31 \times 10^{-10} & 5.09 \times 10^{-13} \end{bmatrix}$$

$$\mathrm{Cov}[\hat{\theta}] = \begin{bmatrix} 9.41 \times 10^{-3} & -3.46 \times 10^{-5} & 2.75 \times 10^{-8} \\ -3.46 \times 10^{-5} & 1.34 \times 10^{-7} & -1.10 \times 10^{-10} \\ 2.75 \times 10^{-8} & -1.10 \times 10^{-10} & 9.18 \times 10^{-14} \end{bmatrix}.$$

We can see that the covariance after the 6th cycle is the same as that obtained with the full dataset.

Script to reproduce results: `ch5_heat_capacity_bayes` □

The Bayes penalty $\frac{1}{2}(\theta - \bar{\theta})\Sigma_{\theta}^{-1}(\theta - \bar{\theta})$ naturally arises from the prior pdf. Now recall that one can also find trade-off points between the likelihood and the penalty function by solving the problem:

$$\min_{\theta} \; -\log f(\mathbf{y}|\theta) \tag{5.128a}$$

$$\text{s.t.} \; \frac{1}{2}(\theta - \bar{\theta})\Sigma_{\theta}^{-1}(\theta - \bar{\theta}) \leq \kappa. \tag{5.128b}$$

As discussed in Chapter 3, the constraint defines an ellipsoid with center point $\bar{\theta}$ and radius κ. Moreover, since the prior is Gaussian, we have that setting $\kappa = \mathbb{Q}(1-\alpha)$ (the $(1-\alpha)$-quantile of $\chi^2(n)$) gives us the ellipsoid that contains θ with probability $1-\alpha$. The Bayesian estimation problem can thus also be written as:

$$\min_{\theta} \; -\log f(\mathbf{y}|\theta) \tag{5.129a}$$

$$\text{s.t.} \; \mathbb{P}(\theta \in \mathcal{E}) = 1 - \alpha. \tag{5.129b}$$

Consequently, by adding prior information, we are implicitly constraining the probability that the estimate $\hat{\theta}$ is in an ellipsoidal confidence region. We can control the size of this region by controlling the value of α (which is more natural than controlling the value of κ). This highlights how the Bayesian estimation approach provides natural statistical interpretations of our estimates.

The Bayes framework can be naturally extended to capture prior information on input data $\mathbf{X}$ and on other parameters such as Σ. Assume that we have uncertainty in the input data, which we capture using the pdf $f(\mathbf{X})$. For example, we might consider that our inputs are $\mathbf{X} \sim \mathcal{N}(\bar{\mathbf{X}}, \Sigma_X)$; here, the reference point (the mean) $\bar{\mathbf{X}}$ is the measurements and Σ_X captures their uncertainty (e.g., precision of the sensor reporting the input data). Given a model $\mathbf{y} = \mathbf{m}(\theta, \mathbf{X})$, we seek to simultaneously estimate the parameters θ and the inputs $\mathbf{X}$.

Bayes' theorem tells us that:

$$f(\theta, \mathbf{X}|\mathbf{y}) \propto f(\mathbf{y}|\theta, \mathbf{X}) f(\theta, \mathbf{X}). \tag{5.130}$$

If we assume that θ and $\mathbf{X}$ are independent, we can write $f(\theta, \mathbf{X}) = f(\theta)f(\mathbf{X})$. Our estimates can thus be obtained by solving the problem:

$$\max_{\theta, \mathbf{X}} \; \log f(\mathbf{y}|\theta, \mathbf{X}) + \log f(\theta) + \log f(\mathbf{X}) \tag{5.131}$$

If we assume that $\theta \in \mathcal{N}(\bar{\theta}, \Sigma_{\theta})$, the Bayesian estimation problem can be expressed as:

$$\min_{\theta, \mathbf{X}} \; \frac{1}{2}(\mathbf{y} - \mathbf{m}(\theta, \mathbf{X}))^T \Sigma^{-1} (\mathbf{y} - \mathbf{m}(\theta, \mathbf{X})) + \frac{1}{2}(\theta - \bar{\theta})^T \Sigma_{\theta}^{-1}(\theta - \bar{\theta}) + \frac{1}{2}(\mathbf{X} - \bar{\mathbf{X}})^T \Sigma_X^{-1}(\mathbf{X} - \bar{\mathbf{X}}). \tag{5.132}$$

This problem is also known as *errors-in-variables* estimation and is particularly useful in reconciling variables to data. For instance, in a chemical process or a power plant, hundreds of flow measurements are collected and these are subject to error; here, we aim to reconcile such flows to a physical model that satisfies mass and energy balances. This problem is analogous to the flow conservation problem that we discussed in Chapter 1.

Example 5.17 Bayesian Estimation for Uncertain Input

Consider a simple system with output flow Y and input flow X. There is no chemical transformation in the system, and we thus know that conservation $Y = X$ must hold. We have a single input–output observation (measurement) that we denote as (x_ω, y_ω). These input and output measurements are subject to Gaussian errors with variances σ_x^2 and σ_y^2. Because of errors in the input and output flow, the observed input and output flow do not satisfy conservation (i.e., $x_\omega \neq y_\omega$). Our goal is to use Bayesian estimation to obtain an estimate of the input flow $\hat{x}$ that reconciles the observed input and output data to the conservation equation (the model).

Bayes' theorem tells us that $f(x|y) \propto f(y|x)f(x)$, and we seek to obtain $\hat{x}$ by finding a value x that maximizes $f(x|y)$. As we have done in our previous discussions, we express the observed output $y_\omega = x_\omega + \epsilon$. This form tells us that any deviation of the output from the input (any deviation from conservation $y_\omega = x_\omega$) is the result of the random error ϵ. The random error is given by $\epsilon \sim \mathcal{N}(0, \sigma_y^2)$, and we thus have $\mathbb{E}[y] = x$ and $\mathbb{V}[y] = \sigma_y^2$. Note that this form is consistent with the general form $y = \theta x + \epsilon$ that we have studied, but in this case, we know that $\theta = 1$. The likelihood $f(y|x)$ is thus the pdf of $y|x \sim \mathcal{N}(x, \sigma_y^2)$. To estimate our input, we assume prior knowledge for x of the form $x \sim \mathcal{N}(x_\omega, \sigma_x^2)$.

The Bayes problem seeks to maximize $f(x|y)$ and can be expressed as:

$$\min_x \quad \frac{1}{2\sigma_y^2}(y_\omega - x)^2 + \frac{1}{2\sigma_x^2}(x - x_\omega)^2. \tag{5.133}$$

This clearly reveals that the estimate $\hat{x}$ seeks to minimize the error from conservation and the deviation from the reference value x_ω (from the observation). In other words, this formulation seeks to "correct" the input observation x_ω in order to close the gap in conservation. How much weight/emphasis is put on the conservation and on the correction is controlled by the variances σ_y^2 and σ_x^2.

To find the solution to the problem, we take the derivative of the objective function with respect to x. This gives us the necessary condition:

$$-\frac{1}{\sigma_y^2}(y_\omega - x) + \frac{1}{\sigma_x^2}(x - x_\omega) = 0. \tag{5.134}$$

From here, we can determine that the estimate $\hat{x}$ is:

$$\hat{x} = \frac{\frac{1}{\sigma_y^2} y_\omega + \frac{1}{\sigma_x^2} x_\omega}{\frac{1}{\sigma_y^2} + \frac{1}{\sigma_x^2}}. \tag{5.135}$$

This can also be written as:

$$\hat{x} = \frac{y_\omega + \frac{\sigma_y^2}{\sigma_x^2} x_\omega}{1 + \frac{\sigma_y^2}{\sigma_x^2}}. \tag{5.136}$$

From this expression, it is clear that, if we trust the output measurement much more than the input measurement, then $\sigma_y^2/\sigma_x^2 \to 0$ and thus $\hat{x} \to y_\omega$. On the other hand, if we trust the input measurement much more than the output measurement, then $\sigma_y^2/\sigma_x^2 \to \infty$ and thus $\hat{x} \to x_\omega$.

Consider that $(x_\omega, y_\omega) = (1,2)$ and $\sigma_y^2/\sigma_x^2 = 0.1$ (we trust the output measurement 10 times more than the input measurement). The estimate is given by $\hat{x} = 1.09$. If we instead consider $\sigma_y^2/\sigma_x^2 = 10$ (we trust the input 10 times more than the output), the estimate is $\hat{x} = 1.90$. □

The approach followed in the previous example can be generalized to reconcile multiple inputs to outputs that satisfy a structural model $\mathbf{y} = \mathbf{m}(\mathbf{X}) + \mathbf{e}$. Moreover, one can simultaneously reconcile the inputs and estimate parameters for a model $\mathbf{y} = \mathbf{m}(\theta, \mathbf{X}) + \mathbf{e}$. Note that, under a Bayesian framework, this process is *exactly equivalent* as to assume that the input data $\mathbf{X}$ are parameters of the model. This consistency is a key benefit of this framework.

5.6 Bayesian Decision-Making

Bayesian decision-making is a powerful paradigm that is used for making decisions under uncertainty. This paradigm involves a:

$$data \to models \to decisions$$

pipeline and thus integrates a broad range of concepts that we have covered so far. Specifically, we have learned how to use data to build predictive models and to quantify the uncertainty of our predictions. We have also emphasized that data collection is essential in obtaining parameter estimates and predictive models that are reliable. Our goal is now to use our models to guide data collection and to make decisions in order to counteract uncertainty.

To explain these concepts in simple terms, consider that we have a linear predictive model of the form $\hat{y} = x^T\hat{\theta}$. Here, the output $\hat{y}$ is a performance metric that we aim to maximize/minimize (e.g., profit, conversion, and energy use), the parameter estimate $\hat{\theta}$ was obtained with an existing dataset to obtain the covariance $\mathrm{Cov}[\hat{\theta}]$, and x is a decision that we have control over (e.g., a new experiment).

Because the model is linear, we have that the covariance of the model prediction is $\mathrm{Cov}[\hat{y}] = x^T\mathrm{Cov}[\hat{\theta}]x$; this covariance allows us to quantify the uncertainty of our prediction, and note that we can influence this uncertainty via our decision x. To simplify notation, we will define the model prediction as $\mu(x) = x^T\hat{\theta}$ and the uncertainty of the prediction as $\sigma^2(x) = x^T\mathrm{Cov}[\hat{\theta}]x$. Note that our predictive model is thus $\mathcal{N}(\mu(x), \sigma^2(x))$.

Now note that we can select our decision x in order to maximize (or minimize) our predicted performance $\mu(x)$; the idea here is that we are exploiting the available information (contained in the model $\mu(x)$) to optimize the system performance. But, what if our prediction has too much uncertainty? In such a case, we face the risk that our prediction will significantly deviate from the true performance. To mitigate this risk, we need to collect more data; this can be done by selecting x that maximizes $\sigma^2(x)$; the idea is that, by maximizing the uncertainty of our prediction, we are exploring or learning something new about the system (we are maximizing information gain). This is the basic principle used in an area of statistics called *Bayesian experimental design* [48]. Once we collect new data, we can refine our predictive model $\mathcal{N}(\mu(x), \sigma^2(x))$ by re-estimating (recalibrating) our parameter estimates $\hat{\theta}$; we can then use this new model to make a new decision that either maximizes performance or maximizes information gain.

There are many variants of the above paradigm that are beyond the scope of this book. As an example, one can define different ways to measure information gain; for instance, in some techniques, one can aim to select new experiments to manipulate the eigenvalues of the covariance of the estimate $\text{Cov}[\hat{\theta}]$ [49]. In addition, when the system performance is a nonlinear function of the inputs ($\mu(x)$ is nonlinear), one needs more sophisticated techniques to build models from data or to combine physical and empirical models. Some of such techniques include Gaussian process modeling and Bayesian Optimization (BO) [50], which we will discuss in Chapters 6 and 7.

5.7 Summary

In this chapter, we have discussed techniques to build predictive models and to quantify the uncertainty of their parameters and their predictions; some of the important take-home messages are the following:

- Predictive models leverage structural relationships between RVs in order to make predictions of some as a function of others. Structural relationships arise from physical connections between variables and manifest as correlations between variables.
- Predictive models can be empirical, physical, or combinations (hybrid). Empirical models have model structures that are non-physical (e.g., linear or polynomial), while physical models have structures dictated by first-principles (e.g., kinetic and thermodynamic models).
- The goal of estimation is to leverage available input–output data to determine the parameters of a model. At the core, estimation aims to separate systematic trends that are contained in the data (what can be predicted or is known) from random effects that are contained in the data (what cannot be predicted or is unknown). We can thus see estimation as a separation process that extracts knowledge (value) from raw data.
- MLE provides a powerful optimization paradigm to determine parameters from data. Estimates obtained from MLE have many important properties; for instance, MLE estimates are unbiased and have optimality guarantees (i.e., there is no better set of parameters).
- For linear models, we can derive analytical expressions to determine the optimal MLE estimates, to quantify the uncertainty of the estimates, and to quantify the uncertainty of model predictions. This provides interesting and important insights; for instance, the Hessian matrix of the MLE problem is directly associated with the covariance of parameter estimates and with the FI matrix.

- For nonlinear models, it is not possible to derive analytical expressions, and one must resort to numerical techniques, such as Newton's method. Nonlinear models can also exhibit parasitic behavior (presence of minimum and maximum points and regions in which the model is not well defined); as such, one has to be careful in providing good initial guesses for the parameter estimates and in validating that the solution makes physical sense.
- One can encode physical or expert knowledge into MLE formulations by using constraints or regularization terms. Constraints aim to filter out regions in the parameter space that do not make physical/practical sense; the constraints can also be seen as a mechanism to control the flexibility of the model. Regularization also aims to control the flexibility of the model by introducing penalty terms in the objective function that penalize movement of the parameters; one can control the degree of flexibility by tuning a trade-off parameter that balances how much weight you place on model fit (the likelihood function) and the penalty term.
- Regularization can be used as a mechanism to reduce model complexity (eliminating parameters or input variables that are not relevant) and with this prevent overfitting. Regularization thus allows us to navigate the trade-off of model complexity vs. model fit; in general, we want models that are simple (contain few parameters), but this might sacrifice fit. Constraints and regularization can also be used to help numerical solvers find solutions.
- There is a broad range of techniques that can be used to assess if the parameter estimates and resulting predictive model are satisfactory. It is particularly important to monitor the residuals (the model errors), as this can provide an indication that the model cannot fully capture trends present in the data (thus suggesting the need for an alternative model). One can also use simple metrics such as the coefficient of determination and the Pearson correlation to determine how well the predicted outputs match the observed outputs. Other important mechanisms include the use of uncertainty models (to quantify how confident we are in our estimates and model predictions) and the use of cross-validation procedures (to determine how well our model generalizes to data not used in the estimation process).
- Bayesian estimation provides another powerful paradigm for parameter estimation. This differs from MLE in that it provides a more rigorous statistical treatment of the parameters. The Bayesian paradigm provides a natural framework to capture the uncertainty of the parameters before (prior) and after (posterior) the estimation process. This facilitates the incorporation of prior knowledge and the derivation of recursive/sequential procedures to deal with large datasets.
- Bayesian decision-making provides a general paradigm to understand how data can be used to build models and how models can be used to make decisions. At the core of this paradigm are uncertainty models that allow us to make decisions on how to collect new data to reduce the uncertainty of our predictions (gain information) and to make decisions to use the predictive model to maximize a desired target (e.g., profit).

The area of parameter estimation has been explored for many years (going back to the seminal work of Gauss). As such, there is a broad range of techniques that have been proposed and that are used in practice. Our goal in this book is to focus on the fundamental principles behind these techniques so that the reader can more easily understand and use advanced techniques. Our goal also is to highlight the deep connections between optimization, linear algebra, and statistics; for instance, we emphasized how the Hessian matrix of MLE problems is influenced by the properties of the input data matrix, how the Hessian is directly connected with measures of information and uncertainty, and how to use eigenvalues to quantify properties of the Hessian. We will continue to highlight these connections in the next chapter.

5.8 Exercises

Exercise 5.1. Water Quality Prediction in Lakes.
State and national organizations are becoming increasingly concerned about the water quality of lakes. Excess phosphorus concentration, in particular, can have significant detrimental impacts on lakes, as it is a primary driver for harmful algae blooms (HABs). High phosphorus concentration is usually the result of fertilizer runoff from agricultural fields. HABs can have serious negative environmental, economic, and health impacts (e.g., release toxins and affect tourism). To monitor lake water quality, samples are frequently taken to measure the total phosphorus concentration (μg/L) and chlorophyll-a (a measure of algae content, in μg/L). You have real data for 283 lakes in the state of Wisconsin, and you would like to establish if phosphorus concentration is indeed linked to chlorophyll-a concentration.

Use the data in `WI_lake_water_quality.csv` (the first column contains the total phosphorus concentration and the second column contains chlorophyll-a concentration) to address the following:

(a) Obtain a model that captures the relationship between the total phosphorus and chlorophyll-a. Obtain a parity plot and the prediction errors (residuals). Compute the summarizing statistics (mean, median, and standard deviation) of the residuals. What can you conclude from your results?
(b) An acceptable model prediction error (absolute value of error) for chlorophyll-a is 10 μg/L; determine the fraction of lakes that have acceptable and unacceptable prediction errors. Is the proposed model reasonable?
(c) Standardize the residuals and obtain their pdf. Conduct a Kolmogorov - Smirnov test to determine whether they follow a standard normal distribution. If these do not follow a standard normal, what can be the causes for this? What could be the possible remedies?

Exercise 5.2. Arrhenius Law.
The Arrhenius law establishes an explicit dependency between the rate constant of a reaction (how quickly it progresses) and temperature and has the following

nonlinear form:

$$k = A \cdot \exp\left(\frac{-E_a}{R \cdot T}\right).$$

Here, k (in 1/second) is the reaction rate constant, A is known as the pre-exponential factor (same units as k), E_a is the activation energy (in J/mol), T is temperature (in K), and R is the universal gas constant (8.314 J/mol · K).

The activation energy and pre-exponential factor for a given reaction can be estimated from experimental data. For instance, one typically runs a reaction in a batch reactor (at a given temperature) and keeps track of the concentration of a reactant or product. One then uses the concentration data to determine the reaction rate and associated constant k. Determining the rate constant is a complicated procedure, and data might contain inherent variability due to a variety of factors that are difficult to control (e.g., sample preparation). As such, it is often necessary to ensure reproducibility by running the experiment at the same temperature multiple times.

You have conducted experiments at four temperatures and repeated the experiment 10 times at each temperature. The outcomes of the experiments are recorded in `arrhenius_data.dat` (first column is the reaction rate constant and second column is temperature).

A standard approach to avoid the nonlinearity of the Arrhenius model is to apply a logarithmic transformation to obtain:

$$\log(k) = \log(A) - \frac{E_a}{R} \cdot \frac{1}{T}.$$

With this, the model can be written as a linear model $Y = \theta_0 + \theta_1 X$ with $Y = \log(k), \theta_0 = \log(A), \theta_1 = -E_a/R$, and $X = 1/T$. Once the parameters θ_0 and θ_1 are estimated, we can obtain estimates for the activation energy and pre-exponential factor.

(a) Determine the maximum likelihood estimate for the parameters of the linear model and obtain the values for the activation energy and pre-exponential factors.
(b) Determine the standard deviation of the noise (which captures the variability of the measurements for k). Comment on the variability observed in the repetition of the experiments; is the experimental procedure reliable?
(c) Determine if the parameters can be estimated uniquely from the available data.
(d) Compare the data against the model predictions (using a k vs. T plot). Does the Arrhenius law match the data well?
(e) Use the predicted and observed model outputs to determine the coefficient of determination and the Pearson correlation. What can you conclude from this?
(f) Determine the precision and 95% confidence intervals of the parameter estimates θ_0, θ_1. What can you conclude from this?
(g) Determine the FI and its eigenvalues. What can you conclude from this?

Exercise 5.3. Why Am I Feeling so Sleepy in This Class?

Humans generate significant amounts of CO_2 through respiration; if an occupied space is not properly ventilated, the generated CO_2 will accumulate and the air can reach harmful concentration levels that affect health (manifests as reduced cognitive capacity and sleepiness). Monitoring and controlling CO_2 levels in occupied spaces such as office buildings and schools is important.

The dynamics of the CO_2 concentration in a classroom can be described by a reactor model of the form:

$$\frac{dC(t)}{dt} = \frac{Q}{V}(C_{in} - C(t)) + \frac{G}{V}n_{oc},$$

where $C(t)$ is the CO_2 concentration in ppm, t is time in min, Q is the airflow rate in cfm = ft^3/min, $C_{in} = 400$ is the CO_2 concentration in the fresh airflow in ppm, $V = 10{,}000$ is the classroom volume in cf = ft^3, G is the CO_2 generation rate per occupant in ppm–cf/min, and $n_{oc} = 40$ is the number of occupants in the classroom. All the units in the model are consistent.

The American Society of Heating, Refrigerating, and Air-Conditioning Engineers (ASHRAE) recommends that a minimum ventilation flow rate of 8 cfm *per occupant* should be maintained at all times. Unfortunately, there is no flow sensor in the classroom to validate this, and thus, the flow rate needs to be estimated from data. Moreover, the CO_2 generation rate per occupant is difficult to determine from first-principles (depends on many factors such as age, metabolic rate, and activity level) and thus we would like to determine it directly from data.

To estimate the ventilation flow rate and generation rate, you have set up the following experiment. You begin with an empty classroom 400 ppm (atmospheric conditions) and start recording the CO_2 concentration after the students come into the classroom. You record the concentration every Δt =5 min for 1 hr (total of $N = 13$ sampling times) and approximate your model using the discrete-time form:

$$\frac{C_{k+1} - C_k}{\Delta t} = \frac{Q}{V}(C_{in} - C_k) + \frac{G}{V}n_{oc}, \quad k = 0, \ldots, N-1,$$

where C_k is the observed CO_2 concentration at time $k \cdot \Delta t$. The recorded CO_2 data is available at `co2_classroom_data.dat` (first column is time and second column is concentration).

(a) Use the discrete-time model to estimate the flow rate and the generation rate per occupant in the classroom.
(b) Determine whether the flow rate per occupant satisfies the requirements of ASHRAE.
(c) Compare the model predictions against the concentration data (use a plot of concentration vs. time for this). What can you conclude from this?
(d) Use the model and estimated parameters to determine how quickly will the CO_2 concentration reach 1,200 ppm (considered an unhealthy level) if the classroom has an initial concentration of $C_0 = 400$ ppm and the classroom has 50 occupants. Assuming that the class lasts 40 min, will the CO_2 concentration reach the unhealthy level within the classroom time?

Exercise 5.4. Monitoring Reactor Performance.

You are a process engineer responsible for monitoring the operation of a chemical reactor. The operator of the facility mentioned that the laboratory samples used to measure the quality of the product indicate that the product is getting close to being off-specification. You do not have a real-time sensor to monitor the product quality; the only data you have available are lab measurements that include operating temperature (T in K), operating pressure (P in bar), and concentration of product in the reactor outlet (y in mol fraction).

The operator wants you to find suitable operating conditions that will prevent the reactor from going off-specification. Initially, you consider building a simulation model of the reactor; however, you soon realize you do not have access to the kinetic data that would be required. You then recall the various modeling techniques you learned in your statistics course and decide to generate an empirical (data-driven) model that predicts quality y as a function of T and P.

Load the data in `reactor_conversion.dat`; the first column contains the reactor temperature, the second column is its pressure, and the third is the mol fraction of the product.

Address the following:

(a) Develop a quadratic polynomial model that predicts y as a function of T and P of the form $y = \theta_0 + \theta_1 T + \theta_2 P + \theta_3 T^2 + \theta_4 P^2 + \theta_5 T\, P$. Determine what parameters you can eliminate from the model and verify that you can obtain an adequate fit.

(b) Evaluate the model in the domain $[T, P] \in [623, 823] \times [200, 300]$ and identify the subdomain that ensures that the outlet stream satisfies the required quality specifications ($y \geq 0.5$).

Exercise 5.5. Predicting Solubility.
Scientists are planning experiments to build a predictive model of paracetamol solubility in ethanol. Data is provided in `solubility_data.dat` (the first column is temperature T in °C and the second column is solubility S in g of solute per kg of solvent). You are interested in testing models of different complexity (see below). Provide parity plots for all models and report goodness of fit in terms of R^2 and RMSE. Recommend the best model structure to use.

Name	Structure
Linear	$S = \theta_0 + \theta_1 T$
Polynomial	$S = \theta_0 + \theta_1 T + \theta_2 T^2$
Exponential	$S = \theta_0 \exp(\theta_1 T)$
Apelblat	$S = \exp(\theta_0 + \frac{\theta_1}{T} + \theta_2 \log(T))$

Exercise 5.6. Predicting CMC Using QSPR.
Surfactants are amphiphiles that are constituted by a hydrophilic head group and a hydrophobic tail group; these are widely used in consumer products such as soaps and detergents [15]. Given their amphiphilic structure, surfactants tend to self-assemble in solution, leading to the formation of spherical micelles where the hydrophilic head groups are oriented toward the aqueous phase and the hydrophobic tail groups are positioned within the micelle core. An important property that

guides the design of a surfactant is the critical micelle concentration (CMC), which is the minimum concentration of surfactant at which self-assembly occurs. The CMC is dependent on the molecular structure of the surfactant; as such, one can predict CMC values based on molecular descriptors (molecular characteristics) using quantitative structure–property relationship (QSPR) models.

Load the data in `cmc_md.mat`. This dataset contains 202 surfactants, each represented as a vector containing six molecular descriptors. The input descriptor data is reported as a matrix $\mathbf{X} \in \mathbb{R}^{202\times 6}$. The name of the descriptors is defined in the `md_colnames` variable. The experimental $\log_{10}$CMC values (output y) for each surfactant are also provided in the dataset. Our goal is to construct simple models to predict $\log_{10}$CMC as a function of the molecular descriptors.

Address the following:

(a) Compute the mean, standard deviation, min value, and max value for each molecular descriptor and report them in a table. Scale each column of $\mathbf{X}$ to obtain values that are between 0 and 1 and construct and scaled matrix $\mathbf{X}_{scaled}$.
(b) Obtain a linear regression model using the scaled input data to predict the output $\log_{10}$CMC. Report the RMSE, R^2, and adjusted-R^2 values. Report the values of the parameters.
(c) Scaling the input data helps interpret the parameters more easily. Based on the magnitude of the parameters, identify the molecular descriptor with the most impact on $\log_{10}$CMC. Use a scatter plot to show the relationship between $\log_{10}$CMC and the most significant descriptor. How does this descriptor influence the CMC?
(d) Obtain a parity plot for the model (with all the descriptors) and comment on the model quality.
(e) Using descriptors to represent a molecule is intuitive, but this method has a few limitations. Can you identify possible drawbacks of this data representation?

Exercise 5.7. Predicting CMC Using SMILES.
In the context of chemical and biological engineering, we can encounter datasets in different forms. For instance, molecules can be represented using molecular graphs, fingerprints, and molecular descriptors (e.g., using QSPR). In this problem, we will focus on a data representation of molecules called SMILES (simplified molecular-input line-entry system) and familiarize ourselves with the one-hot encoding technique to convert SMILES strings into numerical values that can be used for building structural models.

SMILES strings are composed of text characters that represent atoms, bonds, and other structural information following a predefined set of rules. This data representation is widely used in cheminformatics for applications including chemical database searching, screening, and molecular property prediction. In order to handle text strings mathematically, we need to transform these strings into numerical values (matrices). One-hot encoding is a technique that transforms a SMILES string into a binary matrix that captures the chemical structure of a molecule.

In this problem, we revisit the CMC dataset of Exercise 5.6 and construct models that predict $\log_{10}$CMC using one-hot encoded SMILES strings. Load the data in `cmc_ohe.mat`. This dataset contains 202 surfactants, each represented as a SMILES string. The experimental $\log_{10}$CMC values (outputs) are also provided.

To illustrate how one-hot encoding works, consider the sample surfactant 1,2-decanediol:

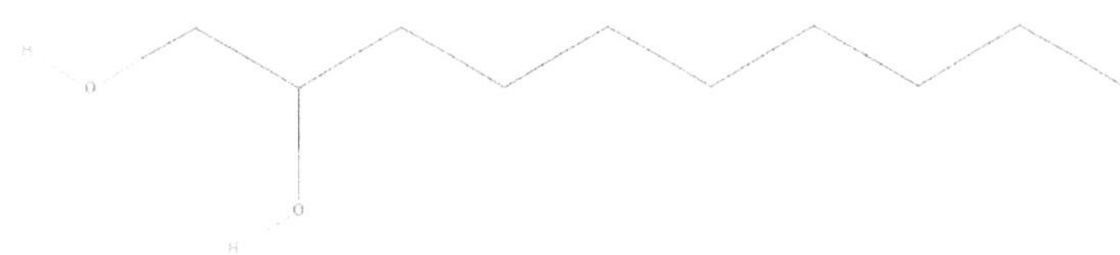

The SMILES string for 1,2-decanediol is "CCCCCCCCC(O)CO," where the carbon atoms are denoted as "C," the oxygen atoms are denoted as "O," the side chain is captured in parentheses, and the hydrogen atoms are omitted (they are implied from the structure). The connectivity between atoms is reflected by the sequence of the characters.

Taking 1,2-decanediol as an example, we go through the process of converting this SMILES string into a one-hot encoded matrix. We define an ordered list for the SMILES alphabet: "C," "O," "(," and ")," which are the unique characters found in the SMILES string. We then define a matrix of zeros in which the number of columns is the number of characters in the SMILES string (in this case 15) and the number of rows is the size of the alphabet (in this case 4). We then populate the matrix as follows: the first row corresponds to the character "C," and we place a value of 1 in the positions in which this character is found in the string. We then go to the second row and place a value of 1 at the positions in which "O" is found; the procedure continues for all the rows. The one-hot encoded matrix for 1,2-decanediol is:

$$\begin{bmatrix} 1 & 1 & 1 & 1 & 1 & 1 & 1 & 1 & 1 & 0 & 0 & 0 & 1 & 0 \\ 0 & 0 & 0 & 0 & 0 & 0 & 0 & 0 & 0 & 0 & 1 & 0 & 0 & 1 \\ 0 & 0 & 0 & 0 & 0 & 0 & 0 & 0 & 0 & 1 & 0 & 0 & 0 & 0 \\ 0 & 0 & 0 & 0 & 0 & 0 & 0 & 0 & 0 & 0 & 0 & 1 & 0 & 0 \end{bmatrix}.$$

In real-world applications, SMILES strings vary in length (depending on molecule complexity). In order to compare molecules, it is necessary to obtain matrices of equal sizes. This is done by adding columns of zeros to match the size of the one-hot encoded matrix with the maximum length in the dataset (This technique is called "zero-padding"). In the CMC dataset, we have identified 28 unique characters in the SMILES alphabet, and the maximum length of the SMILES strings is 93. As such, all matrices need to be of size 28×93.

In order to enable the use of simple structural models, we can convert a matrix into a single vector. In this example, this gives a vector of length $28 \times 93 = 2604$. The SMILES strings for all the surfactants have been preprocessed and are provided as a matrix $\mathbf{X}_{ohe} \in \mathbb{R}^{202\times 2604}$.

Address the following:

(a) Obtain the SMILES string and one-hot encoded matrix for 1,3-undecanediol:

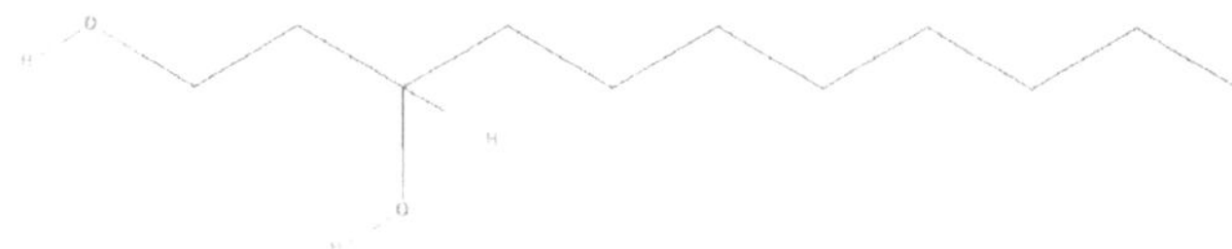

(b) Fit a linear regression model using $\mathbf{X}_{ohe}$ to predict y. Report the RMSE, R^2, and adjusted-R^2 values.
(c) Build a parity plot (predicted vs. experimental $\log_{10}$CMC) and comment on the ability of the model to fit the data.
(d) Determine the number of non-zero parameters in the model. What does this tell you about this model?
(e) Compared with the use of molecular descriptors (Exercise 5.6), do SMILES strings increase or decrease model accuracy? Discuss possible drawbacks of using the SMILES data representation compared to molecular descriptors.

Exercise 5.8. Predicting Atmospheric CO_2.
The concentration of carbon dioxide (CO_2) in the atmosphere has been measured and recorded for several decades, and it has been found that the concentration has been increasing. Load the data in `CO2_overtime.csv`; it contains two columns: the first column is for "Year," and the second column is for "Carbon Dioxide Concentration" (measured in parts per million [ppm]). This data is obtained from the Mauna Loa Observatory public data from the National Oceanic and Atmospheric Administration (NOAA) [16].

Address the following:

(a) Determine if CO_2 has been increasing at a linear, quadratic, or exponential rate over time.
(b) Use your best model predict to CO_2 in the year 2030 and report confidence intervals for your predictions (you can use a software package for obtaining your intervals). Compare your prediction to the CO_2 concentration observed in 2020.
(c) Discuss the limitations of the models and potential ways to improve them.

Exercise 5.9. Estimation of Kinetic Parameters.
A student is investigating the kinetics of a reaction in a continuous stirred-tank reactor (CSTR). The reaction is modeled by the following rate law:

$$r = k \cdot C,$$

where r is the reaction rate (in mol$\text{L}^{-1}\text{s}^{-1}$), k is the rate constant (in s^{-1}), and C is the concentration (in mol$\cdot$ L^{-1}) of the reactant. The rate constant k depends on the temperature by following the Arrhenius law:

$$k = A\exp(-E_a/RT),$$

where A is the pre-exponential factor, E_a is the activation energy for the reaction (in $J \cdot mol^{-1}$), R is the gas constant (in $J \cdot mol^{-1} \cdot K^{-1}$), and T is the temperature (in K).

Load the data in `KineticData.csv`; the first column is temperature (K), the second column is measured reactant concentration ($mol \cdot L^{-1}$), and the third column is measured reaction rate ($molL^{-1}s^{-1}$). The goal is to use the data to estimate the activation energy and pre-exponential factor.

Address the following:

(a) Estimate the reaction constant k for each temperature.
(b) Determine the activation energy E_a and the pre-exponential factor A. Compare the experimental rate constants against those predicted by the Arrhenius model.
(c) Use the model to predict the reaction rate at a temperature of 355 K and a concentration of 0.8 mol/L. Predict how much faster will the reaction proceed if we increase the temperature to 375 K.

Exercise 5.10. Factorial Experimental Design.
You are seeking to develop a linear regression model of the form $Y = \theta_0 + \theta_1 X_1 + \theta_2 X_2$. The model will be used in a process to predict the selectivity of a new catalyst as a function of temperature X_1 and reactant concentration X_2. You need to design a set of experiments to collect data and estimate the parameters.

To understand the effect of temperature, you design experiments that operate the reactor at high, medium-high, medium-low, or low temperature. These settings are represented using X_1 values of $+1, +0.5, -0.5$, and -1, respectively (the temperature range has been scaled by the highest and lowest values). To capture the effect of concentration, you operate the reactor at high, medium-high, medium-low, or low concentration (represented as X_2 values of $+1, +0.5, -0.5$, and -1). The standard deviation of the output noise is assumed to have a value of 1.

(a) Determine the FI matrix for the set of four experiments: $(-1, -1)$, $(-0.5, -0.5)$, $(+0.5, +0.5)$, and $(1, 1)$. If we define the determinant of the FI (product of its eigenvalues) as a measure of the total information content, what is the total information provided by this experimental design? Use your intuition to explain why this is a badly designed set of experiments.
(b) Determine the determinant of the FI for the following set of experiments: $(-1, -1)$, $(-0.5, +0.5)$, $(+0.5, -0.5)$, and $(+1, +1)$. Is this a better design of experiments in terms of total information content? Use your intuition to explain why (or why not) this is a better set of experiments.
(c) Factorial design is a powerful technique to design experiments. This technique suggests that we should ignore the middle points and instead select the extreme points of our experimental space. Factorial design thus recommends the experiments: $(-1, -1)$, $(+1, -1)$, $(-1, +1)$, and $(+1, +1)$. Is information content improved under this design? Use your intuition to explain why this is (or not) a better set of experiments.

(d) Compare the structure of the FI matrix across the different experimental designs; comment on any differences observed and comment on why these differences are relevant from the perspective of the covariance of the parameter estimates.

Exercise 5.11. Box–Behnken and the Response Surface Methodology.
George P. Box is one of the most influential statisticians in history and was the founder of the Department of Statistics at the University of Wisconsin–Madison. Box was a mathematician, but he started his professional career in the chemical industry. He is particularly famous for the quote "All models are wrong, but some are useful." Box and Behnken proposed techniques to design experiments in order to effectively collect data to build models and to use such models to find optimal operating conditions (this technique is known as the response surface methodology [RSM]).

In a recent paper, a team of scientists used RSM to determine optimal operating conditions to extract pectin from sunflower heads using oxalic acid [17]. Pectin is a complex polysaccharide that is contained in plant primary cell walls and is widely used in industry (e.g., food, pharma, and cosmetics). The scientists are seeking to identify the influence of three variables (also known as factors in the design of experiment literature): X_1 (extraction temperature), X_2 (extraction time), and X_3 (oxalic acid concentration) on the extraction yield (denoted as Y). They propose a generalized 2nd-order polynomial model:

$$\begin{aligned} Y = \theta_0 + \theta_1 X_1 + \theta_2 X_2 + \theta_3 X_3 + \theta_4 X_1^2 + \theta_5 X_2^2 + \theta_6 X_3^2 \\ + \theta_7 X_1 X_2 + \theta_8 X_1 X_3 + \theta_9 X_2 X_3. \end{aligned}$$

For this type of model, Box–Behnken recommends collecting data by performing the set of experiments shown in `box_behnken.dat`. The first three columns are the scaled temperatures, extraction times, and oxalic acid concentration, while the fourth column is the corresponding yields (in %). The scaled values (−1, 0, and 1) of factor X_1 (extraction temperature) represent 70, 85, and 100°C. The scaled values (−1, 0, and 1) of factor X_2 (extraction time) represent 1.0, 1.5, and 2.0 hours. The scaled values (−1, 0, and 1) of factor X_3 (oxalic acid concentration) represent 0.1, 0.2, and 0.3%.

Address the following:

(a) Use MLE to determine the best estimate for the parameters of the model.
(b) Determine whether the design of experiments of Box–Behnken yields unique estimates. Is there something particular that you notice about the structure of the FI matrix? What does this tell you about the effectiveness of the experimental design of Box–Behnken?
(c) Use the model with the estimated parameters to determine the optimal extraction conditions. In other words, find the solution to the following problem:

$$\begin{aligned} \max_{X_1, X_2, X_3} \; \theta_0 + \theta_1 X_1 + \theta_2 X_2 + \theta_3 X_3 + \theta_4 X_1^2 + \theta_5 X_2^2 + \theta_6 X_3^2 \\ + \theta_7 X_1 X_2 + \theta_8 X_1 X_3 + \theta_9 X_2 X_3, \end{aligned}$$

subject to the constraints that $-1 \leq X_i \leq 1$ for $i = 1,2,3$. Report the optimal conditions in the actual engineering units for temperature, time, and concentration (as opposed to their scaled values). You can use a simple linear interpolation to obtain these. Report also the optimal yield predicted by the model; is this higher than those identified already experimentally?

Exercise 5.12. DHP Dehydrogenation.
Catalytic upgrading of biomass can be used to produce solvents in a sustainable manner. A multidisciplinary team of scientists has recently explored the production of tetrahydropyran (THP) from furfural, which can be produced from biomass [18]. THP is an ether that is biodegradable and noncarcinogenic; as such, it has the desirable properties of a green solvent. The research team investigated a route to produce THP via the gas-phase hydrogenation of DHP over a solid Ni/SiO_2 catalyst. The scientists hypothesize that the reaction mechanism follows these steps:

$$\begin{aligned} \mathrm{DHP} + \theta &\overset{K_1}{\longleftrightarrow} \mathrm{DHP}_{ads} \\ \mathrm{H}_2 + 2\theta &\overset{K_2}{\longleftrightarrow} 2\mathrm{H}_{ads} \\ \mathrm{DHP}_{ads} + 2\mathrm{H}_{ads} &\overset{k_3}{\longrightarrow} \mathrm{THP}_{ads} + 2\theta \\ \mathrm{THP}_{ads} &\overset{K_4}{\longleftrightarrow} \mathrm{THP} + \theta. \end{aligned}$$

Here, θ denotes catalyst sites at which DHP and hydrogen (H_2) adsorb (equilibrium steps 1 and 2) and react to form THP (reaction step 3). The THP produced subsequently desorbs from the site (equilibrium step 4). The overall reaction rate of these steps can be modeled using a Hougen-Watson (HW) kinetic model:

$$r = \frac{k_3 K_1 p_{\mathrm{DHP}} (K_2 p_{\mathrm{H}_2})^2}{1 + K_1 p_{\mathrm{DHP}} + \sqrt{K_2 p_{\mathrm{H}_2}} + \tilde{K}_4 p_{\mathrm{THP}}}, \tag{5.137}$$

where $\tilde{K}_4 = 1/K_4$. To test the hypothesis that this is indeed the mechanism, the team conducts a set of experiments at different partial pressures for, DHP, H_2, and THP to measure the reaction rate. The data is contained in `hougenwatson_dhp.dat`; the first three columns are the partial pressures of DHP, H_2, and THP, respectively (in psi) and the fourth column is the reaction rate in mmol/(hr· gr_{cat}). All the adsorption constants $K_1, K_2, \tilde{K}_4$ are dimensionless and the reaction constant k_3 has units of mmol/(hr· gr_{cat}· psi^3). The parameter $\tilde{K}_4$ is introduced instead of $1/K_4$ in order to avoid numerical instability (divisions by zero).

Address the following:

(a) Estimate and report the parameters of the HW model ($K_1, K_2, K_3, \tilde{K}_4$). Determine if they provide adequate fit (show a parity plot). Does the proposed mechanism seem to explain the data? **Hint:** Make sure to introduce lower and upper bounds for the parameters and note that all parameters must be non-zero.

(b) Use the Jacobian of the model and the variance of the residuals to determine the covariance matrix of the parameters and their 95% confidence intervals. Are the parameter estimates precise?

(c) Use the Jacobian of the model and the covariance of the parameters to obtain the 95% confidence intervals of the predicted outputs. Display the confidence intervals in a parity plot. Is the prediction uncertainty reasonable?

(d) Determine the eigenvalues of the Hessian matrix. Based on this, comment on whether the available data is sufficient to estimate the parameters in a reliable manner.

Part III

Advanced Concepts

6 Statistical Learning

The previous chapters of the book have focused on how we can use data to conduct a variety of tasks such as computing event probabilities, visualizing trends, modeling random phenomena, and developing predictive models. In this chapter, we will discuss advanced tools from data science and machine learning (ML) that will allow us to preprocess data with the goal of revealing hidden information and facilitating modeling, analysis, and visualization tasks.

Data science and ML are fast-growing fields that provide powerful tools to perform tasks such as:

- Data analysis (e.g., dimensionality reduction, pattern recognition, clustering, and computer vision).
- Predictive modeling (e.g., neural nets, kriging, logistic regression, and support vector machines [SVMs]).
- Artificial intelligence (e.g., data collection, experimentation, learning, and control).

Many advanced tools used in data science and ML are derived from *statistical principles and tools* (as those that we have discussed in the book). Such techniques are often combined with techniques from other areas of mathematics (e.g., geometry, optimization, and linear algebra). As such, the goal of this chapter is to highlight how all the principles that we have learned so far can be used to develop and use advanced data science and ML tools.

Our focus here will not be to provide an extensive review of data science and ML tools (which is a large topic that deserves a book on its own); instead, we will aim to explain fundamental statistical principles that are behind such tools. We will see that understanding the fundamentals is important in explaining the behavior and limitations of these tools, in selecting proper tools, and in developing new tools.

As we have discussed, data is at the core of modeling and decision-making, as data is what allows us to characterize the behavior of systems. However, in many applications of interest, *raw data* is high-dimensional (e.g., a chemical process contains hundreds to thousands of sensors), is noisy, and might appear in complex formats (e.g., molecular structures, time series, images, video, and text). Our goal is to extract knowledge from such large and complex datasets. We will discuss how to achieve this by using *data transformation techniques* that aim to map data from its original domain (also known as "space") to an alternative space; we will see that this transformation can be used to eliminate data redundancies and noise, reduce the dimensionality of data, visualize trends, and extract useful information to make predictions and decisions.

We will discuss empirical (data-driven) modeling techniques that conduct transformations of the data prior to building the model. We will see that such transformations can help reduce the complexity of models or highlight aspects of the data that are hidden in our raw data. We will see how such transformations can be used to build models that do not require parameters (called nonparametric models) to make predictions for highly nonlinear systems. Moreover, we will explore how data transformations are used to build parametric models that are universal (e.g., neural networks [NNs]), in the sense that they can use data to predict virtually any type of complex nonlinear behavior seen in practice. These techniques are flexible and aim to mimic the way the human brain learns from data collected from our sensory systems (e.g., vision and hearing). We will also discuss a new type of predictive model (known as a classification model) that aims to predict outputs that are discrete in nature.

We will discuss how to build empirical models that use data that appears in complex formats such as images, time series, and spatial patterns. We will see that making predictions based on such objects requires preprocessing steps that transform the data into a representation that enables predictions and decisions (e.g., our vision system transforms image data into signals prior to feeding this into our brain for processing). Ultimately, we will see that the way we represent our data can play a key role in our ability to extract useful information. In this respect, we will see that obtaining effective data representations requires a combination of mathematical principles, physical understanding, and creative thinking. We have already alluded to techniques that can be used to represent complex data formats for use in predictive models (e.g., text and one-hot encoding to represent molecules, as discussed in Exercise 5.7).

6.1 Principal Component Analysis

We begin by exploring techniques that help us address the following important questions:

- How can I extract useful information from *high-dimensional* data?
- How can I eliminate (filter out) data that is non-informative?
- How can I reduce (compress) data to facilitate analysis, visualization, storage, and use?

Consider the following *statistical mixing* problem:

We have multiple input random variables (RVs) $X = (X_1, X_2, \ldots, X_n)$ and you want to create a *mixture* of these RVs of the form:

$$T = \sum_{i=1}^{n} w_i X_i = w^T X, \tag{6.1}$$

where $w_i \in \mathbb{R}$ are the mixture proportions/weights (that we collect in vector $w = (w_1, w_2, \ldots, w_n)$) and T is a scalar RV that results from mixing the RVs.

Now consider the following questions:

- What weights w give a product T that contains maximum information about X?
- What weights w give a product T that contains the second most information about X?
- What weights w give a product T that contains minimum information about X?

Think about the analogy of this statistical mixing problem with that of a physical mixing process (see Figure 6.1). In a physical context, we want to mix a set of input flows of different quality in such a way that the product is as valuable as possible. Here, we expect that some inputs will be more valuable than others and some might even be redundant or nonessential. From a statistical standpoint, we can measure quality in terms of *information content*. We thus want the mixture T to encode as much information about the inputs X as possible; in doing so, we need to select the proportions w judiciously. For instance, if we do not select the proportions correctly, the product T might not convey information about the inputs (the information about some inputs might get lost in the mix). This mixing analogy also resembles the situation in which we mix ingredients to obtain a beverage (e.g., an "old-fashioned"); here, the question is: To what extent can we infer the ingredients by tasting the product? (e.g., if one ingredient dominates, it can hide the presence of the others).

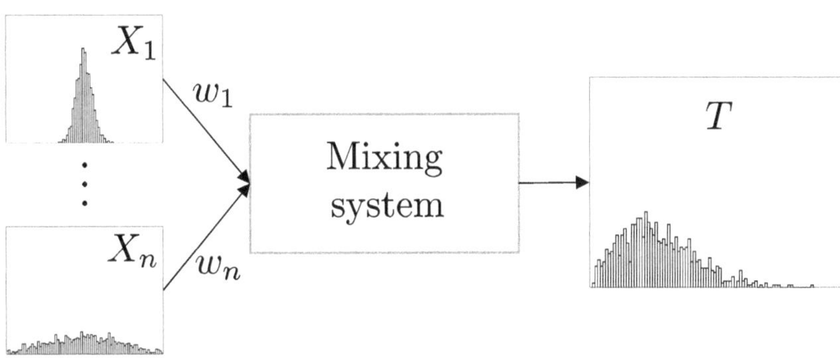

Figure 6.1 Statistical mixing problem to be studied using principal component analysis (PCA).

The mixing problem can be solved by using a technique called *principal component analysis* (PCA) [51], which was proposed by statistician Karl Pearson in 1901. In a nutshell, the goal of PCA is to identify mixtures (also known as principal components) that contain maximum information about the input data. These mixtures contain dominant features of the data and can be used to summarize and analyze high-dimensional data. Principal component analysis implicitly eliminates nonobvious data redundancies (that do not contribute information), and this is useful in enabling data compression. In other words, we will see that PCA aims to transform data from its original space to an information (feature) space that might be easier to analyze and compress.

Principal component analysis is widely used in data analysis and combines concepts of statistics, linear algebra (analysis of matrix eigenvalues), and optimization that will be useful in discussing other techniques. For instance, we will see that PCA can be used as an effective data preprocessing technique that facilitates the construction of predictive models (e.g., linear regression and neural networks). The principles of PCA are also used in the design of advanced ML tools such as autoencoder networks, which also aim to reduce high-dimensional data.

The PCA procedure can be summarized as follows:

- Collect observations $x_\omega \in \mathbb{R}^n$, $\omega \in \mathcal{S}$ for the multivariate RV $X = (X_1, X_2, \dots, X_n)$. Here, we make the implicit assumption that the observations are *random samples* of X (they are i.i.d.).
- Store the observations in the $S \times n$ data matrix:

$$\mathbf{X} = \begin{bmatrix} x_{1,1} & x_{1,2} & \cdots & x_{1,n} \\ x_{2,1} & x_{2,2} & \cdots & x_{2,n} \\ \vdots & \vdots & \vdots & \vdots \\ x_{S,1} & x_{S,2} & \cdots & x_{S,n} \end{bmatrix}. \tag{6.2}$$

- Scale the columns of matrix $\mathbf{X}$ in such a way that their means are zero:

$$\frac{1}{S}\sum_{\omega=1}^{S} \mathbf{X}_{\omega,j} = 0, \qquad j = 1,\dots,n. \tag{6.3}$$

  This can be done by taking each column of $\mathbf{X}$ and subtracting its mean; this approach is also called "data centering." If one conducts this scaling, one can show that the sample covariance matrix of X is:

$$\hat{\text{Cov}}(X) = \mathbf{X}^T\mathbf{X}. \tag{6.4}$$

  Here, the matrix $\mathbf{X}^T\mathbf{X}$ is known as the *kernel* matrix, and we denote this as Σ. We note that the kernel matrix is a square $n \times n$ matrix. We also note that the kernel matrix contains all information about the inputs X; in our context, *variance* is our measure of *information content*. Recall that the kernel matrix appears in the covariance matrix for the parameter estimates of linear models (analyzed in Chapter 5); specifically, we have that $\text{Cov}[\hat{\theta}] = \sigma^2(\mathbf{X}^T\mathbf{X})^{-1}$. We will analyze the relevance of this connection shortly.

- The mixture $T = w^T X$ (known as a principal component) is a univariate RV, and one can show that its sample variance is:

$$\hat{\mathbb{V}}[T] = w^T \Sigma w. \tag{6.5}$$

This establishes a direct connection between the sample variance of T and the sample covariance of X. The variance of the principal component T will be used as a measure of information content of the mixture.

The mixture weights w that contain maximum information about X are those that maximize the variance of T. Consequently, we can find these proportions by solving the optimization problem:

$$\max_w \; w^T \Sigma w \text{ s.t. } \|w\| = 1. \tag{6.6}$$

The constraint $\|w\| = 1$ is a normalization constraint that prevents the weights w from being arbitrarily large (or small). The solution to this optimization problem is denoted as $w_1 \in \mathbb{R}^n$, and one can show that this is the *eigenvector* of Σ associated with its largest eigenvalue $\lambda_1 \in \mathbb{R}$. We recall that an eigenvalue–eigenvector pair must satisfy

$$\Sigma w_1 = \lambda_1 w_1. \tag{6.7}$$

This highlights that the eigenvalue–eigenvector pair aims to summarize/compress the matrix Σ.

The optimal proportions w_1 give the optimal mixture $T_1 = w_1^T X$ that contains maximum information about X. One can show that the eigenvalue–eigenvector pair satisfies $\lambda_1 = w_1^T \Sigma w_1 = \hat{\mathbb{V}}[T_1]$. As such, the eigenvalue measures information content (the variance of T_1).

- One can identify the mixture that contains the second most information by solving the optimization problem:

$$\max_w \; w^T \Sigma w \text{ s.t. } \|w\| = 1, \; w^T w_1 = 0. \tag{6.8}$$

The solution to this problem is denoted as $w_2 \in \mathbb{R}^n$, and this is the eigenvector associated with the second largest eigenvalue $\lambda_2 = w_2^T \Sigma w_2$. The constraint $w_2^T w_1 = 0$ (known as an orthogonality constraint) ensures that the second best mixture $T_2 = w_2^T X$ is not correlated to the best mixture $T_1 = w_1^T X$. This guarantees that the principal components T_2 and T_1 provide *complementary* (nonredundant) information.

- One can continue the above procedure to obtain all eigenvectors $w_k \in \mathbb{R}^n$, $k = 1, \ldots, n$ of the covariance matrix (a matrix of dimension $n \times n$ contains n eigenvectors). Each eigenvector w_k has an associated mixture $T_k = w_k^T X$ and information content $\lambda_k = w_k^T \Sigma w_k$.

Importantly, the family of mixtures obtained from the above procedure are ranked in order of information content:

$$\lambda_1 \geq \lambda_2 \geq \ldots \lambda_n. \tag{6.9}$$

The last eigenvector w_n and the associated mixture $T_n = w^T X$ contain minimum information of X. We also note that, because the covariance matrix Σ is positive semidefinite, all its eigenvalues are nonnegative (positive or zero).

The set of eigenvectors and eigenvalues can be used to decompose the kernel matrix Σ as:

$$\Sigma = \lambda_1 w_1 w_1^T + \lambda_2 w_2 w_2^T + \cdots + \lambda_n w_n w_n^T. \tag{6.10}$$

Truncating this series (by eliminating the smallest eigenvalues) enables compression/reduction of the kernel matrix Σ. We expect that eliminating small eigenvalues will not affect the overall features of the matrix Σ. Specifically, imagine the extreme case in which some eigenvalues are actually zero; this means that those terms do not contribute any information. We will see later that the presence of small eigenvalues is a sign that there is *redundancy in the data*, which might arise from linear combinations of variables or correlations between them.

The eigenvalue decomposition of Σ can also be written in the following compact matrix form:

$$\Sigma = \mathbf{W}\Lambda\mathbf{W}^T, \tag{6.11}$$

where $\mathbf{W} \in \mathbb{R}^{n \times n}$ is the eigenvector matrix of the form:

$$\mathbf{W} = [w_1 | w_2 | \ldots | w_n]. \tag{6.12}$$

We can thus see that the columns of $\mathbf{W}$ are the eigenvectors of Σ. The matrix $\Lambda \in \mathbb{R}^{n \times n}$ is the eigenvalue matrix; this is a diagonal matrix of the form:

$$\Lambda = \text{diag}(\lambda) = \begin{bmatrix} \lambda_1 & & & \\ & \lambda_2 & & \\ & & \ddots & \\ & & & \lambda_n \end{bmatrix}. \tag{6.13}$$

We can thus see that the eigenvalues are the diagonal entries and thus $\Lambda_{k,k} = \lambda_k$.

The mixtures $T_1, T_2, \ldots T_n$ contain all the information about X. We can use the eigenvector matrix $\mathbf{W}$ to *project* the data matrix $\mathbf{X}$ into the space of the principal components by using the linear transformation:

$$\mathbf{T} = \mathbf{X}\mathbf{W}, \tag{6.14}$$

where $\mathbf{T} \in \mathbb{R}^{S \times n}$ is a matrix with entries $\mathbf{T}_{\omega,j}$, $\omega = 1, \ldots, S$, $j = 1, \ldots, n$. Here, $\mathbf{T}_{\omega,j}$ is the ω-th observation of the j-th principal component T_j.

The information content of the matrix Σ (known as the *spectrum*) is typically summarized in the form of a curve that shows the decaying value of the eigenvalues. A curve that decays rapidly indicates that the first few eigenvalues contain most of the information; these matrices are easy to compress, as they contain lots of redundancy (data contains redundant information). A curve that decays slowly indicates that there is a significant amount of information in a large number of eigenvalues; these matrices are more difficult to compress, as they contain little redundancy (data contains unique information). The spectrum of the Σ can also be summarized

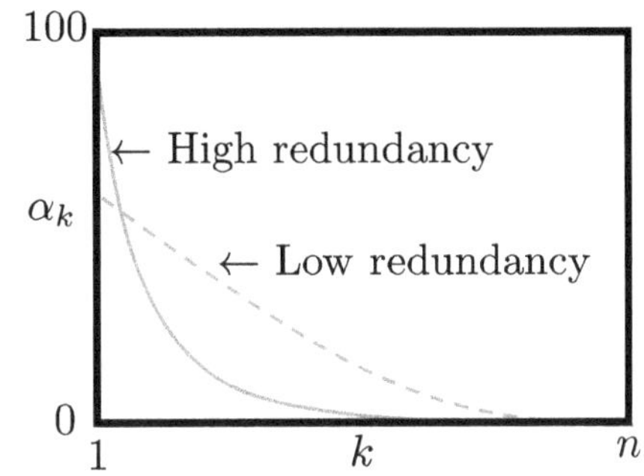

Figure 6.2 Illustration of the spectrum of a couple of covariance matrices with high redundancy (solid) and low redundancy (dashed).

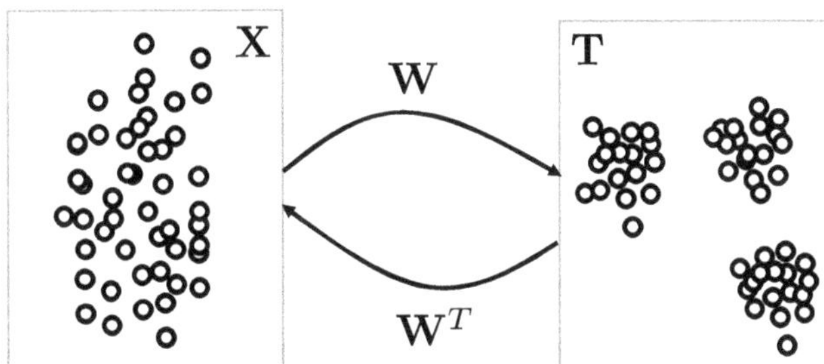

Figure 6.3 Projection of data from physical space to information/latent space (and back) using the eigenvector matrix **W**.

in the form of a curve that shows the proportion/percentage of the total information that each eigenvalue λ_k contains; the total information can be expressed as the sum of the eigenvalues $\sum_{j=1}^{n} \lambda_j$ or as the product of eigenvalues $\prod_{j=1}^{n} \lambda_j$. If total information is the sum, the percentage of the total information that the eigenvalues contain is:

$$\alpha_k = 100 \cdot \frac{\lambda_k}{\sum_{j=1}^{n} \lambda_j}, \quad k = 1, \ldots, n. \tag{6.15}$$

The spectrum of a couple of covariance matrices is illustrated in Figure 6.2.

The projection $\mathbf{T} = \mathbf{X}\mathbf{W}$ *transforms the data* from the original *physical space* (under which $\mathbf{X}$ lives) into an *information space* (under which $\mathbf{T}$ lives). The ability to analyze data in the *information space* can provide deep insights into underlying trends/patterns that are not immediately obvious when analyzing data in a physical space. This is illustrated in Figure 6.3. This is important because we, as engineers, are often used to visualize the behavior of a system in a physical space, and this might hide important features of the system. Principal component analysis thus provides an *alternative lens* or perspective under which we can analyze behavior. The information space is also often known as the *latent* space, and the principal components are known as latent/hidden/score variables.

A key property of PCA is that we can transform data from the information space back to the physical space by using the inverse relationship $\mathbf{X} = \mathbf{T}\mathbf{W}^T$. This property is important, as it provides a mechanism to move back and forth between the latent (information) space and the physical space. This property is also important because it allows us to interpret PCA as a tool that minimizes reconstruction errors (when we move from the latent space to the physical space). Specifically, one can show that the eigenvector matrix is a projection matrix that minimizes the reconstruction error; we will elaborate on this topic in our discussion of singular value decomposition (SVD).

One often selects a few principal components (typically the first two or three T_1, T_2, T_3) to summarize and visualize the high-dimensional data of X in a small-dimensional information space. This reduction procedure is achieved by extracting the first few columns of the principal component matrix $\mathbf{T}$.

From the projection $\mathbf{T} = \mathbf{X}\mathbf{W}$, we can observe that the covariance of the principal components is:

$$\begin{aligned} \hat{\text{Cov}}[T] &= \mathbf{T}^T\mathbf{T} \\ &= \mathbf{W}^T\mathbf{X}^T\mathbf{X}\mathbf{W} \\ &= \mathbf{W}^T\mathbf{W}\Lambda\mathbf{W}^T\mathbf{W} \\ &= \Lambda. \end{aligned} \tag{6.16}$$

Because Λ is a diagonal matrix, we can conclude that the principal components are uncorrelated (there is no redundant information between them). Moreover, we can also conclude that all information on the principal components is contained in the eigenvalues. The above derivation uses the property that the eigenvector matrix is orthonormal (its columns are orthogonal) and thus $\mathbf{W}^T = \mathbf{W}^{-1}$. The orthogonality property is precisely what ensures that the principal components are uncorrelated. We also used this property in the analysis of ellipses for Gaussian variables in Chapter 3. The orthogonality property also has interesting geometric interpretations; for instance, one can show that the projection of the data into an information space is "rotating" the data.

A key property of the principal components is that they retain the *structure* of the original high-dimensional data. This is important because, in practice, we are often tempted to analyze and visualize data by selecting variables that we deem most important (e.g., we analyze pairs of data). Using a variable selection approach can be dangerous, as it might not preserve the structure of the original data and might hide relevant information. Principal component analysis reduces the data by eliminating redundant information; for example, if input variables X_1, X_2 are strongly correlated, then one variable carries significant information about the other (and thus one can be ignored as it is redundant). Redundant information also manifests as data *collinearities*, in which some variables can be obtained as linear combinations of others (linear dependencies in the rows/columns of the data matrix are present).

Principal component analysis can be used to identify dominant trends (clusters) and outliers in the data. This is because dimensionality reduction facilitates visualization. Fundamentally, however, one can think of PCA as a data transformation (preprocessing) technique that uncovers dominant features. We will see that this feature information can be used in a variety of tasks such as in the development of predictive models. We will also see that the principle of eigenvalue decomposition used in PCA can be generalized to extract dominant features from images and general matrices.

It is important to highlight that it is not necessary to recenter data (to obtain a zero mean) in order to apply PCA. Normalization is only required if one seeks to interpret the matrix $\mathbf{X}^T\mathbf{X}$ as the covariance matrix of X. However, one can directly apply PCA on the matrix $\mathbf{X}^T\mathbf{X}$ without recentering to reduce the dimension of the data. Scaling of data is also a common technique used to make all variables of approximately the same magnitude; this can facilitate the interpretation of the results.

Example 6.1 PCA for Gibbs Reactor Data

Consider the Gibbs reactor example; we have a dataset with measurements of pressure, temperature, and conversion (P,T,C). When we visualize the data in a 3D (P,T,C) physical space (see Figure 6.4(a)), we can see that the data separates into a couple of clusters/communities. One of these clusters is times at which the reactor operates under normal mode (cooling system operates normally) and the other is times under which it operates under failing mode (cooling system fails).

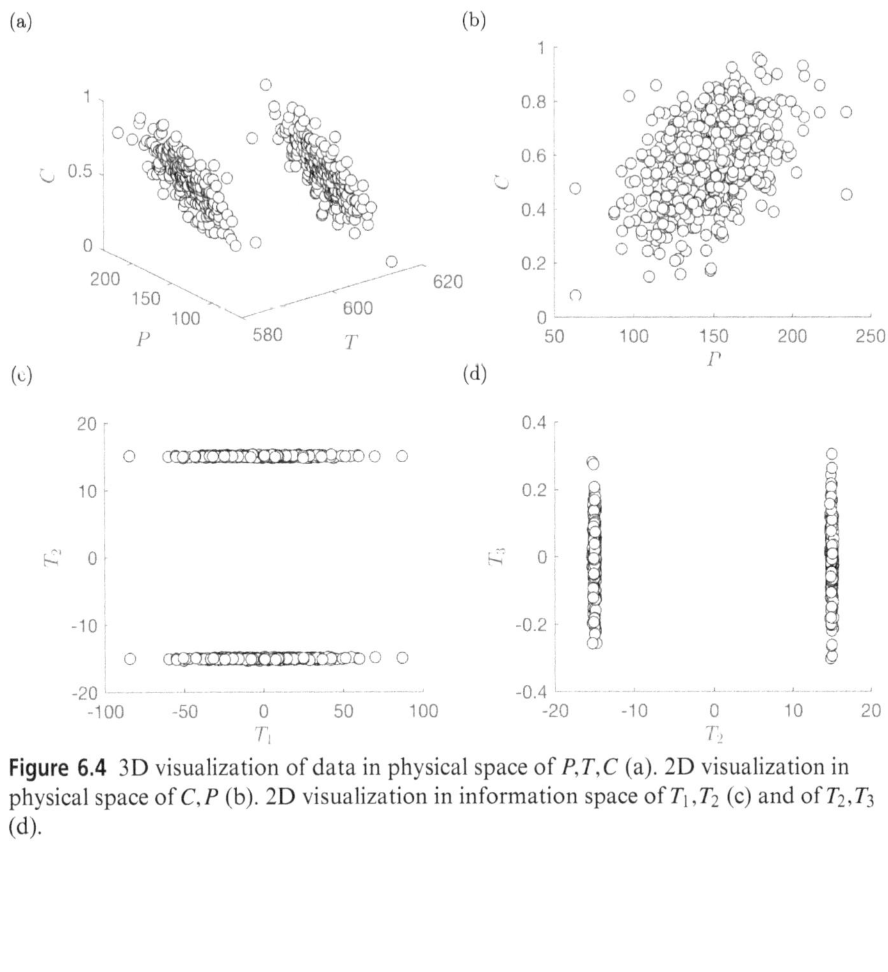

Figure 6.4 3D visualization of data in physical space of P,T,C (a). 2D visualization in physical space of C,P (b). 2D visualization in information space of T_1,T_2 (c) and of T_2,T_3 (d).

Humans can only visualize data in a small number of dimensions (usually in 2D and 3D). Imagine that we visualize our data in 2D by ignoring the dimension associated with temperature; in other words, we visualize in the 2D space of (P,C). We can see that visualizing the data in this reduced space hides the clusters in the data (see Figure 6.4(b)). This is because the failure of the reactor is driven by a failure in the cooling system (affecting temperature). This illustrates how improper reduction of data can miss information and thus one needs to proceed with care. This is a fundamental issue in data analysis, particularly in complex problems that involve many dimensions (hundreds to thousands of variables).

In Figure 6.4(c and d), we visualize the data in 2D but this time in the information space of the principal components (T_1,T_2) and of (T_2,T_3). We can see that this visualization maintains the clusters that are present in the original data and reveals the instances under which the reactor is operating in normal and failure modes. This highlights that PCA reduces the dimension of our data in a way that retains its structure.

The eigenvalues of the covariance $\mathbf{X}^T\mathbf{X}$ are $\lambda = (3.19 \times 10^5, 1.12 \times 10^5, 4)$; we thus have that a couple of eigenvalues are extremely large and one is small. Specifically, the first eigenvalue contains 74% of the total information contained in the data, the second eigenvalue contains 26% of the information, and the third eigenvalue contains negligible information. The presence of a small eigenvalue indicates that there is a redundant dimension. This redundancy arises because of the high correlation between pressure P and conversion C. This indicates that we can throw away either P or C, and we will still be able to maintain most information of our data.

Script to reproduce results: `ch6_gibbs_pca` □

6.1.1 Principal Component Estimation

In Chapter 5, we explored strategies to estimate parameters for a linear regression model of the form $\mathbf{y} = \mathbf{X}\theta$. We also discussed how redundancy in the data leads to rows or columns in the data matrix $\mathbf{X}$ that are linearly dependent, and this can lead to high parameter uncertainty (low precision). Here, we will show how to use PCA to identify and eliminate such data redundancies and with this obtain more precise parameters. This is done using a technique known as principal component regression (or estimation) [52].

The data matrix $\mathbf{X}^T\mathbf{X}$ analyzed in PCA is related to the Hessian matrix of the linear estimation problem as $\mathcal{H}(\hat{\theta}) = (1/\sigma^2)\mathbf{X}^T\mathbf{X}$. Here, we use a simplified version of the Hessian (assume diagonal noise covariance) in order to facilitate the discussion. The covariance matrix of the estimates $\hat{\theta}$ is the inverse Hessian matrix; this can be written in terms of the eigenvectors and eigenvalues of $\mathbf{X}^T\mathbf{X}$ (obtained with PCA) as:

$$\begin{aligned}
\mathrm{Cov}[\hat{\theta}] &= \mathcal{H}(\hat{\theta})^{-1} \\
&= \sigma^2(\mathbf{X}^T\mathbf{X})^{-1} \\
&= \sigma^2(\mathbf{W}\Lambda\mathbf{W}^T)^{-1} \\
&= \sigma^2\mathbf{W}\Lambda^{-1}\mathbf{W}^T \\
&= \sigma^2(\lambda_1^{-1}w_1w_1^T + \lambda_2^{-1}w_2w_2^T + \cdots + \lambda_n^{-1}w_nw_n^T).
\end{aligned} \tag{6.17}$$

This reveals that small eigenvalues of the data matrix $\Sigma = \mathbf{X}^T\mathbf{X}$ lead to large covariance matrices for the parameter estimates (high parameter uncertainty). We also recall that this is a sign of degeneracy (the parameter estimates are nonunique). These issues arise because of data redundancy; in the extreme case in which the matrix $\mathbf{X}^T\mathbf{X}$ is singular (some of the rows or columns are linearly dependent), we have that at least one eigenvalue is *exactly zero* and the estimate has infinite covariance (infinite uncertainty). These linear dependencies typically arise from strong correlations in the input variables.

Motivated by the decomposition of the parameter covariance matrix, it makes sense to try to build a predictive model $\mathbf{y} = \mathbf{X}\theta$ by first *transforming* the input data matrix $\mathbf{X}$. Specifically, we are interested in removing the effect of small eigenvalues associated with non-informative aspects of the data. This is done by partitioning the eigenvector matrix:

$$\begin{aligned} \mathbf{W} &= [\mathbf{W}_1 | \mathbf{W}_2] \\ &= [\underbrace{w_1, w_2, \ldots, w_{n_1}}_{\text{high information}} \mid \underbrace{w_{n_1+1}, w_{n_1+2}, \ldots, w_n}_{\text{low information}}]. \end{aligned} \tag{6.18}$$

We can thus see that $\mathbf{W}_1 \in \mathbb{R}^{S\times n_1}$ contains the eigenvectors associated with the large eigenvalues of Σ (high information) and where the matrix $\mathbf{W}_2 \in \mathbb{R}^{S\times n_2}$ contains the eigenvectors associated with the small eigenvalues (low information). We also have that $n_1 + n_2 = n$, which is the number of columns of $\mathbf{X}$ (the number of input variables).

The eigenvector partitioning also induces a partitioning of the eigenvalue matrix of the form:

$$\Lambda = \left[\begin{array}{c|c} \Lambda_1 & \\ \hline & \Lambda_2 \end{array}\right], \tag{6.19}$$

where $\Lambda_1 \in \mathbb{R}^{n_1\times n_1}$ is a diagonal matrix that contains the n_1 largest eigenvalues (high information) and $\Lambda_2 \in \mathbb{R}^{n_2\times n_2}$ is a diagonal matrix that contains the n_2 smallest eigenvalues (the ones that we wish to eliminate because they have low information).

We project the input data into the information space as:

$$\mathbf{T}_1 = \mathbf{X}\mathbf{W}_1, \tag{6.20}$$

and we note that the projected matrix is $\mathbf{T}_1 \in \mathbb{R}^{S\times n_1}$. We then use the principal component data matrix $\mathbf{T}_1$ to build a *reduced* model of the form:

$$\mathbf{y} = \mathbf{T}_1\gamma, \tag{6.21}$$

where $\gamma \in \mathbb{R}^{n_1}$ are the parameters. We call this a *reduced model* because the projection of the data has reduced the number of input variables (from n in $\mathbf{X}$ to n_1 in $\mathbf{T}_1$), and as a result, we have reduced the number of parameters needed (from n in θ to n_1 in γ). In other words, reducing the data has the effect of reducing the complexity of the model. We recall that, in Chapter 5, we reduced the complexity of the model by penalizing parameters and eliminating input variables. Here, we are reducing complexity by using data projection.

We estimate the parameters γ by solving the problem:

$$\min_{\gamma} \frac{1}{2}\|\mathbf{y} - \mathbf{T}_1\gamma\|_2^2. \tag{6.22}$$

The covariance of the estimate is given by:

$$\mathrm{Cov}[\hat{\gamma}] = \sigma^2(\mathbf{T}_1^T\mathbf{T}_1)^{-1}. \tag{6.23}$$

The estimated parameters $\hat{\gamma}$ live in an information space that is smaller than that of our original physical space of the parameters θ. To obtain an estimate of the parameters in the original physical space, we note that the objective function can be written as $\frac{1}{2}\|\mathbf{y} - \mathbf{X}\mathbf{W}_1\gamma\|_2^2$ and thus:

$$\hat{\theta} = \mathbf{W}_1\hat{\gamma}. \tag{6.24}$$

From this observation, we can also determine that the estimate in the original space is a linear transformation of the estimate in the information space; as such, we have that:

$$\mathrm{Cov}[\hat{\theta}] = \mathbf{X}\mathrm{Cov}[\hat{\gamma}]\mathbf{X}^T. \tag{6.25}$$

One can use the orthogonality properties of the eigenvectors to show that the covariance of the estimate $\hat{\theta}$ is:

$$\begin{aligned} \mathrm{Cov}[\hat{\theta}] &= \sigma^2\mathbf{W}_1\Lambda_1^{-1}\mathbf{W}_1^T \\ &= \sigma^2(\lambda_1^{-1}w_1w_1^T + \lambda_2^{-1}w_2w_2^T + \cdots + \lambda_{n_1}^{-1}w_{n_1}w_{n_1}^T). \end{aligned} \tag{6.26}$$

By comparing Eq. (6.26) with Eq. (6.17), we can see that projecting the data has the effect of eliminating the effect of the small eigenvalues of the kernel $\mathbf{X}^T\mathbf{X}$ from the estimate covariance, and this makes the parameter estimates more precise. We also highlight that data reduction implicitly reduces the number of parameters, and, with this, we can prevent overfitting and avoid parameters (and model predictions) with high uncertainty.

This estimation approach that preprocesses the data is known as principal component estimation (or regression). The name arises from the fact that we build our estimation model by using the principal components of the data (as opposed to the data itself). This helps us eliminate the undesired effects of our data. Another way to think about data pre-processing is that, intuitively, we do not lose significant information by ignoring certain (redundant) elements of the input data. However, we need to be careful about how we reduce our data so as not to lose too much information. For instance, when facing high estimate variability, a common approach to reducing data is to simply ignore certain input variables or observations. This approach (called subset selection) is intuitive but can miss significant information. The reduction of data using PCA also allows us to discover the inherent dimensions of our data; even if our original data is high-dimensional (e.g., hundreds of variables), the data might actually live in an information space of small dimension.

Principal component estimation shares commonalities with other regression techniques that use dimensionality reduction, such as partial least-squares (PLS) regression and principal covariates regression (PCovR) [53, 54]. An introductory comparative analysis of some of these techniques can be found in [30, chapter 4].

Example 6.2 PCA Estimation for Gibbs Reactor

We revisit the estimation problem of Example 5.14. Our goal is to develop a linear model for the Gibbs reactor that predicts conversion as a function of different operating variables. The model we consider has the form:

$$Y = \theta_0 + \theta_1 X_1 + \theta_2 X_2 + \theta_3 X_3 + \theta_4 X_4 + \theta_5 X_5 + \epsilon. \tag{6.27}$$

Here, Y is the conversion, X_1 is the pressure, X_2 is the output flow of CO, X_3 is the output flow of H_2, X_4 is the output flow of CH_3OH, and X_5 is the temperature. We expect that some of the data is redundant because some of the variables are strongly correlated (see Example 3.15). In Example 5.14, we also found evidence that one can significantly reduce the complexity of the model by eliminating variables. In this example, we will explore this issue further by using PCA techniques.

We begin by applying a standard linear estimation procedure and solving a least-squares problem to obtain the estimates $\hat{\theta}$. The fit obtained with such parameters is shown in Figure 6.5(a); we can see that these parameters give a good fit. Specifically, the mean squared error (MSE) is 2.44×10^{-3}; however, we find that the covariance matrix $\mathrm{Cov}[\hat{\theta}]$ for the parameters is singular (contains zero eigenvalues). As such, it is not possible to compute the parameter variances and their confidence intervals. Upon inspection, we find that the eigenvalues of the data matrix $\mathbf{X}^T\mathbf{X}$ are:

$$\lambda = (1.45 \times 10^8, 2.36 \times 10^5, 1.97 \times 10^3, -5.65 \times 10^{-8}, -1.23 \times 10^{-9}, 2.18 \times 10^{-12}).$$

We thus see that there are three eigenvalues that are effectively zero. This tells us that there are three parameters that cannot be estimated reliably; this is the result of having data redundancy (3 input variables are not necessary). Instead of eliminating variables, we project our data into an information space. To do so, we obtain the eigenvectors $\mathbf{W}$ of the matrix $\mathbf{X}^T\mathbf{X}$ and eliminate the three columns corresponding to the three smallest eigenvalues to obtain the matrix $\mathbf{W}_1$. We then reduce the input data by using the projection $\mathbf{T}_1 = \mathbf{X}\mathbf{W}_1$ and solve the estimation problem to obtain a new estimate $\hat{\gamma}$ and $\hat{\theta} = \mathbf{W}_1\hat{\gamma}$. Note that $\hat{\gamma}$ only contains three parameters (as we have eliminated 3).

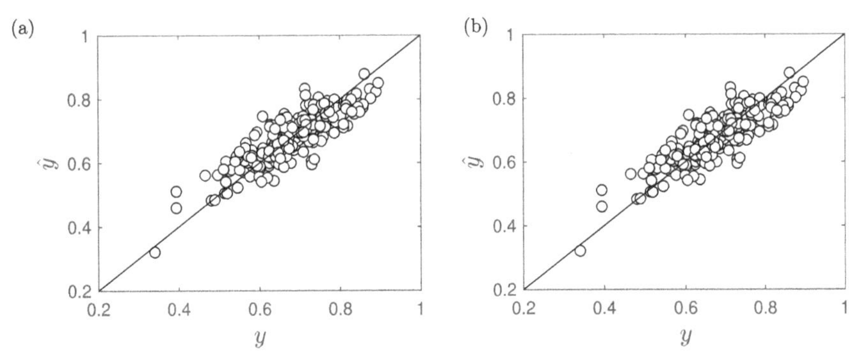

Figure 6.5 Fit of model under standard estimation (a) and PCA estimation (b).

The fit obtained with such parameters is shown in Figure 6.5(b). Note that the fit is virtually the same. However, we find that the new parameter covariance matrix is nonsingular and the variances of the parameters are:

$$\mathbb{V}(\hat{\theta}) = (7.34 \times 10^{-13}, 3.19 \times 10^{-7}, 1.77 \times 10^{-7}, 2.49 \times 10^{-7}, 2.56 \times 10^{-7}, 2.49 \times 10^{-7}).$$

From this, we see that the parameters are highly precise. This illustrates how processing or "cleaning" of input data can eliminate redundancies and improve our models.

Script to reproduce results: `ch6_gibbs_pca_estimation` □

6.1.2 Sparse Principal Component Analysis

The eigenvectors w obtained with PCA contain the proportions of our mixture $T = w^T X$. The magnitude of the i-th component w_i (given by $|w_i|$) implicitly tells us the contribution of the input variable X_i in the mixture. In other words, the magnitudes of the entries of w reveal the key (dominant) input variables (ingredients) in our mixture. A problem that often arises in complex applications is that the number of input variables is quite large, and we only want to find a handful of key variables. These key variables can be found by using a technique called sparse principal component analysis (sparse PCA) [55].

The idea would be to make zero as many entries of the eigenvector w as possible (ignore elements of the eigenvector) while making sure that the new sparse eigenvector is close to the original one. Intuitively, one could simply discard entries with small enough magnitudes $|w_j|$; however, finding an appropriate threshold value is not always obvious.

Sparse PCA is based on the observation that the eigenvector w_j can be recovered as the solution of the least-squares optimization problem:

$$\min_{w} \frac{1}{2} \|\mathbf{X} w_j - \mathbf{X} w\|_2^2. \tag{6.28}$$

The objective function here is to find a vector w that minimizes the distance between the data projection obtained with the eigenvector w_j and that obtained with w. A sparse eigenvector is found by minimizing the number of nonzero entries of w; this is done by adding a Lasso penalty function to the objective function as:

$$\min_{w} \frac{1}{2} \|\mathbf{X} w_j - \mathbf{X} w\|_2^2 + \kappa \|w\|_1, \tag{6.29}$$

where $\|w\|_1 = \sum_{i=1}^{n} |w_i|$ and $\kappa \in \mathbb{R}_+$ is a parameter that aims to trade-off sparsity of the sparse eigenvector $\tilde{w}_j$ and proximity of this to the original eigenvector w_j. Specifically, minimizing the penalty $\|w\|_1$ reduces the number of nonzero entries (i.e., eliminates variables). The sparse eigenvector is made to satisfy $\|\tilde{w}_j\|_2 = 1$ by scaling the vector as $\tilde{w}_j \leftarrow \tilde{w}_j / \|\tilde{w}_j\|_2$. The sparsification approach is applied to all eigenvectors to obtain a sparse eigenvector matrix $\tilde{\mathbf{W}}$. In the limit when κ is large, every sparse eigenvector will only contain a single variable. One can interpret such variables as the dominant variables associated with each principal component.

Sparse eigenvectors are easier to interpret than the dense eigenvectors obtained with the direct use of PCA. Specifically, they reveal key variables that drive our data reduction procedures. For example, in the context of PCA estimation, sparse eigenvectors can reveal variables that are responsible for the small eigenvalues of the kernel matrix $\mathbf{X}^T\mathbf{X}$. This information can be used to eliminate data for problematic variables prior to performing estimation.

Example 6.3 Sparse PCA Estimation for Gibbs Reactor

In our previous example, we applied PCA to reduce the data, and we saw that this improves the precision of the estimates. However, such a procedure did not provide much insight into which input variables might be responsible for the small eigenvalues of the kernel matrix $\mathbf{X}^T\mathbf{X}$. Specifically, the eigenvector matrix of the kernel matrix is given by:

$$\mathbf{W} = \begin{bmatrix} -0.00 & -0.00 & -0.00 & -0.00 & 0.00 & -0.00 \\ -0.19 & -0.84 & 0.50 & -0.20 & 0.01 & -0.01 \\ -0.04 & 0.23 & 0.38 & -0.03 & -0.43 & -0.79 \\ -0.61 & 0.40 & 0.45 & -0.59 & 0.53 & 0.06 \\ -0.09 & -0.25 & -0.45 & -0.04 & 0.59 & -0.62 \\ -0.76 & -0.09 & -0.45 & -0.78 & -0.42 & 0.02 \end{bmatrix}.$$

The last three eigenvectors (last three columns) are responsible for the small eigenvalues but do not provide a clear indication of which variable (which row) might be driving such behavior. Having such information can help us figure out which input variables might be problematic, and we can explicitly eliminate them from the input data matrix. To identify the key problematic variables, we apply sparse PCA with a large value of κ and obtain the sparse eigenvector matrix:

$$\tilde{\mathbf{W}} = \begin{bmatrix} 0 & 0 & 0 & 0 & 0 & 0 \\ 0 & 1.00 & 0 & 0 & 0 & 0 \\ 0 & 0 & 0 & 0 & 1.00 & 0 \\ 0 & 0 & 1.00 & 0 & 0 & 0 \\ 0 & 0 & 0 & 0 & 0 & 1.00 \\ 1.00 & 0 & 0 & 1.00 & 0 & 0 \end{bmatrix}.$$

Note that every eigenvector has exactly one entry, and from the last three columns, we see that the 6th, 3th, and 5th rows have nonzero entries. Recall that the first row is associated with the parameter θ_0 and thus the rows with nonzero entries correspond to the input variables X_5 (temperature), X_2 (CO flow), and X_4 (CH_3OH flow). By comparing the dense and sparse eigenvectors, we see that such a conclusion cannot be reached by simply identifying entries with large magnitudes. Specifically, recall that sparse PCA is identifying sparse eigenvectors that are closest to the dense ones and the solution is nonobvious.

As suggested by sparse PCA, we eliminate the data associated with input variables X_5, X_2, X_4. The reduction leads to a model with only three parameters. Upon estimating these parameters, we find that the fit obtained is the same as that achieved with PCA estimation (see Figure 6.6). Moreover, upon inspecting the variances, we find that these are small and given by:

$$\mathbb{V}(\hat{\theta}) = (2.78 \times 10^{-1}, 3.19 \times 10^{-7}, 9.27 \times 10^{-7}).$$

From this, we conclude that the variables identified by sparse PCA indeed seem to be the problematic ones. We can verify that if we keep the data input variables X_5, X_2, X_4 and eliminate the rest, we obtain an estimate with poor fit and that the kernel matrix is singular. This confirms our conclusion that the input variables X_5, X_2, X_4 are indeed problematic. We will explore yet another approach to detect these problematic variables in the next section.

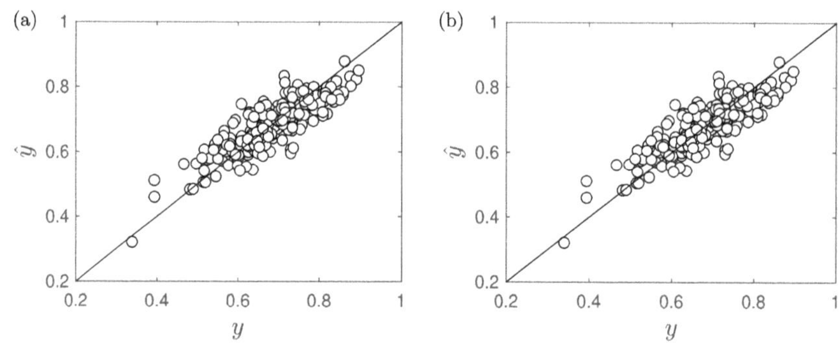

Figure 6.6 Fit of a linear model under PCA estimation (a) and fit using subset of variables identified with sparse PCA (b).

Script to reproduce results: `ch6_gibbs_pca_estimation_sparse` □

6.2 Singular Value Decomposition

Principal component analysis relies on an eigenvalue decomposition of the kernel matrix of the form $\Sigma = \mathbf{W}\Lambda\mathbf{W}^T$. Singular value decomposition (SVD) is a generalization of eigenvalue decomposition; specifically, eigenvalue decomposition can only be applied to square matrices (such as Σ), while SVD can be applied to rectangular matrices (such as $\mathbf{X}$). As such, SVD is more broadly applicable and is widely used for the compression of different forms of data (e.g., image analysis).

To explain how SVD works, it will be convenient to think about this technique in the context of PCA. One can implement PCA by performing an SVD of the data matrix $\mathbf{X} \in \mathbb{R}^{S \times n}$ (as opposed to performing an eigendecomposition on $\mathbf{X}\mathbf{X}^T$). Here, we will assume that $n \leq S$, which is the most common case found in applications (we have fewer variables than observations). Under this assumption, the *economy-size* SVD decomposition is given by:

$$\mathbf{X} = \mathbf{U}\mathbf{S}\mathbf{V}^T, \tag{6.30}$$

where $\mathbf{U} \in \mathbb{R}^{S\times n}$ is called the left eigenvector matrix and $\mathbf{V} \in \mathbb{R}^{n\times n}$ is called the right eigenvector matrix. These matrices have the corresponding eigenvectors as columns:

$$\mathbf{U} = [u_1|u_2|\ldots|u_n] \tag{6.31a}$$

$$\mathbf{V} = [v_1|v_2|\ldots|v_n], \tag{6.31b}$$

where $u_j \in \mathbb{R}^S$ and $v_j \in \mathbb{R}^n$ for all $j = 1,\ldots,n$. The singular value matrix $\mathbf{S} \in \mathbb{R}^{n\times n}$ is diagonal and contains the singular values $s_j \in \mathbb{R}_+,\ j = 1,\ldots,n$ in its diagonal. The singular values are ordered in magnitude $s_1 \geq s_2 \geq \cdots \geq s_n$, and all of these values are nonnegative (either positive or zero).

A key property is that the number of nonzero singular values reveals the *rank* of the matrix **X**. We recall that the rank of a matrix is the maximum number of linearly independent rows (equivalent to the number of linearly independent columns). Since we have assumed that $n \leq S$, we say that **X** is *regular* if the rank of the matrix is n.

When a matrix is not regular, it means that there are redundant rows or columns; in our context, this means that there are redundant observations or variables. To be precise, redundancies arise because some rows are linear combinations of other rows (or some columns can be obtained from linear combinations of other columns). As a trivial example, imagine that one of our variables is length (in meters) and the other is length (in feet); these variables are linearly dependent, as one is a scaled version of the other. Redundancies are common but are not always spotted easily; as we will see shortly, these can arise from structural dependencies and correlations between variables. Singular values are an important tool that will help us detect and eliminate redundancies in the data matrix **X**.

The matrices **U**, **V** obtained from SVD have the important property that they are orthogonal and thus satisfy $\mathbf{U}^T\mathbf{U} = \mathbf{I}$ and $\mathbf{V}^T\mathbf{V} = \mathbf{I}$; consequently, we can observe that:

$$\begin{aligned}\mathbf{X}^T\mathbf{X} &= (\mathbf{U}\mathbf{S}\mathbf{V}^T)^T(\mathbf{U}\mathbf{S}\mathbf{V}^T)\\ &- \mathbf{V}\mathbf{S}^T\mathbf{S}\mathbf{V}^T.\end{aligned} \tag{6.32}$$

By defining $\mathbf{W} = \mathbf{V}$ and $\Lambda = \mathbf{S}^T\mathbf{S}$, we obtain standard PCA (based on an eigendecomposition). This also reveals that the eigenvalues of $\mathbf{X}^T\mathbf{X}$ and the singular values of **X** are related as $\lambda_j = s_j^2,\ j = 1,\ldots,n$.

The SVD of **X** can be written as an expansion of the form:

$$\mathbf{X} = \sum_{j=1}^{n} s_j u_j v_j^T. \tag{6.33}$$

As in the case of the eigenvalue decomposition, one can truncate this expansion to obtain a compressed/reduced representation of the data matrix **X** (we eliminate the entries with small singular values).

Note that we have assumed that the dimensions of **X** are such that $n \leq S$. For situations in which $S \leq n$, the concepts are analogous, but the dimensions of the eigenvector matrices change; specifically, we have that $\mathbf{U} \in \mathbb{R}^{S \times S}$ and $\mathbf{V} \in \mathbb{R}^{n \times S}$. Moreover, if $S \leq n$, we have that the matrix is regular if its rank is S.

Example 6.4 SVD for Gibbs Reactor Data

We analyze the data matrix $\mathbf{X}$ for the Gibbs problem using SVD. This is a matrix with $S = 250$ rows (number of observations) and $n = 6$ columns (number of inputs); we thus have that $n \leq S$ and thus the matrix is deemed regular if its rank is $n = 6$. We perform SVD to this matrix and find that the singular values are:

$$\text{diag}(\mathbf{S}) = (12059.49, 486.03, 44.43, 0.00, 0.00, 0.00).$$

From this, we see that the matrix has a rank of 3 (there are three nonzero singular values). This indicates that there are three linearly dependent columns and that the matrix is not regular (otherwise, it would have a rank of 6).

We inspect the first six rows of the matrix:

$$\mathbf{X} = \begin{bmatrix} 1.00 & 166.98 & 25.56 & 451.12 & 74.44 & 583.15 \\ 1.00 & 170.71 & 24.75 & 449.51 & 75.25 & 583.15 \\ 1.00 & 141.17 & 32.22 & 464.44 & 67.78 & 583.15 \\ 1.00 & 146.63 & 30.64 & 461.28 & 69.36 & 583.15 \\ 1.00 & 164.67 & 26.08 & 452.16 & 73.92 & 583.15 \\ 1.00 & 157.43 & 27.79 & 455.59 & 72.21 & 583.15 \end{bmatrix}.$$

From this, we can see that the first column (corresponding to parameter θ_0) and the sixth column (corresponding to temperature) have constant values. As such, one can always obtain the sixth column by scaling the first column, and therefore, these columns are linearly dependent. This tells us that we can either choose to preserve the first or sixth column (but not both). Identifying the rest of the dependencies is not straightforward, but physical intuition tells us that one can infer the flow of a species from the flow of others (due to mass conservation).

To identify the rest of the redundancies, we compute the rank of matrix $\mathbf{X}$ by adding one column at a time and see when the rank increases (indicating that the new column is independent of the rest) or not (indicating that the new column is dependent on the rest). For example, we compute the rank of $\mathbf{X}[:, 1] = [x_1]$, $\mathbf{X}[:, 1, 2] = [x_1 \mid x_2]$, $\mathbf{X}[:, 1, 2, 3] = [x_1 \mid x_2 \mid x_3]$, and so on. By applying this procedure we find that:

$$\begin{aligned} \text{rank}(\mathbf{X}[:, 1]) &= 1 \\ \text{rank}(\mathbf{X}[:, 1, 2]) &= 2 \\ \text{rank}(\mathbf{X}[:, 1, 2, 3]) &= 3 \\ \text{rank}(\mathbf{X}[:, 1, 2, 3, 4]) &= 3 \\ \text{rank}(\mathbf{X}[:, 1, 2, 3, 4, 5]) &= 3 \\ \text{rank}(\mathbf{X}[:, 1, 2, 3, 4, 5, 6]) &= 3. \end{aligned}$$

From this, we can see that the 4th, 5th, and 6th columns are dependent on the 1st, 2nd, and 3rd columns (they can be obtained as a linear combination of these). As such, we can safely remove the last three columns from the data matrix (eliminate the variables X_3, X_4, X_5 and preserve X_1, X_2 and the column of ones). Note that this selection is not unique; for example, we have that $\text{rank}(\mathbf{X}[:, 3, 5, 6]) = 3$ is also viable; this corresponds to the variable selection of X_2, X_4, X_5 (and elimination of X_1, X_3 and column of ones) used in Example 6.3. The above procedure can thus help us identify all combinations that lead to degeneracy/redundancy.

Script to reproduce results: `ch6_gibbs_pca_estimation_svd` □

Example 6.5 Images as Matrices

Any matrix $\mathbf{X}$ can be represented as a *grayscale image* in which each matrix entry corresponds to a pixel and the number at such entry is the intensity of the pixel (i.e., how dark or bright the pixel is). This indicates that numerical *patterns* encoded in a matrix can be visualized using images. Similarly, any grayscale image can be represented as a matrix $\mathbf{X}$, and this is the basic property that enables image processing (e.g., compression and filtering). There is thus an inherent duality between grayscale images and matrices; we have already used this idea to visualize matrices in Chapter 3. We will see shortly that we can leverage this duality to find alternative representations of data that facilitate visualization and processing.

To illustrate the connections between matrices and images, consider the matrix shown below; here, we can see that the matrix has some obvious numerical patterns. Specifically, it is symmetric and the magnitude of the entries increases as we approach the center.

$\mathbf{X} =$

0	0	0	0	0	0	0	0	0	0
0	0	0	1	1	1	1	0	0	0
0	0	1	2	4	4	2	1	0	0
0	1	2	6	9	9	6	2	1	0
0	1	4	9	14	14	9	4	1	0
0	1	4	9	14	14	9	4	1	0
0	1	2	6	9	9	6	2	1	0
0	0	1	2	4	4	2	1	0	0
0	0	0	1	1	1	1	0	0	0
0	0	0	0	0	0	0	0	0	0

Now imagine that we perturb this matrix by adding random noise as $\mathbf{X}_\epsilon = \mathbf{X} + \epsilon$; this gives the matrix shown below. Here, we can see that the overall pattern persists, but some entries are disrupted.

$$\mathbf{X}_\epsilon = \begin{array}{|c|c|c|c|c|c|c|c|c|c|} \hline 0 & 0 & 0 & 0 & 0 & 2 & 0 & 0 & 0 & 0 \\ \hline 0 & 1 & 0 & 0 & 1 & 2 & 0 & 0 & 1 & 0 \\ \hline 0 & 1 & 1 & 3 & 7 & 5 & 1 & 1 & 0 & 0 \\ \hline 0 & 1 & 0 & 6 & 9 & 8 & 5 & 4 & 1 & 0 \\ \hline 2 & 2 & 2 & 9 & 14 & 14 & 9 & 4 & 0 & 2 \\ \hline 0 & 0 & 4 & 10 & 14 & 12 & 12 & 3 & 3 & 1 \\ \hline 0 & 1 & 4 & 5 & 8 & 10 & 4 & 2 & 1 & 0 \\ \hline 0 & 2 & 2 & 3 & 3 & 2 & 4 & 2 & 0 & 0 \\ \hline 1 & 0 & 2 & 0 & 1 & 0 & 0 & 1 & 0 & 0 \\ \hline 0 & 0 & 0 & 0 & 0 & 0 & 0 & 0 & 0 & 0 \\ \hline \end{array}$$

We visualize the previous matrices as images in Figure 6.7(a and b). These images contain a total of $10 \times 10 = 100$ pixels (corresponding to all the entries of their matrices). The images are 2D fields that capture the *spatial pattern* of the matrix. To the human eye, it is easier to visualize the patterns using the image representation (particularly when the matrices become large). However, the numerical representation of the image allows us to analyze it and to conduct computer/mathematical operations.

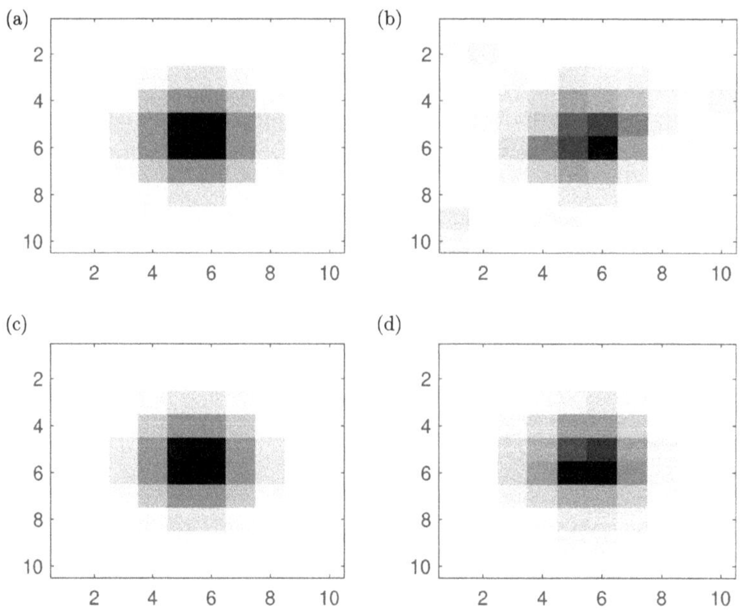

Figure 6.7 Image corresponding to matrix $\mathbf{X}$ (a) and to perturbed matrix $\mathbf{X}_\epsilon$ (b). Image corresponding to compression of $\mathbf{X}$ (c) and of the perturbed matrix $\mathbf{X}_\epsilon$ (d).

For example, imagine that we perform an SVD of the image to obtain $\mathbf{X} = \mathbf{USV}^T$. We recall that this allows us to express the matrix as the expansion:

$$\mathbf{X} = \sum_{j=1}^{10} s_j u_j v_j^T$$

where u_j and v_j are vectors of dimension 10. From the SVD, we find that the image has four nonzero singular values, and thus, the image has a rank of 4 (indicating that there are six linearly dependent rows or columns). The nonzero singular values of the image are:

$$(41.8629, 1.5625, 0.8498, 0.5757).$$

From this, we see that the first singular value dominates the rest by an order of magnitude. This indicates that the image (a 10×10 matrix) can be approximated by using only the first component of the expansion:

$$\mathbf{X} \approx s_1 u_1 v_1^T.$$

This approximation is visualized in Figure 6.7(c). By comparing to the original image (Figure 6.7(a)), we can see that these appear to be identical (at least to the human eye). It is possible to obtain this approximation with just one singular value because the image is symmetric in all directions (there is significant redundancy). Think about the relevance of this result; this indicates that we can compress the original image (comprising a total of $10 \times 10 = 100$ numerical values) by storing only the vectors u_1 and v_1 and the scalar s_1 (a total of $10 + 10 + 1 = 21$ numerical values). In other words, we have reduced the amount of data that we need to store by 80%. In this simple example, this might not seem too significant, but more complex (high-resolution) images can contain millions of pixels, and thus, compression is essential.

Now consider that we perform an SVD on the noisy image $\mathbf{X}_\epsilon = \mathbf{U}\mathbf{S}\mathbf{V}^T$. Here, we find that this image has nine nonzero singular values (has a rank of 9). This indicates that adding noise adds information to the image. However, by inspecting the four largest singular values

$$(41.6533, 5.6010, 4.0541, 3.2470),$$

we again find that the first singular value dominates the rest by an order of magnitude. As such, the noise is not contributing significant information. We approximate the image by using only the first term in the SVD expansion $\mathbf{X}_\epsilon \approx s_1 u_1 v_1^T$ and visualize this in Figure 6.7(d). Here, it is clear that the singular value preserves the structure of the original (noise-free) matrix $\mathbf{X}$. This indicates that eliminating principal components has the effect of "filtering" out the noise. These results highlight how understanding the interplay between images and matrices can help us gain insight into our data. In Section 6.3, we will discuss more sophisticated image-processing techniques.

Script to reproduce results: `ch6_svd_image_field` □

Example 6.6 Image Compression Using SVD

In this example, we decompose a complex image using SVD as $\mathbf{X} = \sum_{j=1}^{10} s_j u_j v_j^T$ and illustrate how the elements of expansion contain different features of the image. In general, we will see that large components of the series tend to contain broad (global) features of the image, while small components contain granular (local) features such as noise and textures. As discussed in the previous example, by truncating the expansion to compress the image, we ignore (filter out) granular behavior.

In Figure 6.8(a), we present a micrograph of a liquid crystal (LC) droplet; this image is a matrix $\mathbf{X}$ of dimension 199×199 and has a total of 39,601 pixels. Upon applying SVD, we find that the matrix has a rank of 199 (it is regular). In Figure 6.8(b), we present approximate images obtained using the truncated series with $n = 1, 3, 5, 20$ components. Note how coarse features develop quickly (see image for $n = 3$), while granular features develop more slowly. This indicates that different elements $s_j u_j v_j^T$ of the series encode different "features" of the image. We also observe that a nearly perfect image is obtained with $n = 20$. This indicates that truncating the series filters out granular features of the image that are *imperceptible to the human eye*. These features are often associated with "frequencies" (we will formally define this shortly) that our eyes cannot perceive. This is analogous to our inability to perceive certain sound frequencies (dogs can perceive sounds that we cannot), highlights inherent limitations of our sensory systems, and highlights the need for techniques that can overcome such limitations.

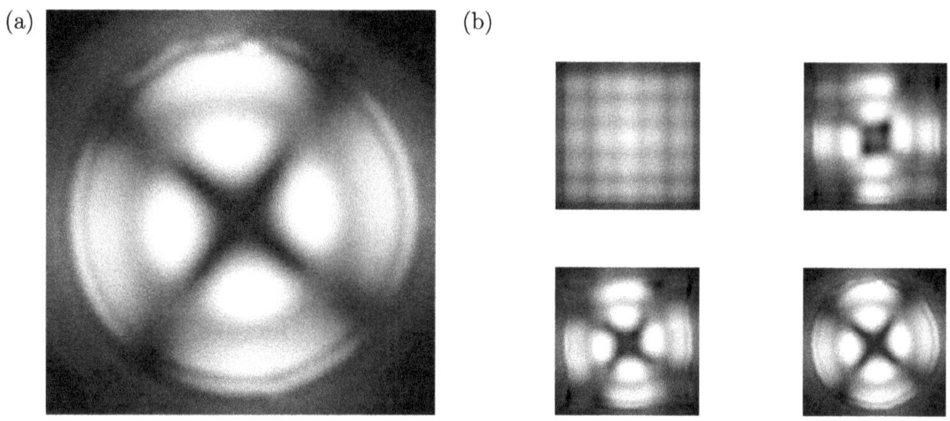

Figure 6.8 Original image (a) and compressions using $n = 1, 3, 5, 20$ singular values (b).

Script to reproduce results: `ch6_svd_reconstruction` □

6.3 Signal Processing

Signal processing is a field that provides tools for analyzing and transforming data [56]. Data often appears as temporal signals (e.g., temperature, vibration, and sound) or spatial signals (e.g., images, tomography, and micrographs). Here, we will focus our attention on a fundamental signal processing task known as *filtering*; as the name suggests, filtering seeks to remove unwanted features from the data with the goal of extracting (uncovering) dominant patterns or features. Moreover, filtering is a powerful way to preprocess raw data, and such preprocessing can be beneficial in performing subsequent tasks (e.g., estimation, predictive modeling, visualization). These capabilities are similar in spirit to those provided by PCA and SVD, but here, we will use techniques that extract other types of information (i.e., not in the form of eigenvalues).

Convolution is a fundamental mathematical operation behind filtering approaches used in signal processing. We will see that this operation is nothing but a *weighted sum of data points* over a local subdomain that enables the extraction of patterns and the transformation of data into diverse forms. We will also see that we can construct a broad range of weighting schemes by using probability density functions (pdfs) and that different pdfs induce different types of data transformations (e.g., averaging and differentiation).

We will see that convolution operations can be used to determine the properties of RVs. For example, we will use convolution to obtain pdfs for sums of RVs and to gain intuition into why the central limit theorem holds. Specifically, convolution will reveal that the central limit theorem results from recurrent smoothing (averaging) of a pdf and that such smoothing eventually converges to a Gaussian pdf.

We will also explore a powerful data transformation/processing technique known as the *Fourier transform* (FT). The FT transforms the data from an original physical space into an information space, in which information can be measured in terms of frequencies. In the context of signal processing, we will see that the FT facilitates data compression (e.g., by eliminating non-informative frequencies). Moreover, we will see that the FT facilitates the computer implementation of convolution operations. We will also see that information extracted from convolutions can be used to characterize and classify images and to predict properties associated with 1D, 2D, and 3D patterns encoded in audio, images, and video data; this is the basic principle behind convolutional neural networks (CNNs; see Section 6.7.2).

6.3.1 Convolutions

Consider a *continuous signal* represented by a continuous function f and consider a continuous weighting (filter) function represented by g. The convolution of the signal and the filter is a function φ given by:

$$\varphi(t) = \int_{-\infty}^{\infty} f(t-\tau)\cdot g(\tau)d\tau, \quad t \in (-\infty,\infty). \tag{6.34}$$

The convolved signal is often expressed compactly as $\varphi = f * g$. We note that, in this definition, we assume that, the domain of the signal and filter functions is continuous and given by $(-\infty,\infty)$. Note also that this domain is symmetric and centered at zero. One can also define functions over asymmetric and finite domains such as $t \in [0,N]$ (typically found in practice). We do not consider such settings here, as this will unnecessarily complicate the explanation; in addition, the concepts discussed here apply to such settings as well.

Consider now a *discrete signal* given by the sequence $f_j, j = -\infty,\dots,\infty$ and consider a filter represented by the sequence $g_j,\ j = -\infty,\dots,\infty$. The discrete convolution of the signal and the filter is a sequence $\varphi_k,\ k = -\infty,\ldots,\infty$ given by:

$$\varphi_k = \sum_{j=-\infty}^{\infty} f_{k-j} \cdot g_j,\ k = -\infty,\ldots,\infty. \tag{6.35}$$

The convolution function is expressed as $\varphi = f * g$. The continuous representation of the convolution signal is often used to establish analytical results, while the discrete representation is often used for computer implementation.

The convolution operation can be seen as a *matching* operation between the signal and the filter. This operation seeks to extract/highlight dominant patterns (features) from the signal, and the patterns that we search for are defined by the filter. This extraction process can also be interpreted as a process that filters out features that are not present in the filter (hence the origin of the term "filter"). The convolution operation can also be seen as a transformation of the signal, and the filter dictates the type of transformation applied.

Filters can be strategically *designed* to extract different types of features from a signal (or to transform the signal in desirable ways). For example, the moving average filter is a function of the form:

$$g(t) = \begin{cases} 0 & -\infty \leq t < a \\ w(t) & a \leq t \leq b \\ 0 & b < t \leq \infty \end{cases}. \tag{6.36}$$

Here, $w(t) = 1/(b-a), t \in [a,b]$ is the weight function, and this has the property that $\int_a^b w(t)dt = 1$. We note that this filter is the pdf of a uniform RV $\mathcal{U}_c(a,b)$. From this structure, it is not immediately obvious why this filter is called a moving average filter. To see this more clearly, consider the discrete version of this filter, which has the form:

$$g_j = \begin{cases} 0 & -\infty \leq j < a \\ w_j & a \leq j \leq b \\ 0 & b < j \leq \infty \end{cases}. \tag{6.37}$$

Here, we have that $w_j = 1/(b-a+1),\ j = a,\ldots,b$ are weights and note that these have the property that $\sum_{j=a}^{b} w_j = 1$.

As an example, design a filter with $a = -1$ and $b = +1$ so that $b - a + 1 = 3$ with weights $w_j = 1/3$ for $j = -1, 0, +1$. Now consider the weighted average of a signal f over the small window $k = -1, 0, 1$:

$$\begin{aligned} \varphi_k &= w_{+1} \cdot f_{k-1} + w_0 \cdot f_k + w_{-1} \cdot f_{k+1} \\ &= \sum_{j=-1}^{+1} w_j \cdot f_{k-j}. \end{aligned} \tag{6.38}$$

This is a weighted average of the signal over a symmetric window that is centered at k and with weight vector $w = (1/3, 1/3, 1/3)$. This window is also known as the *scanning* window.

k

1 1 0 0 0 [1 1 1] 0 1 0 0 0 1 0 1 0 1 0 1 1

$f_k = (1,1,1) \implies \varphi_k = (1/3)\cdot 1 + (1/3)\cdot 1 + (1/3)\cdot 1 = 1$

$k+1$

1 1 0 0 0 1 [1 1 0] 1 0 0 0 1 0 1 0 1 0 1 1

$f_{k+1} = (1,1,0) \implies \varphi_{k+1} = (1/3)\cdot 1 + (1/3)\cdot 1 + (1/3)\cdot 0 = 2/3$

Figure 6.9 Illustration of the application of moving average filter to a signal.

Now consider a general symmetric window with N entries, obtained by setting $a = -(N-1)/2$ and $b = (N-1)/2$. The weighted average over a window centered at k is given by:

$$\begin{aligned}\varphi_k &= \sum_{j=-(N-1)/2}^{(N-1)/2} w_j \cdot f_{k-j} \\ &= \sum_{j=-\infty}^{\infty} f_{k-j} \cdot g_j.\end{aligned} \tag{6.39}$$

We thus have that the weighted average can be expressed as a convolution of the signal and the filter. The convolution value φ_k is obtained by centering the symmetric window at f_k and computing its weighted average. The next value φ_{k+1} is obtained by shifting (sliding) the scanning window forward, centering it at f_{k+1}, and computing the weighted average (hence the use of the term "moving average filter"). In Figure 6.9, we illustrate the application of the moving average filter to a signal f that contains a binary pattern.

The moving average filter g helps smooth out signals that have noise (e.g., they are too "jumpy"), and this helps better reveal trends and patterns in the signal f. We will discuss a bit later why the moving average indeed has a smoothing effect.

The continuous representation of a discrete filter is usually called the *kernel function* (or generating function) of the filter. The kernel function can be seen as a function that generates a family of filters with similar characteristics (for different values of its parameters). For example, we can change the parameters b and a to generate different types of moving average filters (e.g., that average over symmetric and nonsymmetric windows or over large and small windows). Moving average filters are often expressed as simple vectors of weights; for example, for a $N = 3$ window, we have that the filter is $g = (1/3, 1/3, 1/3)$.

Example 6.7 Applying a Moving Average Filter to a Pulse Signal

In Figure 6.10, we show a rectangular step signal f, a moving average filter g, and their convolution $\varphi = f * g$. Both the signal and the filter have $T = 100$ entries. The moving average filter is constructed by defining an averaging window with $N = 20$

entries, and each weight in the window is given by $1/20 = 0.05$ (the rest of the entries outside the window have zero values).

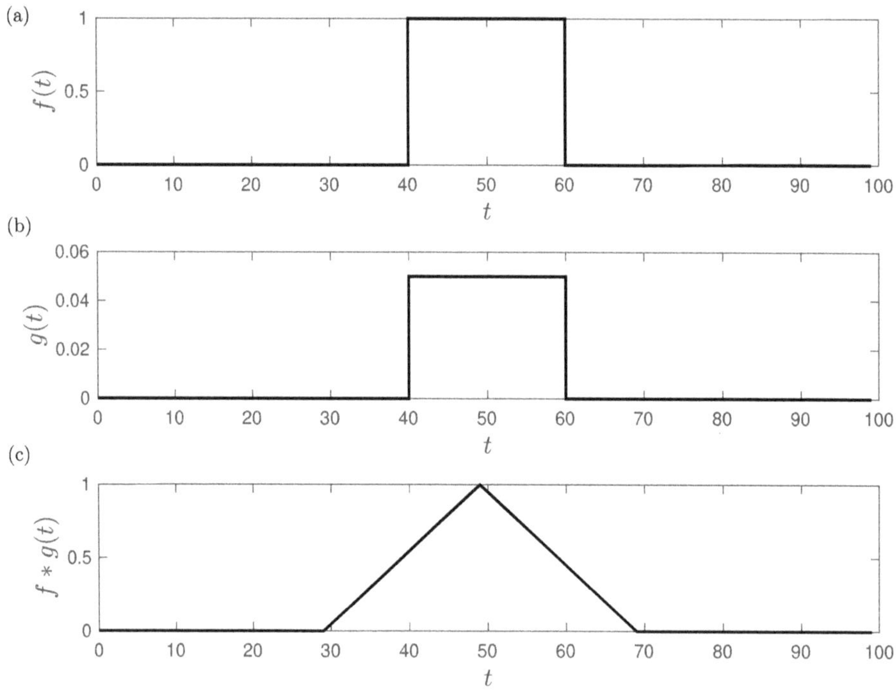

Figure 6.10 Signal (a), filter (b), and convolution (c) for a uniform moving average filter.

The convolution is performed by starting at $k = 0$ using f_0 as the center point, and we obtain the weighted average over the window $-(N-1)/2, \ldots, (N-1)/2$. Because the signal f is not defined for $k < 0$ (this is outside of its domain), we can simply ignore those entries. The convolution is computed by progressively increasing k (using the center point f_k) to obtain all values φ_k, $k = 0, \ldots, T-1$. As we approach the right boundary of the signal, a part of the window will lie outside the domain of the signal; in such a case, we also simply ignore the entries outside of the domain.

We note that the filter and the signal have a similar shape, and we note that the convolution signal has a triangular shape that has a peak in the center point. This indicates that, at the center point, the shape of the signal and of the filter have a strong match. The strength of this match decreases as we move away from the center point. This highlights that the process of convolution can be interpreted as a pattern-matching process; specifically, this reveals regions over which the signal pattern matches the filter pattern (and those regions over which it does not). One can also interpret convolution as a process that transforms the input signal; in this case, the moving average filter smooths out the rectangular signal.

Script to reproduce results: `ch6_example_convolutions` □

Example 6.8 Applying a Moving Average Filter to a Noisy Signal

In Figure 6.11, we show a signal f and the convolution $\varphi = f * g$ using a moving average filter $g = (1/4, 1/4, 1/4, 1/4)$. From the structure of the filter, we see that this has a moving window of length $N = 4$ and that the weights are uniform. This indicates that the filter takes the average of the four values in a given window. From Figure 6.11(b), we observe that the convolution has the effect of eliminating (filtering out) the noise, and this more clearly reveals the periodic structure of the signal.

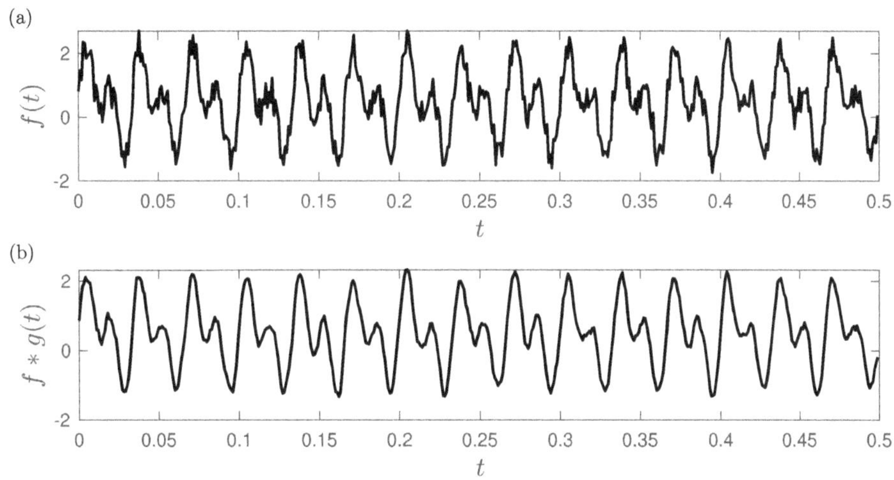

Figure 6.11 Signal (a) and filtered signal obtained from convolution (b) using a uniform moving average filter.

Script to reproduce results: `ch6_example_convolutions` □

Another popular filter used in signal processing is the *Gaussian filter*, which has a continuous representation:

$$g(t) = \frac{1}{\sqrt{2\pi\sigma^2}} e^{-\frac{t^2}{2\sigma^2}}, \ t \in (-\infty, \infty). \tag{6.40}$$

Here, σ^2 is a tunable parameter of the kernel function. The filter is the pdf of $\mathcal{N}(0, \sigma^2)$. Now imagine that we partition the continuous domain into intervals to derive the discrete Gaussian filter:

$$g_j = \frac{1}{\sqrt{2\pi\sigma^2}} e^{-\frac{1}{2}\frac{j^2}{\sigma^2}}, \ j = -\infty, \ldots, \infty. \tag{6.41}$$

The Gaussian filter also induces an averaging effect, but this effect is smoother than that induced by the moving average filter. Specifically, note that the weights of the moving average filter are held constant over the averaging window and then are decreased abruptly to zero outside the window. For the Gaussian filter, on the other hand, the weights are progressively/exponentially reduced inside the averaging window, and this creates a smoother transition outside the window. The decaying weights naturally put less emphasis on the values of the signals as we move away from the middle point. The speed at which the weights decay is controlled by the parameter σ; the smaller this parameter, the faster the decay and thus the smaller the averaging window. The weights are simply the density values of the pdf. Note that there is no need to specify a nonzero mean for the Gaussian filter because the filter is naturally shifted as we slide the convolution window.

One often expresses a discrete Gaussian filter over a symmetric window (as done for the moving-average filter). For example, consider a window of length N, this creates a symmetric domain around zero of the form $-(N-1)/2,\ldots,0,\ldots,(N-1)/2$. With this, we can construct a discrete Gaussian filter of the form:

$$g_j = e^{-j^2/\sigma^2},\ j = -(N-1)/2,\ldots,(N-1)/2. \tag{6.42}$$

with $\sigma = (N-1)/(2\alpha)$. Here, α is called the width factor; for a fixed window length N, increasing the width factor α has the effect of decreasing σ (which leads to a faster decay of the filter weights). This filter is defined in terms of the width factor because it is often more intuitive to control the width factor than controlling the standard deviation. In this filter, we note that we do not include the pre-exponential factor $1/\sqrt{2\pi\sigma^2}$ because this does not alter the overall behavior induced by the filter. Adding this term, however, is necessary if one wants to attribute statistical meaning to the weights.

In the above filter, we also note that the weights are implicitly assumed to be zero for any value outside of the window. As such, Gaussian filters are also often expressed as a compact/small weight vector; for example, for an $N = 5$ window with $\alpha = 2$ (corresponding to $\sigma = 1$), we have that $g = (0.13, 0.6, 1, 0.6, 0.13)$. Note that the weight vector is symmetric and has the highest entry at the middle point (corresponding to the mean). Note also that the weights decay quickly as one moves away from the middle point.

Example 6.9 Applying a Gaussian Filter to a Pulse Signal

In Figure 6.12, we show a rectangular pulse signal f, a Gaussian filter g, and their convolution $\varphi = f * g$. The Gaussian filter is constructed by defining a window with $N = 20$ and by setting $\alpha = 4$ (corresponding to $\sigma = 2.4$).

Note that the filter and the signal now have a different shape (compared to our previous example). From the convolution of the signals, we see that the shape of the signal and the filter has a strong match around the center point, and this results in large values of the convolution signal. The strength of this match decreases smoothly

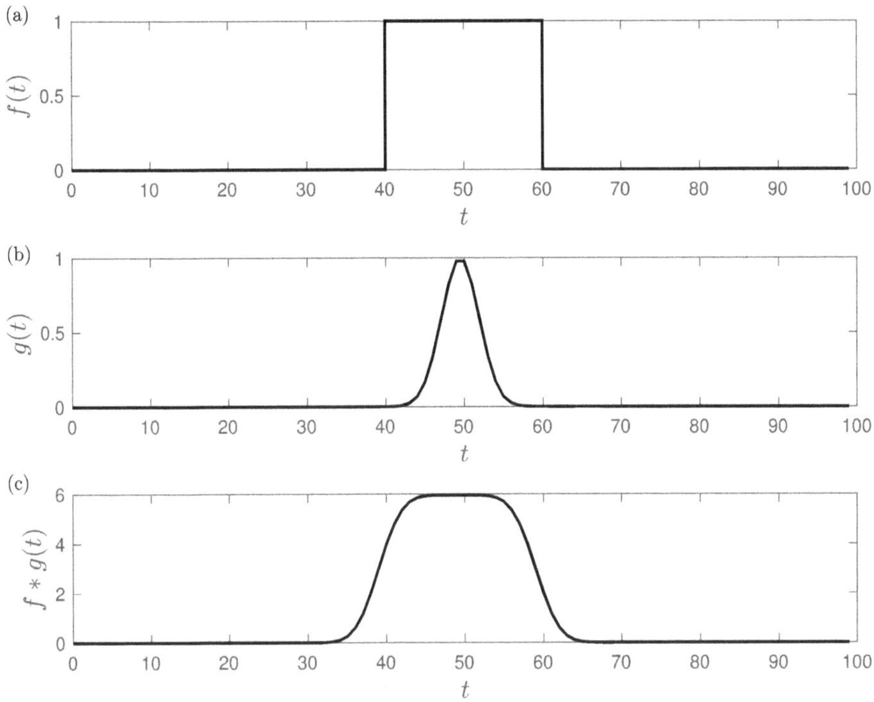

Figure 6.12 Signal (a), filter (b), and convolution (c) for a Gaussian filter.

as we move away from the center point. We can also see that applying the Gaussian filter has the effect of smoothing out the edges of the rectangular pulse, leading to a signal with a bell-like shape. This highlights that convolution transforms the signal (in this case smoothing) and that the nature of transformation is dictated by the type of filter used.

Script to reproduce results: `ch6_example_convolutions` □

Example 6.10 Applying a Gaussian Filter to a Noisy Signal

In Figure 6.13, we show a signal f and the convolution $\varphi = f * g$ using a Gaussian filter with weight vector $g = (0.60, 0.94, 0.94, 0.60)$. The filter was obtained using a window of length $N = 4$ and $\alpha = 1$ (corresponding to $\sigma = 1.5$). This indicates that the filter takes the average of the four values in a given window, but the weights are not uniform. For comparison, we also perform a convolution using a Gaussian filter with $N = 4$ and $\alpha = 4$ (corresponding to $\sigma = 0.37$), which results in a weight vector $g = (0.0003, 0.41, 0.41, 0.003)$. Clearly, the second filter has weights that are more concentrated in the center point and decay faster.

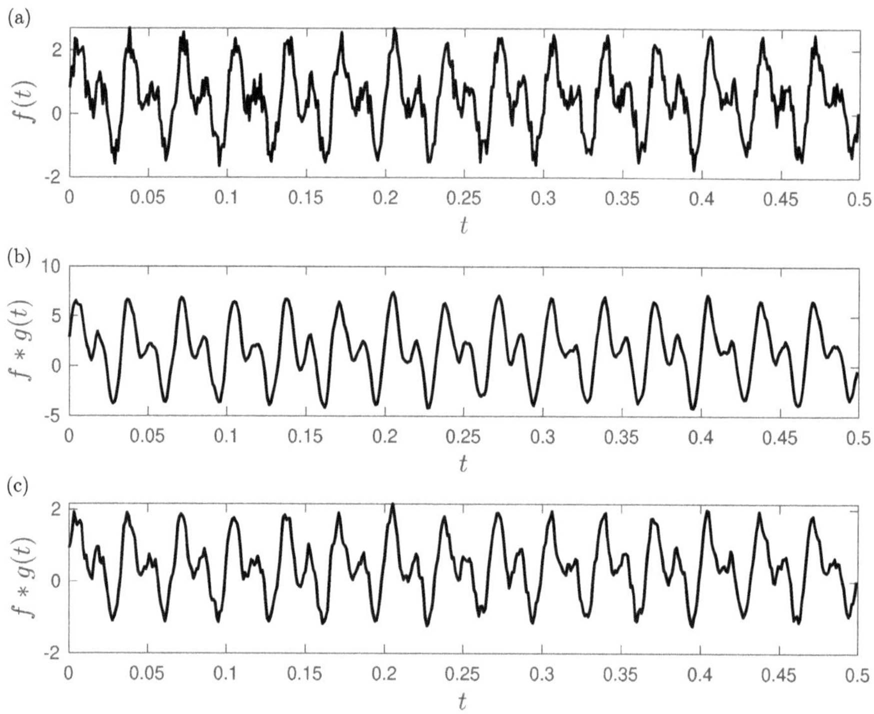

Figure 6.13 Signal (a), convolution with Gaussian filter with $\sigma = 1.5$ (b), and convolution with Gaussian filter with $\sigma = 0.37$ (c).

From Figure 6.13(b), we observe that the convolution with the first Gaussian filter has the effect of filtering out the noise (as observed for the uniform moving average filter), and this more clearly reveals the periodic structure of the signal. From Figure 6.13(c), we observe that the smoothing effect induced by the second filter is less strong. On the other hand, the first filter affects the overall magnitude of the signal, while the second one maintains similar magnitudes. This highlights that "tuning" the Gaussian filter is needed to achieve the behavior that is desired.

Script to reproduce results: `ch6_example_convolutions` □

We can use convolution to highlight characteristics that define the *shape* of a signal (e.g., maximum points, minimum points, flat points, or discontinuities). Such characteristics are *geometrical* features of the signal and are extremely useful for data analysis. For example, imagine that you want to design a filter that computes the first derivative of the signal $f(t)$ at a given point t:

$$\frac{df(t)}{dt} \approx \frac{f(t + \Delta t) - f(t - \Delta t)}{2 \cdot \Delta t}. \tag{6.43}$$

Define $\frac{df(t)}{dt} = \varphi_k$, $f(t) = f_k$, $f(t + \Delta t) = f_{k+1}$, and $f(t - \Delta t) = f_{k-1}$; define also the weight vector:

$$g_{-1} = 1/2 \cdot \Delta t, \; g_0 = 0, \; g_{+1} = -1/2 \cdot \Delta t. \tag{6.44}$$

Now note that the derivative can be written as the weighted sum:

$$\varphi_k = g_{+1} \cdot f_{k-1} + g_0 \cdot f_k + g_{-1} \cdot f_{k+1}. \tag{6.45}$$

This has the same structure as a weight filter with weight vector $g = (1, 0, -1)$. Here, we have eliminated the term $\frac{1}{2}\Delta t$ because this does not alter the behavior of the filter.

The key observation is that the derivative of f is simply the convolution of the signal and the filter $\varphi = f * g$. The filter $g = (1, 0, -1)$ is a derivative filter or *gradient filter*. Note that the weight vector of the gradient filter has positive and negative values (the Gaussian and uniform filters have positive weights). The weights of the gradient filter correspond to the coefficients of the central finite difference approximation of the derivative (6.43). Interestingly, there is a wide range of finite difference approximations, and we therefore can envision many filter designs. For instance, another popular finite difference approximation has weights of the form $g = (1, -2, 1)$, and this corresponds to:

$$\frac{d^2 f(t)}{dt^2} \approx \frac{f(t + \Delta t) - 2f(t) + f(t - \Delta t)}{(\Delta t)^2}. \tag{6.46}$$

Applying a convolution with filter $g = (1, -2, 1)$ thus computes the second derivative of the signal f. As we have seen in our discussions about optimization, computing the first derivative highlights the stationary points of a function (but makes no distinction between minima, maxima, and sharp/flat points); the second derivative highlights the nature of the stationary points. We will see shortly that highlighting these shape features is important in many applications such as image processing. Derivatives can also be used to detect locations in which the signal is discontinuous (as the derivative is not defined at such points).

Example 6.11 Applying a Derivative Filter to a Signal

We know that the pdf of a random variable is the derivative of the corresponding cdf. Now imagine that we want to verify this using convolution; to do so, we construct a signal $f(t)$ that contains values of the cdf for $\mathcal{N}(0, 1)$ at different values in $t \in (-10, 10)$ and construct a derivative filter with weight vector $g = (1, 0, -1)$.

The cdf signal f and the convolution $\varphi = f * g$ are shown in Figure 6.14. Here, we can see that the convolution exactly matches the shape of the pdf of $\mathcal{N}(0, 1)$, and we thus confirm that the convolution with the derivative filter correctly computes the derivative of the cdf. We also use this result to reinforce our idea that convolution can be interpreted as a transformation that highlights interesting aspects of a signal. For instance, here, the convolution tells us at which locations of the domain $(-10, 10)$ the cdf exhibits fast, slow, and no growth (this helps us characterize the shape of the cdf).

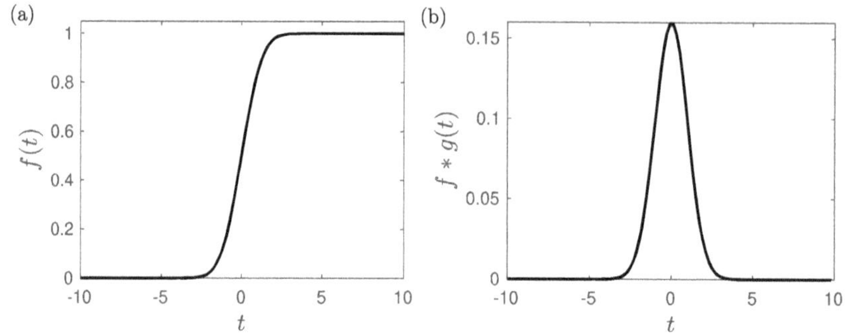

Figure 6.14 Cumulative density function (a) and its derivative obtained via convolution (b).

Script to reproduce results: `ch6_derivative_filter` □

6.3.2 Fourier Transforms

The Fourier Transform (FT) is a powerful tool that is widely used for data analysis. The FT is different from other data analysis tools such as PCA and SVD in that information is quantified in terms of *frequency content*. We will see that convolution can be performed by using the FT and that this facilitates computer implementation and provides important insights into the properties of convolution filters.

We recall that the FT of a continuous function f is given by:

$$\begin{aligned}\hat{f}(\omega) &= \int_{-\infty}^{\infty} f(\tau)\cdot e^{-i(2\pi\omega\tau)}d\tau \\ &= \int_{-\infty}^{\infty} f(\tau)\cdot(\cos(2\pi\omega\tau) - i\sin(2\pi\omega\tau))\,d\tau \\ &= \left(\int_{-\infty}^{\infty} f(\tau)\cos(2\pi\omega\tau)d\tau\right) - i\left(\int_{-\infty}^{\infty} f(\tau)\sin(2\pi\omega\tau)d\tau\right),\end{aligned} \tag{6.47}$$

for all $\omega \in (-\infty, \infty)$. The FT of signal f is often expressed compactly as $\hat{f} = \mathcal{F}\{f\}$. In the above expression, we have that $i = \sqrt{-1}$ and ω is known as the *frequency*. The FT performs a transformation of a signal from its original domain of t (often a physical domain such as time or spatial location) to the frequency (information) domain of ω.

The FT returns a complex function; specifically, each value $\hat{f}(\omega)$ is a complex number with real and imaginary components:

$$\hat{f}_R(\omega) = \left(\int_{-\infty}^{\infty} f(\tau)\cos(2\pi\omega\tau)d\tau\right) \tag{6.48a}$$

$$\hat{f}_I(\omega) = \left(\int_{-\infty}^{\infty} f(\tau)\sin(2\pi\omega\tau)d\tau\right). \tag{6.48b}$$

The magnitude of the complex number $\hat{f}(\omega)$ is given by:

$$|\hat{f}(\omega)| = \sqrt{\hat{f}_R(\omega)^2 + \hat{f}_I(\omega)^2}. \tag{6.49}$$

The transformation from the exponential to the sinusoidal form of the FT uses the so-called Euler's identity. Typically, it is preferred to express the FT in its exponential form because this is more compact and facilitates analysis. However, the sinusoidal form provides interesting insights into the interpretation of the FT.

The *inverse FT* recovers the function $f(t)$ from $\hat{f}(\omega)$ and is given by:

$$f(t) = \int_{-\infty}^{\infty} \hat{f}(\omega)e^{i(2\pi\omega\tau)}d\omega \;\; t \in (-\infty, \infty). \tag{6.50}$$

The inverse FT is expressed as $f = \mathcal{F}^{-1}\{\hat{f}\}$. The FT thus has the key property (similar to PCA) that we can project back and forth between the physical and information spaces.

The *discrete FT* of a sequence of N real numbers f_j, $j = 0, \ldots, N-1$ is a sequence of complex numbers given by:

$$\begin{aligned} \hat{f}_k &= \sum_{j=0}^{N-1} f_j e^{-i(2\pi kj/N)} \\ &= \sum_{j=0}^{N-1} f_j \left(\cos\left(2\pi kj/N\right) - i \sin\left(2\pi kj/N\right)\right), \end{aligned} \tag{6.51}$$

for $k = 0, \ldots, N-1$. Here, every index k has an associated frequency.

The discrete FT is expressed compactly as $\hat{f} = \mathcal{F}\{f\}$, and the inverse FT $f = \mathcal{F}^{-1}\{\hat{f}\}$ is given by:

$$f_k = \frac{1}{N}\sum_{j=0}^{N-1} \hat{f}_j \cdot e^{i(2\pi kj/N)} \;\; k = 0, \ldots, N-1. \tag{6.52}$$

The FT returns a series of complex numbers $\hat{f}_k$, $k = 0, \ldots, N-1$. Each of these numbers corresponds to a given frequency ω_k. The "power" of each frequency component is obtained from the magnitude $|\hat{f}_k|$; the power highlights how strong the presence of a frequency is in a given signal.

One often visualizes the frequency content of a signal by using a plot of frequency ω_k and its intensity (power) $|\hat{f}_k|$. This plot is also often known as the *frequency spectrum* of the signal. It is also important to note that the components are symmetric and thus $|\hat{f}_k| = |\hat{f}_{-k}|$, and this is why we only consider nonzero values of k (negative values are redundant).

The key observation that guides the Fourier decomposition of a signal into frequencies is that *any* smooth signal can be obtained by combining sinusoidal (periodic) functions of different frequencies. This is known as a *Fourier series expansion*. For example, an audio signal (e.g., a song) can be decomposed into periodic signals of different frequencies. This is analogous to the decompositions we used in PCA and SVD, but, in such cases, the signal is decomposed in terms of eigenvalues and singular values (in the FT, the signal is decomposed in terms of frequencies). The FT is widely used in the analysis of rheological and vibration data, as this often involves periodic signals that result from periodic excitation.

Example 6.12 Fourier Transform of a Periodic Signal

We construct a signal of the form:

$$f(t) = \sin(2\pi\omega_1 t) + \sin(2\pi\omega_2 t), \tag{6.53}$$

with frequencies $\omega_1 = 30$ and $\omega_2 = 60$. The signal results from the superposition of a couple of periodic signals that oscillate with different frequencies (the second one oscillates with a higher frequency). In Figure 6.15, we show the signal $f(t)$ and its components (which we denote as $f_1(t)$ and $f_2(t)$). We can see that the addition of simple sinusoidal signals creates a complex signal; the low-frequency component (associated with ω_1) is responsible for the slow-moving trend, while the high-frequency component (associated with ω_2) is responsible for a fast-moving trend.

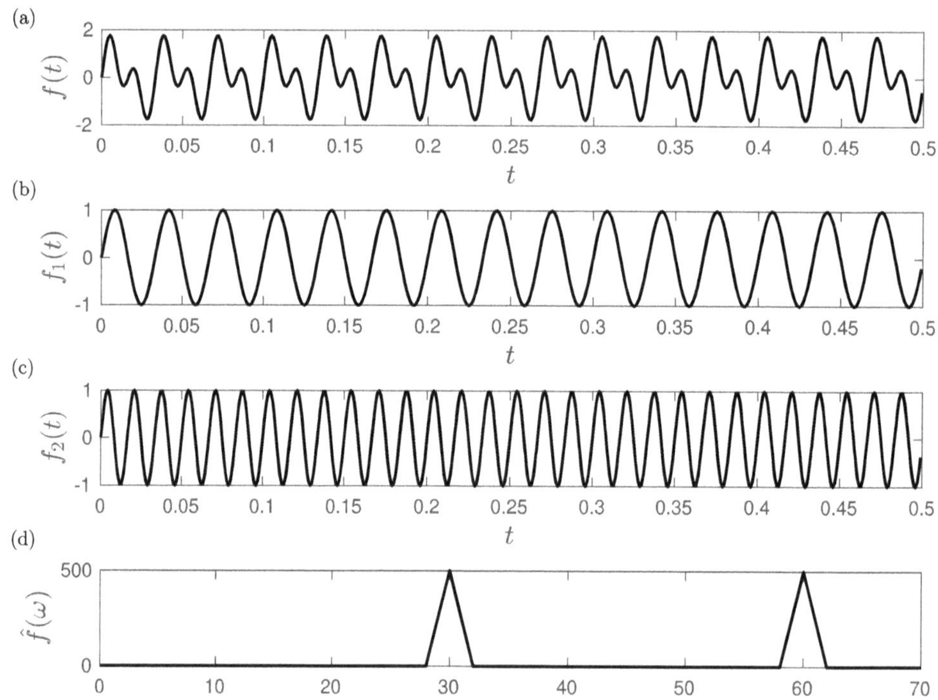

Figure 6.15 Signal $f(t)$ (a) obtained by superimposing periodic signals $f_1(t)$ and $f_2(t)$ (b and c) and corresponding frequency spectrum (d).

We compute the FT for the signal $f(t)$ to obtain $\hat{f}(\omega)$. In Figure 6.15(d), we show the frequency spectrum; this reveals that the signal contains two dominant components at $\omega_1 = 30$ and $\omega_2 = 60$. We can thus see that the FT reveals the dominant frequencies of the signal.

Script to reproduce results: `ch6_fourier_transform` □

Example 6.13 Fourier Transform of a Pulse and of a Gaussian Signal

In Figure 6.16, we show the Fourier transform for a rectangular pulse function and for the Gaussian function (with $\sigma = 1$ and $\sigma = 0.4$). From Figure 6.16(a), we see that the rectangular function has a strong information content at low frequencies and that information content decays rapidly (exponentially) at high frequencies. This indicates that a rectangular pulse is composed of multiple sinusoidals of low oscillation frequency; however, we note that the pulse also contains a non-negligible amount of high-frequency components. This is one of the reasons why rectangular signals are used to excite dynamical systems, as they introduce a wide spectrum of frequencies; this is a topic that is widely discussed in the areas of system identification and process control (we will discuss this in more detail shortly).

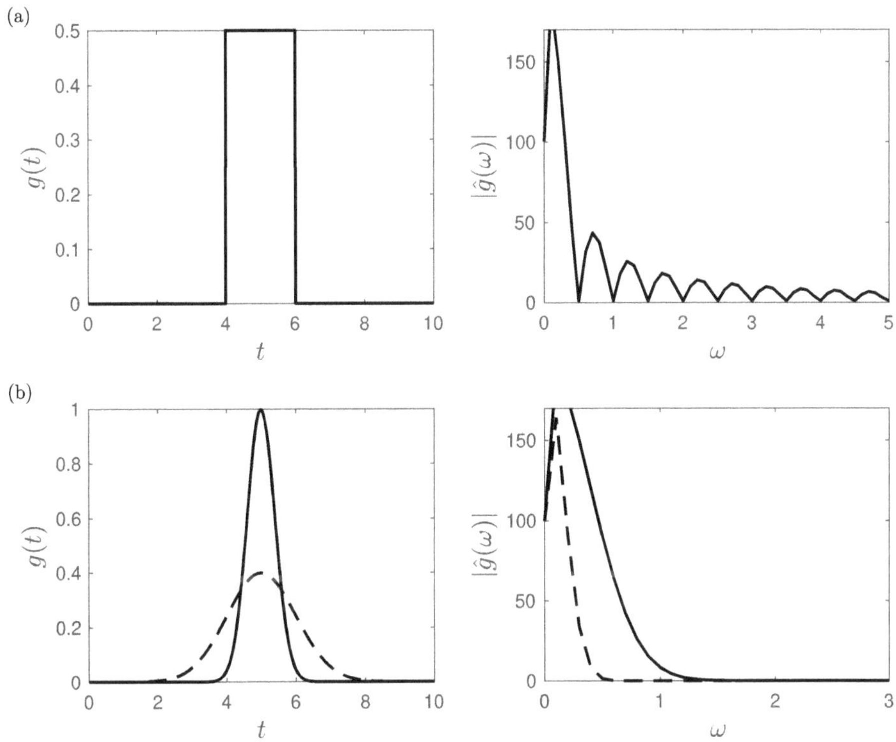

Figure 6.16 Rectangular pulse and Fourier transform (a) and Gaussian function and Fourier transform (b).

From 6.16(b), we see that the Gaussian function also has a high information content at low frequencies and that information decays exponentially at high frequencies. Compared to the rectangular pulse, however, the Gaussian function has a negligible content at high frequencies. We also observe that the speed at which the frequency content decays is controlled by σ (larger σ leads to a faster decay). This makes sense because, as σ is increased, the Gaussian function exhibits slower decay, and thus, the frequencies of the sinusoidal terms that approximate the Gaussian function have low frequencies. On the other hand, as σ is decreased, sinusoidal functions with high frequencies are needed to approximate the fast-growing Gaussian function. These results also tell us that we can control the frequency content of the Gaussian by tuning σ.

Script to reproduce results: `ch6_fourier_transform_filters` □

The FT facilitates the analysis and computer implementation of convolution operations. This is enabled by a fundamental result of mathematics known as the *convolution theorem*. This theorem states that:

$$\mathcal{F}\{f * g\} = \mathcal{F}\{f\} \cdot \mathcal{F}\{g\}. \tag{6.54}$$

This can also be written as:

$$f * g = \mathcal{F}^{-1}\{\mathcal{F}\{f\} \cdot \mathcal{F}\{g\}\}. \tag{6.55}$$

This indicates that, if you have the FTs of the signal f and of the filter g, then you can obtain their convolution by *multiplying* their FTs and then applying the inverse FT to their product. This is important because computing a convolution involves a complex integral in the original domain, but it only involves a product in the frequency domain.

The convolution theorem is also relevant because one can show that the product of the FTs has the effect of preserving frequencies of f that are also in g (while eliminating the rest). This is the basic principle of filtering and tells us that one can strategically remove frequencies from a signal f by properly selecting a filter g; as such, it is important to understand the frequency content that different signals and filters contain.

Example 6.14 Applying Convolution Theorem to Filter Signal

We consider a signal that has two frequencies $\omega_1 = 30$ and $\omega_2 = 100$. The signal f and its FT $\mathcal{F}\{f\}$ are shown in Figure 6.17(a). Here, it is clear that the frequency spectrum

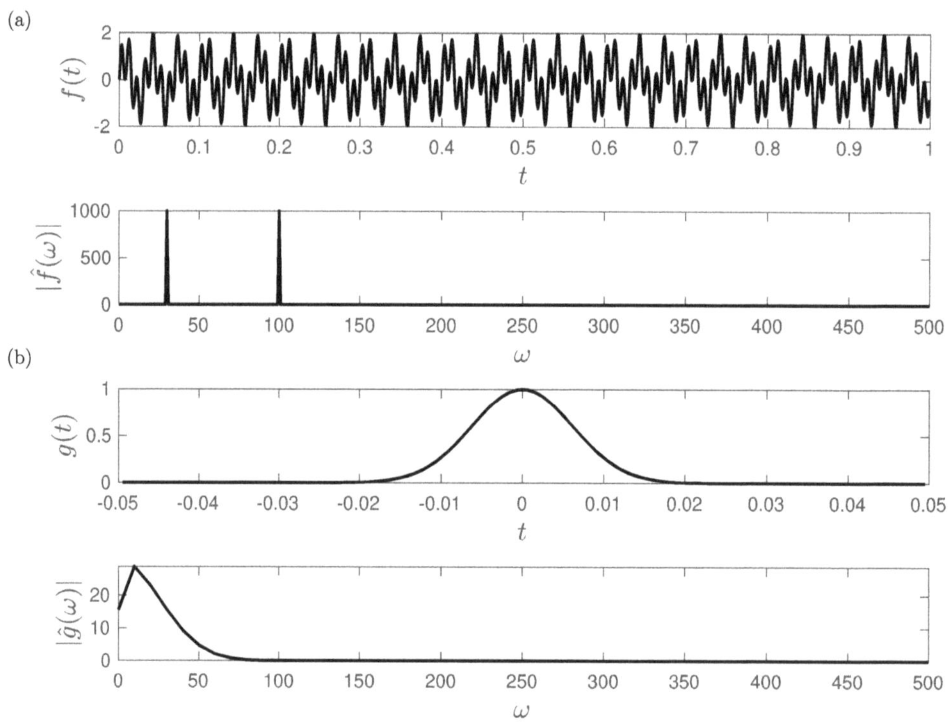

Figure 6.17 Signal and Fourier transform (a) and Gaussian filter and Fourier transform (b).

of the signal contains only a couple of components. We would like to filter the high-frequency component from the signal using a Gaussian filter. The filter g and its FT $\mathcal{F}\{g\}$ are shown in Figure 6.17(b). Here, we can see that the Gaussian filter has a strong content at frequencies below a value of 50 and has negligible content at frequencies above 50. As such, computing the product $\mathcal{F}\{f\} \cdot \mathcal{F}\{g\}$ will eliminate any frequencies above a value of 50 from f and preserve the rest (which is precisely what we want). We corroborate this result by computing the convolution $\varphi = f * g$. The signal f and the convolution $f * g$ are shown in Figure 6.18; here, we confirm that the high-frequency component has been eliminated, and the low-frequency component is maintained.

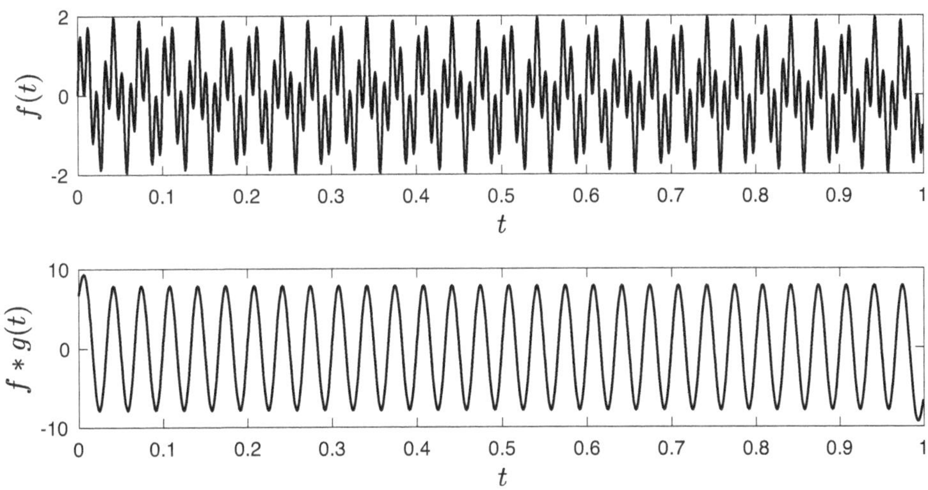

Figure 6.18 Original signal f and filtered signal $f * g$ using a Gaussian filter.

Script to reproduce results: `ch6_convolution_filters` □

6.3.3 System Identification

Convolution and the FT are essential tools that are used in the modeling of dynamical systems from time-dependent data (an area called system identification). The development of such models is essential for the design of feedback controllers (e.g., your home thermostat or a controller for a chemical process).

To explain the basic principles of system identification, we consider a first-order dynamical system of the form:

$$\tau \frac{dx(t)}{dt} + x(t) = Ku(t), \; t \in [0, \infty). \tag{6.56}$$

Here $t \in [0, \infty)$ is time, $x(t)$ is the system state (e.g., the temperature of your home), and $u(t)$ is the control signal (e.g., the ventilation rate of your home). To simplify the discussion, we assume that the initial condition of the state is $x(0) = 0$. The parameters of the model are τ (the time constant or characteristic time of the system) and K (the gain of the system). The time constant captures how slowly or fast the system responds to changes in the control (e.g., your home temperature has a slow dynamic response to ventilation on the order of hours). The gain captures in what direction and how strongly the system responds to changes in the control (e.g., the temperature of your home will decrease when we increase the ventilation rate).

The parameters K,τ are usually determined from time-dependent data. The key idea is to design an input signal $\tilde{u}(t), t \in [0,T]$ to excite the system and obtain a dynamic response $\tilde{x}(t),\ t \in [0,T]$. The design of the input signal plays a key role in the information that we gain from the response of the system; as such, the input signal plays a key role in our ability to obtain precise parameter estimates. For instance, if the input signal is a constant (e.g., we keep your home at a constant ventilation rate), then the system will not be excited and will not provide information. Note how the design of the input signal is analogous to the design of an experiment (the input signal is our design). There are different types of input signals that one might use, such as step signals, rectangular signals, sinusoidal signals, or random signals.

A straightforward strategy to estimate parameters is the following:

- Excite the system using the input signal $\tilde{u}(t),\ t \in [0,T]$ to obtain the empirical response (data) $\tilde{x}(t)$ at a set of times t_k for $k = 0,\ldots,N$.
- Solve the following maximum likelihood estimation (MLE) problem:

$$\min_{K,\tau} \sum_{k=0}^{N} (x(t_k|K,\tau) - \tilde{x}(t_k))^2, \tag{6.57}$$

  where $x(t_k|K,\tau)$ is the model trajectory obtained from simulating the differential equation (6.56) for given values of the parameters K,τ (e.g., by obtaining the analytical solution or by using a numerical scheme such as Euler's method or a Runge–Kutta method). Therefore, the optimization solver will aim to iteratively find K,τ that match the model dynamic response to the empirical response.

While the previous approach appears to be straightforward, we note that it might require repetitive simulations of the model, which can be time-consuming. Moreover, this approach might require storing a lot of data (the full trajectory $\tilde{x}(t)$ to solve the estimation problem).

A key observation made in system identification is that one can use the FT to obtain a *dual representation* of the time-domain model (6.56) in the frequency domain. This dual representation has the form:

$$\hat{x}(\omega) = \hat{h}(\omega) \cdot \hat{u}(\omega),\ \omega \in [0,\infty). \tag{6.58}$$

Here, $\hat{x}(\omega)$ is the system response signal in the frequency domain, $\hat{u}(\omega)$ is the system input signal in the frequency domain, and $h(\omega)$ is the response model in the frequency domain (known as a transfer function). One can obtain the dual form of the model by applying the FT to Eq. (6.56). Specifically, we note that:

$$\begin{aligned}\mathcal{F}\left\{\tau\frac{dx(t)}{dt}\right\} &= \int_0^\infty \tau e^{-i\omega t}\frac{dx(t)}{dt}dt\\ &= \tau\mathcal{F}\left\{\frac{dx(t)}{dt}\right\}\\ &= \tau\cdot(i\omega)\cdot\hat{x}(\omega).\end{aligned}$$

Moreover, $\mathcal{F}\{x(t)\} = \hat{x}(\omega)$ and $\mathcal{F}\{Ku(t)\} = K\cdot\hat{u}(\omega)$. This allows us to conclude that:

$$\hat{x}(\omega) = \underbrace{\frac{K}{\tau\cdot(i\omega)+1}}_{\hat{h}(\omega)}\hat{u}(\omega), \quad \omega\in[0,\infty). \tag{6.59}$$

We recall that these are complex numbers and thus have real and imaginary components. The magnitude (power) of a complex number captures the size of these components. We also note that simulating the dual form of the model in the frequency domain involves computing products, while simulating the model in the time domain involves integration or time-stepping schemes. As such, it is easier to conduct operations in the frequency domain. Specifically, if we have the FTs $\hat{u}(\omega)$ and $\hat{h}(\omega)$, then we can obtain the FT $\hat{x}(\omega)$ by simple product.

The dual model can be written in the following form:

$$|\hat{h}(\omega)| = \frac{|\hat{x}(\omega)|}{|\hat{u}(\omega)|}, \quad \omega\in[0,\infty). \tag{6.60}$$

This shows that the spectrum of the transfer function is the ratio of the power of the state and input; this ratio is known as the amplitude ratio and captures how strongly a system responds to inputs. One can show that the *model spectrum* is:

$$y(\omega|K,\tau) = |\hat{h}(\omega)| = \frac{K}{\sqrt{(\imath\omega)^2+1}}, \quad \omega\in[0,\infty). \tag{6.61}$$

Moreover, we can compute the *empirical spectrum* by obtaining the magnitude of the trajectory data of state and inputs. Specifically:

$$\begin{aligned}|\tilde{h}(\omega)| &= \frac{|\tilde{x}(\omega)|}{|\tilde{u}(\omega)|}\\ &= \frac{|\int_0^T x(t)e^{-i\omega t}dt|}{|\int_0^T u(t)e^{-i\omega t}dt|}\\ &= \frac{|\int_0^T x(t)\cos(\omega t)dt - i\int_0^T x(t)\sin(\omega t)dt|}{|\int_0^T u(t)\cos(\omega t)dt - i\int_0^T u(t)\sin(\omega t)dt|}.\end{aligned} \tag{6.62}$$

Writing in a compact form, we obtain:

$$\tilde{y}(\omega) = |\tilde{h}(\omega)| = \frac{\sqrt{A_1(\omega)^2 + A_2(\omega)^2}}{\sqrt{B_1(\omega)^2+B_2(\omega)^2}}, \quad \omega\in[0,\infty), \tag{6.63}$$

where:

$$\begin{aligned} A_1(\omega) &= \int_0^T \tilde{x}(t)\cos(\omega t)dt \\ A_2(\omega) &= \int_0^T \tilde{x}(t)\sin(\omega t)dt \\ B_1(\omega) &= \int_0^T \tilde{u}(t)\cos(\omega t)dt \\ B_2(\omega) &= \int_0^T \tilde{u}(t)\sin(\omega t)dt. \end{aligned} \tag{6.64}$$

These constants can be obtained from the trajectory data for a set of frequencies ω_k, $k = 0, \ldots, N$.

The dual form of the model facilitates parameter estimation; the key idea will be to transform our data $\tilde{u}(t)$ and $\tilde{x}(t)$ from the time domain to the frequency domain to obtain the empirical spectrum $\tilde{y}(\omega)$; we then find the parameters K, τ that match the model spectrum $y(\omega \mid K, \tau)$ to the empirical spectrum $\tilde{y}(\omega)$. This can be done by solving the estimation problem:

$$\min_{K,\tau} \sum_{k=0}^{N} (y(\omega_k|K,\tau) - \tilde{y}(\omega_k))^2. \tag{6.65}$$

Note that this problem does not require the simulation of the dynamical model (it is much easier to solve).

In order to obtain a dynamic response of the system in the time domain, we can use the convolution theorem (6.55). Specifically, this tells us that:

$$\begin{aligned} x(t) &= \mathcal{F}^{-1}\{\hat{x}(\omega)\} \\ &= \mathcal{F}^{-1}\{\hat{h}(\omega) \cdot \hat{u}(\omega)\} \\ &= h(t) * u(t), \end{aligned} \tag{6.66}$$

where $h(t) = \mathcal{F}^{-1}\{\hat{h}(\omega)\}$ (this is the time-domain representation of the transfer function). This tells us that we can determine the system response in the time domain $x(t)$ via convolution of h and u.

We can thus see that the FT and convolution are useful tools that can help analyze dynamical data; the key idea is to transform the data from the original physical space (time) to the frequency domain.

Example 6.15 System Identification for CO_2 Dynamics

Consider the CO_2 monitoring problem of Exercise 5.3. The dynamics of the CO_2 concentration in a classroom can be described by:

$$\frac{dC(t)}{dt} = \frac{Q}{V}(C_{in} - C(t)) + \frac{G}{V} n_{oc}(t),$$

where $C(t)$ is the CO_2 concentration in ppm, t is the time in min, Q is the airflow rate in cfm = ft^3/min, $C_{in} = 400$ is the CO_2 concentration in the fresh airflow in ppm, $V = 10{,}000$ is the classroom volume in cf = ft^3, G is the CO_2 generation rate per occupant in ppm–cf/min, and $n_{oc}(t)$ is the time-varying number of occupants in the classroom. Here, we assume that the airflow rate Q is constant.

You have collected data for the CO_2 concentration $C(t)$ and for the number of occupants $n_{oc}(t)$ that captures the dynamics of the classroom as occupants come and go throughout the day (see Figure 6.19). You would like to use this to estimate the flowrate Q and the generation rate G. We can reformulate this model in the standard form equation (6.56) by defining $x(t) = C_{in} - C(t)$, $u(t) = n_{oc}(t)$, $\tau = V/Q$, and $K = G/Q$.

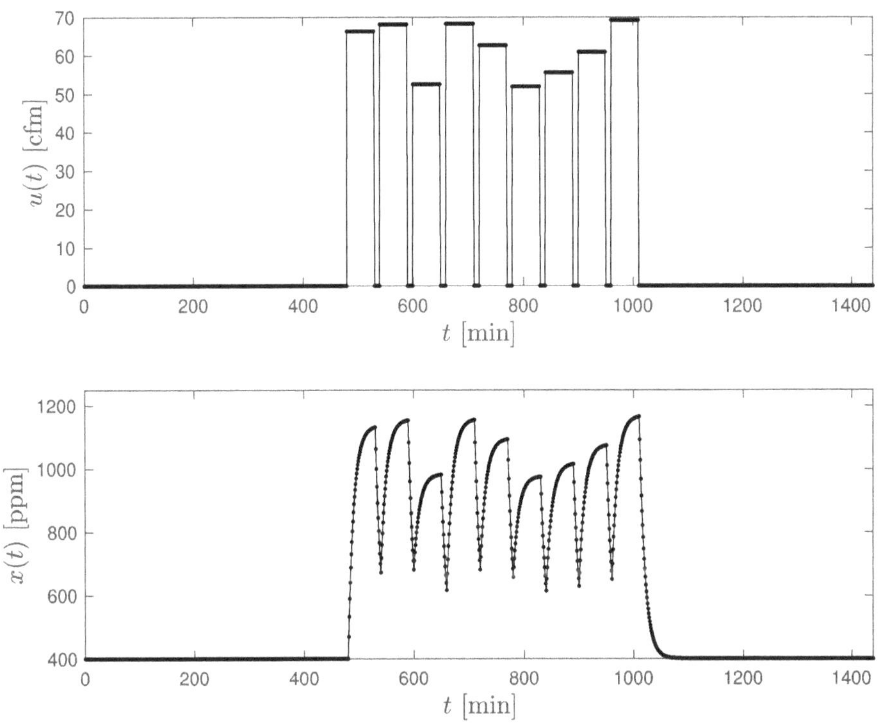

Figure 6.19 Data for occupancy dynamics $n_{oc}(t)$ and resulting CO_2 concentration.

We convert the available data from the time domain to the frequency domain to obtain the empirical spectrum of the response $\tilde{x}(\omega)$, the input $\tilde{u}(\omega)$, and the amplitude ratio $\tilde{y}(\omega)$. The spectra are shown in Figure 6.20; here, we can see that the input and output signals contain a broad range of frequencies; this is because the input signal follows a rectangular shape. This indicates that the data contains rich information to estimate the parameters. Most of the information is in the low-frequency range (frequencies 10^{-2} to 10^{-3}); this frequency range corresponds to times in the range of 100–1,000 min. This is because periods of occupancy are occurring in the timescale of hours.

We formulate an estimation problem to match the model amplitude ratio $y(\omega|K, \tau)$ to the empirical ratio. From this, we determine that the parameters are $K = 11.1$ and $\tau = 9.92$. We can use these values to backcalculate the ventilation and generation rates. From Figure 6.20(c), we can see that the model amplitude ratio matches the empirical counterpart. This type of plot is known as a Bode plot and is widely used in system identification and control to visualize the spectrum of a dynamical system.

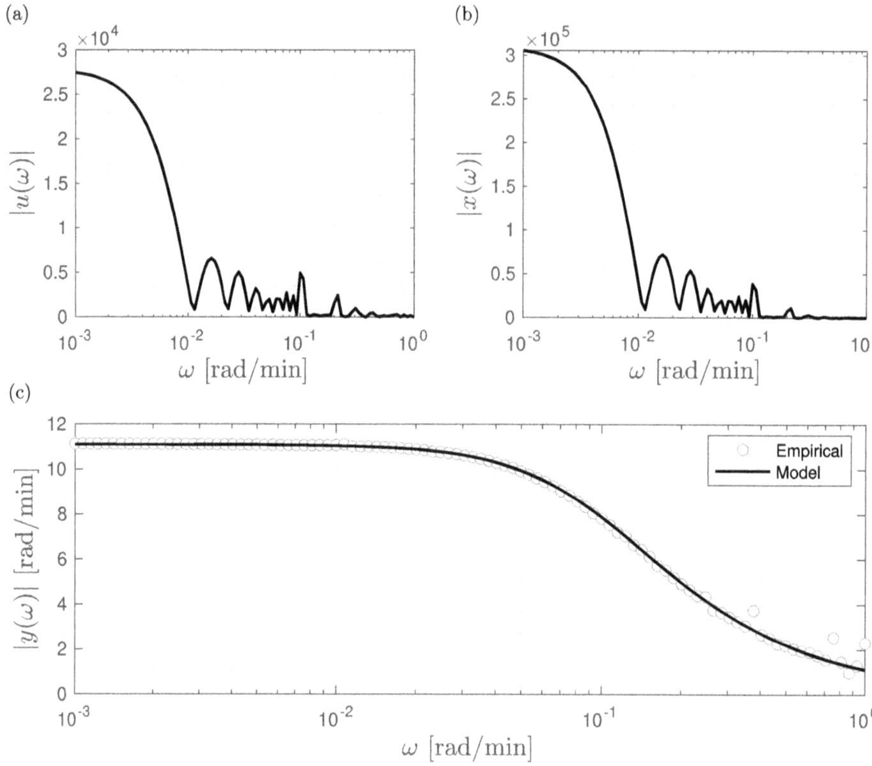

Figure 6.20 Spectrum of input signal (a) and output signal (b). Fit of model amplitude ratio to empirical counterpart (c).

Script to reproduce results: `ch6_sysid_co2` and `ch6_sysid_co2_est` □

6.3.4 Convolution and the Central Limit Theorem

A surprising result of statistics is that the pdf of a sum of independent RVs X_1 and X_2 (of *any* kind) is the convolution of their pdfs f_{X_1} and f_{X_2}. Specifically, the pdf of $Y = X_1 + X_2$ is given by $f_Y = f_{X_1} * f_{X_2}$.

This result is obtained by first noticing that the pdf of $Y = X_1 + X_2$ can be written as:

$$f_Y(y) = \int_{-\infty}^{\infty} f_{X_1,X_2}(x_1, y - x_1)dx_1, \tag{6.67}$$

where f_{X_1,X_2} is the joint pdf of X_1, X_2. Now note that, if X_1, X_2 are independent, we have that $f_{X_1,X_2} = f_{X_1}f_{X_2}$ and thus:

$$f_Y(y) = \int_{-\infty}^{\infty} f_{X_1}(x_1)f_{X_2}(y - x_1)dx_1, \tag{6.68}$$

which is precisely the definition of the convolution $f_{X_1} * f_{X_2}$. One can apply this result *recursively* to show that the convolution of the pdfs of a sum of independent

RVs $Y = X_1 + X_2 + \ldots + X_n$ approaches the pdf of a Gaussian RV as $n \to \infty$. In other words, we have that:

$$f_Y = f_{X_n} * \cdots * (f_{X_3} * (f_{X_1} * f_{X_2})), \tag{6.69}$$

approaches the pdf of a Gaussian RV. Note that this is exactly the *central limit theorem* that we have already discussed in Chapter 4 but stated in terms of the convolution of pdfs.

Example 6.16 Recursive Application of Convolution

Consider a set of independent RVs $X_1, X_2, X_3, X_4 \sim \mathcal{U}_c(4, 6)$ with associated pdf $f(x)$. We want to obtain the pdf of X_1 (given by $\varphi = f$), $X_1 + X_2$ (given by $\varphi = f * f$), $X_1+X_2+X_3$ (given by $\varphi = f*(f*f)$), and $X_1+X_2+X_3+X_4$ (given by $\varphi = f*(f*(f*f))$). With this, we seek to confirm that the pdf of the sum of the RVs approaches that of a Gaussian RV.

The results are shown in Figure 6.21; here we see that the pdf of X_1 is that of $\mathcal{U}_c(4, 6)$ (given by a rectangular pulse). After applying a convolution to itself $f * f$, we obtain the pdf of $X_1 + X_2$, and we see that this is given by a triangular pulse. Applying an additional convolution $f * (f * f)$ gives the pdf of $X_1 + X_2 + X_3$, and we see that this looks like a pdf of a Gaussian RV. Applying an additional convolution does not alter this shape in a perceptible way. Here, we scale the functions so that their maximum value is always one (this facilitates the comparison). This result confirms the central limit theorem.

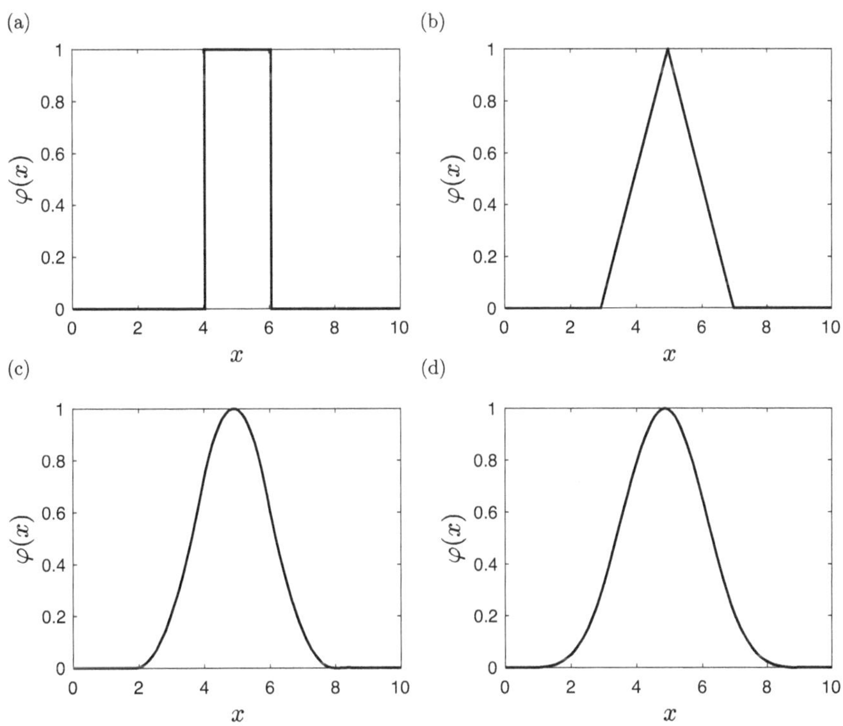

Figure 6.21 Pdf of X_1 (a), $X_1 + X_2$ (b), $X_1 + X_2 + X_3$ (c), and $X_1 + X_2 + X_3 + X_4$ (d).

One can think of this result as follows: a recursive application of convolution induces a progressive smoothing of the pdf, and the resulting smoothed function approaches a Gaussian function.

Script to reproduce results: `ch6_clt_convolution` □

6.3.5 Convolutions and Fourier Transforms

Convolution operations can be performed on 2D and 3D signals (or even higher-dimensional signals), and this enables the analysis and processing of images, spatial fields, videos, and other complex data objects that can be represented as matrices and tensors.

Consider a 2D discrete signal represented by the 2D field:

$$f_{j_1,j_2}, \quad j_1, j_2 = -\infty, \ldots, \infty, \tag{6.70}$$

and consider a filter represented by the 2D field:

$$g_{j_1,j_2}, \quad j_1, j_2 = -\infty, \ldots, \infty. \tag{6.71}$$

The convolution of the signal and the filter is a 2D field given by:

$$\varphi_{k_1,k_2} = \sum_{j_1=-\infty}^{\infty} \sum_{j_2=-\infty}^{\infty} g_{j_1,j_2} \cdot f_{k_1-j_1,k_2-j_2}, \quad k_1, k_2 = -\infty, \ldots, \infty. \tag{6.72}$$

As in the 1D case, one can think of a 2D convolution φ_{k_1,k_2} as a weighted sum (with coefficients given by g) in a given neighborhood of the signal f that is centered around f_{k_1,k_2}. To see this, consider a $N{\times}N$ neighborhood of the signal f (centered at f_{k_1,k_2} and with $N = 3$). This neighborhood (window) is a mesh/matrix with $N^2 = 9$ entries. Our goal is to compute a weighted average of all the entries in the window; to do so, we associate a weight to each signal value in the window, as shown in Figure 6.22.

The convolution of the signal f with the weight g over the window is given by an element-by-element multiplication and the addition of the entries:

$$\begin{aligned}\varphi_{k_1,k_2} = {} & f_{k_1-1,k_2+1} \cdot g_{+1,-1} + f_{k_1,k_2+1} \cdot g_{0,-1} + f_{k_1+1,k_2+1} \cdot g_{-1,-1} \\ & f_{k_1-1,k_2} \cdot g_{+1,0} + f_{k_1,k_2} \cdot g_{0,0} + f_{k_1+1,k_2} \cdot g_{-1,0} \\ & f_{k_1-1,k_2-1} \cdot g_{+1,+1} + f_{k_1,k_2-1} \cdot g_{0,+1} + f_{k_1+1,k_2-1} \cdot g_{-1,+1}.\end{aligned}$$

This can be expressed compactly:

$$\varphi_{k_1,k_2} = \sum_{j_1=-1}^{+1} \sum_{j_2=-1}^{+1} g_{j_1,j_2} \cdot f_{k_1-j_1,k_2-j_2}. \tag{6.73}$$

(a)

f_{k_1-1,k_2+1}	f_{k_1,k_2+1}	f_{k_1+1,k_2+1}
f_{k_1-1,k_2}	f_{k_1,k_2}	f_{k_1+1,k_2}
f_{k_1-1,k_2-1}	f_{k_1,k_2-1}	f_{k_1-1,k_2+1}

(b)

$g_{+1,-1}$	$g_{0,-1}$	$g_{-1,-1}$
$g_{+1,0}$	$g_{0,0}$	$g_{-1,0}$
$g_{+1,+1}$	$g_{0,+1}$	$g_{-1,+1}$

Figure 6.22 Signal f in 3×3 window centered at (k_1, k_2) (a) and weights g (b).

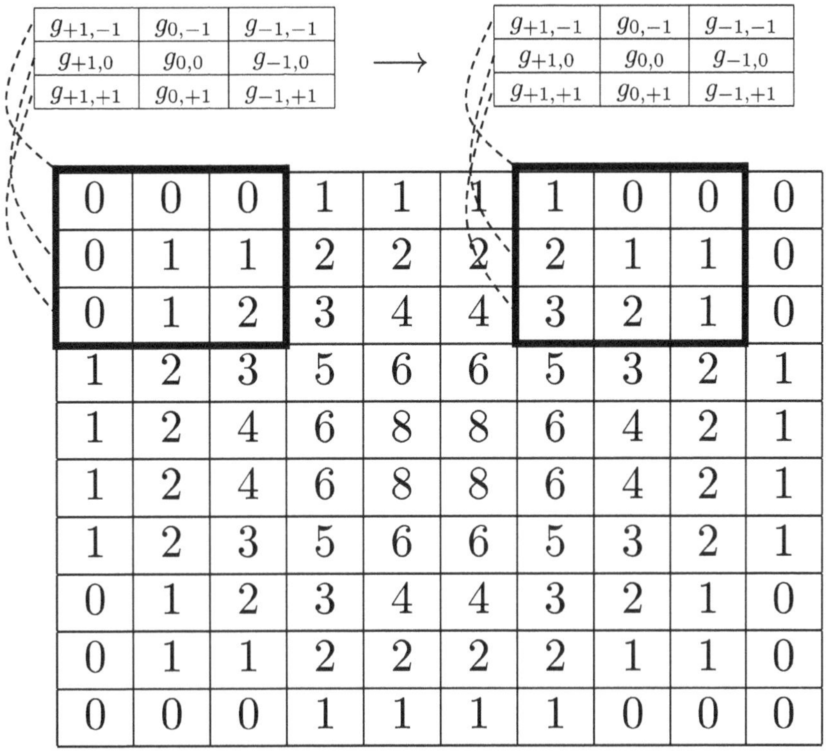

Figure 6.23 Example of convolution of a 2D field by using a moving window.

Consequently, the weighted sum is the convolution of the signal and the weight over the window. We denote the value that results from the convolution (the weighted sum) as $\varphi_{k_1,k_2} = (f * g)_{k_1,k_2}$. The weighted sum can be generalized to any symmetric neighborhood of size $N \times N$ as:

$$\varphi_{k_1,k_2} = \sum_{j_1=-(N-1)/2}^{(N-1)/2} \sum_{j_2=-(N-1)/2}^{(N-1)/2} g_{j_1,j_2} \cdot f_{k_1-j_1,k_2-j_2}, \tag{6.74}$$

for $k_1, k_2 = -(N-1)/2, \ldots, (N-1)/2$. As in the 1D case, convolutions are applied to a signal by applying the filter over a moving window. In the 2D case, the window is shifted over the horizontal axis by increasing index k_2 (for a fixed value of k_1) and then by shifting vertically by increasing index k_1. The 2D convolution is a process that extracts a pattern from a matrix (the filter defines the pattern). The process is illustrated in Figure 6.23.

Example 6.17 Convolution of Matrix with Filter

Consider the convolution of f (a 5×5 matrix) and a filter g (a 3×3 matrix):

$$\underbrace{\begin{bmatrix} 1 & 0 & 1 & 0 & 0 \\ 0 & 1 & 0 & 1 & 0 \\ 1 & 0 & 1 & 1 & 1 \\ 0 & 0 & 1 & 1 & 0 \\ 0 & 1 & 1 & 0 & 0 \end{bmatrix}}_{f} * \underbrace{\begin{bmatrix} 1 & 0 & 1 \\ 0 & 1 & 0 \\ 1 & 0 & 1 \end{bmatrix}}_{g} = \underbrace{\begin{bmatrix} 5 & 1 & 4 \\ 1 & 4 & 2 \\ 3 & 3 & 4 \end{bmatrix}}_{\varphi}.$$

Note that there are two ways of performing the convolution. In the first approach, we center the filter at every entry of the signal matrix but in such a way that the window is always inside the signal matrix (we avoid the boundary entries); this will return a convolution $f * g$ that is a 3×3 matrix. In the second approach, we center the filter at every entry of the signal matrix and allow for some entries of the window to be outside the signal matrix; we simply ignore any filter entries that are outside. This approach is often called *zero-padding* (as it is equivalent to adding artificial zeros to entries outside the signal). In our example, applying this approach will return a convolution $f * g$ that is a 5×5 matrix. One can use any approach, depending on what type of information is needed. Here, we will use the first approach, which returns a 3×3 convolution matrix.

To obtain the $\varphi_{1,1}$ entry of the convolution matrix, we center the filter matrix at the $f_{2,2}$ entry of the signal matrix (see the left matrix in (6.75) with center highlighted). Since the matrix and filter have entries of zeros and ones, the value $\varphi_{1,1} = 5$ entry of the convolution matrix is the sum of all the one entries in the filter that correspond to the one entries of the signal in the window. Note that this indicates that the pattern of the filter perfectly matches the pattern of the signal over the defined window. To obtain the $\varphi_{1,2}$ entry of the convolution matrix, we center the filter at the $f_{2,3}$ entry of the signal (see the right matrix in Eq. (6.75) with center highlighted). The entry value of $\varphi_{1,2} = 1$ indicates that only one value of the filter matrix coincides with that of the signal matrix in the new window.

$$\left[\begin{array}{ccc|cc} 1 & 0 & 1 & 0 & 0 \\ 0 & \mathbf{1} & 0 & 1 & 0 \\ 1 & 0 & 1 & 1 & 1 \\ \hline 0 & 0 & 1 & 1 & 0 \\ 0 & 1 & 1 & 0 & 0 \end{array}\right] \quad \left[\begin{array}{c|ccc|c} 1 & 0 & 1 & 0 & 0 \\ 0 & 1 & \mathbf{0} & 1 & 0 \\ 1 & 0 & 1 & 1 & 1 \\ \hline 0 & 0 & 1 & 1 & 0 \\ 0 & 1 & 1 & 0 & 0 \end{array}\right]. \tag{6.75}$$

We continue the process by sliding the window to the entry $f_{2,4}$, then $f_{3,2}$, $f_{3,3}$, $f_{3,4}$, and so on. From the resulting convolution matrix φ we can see that there are four locations in the signal matrix in which their pattern strongly resembles that of the filter pattern. These correspond to the large values in the convolution matrix (say 5 and 4). The entries with small values in the convolution matrix (say 1) reveal regions in the signal matrix where the pattern weakly resembles that of the filter. This example highlights how convolutions can be used as a pattern-matching process that extracts features from a signal. □

We can define 2D kernel functions to generate discrete filters; for example, the kernel function of a 2D Gaussian filter is given by:

$$g(x,y) = \frac{1}{\sqrt{2\pi\sigma^2}} e^{-\frac{x^2+y^2}{2\sigma^2}}. \tag{6.76}$$

This is the pdf of $\mathcal{N}(0,\Sigma)$ with zero mean and covariance matrix Σ with entries $\Sigma_{1,1} = \Sigma_{2,2} = \sigma^2$ and $\Sigma_{1,2} = \Sigma_{2,1} = 0$. One can generalize this filter by allowing for general covariance matrices. The pdf projected on a 2D field defines concentric ellipses (circles if there is no correlation). The 2D Gaussian kernel can be used to generate discrete Gaussian filters (by tuning the parameter σ). As in the 1D case, decreasing σ has the effect of decreasing the weight that one places on values that are away from the center of the filter.

Example 6.18 Weight Matrix for 2D Gaussian Filter

In Figure 6.24, we present the weighting matrix of a 2D Gaussian filter defined over an $N \times N$ window with $N = 10$ and $\sigma = \sqrt{2}$. Here, we also visualize the filter as an image (a 2D field). Note that the filter is symmetric around the center point; the center point has the highest value and values decay quickly as one moves away from the center point.

(a)

0	0	0	1	1	1	1	0	0	0
0	1	1	2	2	2	2	1	1	0
0	1	2	3	4	4	3	2	1	0
1	2	3	5	6	6	5	3	2	1
1	2	4	6	8	8	6	4	2	1
1	2	4	6	8	8	6	4	2	1
1	2	3	5	6	6	5	3	2	1
0	1	2	3	4	4	3	2	1	0
0	1	1	2	2	2	2	1	1	0
0	0	0	1	1	1	1	0	0	0

(b)

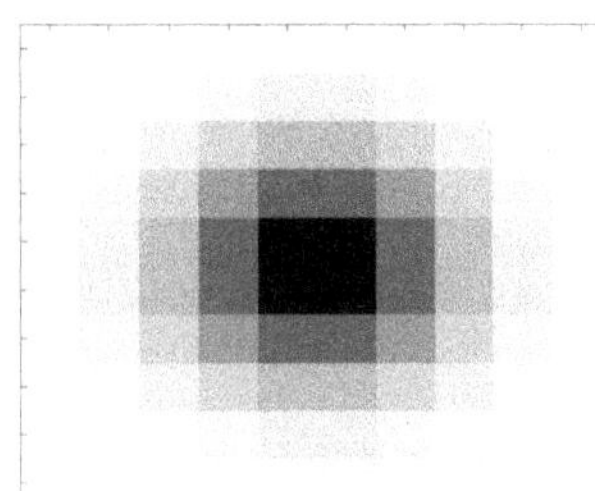

Figure 6.24 Weight matrix for a 2D Gaussian filter (a) and visualization of the filter as a 2D field (b).

Script to reproduce results: `ch6_gauss_filter_2D` □

Derivative filters are widely used in image analysis to extract contrasts (gradients) of light. This information is often used, for instance, to reveal object boundaries. Consider that you want to compute the second derivative:

$$\nabla^2 f(x,y) = \frac{\partial^2 f(x,y)}{\partial x^2} + \frac{\partial^2 f(x,y)}{\partial y^2}. \tag{6.77}$$

This derivative is known as the Laplace operator and can be approximated using finite differences as:

$$\begin{aligned}\nabla^2 f(x,y) &= \frac{f(x-\Delta,y) - 2f(x,y) + f(x+\Delta,y)}{\Delta^2} + \frac{f(x,y-\Delta) - 2f(x,y) + f(x,y+\Delta)}{\Delta^2} \\ &= \frac{f(x-\Delta,y) + f(x+\Delta,y) - 4f(x,y) + f(x,y-\Delta) + f(x,y+\Delta)}{\Delta^2}.\end{aligned} \tag{6.78}$$

If we define:

$$
\begin{aligned}
f_{k_1-1,k_2} &= f(x-\Delta,y)\\
f_{k_1+1,k_2} &= f(x+\Delta,y)\\
f_{k_1,k_2+1} &= f(x,y+\Delta)\\
f_{k_1,k_2-1} &= f(x,y-\Delta)\\
f_{k_1,k_2} &= f(x,y)\\
f_{k_1+1,k_2+1} &= 0\\
f_{k_1+1,k_2-1} &= 0\\
f_{k_1-1,k_2-1} &= 0\\
f_{k_1-1,k_2+1} &= 0,
\end{aligned}
\tag{6.79}
$$

the 2D derivative can be written as a convolution:

$$
f*g = \begin{bmatrix} f_{k_1-1,k_2+1} & f_{k_1,k_2+1} & f_{k_1+1,k_2+1} \\ f_{k_1-1,k_2} & f_{k_1,k_2} & f_{k_1+1,k_2} \\ f_{k_1-1,k_2-1} & f_{k_1,k_2-1} & f_{k_1+1,k_2-1} \end{bmatrix} * \begin{bmatrix} 0 & 1 & 0 \\ 1 & -4 & 1 \\ 0 & 1 & 0 \end{bmatrix}. \tag{6.80}
$$

The second matrix is the weighting matrix corresponding to the Laplace operator. This matrix is also known as the stencil (in the study of partial differential equations) or as the Laplacian filter (in image processing).

Another set of popular derivative filters is the so-called Sobel filters; the vertical and horizontal Sobel filters are:

$$
\begin{bmatrix} 1 & 2 & 1 \\ 0 & 0 & 0 \\ -1 & -2 & -1 \end{bmatrix} \quad \begin{bmatrix} 1 & 0 & -1 \\ 2 & 0 & -2 \\ 1 & 0 & -1 \end{bmatrix}. \tag{6.81}
$$

Example 6.19 Applying 2D Filters to an Image

We show how to apply a Gaussian filter and a Sobel filter to images. In Figure 6.25(a) we present a picture of an LC droplet. The shape (spatial light pattern) of the droplet results from the attachment of a contaminant to its boundary layer. In Figure 6.25(b), we show the image after applying a horizontal Sobel filter. Note how the resulting image highlights locations where the light changes rapidly (the edges). Specifically, areas with high gradients are highlighted here in dark. In Figure 6.25(c) we apply a Gaussian filter with $\sigma = 1$; we note that this smooths out the image (although this is difficult to perceive with our eyes). This is because this filter is removing only granular (high-frequency) features from the image that we cannot easily detect with our eyes. To better highlight the smoothing effect, we apply the Gaussian filter with a value of $\sigma = 10$. The smoothing is now perceptible to our eyes; this is because the filter removes coarse (low-frequency) features from the image that we can detect. □

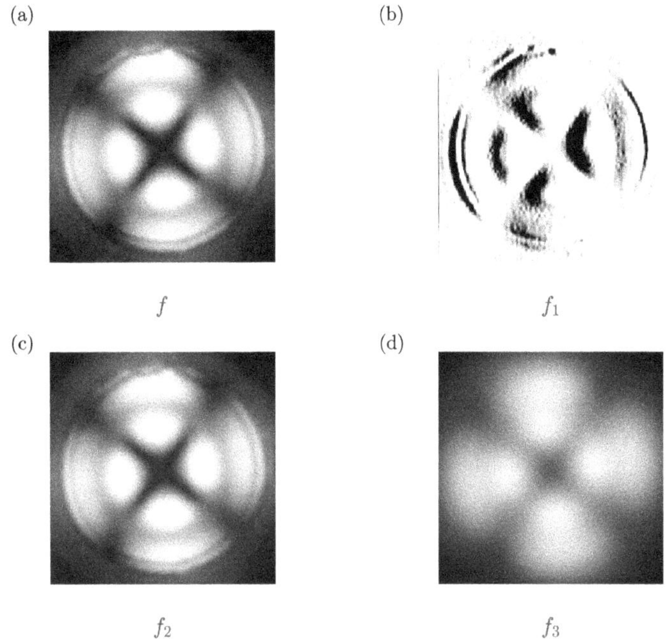

Figure 6.25 General image (a), image transformed with a Sobel filter (b), and image filtered with a Gaussian filter with small σ (c) and large σ (d).

Script to reproduce results: `ch6_image_filters_2D` □

FTs for 2D signals are commonly used to perform convolution operations. The discrete FT of a 2D field is:

$$\hat{f}_{k_1,k_2} = \frac{1}{N}\sum_{j_1=0}^{N-1}\sum_{j_2=0}^{N-1} f_{j_1,j_2} e^{-i2\pi\left(\frac{j_1 k_1}{N}+\frac{j_2 k_2}{N}\right)}, \; k_1 = k_2 = 0,\ldots,N-1. \tag{6.82}$$

The 2D FT returns a 2D field in the frequency domain. The FT allows us to characterize the frequency content of 2D signals and filters. Also, as in the 1D case, one can use the convolution theorem to compute the convolution as $f * g = \mathcal{F}^{-1}\{\mathcal{F}\{f\}\cdot\mathcal{F}\{g\}\}$.

Example 6.20 2D Signals as Superposition of 2D Waves

We have discussed that any 1D signal is the superposition of multiple periodic signals (each with different frequencies). A signal of high frequency exhibits fast oscillations, while a signal with low frequency exhibits slow oscillations. In the 2D case, we have a similar situation; a 2D field (e.g., an image) is a superposition of many different 2D fields (each with a different frequency). Think of a 2D field with a given frequency as a sheet with wave patterns defined by the frequency (higher frequencies show finer waves). In Figure 6.26(a and b), we show the 2D function:

$$f(x_1,x_2) = \sin\, 2\pi\omega_1(x_1+x_2) - \sin\, 2\pi\omega_2(x_1+x_2),$$

with frequencies $\omega_1 = 20$ and $\omega_2 = 60$. Figure 6.26(a, c, and e) shows the function in 3D, and we can see that this forms a wavy field (sheet) and that the sheet has two types of waves (corresponding to the low and high frequencies). Figure 6.26(b, d, and f) shows the function as a projection in 2D in which the darker areas correspond to planes in which the function has higher values and light places in which it has lower values. We see that this 2D field appears as diagonal stripes (corresponding to the peaks and valleys of the function), and we see that there are fast and complex changes in brightness (corresponding to places in which the waves decay faster).

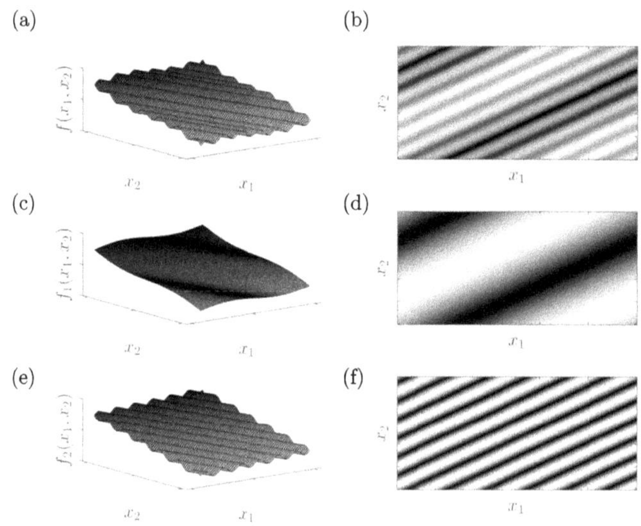

Figure 6.26 2D function $f(x_1,x_2)$ shown in 3D and as a projection in 2D (a and b). High-frequency component (c and d) and low-frequency component (e and f).

The 2D field is what you would see if you observed the wavy sheet from the top. In Figure 6.26(c–f), we show the two components of the function $f_1(x_1,x_2) = \sin 2\pi\omega_1(x_1 + x_2)$ and $f_2(x_1,x_2) = \sin 2\pi\omega_2(x_1 + x_2)$. Here, we can see that the low-frequency component forms a field with coarse waves, while the high-frequency component forms sheets with fine waves. This illustrates how simple shapes (fields) can be used to form complex patterns. This is precisely the principle behind image analysis; any image is the superposition of many wave fields of different frequencies. Identifying the frequencies associated with such waves allows us to design filters that can effectively eliminate undesired aspects of the image.

Script to reproduce results: `ch6_fourier_transform_2D` □

Example 6.21 2D Fourier Transform Applied to Filtering

In this example, we analyze the frequency content of a 2D uniform average filter (defined by a 2D step function) and of a 2D Gaussian filter (defined by a 2D Gaussian kernel). The results are shown in Figure 6.27. In Figure 6.27(a), we see the uniform filter shown as a 2D field and note that this is a black rectangle with a white space. If we were to visualize this filter as a 3D function (with the third dimension being the magnitude of the entry in the field), we would see a rectangular block (which is the 2D version of the rectangle pulse seen in 1D). In Figure 6.27(b), we see the FT of the filter (shown as a field), and Figure 6.27(c) we show this as a 3D function. We see that the 2D frequency field is symmetric, with the low frequencies concentrated in the middle region, and the high frequencies in the regions toward the edges. Here, we see that the uniform filter has a strong content of low frequencies and that the strength decays at higher frequencies. This decay is oscillatory and extends over the entire region (this is similar to the behavior that observed in the 1D case). In Figure 6.27(d–f), we show the 2D Gaussian filter, its frequency content shown as a 2D field, and its frequency content shown as a 3D function. We can see that the frequency content remains highly concentrated at low frequencies and decays rapidly to virtually zero at high frequencies (this is similar to the behavior that observed in the 1D case). From this analysis, we can see that the frequency content of filters generalizes from 1D to 2D.

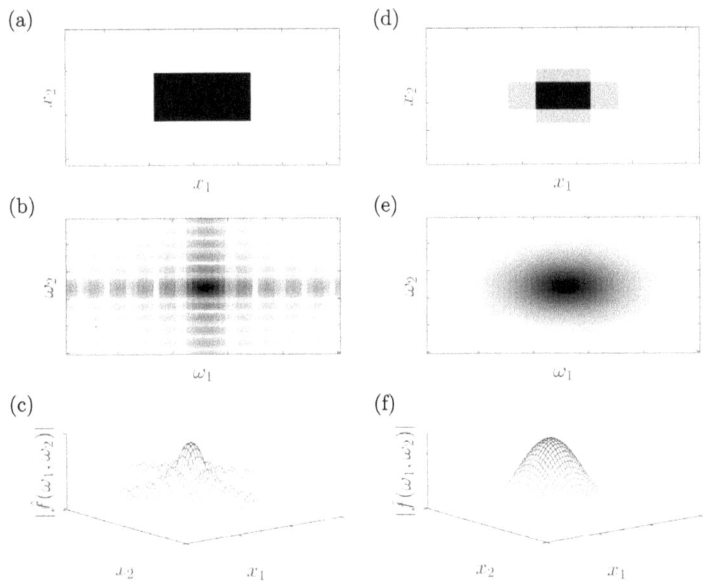

Figure 6.27 Uniform filter (a), Fourier transform in 2D (b), and Fourier transform in 3D (c). Gaussian filter (d), Fourier transform in 2D (e), and Fourier transform in 3D (f).

We now explore the properties of FTs and of the convolution theorem to apply filters. We take our original image f and obtain its 2D FT $\hat{f}$. We then construct a Gaussian filter with $\sigma = 1$ that we denote as g_1 and obtain its FT $\hat{g}_1 = \mathcal{F}\{g_1\}$. We also construct a Gaussian filter with $\sigma = 10$ that we denote as g_2 and obtain its FT $\hat{g}_2 = \mathcal{F}\{g_2\}$. We note that the value of σ for the first filter is much smaller than for the second one; this indicates that the first filter is designed to only eliminate high frequencies, while the second one eliminates high and low frequencies. We obtain filtered images by using the convolution theorem; specifically, the filtered images in the frequency domain are $\hat{f}_1 = \hat{f} \cdot \hat{g}_1$ and $\hat{f}_2 = \hat{f} \cdot \hat{g}_2$. We obtain the filtered images in the original domain by using the inverse FT as $f_1 = \mathcal{F}^{-1}\{\hat{f}_1\}$ and $f_2 = \mathcal{F}^{-1}\{\hat{f}_2\}$.

In Figure 6.28(a–c), we present the original image f and the filtered images f_1, f_2. Here, it is clear that the second filter eliminates high and low frequencies and blurs the image, while the first filter eliminates high frequencies that are less perceptible to the human eye. In Figure 6.28(d–f), we show the frequency spectrum of the original and filtered images. From Figure 6.28(d), we see that the original image has a highly complex spectrum that involves low and high frequencies. This highlights the fact that images are often highly complex objects that result from the superposition of many frequency fields. From Figure 6.28(e), we see that the first filter only eliminates a small portion of the high frequencies (dark regions in the edges) and thus preserves most of the frequencies. From Figure 6.28(f), we see that the filter eliminates most frequencies and only preserves low frequencies (the bright region in the middle). This explains why the second filter only preserves coarse features of the image. We highlight, however, that it is rather remarkable that a small portion of the spectrum is sufficient to capture coarse features of the image.

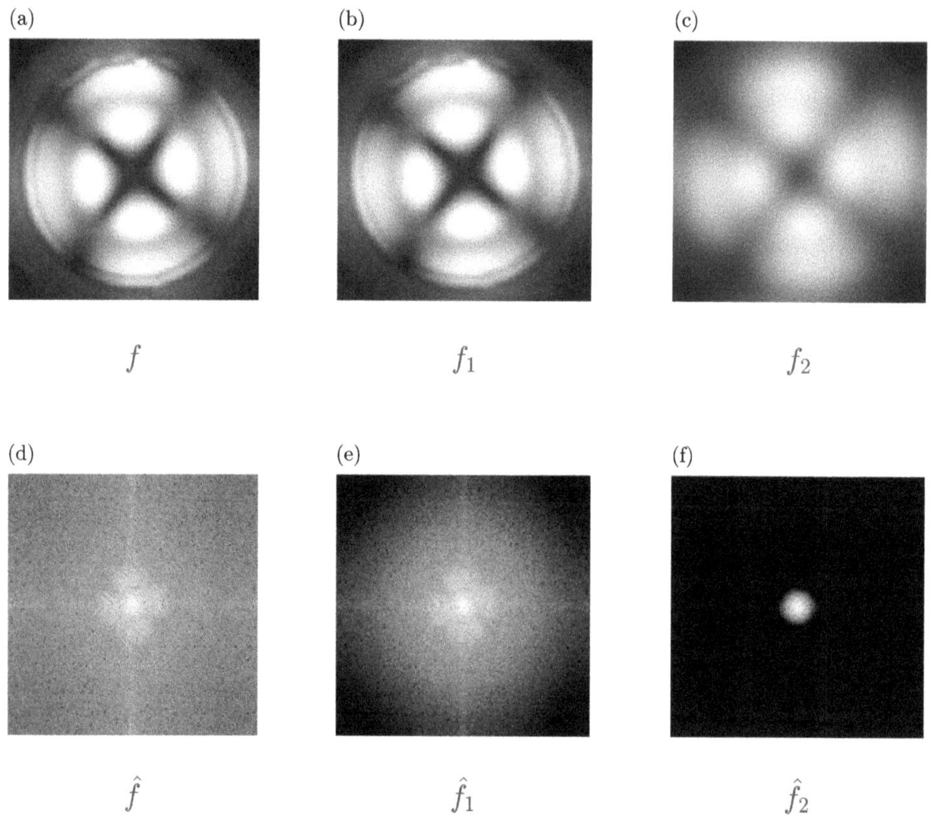

Figure 6.28 (a–c) Original and filtered images. (d–f) Frequency spectrum of original and filtered images.

Script to reproduce results: `ch6_fourier_transform_filters_2D` and `ch6_image_filters_2D` □

6.4 Classification Models

Statistical learning provides many tools to construct predictive models from data. These predictive models can be seen as generalizations of the models that we have already analyzed. Specifically, we will see that statistical learning techniques can construct models that capture an arbitrarily large number of input variables and that capture complex (nonlinear) relationships between variables. We will also explore a new type of predictive model in which the output of the model is a binary variable (known as classification models) and predictive models that admit images as input data. In studying the estimation and modeling techniques of the upcoming sections, we will make ample use of the concepts that we have learned in previous chapters. This will reveal how fundamentals of statistics become valuable in making sense of complex techniques.

Consider the following *classification problem*; imagine we have input variables $X = (X_1, \dots, X_n)$ with observations x_ω that lie in the continuous domain $\mathcal{D}_X \subseteq \mathbb{R}^n$. We have an output variable Y with scalar observations y_ω that lie in the *binary* domain $\mathcal{D}_Y = \{0, 1\}$. We postulate that there exists a relationship between X and Y of the form:

$$Y = m(X, \theta) + \epsilon, \tag{6.83}$$

where $\theta \in \mathbb{R}^n$ are parameters, $m(X, \theta)$ is the model, and ϵ is a noise RV that captures the unknown aspects of the model. Our objective is to find the model that best explains our data:

$$y_\omega = m(x_\omega, \theta) + \epsilon_\omega, \; \omega \in \mathcal{S}. \tag{6.84}$$

In machine learning, the inputs X are often referred to as *features or descriptors* and the output Y is referred to as the *label or class*. What is different about classification models (compared to the predictive models that we have studied) is the binary nature of the output Y. Specifically, in a standard predictive model, the output Y is assumed to be continuous and thus $\mathcal{D}_Y \subset \mathbb{R}$. The binary nature of the output restricts the type of models $m(X, \theta)$ that we can use. To make the distinction clear, predictive models that have discrete outputs are referred to as classification models, while predictive models that have continuous outputs are referred to as regression models.

Classification models have a wide range of applications, and their binary nature makes them intuitive to analyze. Think about, for example, a chemical classification problem: Given the features of a chemical (e.g., its molecular structure), can we predict if this is toxic or not? Classification models are also widely used to detect anomalies in systems. Think about, for example, a fault detection problem: Given a set of operating conditions for a reactor (the features), can we predict if this is malfunctioning or not? Similarly, think about that you want to predict if a given melanoma is benign or malign given its features (color and texture) or that you want to predict if you will get sick from data on blood pressure and glucose levels.

Classification models might also be able to tell us what are the key features of our data (or if some features are not relevant).

To develop our classification model, we will use a technique called *logistic regression* (later we will see that classification models can also be developed using neural networks). In logistic regression, we postulate a model of the form:

$$m(x, \theta) = \frac{1}{1 + e^{-\theta^T x}}. \tag{6.85}$$

The feature mixture $\theta^T x = \theta_0 + \sum_{i=1}^{n} \theta_i x_i$ is known as the *evidence* and the parameters attribute importance (weights) to different features. The postulated model is known as the logistic function and seeks to capture binary 0-1 logic. Specifically, the logistic function is a sigmoidal function that has the following behavior:

- $m(x, \theta) \to 1$ as $\theta^T x \to +\infty$ (large evidence)
- $m(x, \theta) \to 0$ as $\theta^T x \to -\infty$ (small evidence)
- $m(x, \theta) = \frac{1}{2}$ if $\theta^T x = 0$ (no evidence).

Figure 6.29 provides a visualization of the logistic function for different values of θ and for a scalar x. Note that reversing the sign of the parameter has the effect of reversing the logic.

The parameter θ_0 has an interesting interpretation, as this does not attribute weight to any input. In machine learning, this parameter is known as the *bias*; this name arises from the fact that the parameter introduces evidence that has no basis in data. This resembles how humans can make predictions/judgments with no basis in actual data.

The logistic function $m(x, \theta)$ can be used to model the conditional probabilities $\mathbb{P}(Y = 1 \mid x, \theta)$ and $\mathbb{P}(Y = 0 \mid x, \theta) = 1 - \mathbb{P}(Y = 1 \mid x, \theta)$. Specifically, we want to capture the following logic:

- If the evidence $\theta^T x$ is large, there is a high probability that $Y = 1$.
- If the evidence $\theta^T x$ is small, there is a high probability that $Y = 0$.
- If there is no evidence ($\theta^T x = 0$), there is ambiguity (it is equally probable that $Y = 0$ or $Y = 1$).

This logic can be captured by defining a conditional pdf of the form:

$$f(y \mid x, \theta) = \mathbb{P}(Y = y \mid x, \theta) = m(x, \theta)^y (1 - m(x, \theta))^{1-y}. \tag{6.86}$$

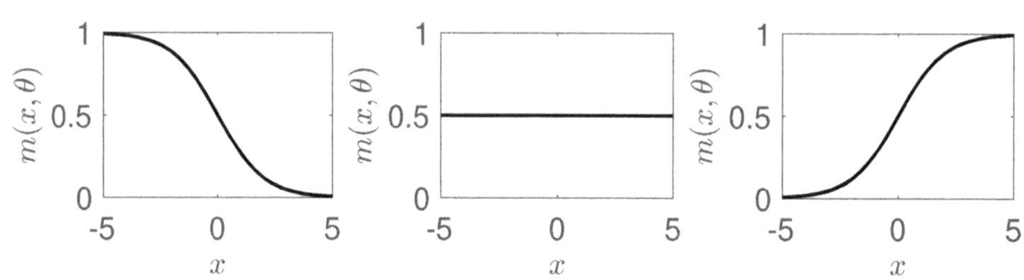

Figure 6.29 Behavior of logistic function for $\theta = -1$, $\theta = 0$, and $\theta = +1$.

Note that this is precisely a likelihood function that tells us the probability of the outputs given inputs and parameters. We thus apply an MLE approach to estimate the parameters θ.

The goal is to find the estimate $\hat{\theta}$ that maximizes the logarithm of the joint probability $\prod_{\omega \in \mathcal{S}} f(y_\omega \mid x_\omega, \theta)$. After applying a logarithmic transformation, this gives the optimization problem:

$$\max_{\theta} \sum_{\omega \in \mathcal{S}} y_\omega \cdot \log(m(x_\omega,\theta)) + (1 - y_\omega) \cdot \log(1 - m(x_\omega,\theta)). \tag{6.87}$$

We solve this problem to obtain the estimate $\hat{\theta}$; we can then use the classification model $m(x,\hat{\theta})$ to predict the class of a given input defined by features x. In machine learning, one refers to the process of estimating the model parameters from data as model training or model learning.

An important observation is that the logistic regression approach does not minimize the sum of squared errors (SSE), as done in standard MLE. The objective function used in logistic regression is called the *log loss*. This function is preferred because the definition of the conditional pdf (6.86) does not require us to make assumptions about the nature of the output error ϵ (this can be done because the output is binary). We recall that, in standard MLE, we assume that the output error is Gaussian, and this is what makes the conditional pdf Gaussian and the MLE objective reduces to SSE. Moreover, the logarithmic transformation used in logistic regression results in an objective that is better behaved numerically (compared to the squared SSE form used in standard MLE). This results from the specific structure of the model $m(x,\theta)$ (which has a sigmoidal form); for instance, note that:

$$\begin{aligned} \log(m(x,\theta)) &= -\log(1 + e^{-\theta^T x}) \\ \log(1 - m(x,\theta)) &= \log(e^{-\theta^T x}) - \log(1 + e^{-\theta^T x}). \end{aligned} \tag{6.88}$$

These functions remain well behaved when the evidence is large and small (this is not the case for the quadratic terms used in SSE).

Because the outputs in classification are discrete, the model quality is not measured in terms of residuals (e.g., MSE or RMSE). Instead, one often uses a measure known as an F1-score; this score is defined as:

$$F_1 = \frac{2 \cdot TP}{2 \cdot TP + FP + FN}. \tag{6.89}$$

Here, TP is the number of true positives; this is the number of observations in which the model correctly predicts a positive class (predicted label of "1"). FN is the number of false negatives; this is the number of observations in which the model predicts a negative class (predicted label of "0"), but the true class is positive. FP is the number of false positives; this is the number of observations in which the model predicts a positive class, but the true class is negative. The number of true negatives (TN) is the number of times that the model predicts a negative class correctly; the TN can be inferred from the rest of the quantities. Ideally, we want an F_1 score of 1.

To get a more comprehensive picture, the quality of a classification model is often visualized in the form of a *confusion matrix*, as illustrated in Figure 6.30. Ideally,

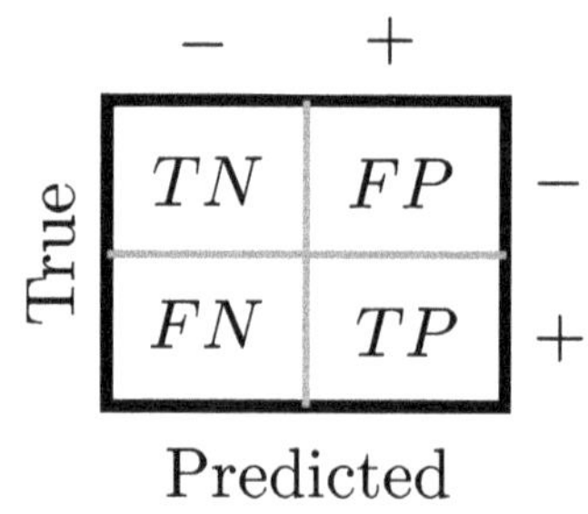

Figure 6.30 Illustration of a confusion matrix for a classification model.

note that the confusion matrix should be diagonal ($FP = 0$ and $FN = 0$). The term "confusion" highlights how the model can get confused in predicting classes.

Other important measures of performance of classification models are the *accuracy*, *precision*, and *recall*, which are defined as:

$$A = \frac{TP + TN}{TP + TN + FP + FN} \tag{6.90a}$$

$$P = \frac{TP}{TP + FP} \tag{6.90b}$$

$$R = \frac{TP}{TP + FN}. \tag{6.90c}$$

Note that the accuracy is simply the fraction of the total number of correct predictions relative to the total number of observations. Precision and recall only focus on the number of true positives. One can show that the F_1 score is the harmonic mean of prediction and recall:

$$F_1 = \frac{2}{\frac{1}{P} + \frac{1}{R}}. \tag{6.91}$$

The discussion that we present here is for binary classification models, but the concepts are easy to generalize to models that make multi-label predictions. We also emphasize that it is important for the modeler to ensure that the classification model generalizes well outside the data used for model training (e.g., by implementing cross-validation).

Example 6.22 Classification for Fault Detection in Reactor

Imagine that we have data $(P_\omega, C_\omega, T_\omega)$ for the Gibbs reactor. The data is separated into a couple of classes; one class represents operation when the reactor is in *normal operation* mode (when T_ω is low), and we define the labels for data entries in such a class as $y_\omega = 1$ (positive class). The other class represents operation under *failure mode* (when T_ω is high), and we define labels for data entries in such class as $y_\omega = 0$ (negative class). The dataset is visualized in Figure 6.31; note that there is a clear separation between the classes, but some points overlap. This type of problem is known in industry as *fault detection*.

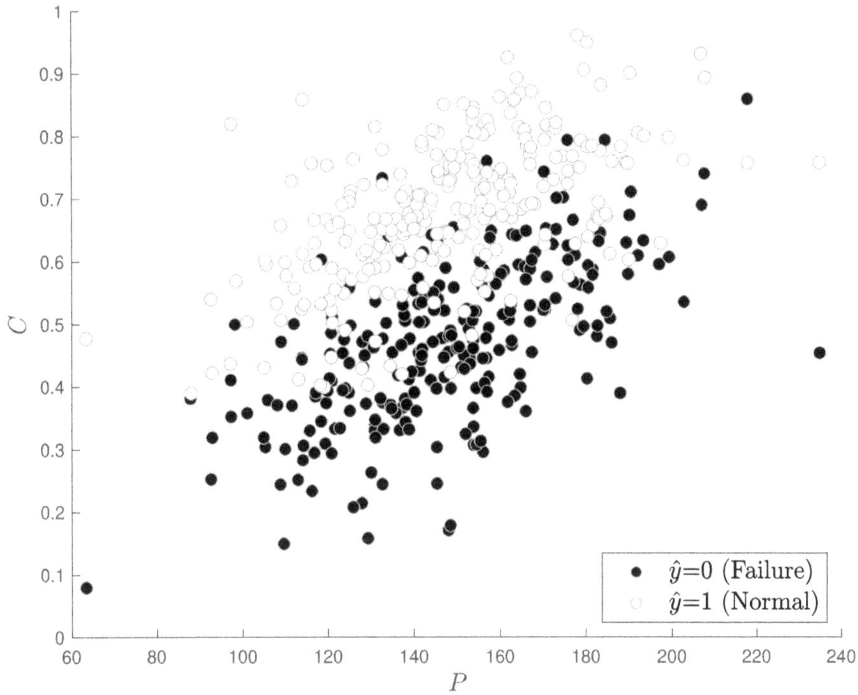

Figure 6.31 Gibbs reactor data showing separation between normal and failure operating modes.

Our goal is to develop a classification model that predicts the mode of operation by using data on pressure and conversion $x_\omega = (P_\omega, C_\omega)$. In other words, the model should be able to tell us if the reactor is failing (or not) by only looking at a data pair of pressure and concentration. The dataset tells us that this might be achievable due to strong class separation observed; specifically, the data points with high pressure and low conversion are usually associated with failure mode; this is expected if we use our physical knowledge because the reaction benefits from high pressures and low temperatures. Consequently, if we observe high pressure and low conversion, this indicates that the temperature might be high. Note also that, for some data points (where the classes overlap), it will be difficult to distinguish the operating mode. This indicates that, for such observations, evidence based on pressure and conversion alone will be insufficient to infer the operating mode.

After estimating the parameters of the classification model, we find that the model correctly predicts classes 85% of the time (428 out of 500 total observations); this is our measure of accuracy A. Logistic regression also gives us an estimate of the probability that predictions are in a given class or in a "fuzzy" (confusion) region. We are also interested in determining if the classification model provides false positives (predicted a failure that did not actually occur) or false negatives (predicted no failure and a failure actually occurred). This information is presented in Figure 6.32; in 6.32(a), we present a confusion matrix, which tells us that the model predicted 214 times that the reactor was in failure mode (label of zero) and that this mode was also the correct one (TN). The confusion matrix also tells us that the model predicted 36 times that the reactor was in failure mode when in fact it was in normal mode (it raised a false alarm). The confusion matrix in this particular example shows that the number of true positives is the same as the number of true negatives and that the number of false positives is the number of false negatives. From these numbers, we conclude that the F_1 score is 0.856.

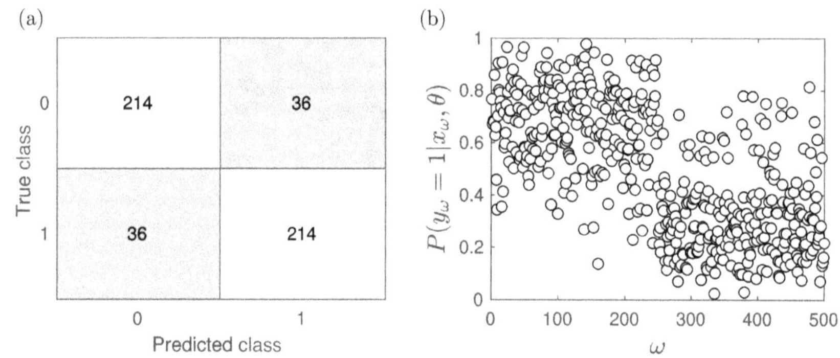

Figure 6.32 Confusion matrix (a) and classification probabilities (b) for Gibbs reactor data.

The Figure 6.32(b) shows us the probabilities that the reactor is in normal mode for all observations; we can see that some observations have high probabilities (close to one), while others have low probabilities (close to zero), and these points are easy to classify. Here, we can also see that a fraction of the data points have probabilities that are close to 0.5; these points are difficult to classify (it is equally likely that they are in either class, given the available evidence). These points are associated with points that lie in an overlapping region between classes. The results presented here show the behavior of the model assuming that all data is used for training; however, one should use cross-validation to verify that the model generalizes well to unseen data (check the accompanying script to see how this is done).

Script to reproduce results: `ch6_gibbs_logistic` □

6.5 Kernel Models

Kernel models are predictive models that are generalizations of the predictive models that we have explored so far. These models have the surprising properties that they require few parameters to make predictions (regardless of the number of inputs) and that they can capture complex nonlinearities (without the need to postulate an input–output model). The key principle of kernel methods is that the covariance of our data can be represented as a matrix function (rather than a fixed matrix). Moreover, such a function (known as a kernel function) can be used to capture complex nonlinear relationships that are present in our data.

To explain the origin and properties of kernel methods, it is useful to discuss a family of predictive models that *mix* different model types (these models are also known as *basis function* models). Specifically, assume that the output Y and the inputs $X = (X_1, \ldots, X_n)$ follow a relationship of the form:

$$Y = \sum_{j=1}^{m} \theta_j \cdot \phi_j(X) + \epsilon, \tag{6.92}$$

where $\phi_j(X),\ j = 1, \ldots, m$ is a collection of basic models (known as *basis functions*). Here, the output and input have continuous domains $\mathcal{D}_Y \subseteq \mathbb{R}$ and $\mathcal{D}_X \subseteq \mathbb{R}^n$ (this is a regression model).

The model is the *mixture* of basis functions:

$$m(X, \theta) = \sum_{j=1}^{m} \theta_j \cdot \phi_j(X). \tag{6.93}$$

For convenience, we define the basis function vector:

$$\phi(X) = (\phi_1(X), \phi_2(X), \ldots, \phi_m(X)), \tag{6.94}$$

and we can thus write the predictive model in compact form as $m(X, \theta) = \theta^T \phi(X)$. The basis functions $\phi(X)$ can be nonlinear (e.g., polynomial, sigmoidal, logarithmic, and exponential) or linear (in which case $\phi(X) = X$). The parameters $\theta \in \mathbb{R}^m$ are *mixing coefficients* for the basis functions, and these will highlight which basis models best explain the output. One can also include physical models in the mixture; for instance, imagine we have multiple alternative physical explanations of a phenomenon (e.g., different kinetic mechanisms); we can incorporate these models in the mixture.

The basis functions can also be interpreted as *data transformations* that seek to highlight different *features* from the input data; for instance, a logarithmic or polynomial transformation scales the input data in different ways. Here, we also note that the number of possible transformations (number of basis functions m) can be significantly larger than the number of input variables n. Consequently, $\phi(x_\omega)$ transforms the input observation x_ω into a high-dimensional space (known as the feature space). As such, the feature space is expected to contain more information than the input data itself. Note that the number of parameters is the same as the number of features; as such, our intuition would tell us that one needs to properly regularize the estimation problem to prevent overfitting.

The following discussion is an adaptation of the analysis made in Bishop [57, chapter 6]. Consider that we estimate the parameter vector $\hat{\theta}$ of the mixture model by solving the problem:

$$\min_{\theta} \frac{1}{2} \sum_{\omega \in \mathcal{S}} (y_\omega - \theta^T \phi(x_\omega))^2 + \frac{1}{2} \lambda\, \theta^T \theta. \tag{6.95}$$

Note that this is a *Bayesian estimation* problem with a Gaussian prior $\theta \sim \mathcal{N}(0, \lambda \mathbf{I})$ (we can also see this problem as an MLE problem with a ridge penalty regularization). We express this problem in a compact form as:

$$\min_{\theta} \frac{1}{2}(\mathbf{y} - \Phi\theta)^T(\mathbf{y} - \Phi\theta) + \frac{1}{2}\lambda\,\theta^T\theta, \tag{6.96}$$

where $\mathbf{y} \in \mathbb{R}^S$ is the output data vector with entries y_ω, $\omega = 1,\ldots,S$. We refer to $\Phi \in \mathbb{R}^{S\times m}$ as the *feature data matrix*; this has entries given by:

$$\Phi_{\omega,j} = \phi_j(x_\omega),\ \omega = 1,\ldots,S\ j = 1,\ldots,m. \tag{6.97}$$

We note that the feature matrix Φ is a transformation of the original input data $\mathbf{X}$ and that the estimation problem is a linear regression problem. If $\Phi = \mathbf{X}$, we obtain a standard linear regression problem (with no data transformation). As such, the only difference in the basis model is that we preprocess (transform) the input data using the basis functions.

We define the objective function of the estimation problem as $S(\theta)$; the necessary condition for optimality of the estimate $\hat{\theta}$ is that $\nabla_\theta S(\hat{\theta}) = 0$. The estimate thus needs to satisfy:

$$-\Phi^T(\mathbf{y} - \Phi\theta) + \lambda\theta = 0. \tag{6.98}$$

This condition can be written as:

$$\theta = \frac{1}{\lambda}\Phi^T(\mathbf{y} - \Phi\theta). \tag{6.99}$$

This condition can in turn be expressed in terms of the residuals as:

$$\theta = \Phi^T\mathbf{r}(\theta), \tag{6.100}$$

where:

$$\mathbf{r}(\theta) = \frac{1}{\lambda}(\mathbf{y} - \Phi\theta). \tag{6.101}$$

By substituting Eq. (6.100) in the objective function $S(\theta)$, we obtain:

$$S(\theta) = \frac{1}{2}\mathbf{r}(\theta)^T\Phi\Phi^T\Phi\Phi^T\mathbf{r}(\theta) - \mathbf{r}(\theta)^T\Phi\Phi^T\mathbf{y} + \frac{1}{2}\mathbf{y}^T\mathbf{y} + \frac{\lambda}{2}\mathbf{r}(\theta)^T\Phi\Phi^T\mathbf{r}(\theta). \tag{6.102}$$

Note that the feature matrix Φ always appears as the pair $\Phi\Phi^T$; we can thus define $\mathbf{K} = \Phi\Phi^T \in \mathbb{R}^{S\times S}$ and simplify:

$$S(\theta) = \frac{1}{2}\mathbf{r}(\theta)^T\mathbf{K}\mathbf{K}\mathbf{r}(\theta) - \mathbf{r}(\theta)^T\mathbf{K}\mathbf{y} + \frac{1}{2}\mathbf{y}^T\mathbf{y} + \frac{\lambda}{2}\mathbf{r}(\theta)^T\mathbf{K}\mathbf{r}(\theta). \tag{6.103}$$

We now make the important observation that the parameters θ only enter implicitly through the residual $\mathbf{r}(\theta)$. As such, we can write the objective function as:

$$S(\mathbf{r}) = \frac{1}{2}\mathbf{r}^T\mathbf{K}\mathbf{K}\mathbf{r} - \mathbf{r}^T\mathbf{K}\mathbf{y} + \frac{1}{2}\mathbf{y}^T\mathbf{y} + \frac{\lambda}{2}\mathbf{r}^T\mathbf{K}\mathbf{r}. \tag{6.104}$$

This allows us to estimate the parameters by using this strategy:

- Solve the following problem to find the optimal *residuals* $\hat{\mathbf{r}}$:

$$\min_{\mathbf{r}} S(\mathbf{r}). \tag{6.105}$$

- Recover *parameter estimate* from $\hat{\theta} = \Phi^T\hat{\mathbf{r}}$.

In other words, this approach first finds the best possible residuals and then determines the parameters associated with those residuals. The formulation (6.105) is known as the *dual representation* of the estimation problem.

It turns out, however, that the parameter estimates $\hat{\theta}$ are *not needed at all* to make predictions with our mixture model. This observation is by no means obvious; to see this, we note that $\hat{\mathbf{r}}$ satisfies the necessary condition $\hat{\mathbf{r}} = (\mathbf{K} + \lambda \mathbf{I})^{-1}\mathbf{y}$ for problem (6.105). The optimal model prediction is thus:

$$\begin{aligned} \hat{\mathbf{y}} &= \Phi\hat{\theta} \\ &= \Phi\Phi^T\hat{\mathbf{r}} \\ &= \mathbf{K}(\mathbf{K} + \lambda \mathbf{I})^{-1}\mathbf{y}. \end{aligned} \tag{6.106}$$

Consequently, the optimal prediction only depends on the input data (encoded in the matrix $\mathbf{K}$) and on the output data (encoded in $\mathbf{y}$). We thus conclude that the parameters $\hat{\theta}$ are not needed to make predictions; this indicates that the parameters are just *intermediary variables* (a means to an end). All that matters is how the input data propagates through the model to predict the output data.

The matrix $\mathbf{K}$ plays a critical role in making predictions; this matrix is also known as the *kernel matrix* (or Gram matrix) and captures interactions/connections between the input variables X. To see this, we note that the entries of this matrix are given by:

$$\mathbf{K}_{\omega,\omega'} = \phi(x_\omega)^T\phi(x'_\omega),\ \omega,\omega' \in \mathcal{S}. \tag{6.107}$$

We can thus see that the entry (ω,ω') captures the interaction of the transformed input data point x_ω and the transformed input data point $x_{\omega'}$. Motivated by this observation, we can consider building the kernel matrix by using different forms of interaction functions (known as kernel functions). Specifically, we can construct the kernel matrix as:

$$\mathbf{K}_{\omega,\omega} = k(x_\omega, x_{\omega'}),\quad \omega,\omega' \in \mathcal{S}, \tag{6.108}$$

where $k(x_\omega, x_{\omega'})$ is the *kernel function*. The kernel function generalizes several strategies that we have discussed; for instance, the case for linear regression corresponds to defining the kernel function as:

$$k(x_\omega, x_{\omega'}) = x_\omega^T x_{\omega'},\ \omega,\omega' \in \mathcal{S}. \tag{6.109}$$

In this case, note that the kernel matrix is simply $\mathbf{K} = \mathbf{X}\mathbf{X}^T$; note how this kernel function captures linear interactions of the data. The case of mixtures of basis functions discussed in this section corresponds to defining a kernel function of the form:

$$k(x_\omega, x_{\omega'}) = \phi(x_\omega)^T\phi(x_{\omega'}),\ \omega,\omega' \in \mathcal{S}. \tag{6.110}$$

Note how this function captures nonlinear interactions in the data (via the definition of the basis functions).

Now that we have a way to generalize our estimation approach, we can start asking ourselves what other types of kernel/interaction functions we might consider. A widely used kernel function used in ML is the radial basis function (RBF):

$$k(x_\omega, x_{\omega'}) = \exp(-\gamma \|x_\omega - x_{\omega'}\|^2), \; \omega, \omega' \in \mathcal{S}, \tag{6.111}$$

where $\gamma \in \mathbb{R}_+$ is a tunable parameter. This kernel function is also known as the Gaussian kernel; the Gaussian kernel measures interaction in terms of distance between the input data vectors $x_\omega \in \mathbb{R}^n$ and $x_{\omega'} \in \mathbb{R}^n$ (the higher the distance, the less the interaction). The parameter γ determines how quickly the interaction decays with distance. This kernel function is widely used in geostatistics, as typically interactions between locations decay exponentially with distance.

It is important to note that the use of the Gaussian kernel does not require that we specify a basis function vector $\phi(X)$ (we do not need to decide what basis functions to use). All that we need in kernel methods is the ability to model the interactions between the variables. Note also that these models are completely data-driven (they only require input–output data).

Example 6.23 Kernel Estimation for Hougen–Watson (HW) Reaction

As we have seen, the structural Hougen–Watson model:

$$Y = \frac{(\theta_0 X_2 - X_3/\theta_4)}{(1 + \theta_1 X_1 + \theta_2 X_2 + \theta_3 X_3)},$$

fits experimental data well. This model, however, has five parameters that might be difficult to estimate from data due to the high nonlinearity of the model. Moreover, one often needs to propose model variants (e.g., for different reaction mechanisms) and identify one that works best.

Instead of using a physical model to make predictions, we can also use data (x_ω, y_ω) to construct a data-driven kernel model. To do so, we use a Gaussian kernel $k(x_\omega, x_{\omega'}) = \exp(-\gamma \|x_\omega - x_{\omega'}\|^2)$ and compare performance with $\gamma = 0.1, 0.001, 0.0001$.

In Figure 6.33, we show the fit of the kernel model to the data for all values of γ. Note that the kernel model can capture the high nonlinearity of the reaction and give an almost perfect fit for $\gamma = 0.1$ and $\gamma = 0.001$. It is rather surprising that such a simple kernel model captures the high nonlinearity of the model with just one parameter (as

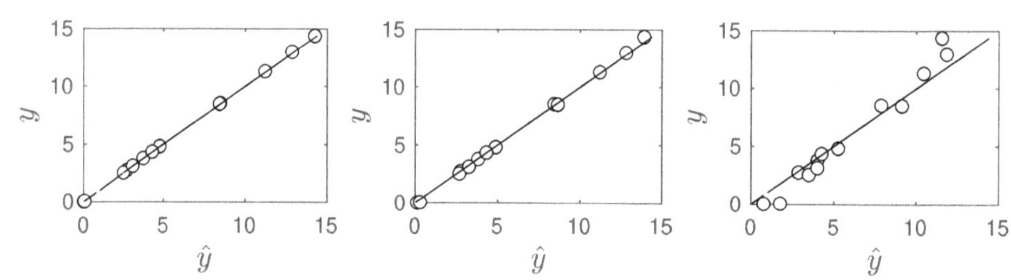

Figure 6.33 Fit of Hougen–Watson reaction data using the Gaussian kernel model for parameter values $\gamma = 0.1, 0.001, 0.0001$.

opposed to five used in the HW model). It is also surprising that the model is rather insensitive to the selection of γ. We note, however, that performance degrades when $\gamma = 0.0001$; in this case, the kernel model does not capture the interactions between the input data points well. It is thus necessary to tune the kernel parameter to achieve adequate performance.

Script to reproduce results: `ch6_hougen_watson_kernel` □

A surprising observation from the previous example is that the Gaussian kernel captures *highly nonlinear behavior*. This is absolutely not obvious from the simple structure of the kernel function, which involves a simple exponential and a quadratic term. We can show, however, that the Gaussian kernel is in fact an *infinite mixture of polynomial* functions. To see this, consider the case where $x_\omega, x_{\omega'}$ are scalars and notice that the Gaussian kernel can be written as:

$$\begin{aligned}
&\exp(-\gamma(x_\omega - x_{\omega'})^2) \\
&= \exp(-\gamma x_\omega^2 + 2\gamma x_\omega x_{\omega'} - \gamma x_{\omega'}^2) \\
&= \exp(-\gamma x_\omega^2)\exp(-\gamma x_{\omega'}^2)\exp(2\gamma x_\omega x_{\omega'}).
\end{aligned} \tag{6.112}$$

Note how the interaction term $x_\omega x_{\omega'}$ is implicitly captured via the exponential term. Now note that the expression $\exp(2\gamma x_\omega x_{\omega'})$ can be represented as a Taylor series to obtain:

$$\begin{aligned}
&\exp(-\gamma(x_\omega - x_{\omega'})^2) \\
&= \exp(-\gamma x_\omega^2)\exp(-\gamma x_{\omega'}^2)\left(1 + \frac{2\gamma x_\omega x_{\omega'}}{1!} + \frac{(2\gamma x_\omega x_{\omega'})^2}{2!} + \ldots\right) \\
&= \exp(-\gamma x_\omega^2)\exp(-\gamma x_{\omega'}^2)\left(1 \cdot 1 + \left(x_\omega\sqrt{2\gamma/1!}\right)\cdot\left(x_{\omega'}\sqrt{2\gamma/1!}\right)\right. \\
&\quad \left. + \left(x_\omega^2\sqrt{(2\gamma)^2/2!}\right)\cdot\left(x_{\omega'}^2\sqrt{(2\gamma)^2/2!}\right) + \ldots\right).
\end{aligned} \tag{6.113}$$

We thus have that Gaussian kernel can be written as:

$$\exp(-\gamma(x_\omega - x_{\omega'})^2) = \phi(x_\omega)^T\phi(x_{\omega'}), \tag{6.114}$$

where $\phi(x)$ is an *infinite* collection of polynomial functions of the form:

$$\phi(x) = e^{-\gamma x^2}(1,\, x\sqrt{2\gamma/1!},\, x^2\sqrt{(2\gamma)^2/2!},\, x^3\sqrt{(2\gamma)^3/3!}, \ldots). \tag{6.115}$$

Note how a *single parameter* γ controls the coefficients of these polynomial functions. As such, the kernel function can generate a wide range of nonlinear behavior by manipulating γ. This also highlights that a kernel function can be seen as a *generator of basis functions*.

From our discussion so far, we can make the following important observations:

- In kernel methods, we do not need to postulate an input–output model (e.g., $Y = \theta^T\phi(X)$) and estimate its parameters (θ). Instead, we postulate a kernel model that captures interactions between variables and tune its parameters (e.g., γ).
- A kernel function can be interpreted as a generator of basis functions (hence the name kernel), and this can capture highly nonlinear behavior.

- The kernel function often involves few parameters and this is a key advantage over standard estimation approaches because overparameterization issues are avoided (regardless of the number of inputs).
- There is a wide range of kernel functions that can be used to capture interactions between variables. This provides a mechanism to capture a rich set of possible nonlinear behavior.

Data transformations using kernel functions can also be used for conducting PCA (this is done using a technique called *kernel PCA*). The ability to transform data into a feature space can help better reveal clusters and trends in the information space of PCA. Mixtures of basis functions are also used in a technique called *symbolic regression*, in which one aims to identify different basic models (or combinations of them) that best explain the behavior of a system; this type of technique has been recently used to discover physical models from data.

6.6 Gaussian Process Modeling

Because there are typically a few number of parameters in the kernel function, one can try to tune them by hand (e.g., by trial and error) or by using simple heuristics. However, questions arise here:

- What are we trying to achieve as we tune the kernel function parameters?
- How do we estimate optimal kernel function parameters?

These questions can be addressed in a general setting by using a technique called *kriging* (also known as *Gaussian process [GP] modeling*). This method was originally proposed by Danie Krige, a South African statistician; the method has origins in geostatistics (a field of statistics that analyzes spatial phenomena).

In GP modeling, we postulate a model of the form:

$$y_\omega = m(x_\omega) + \epsilon_\omega,\ \omega \in \mathcal{S}. \tag{6.116}$$

In compact vector form:

$$\mathbf{y} = \mathbf{m} + \epsilon. \tag{6.117}$$

Here, $\mathbf{y} \in \mathbb{R}^S$ contains the output observations y_ω and $\mathbf{m} \in \mathbb{R}^S$ contains the model evaluated at the input observations $m(x_\omega)$. If we assume that $\epsilon \sim \mathcal{N}(0, \sigma^2\mathbf{I})$, we have that the conditional $f(\mathbf{y}|\mathbf{m})$ is given by $\mathcal{N}(\mathbf{m}, \sigma^2\mathbf{I})$.

The model $\mathbf{m}$ is assumed to be a multivariate Gaussian RV of the form:

$$\mathbf{m} \sim \mathcal{N}(0, \mathbf{K}(\gamma)), \tag{6.118}$$

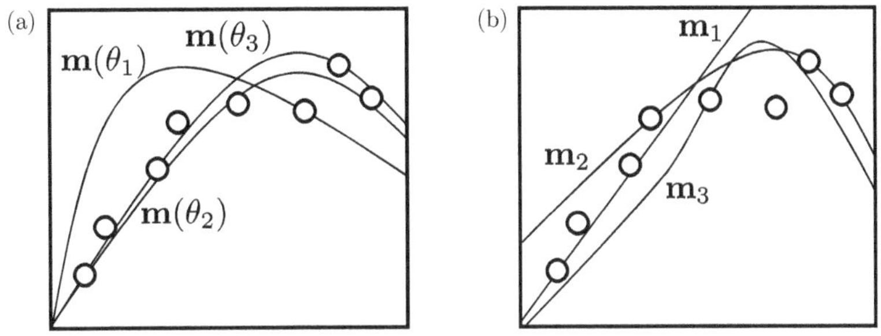

Figure 6.34 Generation of alternative models for parametric model $\mathbf{m}(\theta)$ using different values of θ (a). Generation of alternative models for GP using different realizations of $\mathbf{m} \sim \mathcal{N}(0, \mathbf{K}(\gamma))$ (b).

with associated pdf $f(\mathbf{m}|\gamma)$. Here, $\mathbf{K}(\gamma) = \text{Cov}[\mathbf{m}] \in \mathbb{R}^{S \times S}$ is the model *covariance matrix* and γ are its tunable parameters. The covariance matrix is built by using a kernel function:

$$\mathbf{K}_{\omega,\omega'}(\gamma) = k(x_\omega, x_{\omega'}, \gamma), \; \omega, \omega' \in \mathcal{S}. \tag{6.119}$$

What is different about GP modeling is that the model $\mathbf{m}$ is treated as a *random function*; note also that the model $\mathbf{m}$ has no parameters (hence it is called a *nonparametric* model). This is thus different from the techniques that we have discussed previously, in which we consider parametric models of the form $\mathbf{m}(\theta)$.

A fundamental difference between GP and parametric models is the way in which we generate alternative models to fit the data (see Figure 6.34). In parametric models, we generate alternative models $\mathbf{m}(\theta)$ by testing different parameters θ; from these alternatives, we select the one that fits our data best. In GP, we generate alternative models by sampling $\mathcal{N}(0, \mathbf{K}(\gamma))$ (to generate different realizations of $\mathbf{m}$); from these realizations, we select the one that best fits our data. The covariance matrix $\mathbf{K}(\gamma)$ plays a key role here, as this influences the types of model realizations that are generated. This matrix can be tuned (by tuning the parameters γ) to optimize the types of model realizations obtained; in addition, we can select alternative kernel functions to generate different types of model realizations.

The covariance matrix has tunable parameters; as such, the term nonparametric sometimes is misleading. The term nonparametric means that the model itself does not contain parameters. However, we still require techniques to tune the parameters of the covariance matrix.

In order to derive a procedure to tune the covariance parameters, we recall that the joint pdf of the output $\mathbf{y}$ and the model $\mathbf{m}$ is given by:

$$f(\mathbf{y}, \mathbf{m}|\gamma) = f(\mathbf{y}|\mathbf{m}, \gamma) f(\mathbf{m}|\gamma). \tag{6.120}$$

We thus have that the marginal pdf of $\mathbf{y}$ is:

$$\begin{aligned} f(\mathbf{y}|\gamma) &= \int_{\mathbf{m} \in \mathcal{D}_\mathbf{m}} f(\mathbf{y}, \mathbf{m}|\gamma) d\mathbf{m} \\ &= \int_{\mathbf{m} \in \mathcal{D}_\mathbf{m}} f(\mathbf{y}|\mathbf{m}, \gamma) f(\mathbf{m}|\gamma) d\mathbf{m}. \end{aligned} \tag{6.121}$$

Note that the marginal pdf $f(\mathbf{y}|\gamma)$ captures the likelihood of the outputs $\mathbf{y}$ given the covariance parameters γ (which is what we are trying to estimate). One can show that the marginal pdf corresponds to the pdf of the Gaussian $\mathcal{N}(0, \mathbf{C}(\gamma))$; the entries of the covariance matrix $\mathbf{C}(\gamma)$ are:

$$\mathbf{C}_{\omega,\omega'}(\gamma) = \mathbf{K}_{\omega,\omega'}(\gamma) + \sigma^2 \cdot \delta_{\omega,\omega'}, \quad \omega, \omega' \in \mathcal{S}. \tag{6.122}$$

Here, $\delta_{\omega,\omega'} = 1$ if $x_\omega = x_{\omega'}$ and zero otherwise. In other words, the covariance of the marginal is the covariance of the model plus the covariance of the output noise.

Once we have a likelihood $f(\mathbf{y}|\gamma)$, one can estimate $\hat{\gamma}$ by using MLE. Accordingly, we obtain the estimate by maximizing the log-likelihood $\log f(\mathbf{y}|\gamma)$:

$$\max_{\gamma} \; -\frac{1}{2}\log|\mathbf{C}(\gamma)| - \frac{1}{2}\mathbf{y}^T\mathbf{C}(\gamma)\mathbf{y} - \frac{S}{2}\log 2\pi. \tag{6.123}$$

This provides a systematic approach to estimate the parameters $\hat{\gamma}$.

The MLE optimization problem solved for training the GP model has few parameters, but it is challenging to solve numerically. This is because the matrix $\mathbf{C}(\gamma)$ depends on the parameters in a complex nonlinear manner. Moreover, note that evaluating the likelihood function involves computing the logarithm of the determinant of $\mathbf{C}(\gamma)$; this can be computationally intensive. The solution to the GP MLE problem requires advanced optimization techniques, but modern computing packages have capabilities for solving this problem.

As a kernel method, in GP, one computes predictions by using the kernel function directly (which only depends on $\hat{\gamma}$ and the input data). Model predictions are obtained from the conditional pdf $f(y(x)|\mathbf{y}, \gamma)$, where $y(x)$ is the predicted output (at a new input x of interest) and $\mathbf{y}$ are the observations used to determine $\hat{\gamma}$ (observations used for training the model).

We established that the marginal $f(\mathbf{y}|\gamma)$ is Gaussian and one can easily show that the joint pdf $f(y(x), \mathbf{y})|\gamma)$ is Gaussian. From the joint pdf, we can obtain the conditional $f(y(x)|\mathbf{y}, \gamma)$ and show that this corresponds to the Gaussian $\mathcal{N}(\mu(x), \sigma^2(x))$ with mean and variance:

$$\begin{aligned} \mu(x) &= \mathbf{k}(x, \hat{\gamma})^T\mathbf{C}(\hat{\gamma})^{-1}\mathbf{y} \\ \sigma^2(x) &= c(x, \hat{\gamma}) - \mathbf{k}(x, \hat{\gamma})^T\mathbf{C}(\hat{\gamma})\mathbf{k}(x, \hat{\gamma}), \end{aligned} \tag{6.124}$$

where:

$$\begin{aligned} c(x, \hat{\gamma}) &= k(x, x, \hat{\gamma}) \\ \mathbf{k}_\omega(x, \hat{\gamma}) &= k(x_\omega, x, \hat{\gamma}), \; \omega \in \mathcal{S}. \end{aligned} \tag{6.125}$$

From this, we see that the mean prediction of the model is $\mathbb{E}[y(x)] = \mu(x)$ and its variance/uncertainty is $\mathbb{V}[y(x)] = \sigma^2(x)$. We thus have that GP models naturally provide uncertainty information for predictions; notably, this does not require uncertainty information of the parameters. This highlights that, when we are building predictive models, quantifying the uncertainty of parameters is really not our end goal; instead, our end goal is to quantify the uncertainty of our predictions.

GP models naturally highlight regions in the input domain that are associated with high prediction uncertainty (e.g., regions that have not been explored well). This information can be exploited to recommend experiments that mitigate prediction uncertainty, as we discussed for parametric models in Section 5.6.

There is a wide range of kernel functions that can be used in kriging. For instance, the generalized Gaussian kernel takes the form:

$$k(x_\omega, x_{\omega'}, \gamma) = \gamma_0 \exp\left(-\frac{\gamma_1}{2}\|x_\omega - x_{\omega'}\|^2\right) + \gamma_2, \tag{6.126}$$

where $\gamma = (\gamma_0, \gamma_1, \gamma_2)$ are the parameters. Other popular kernel functions include the Matérn kernel.

We have argued that a key benefit of GP is that we do not need to specify a parametric model to capture relationships between the inputs and outputs, but what if we actually have some knowledge on how they are related? Specifically, what if we have physical knowledge of how the inputs affect the outputs? (e.g., they satisfy the Hougen–Watson equation). To deal with this, it is possible to build *hybrid models* that combine parametric models (e.g., physical models) and nonparametric models (e.g., GP). This can be done by defining the conditional $f(\mathbf{y}|\mathbf{m},\theta)$ as $\mathcal{N}(h(\mathbf{x},\theta) + \mathbf{m}, \sigma^2\mathbf{I})$ where $h(\mathbf{x},\theta)$ is our parametric model that captures what we know about our system. The nonparametric model $\mathbf{m} \sim \mathcal{N}(0, \mathbf{K}(\gamma))$ captures what we do not know about the system and can be seen as a *residual model* that captures behavior that cannot be explained with our physical model. Moreover, it is possible to easily extend GP models to account for uncertainty in our input data. We can thus see that GP provides a powerful framework to model complex systems.

Example 6.24 GP Model for Gibbs Reactor

Our goal is to compare the prediction accuracy of a GP (nonlinear) model against that of a linear model (to demonstrate the versatility of GP models). Here, we use data for the Gibbs reactor, as we have shown that this exhibits nonlinearity. Specifically, we gather data for the inputs (pressure and temperature) $X = (P, T)$ and output (conversion) $Y = C$.

We build the GP model using a Gaussian kernel function of the form (6.126). In Figure 6.35, we present parity plots for the kriging and linear models; it is clear that the kriging model achieves a much better fit over the entire range of outputs; the linear model predictions drift at low and high conversions (where the nonlinear behavior is stronger). It is important to note that both the kriging and linear models have the same number of parameters (three). In the case of the linear model, the parameters are for the input–output model, while for GP, they are for the kernel model.

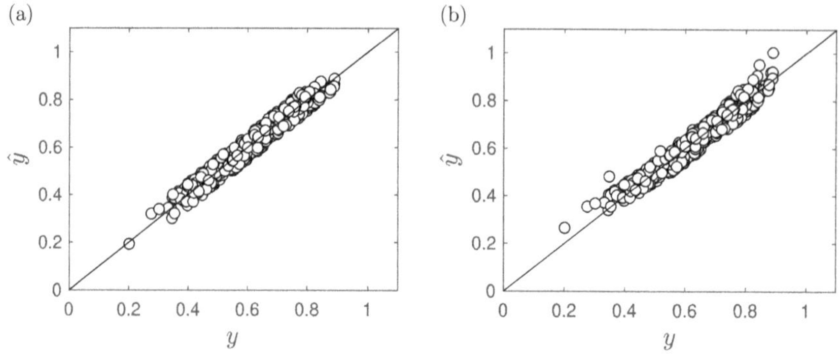

Figure 6.35 Predictions for GP (a) and linear (b) models.

An important feature of GP is that it provides uncertainty information for the output predictions; the uncertainty associated with the predictions (95% confidence intervals) is shown in Figure 6.36. The confidence intervals are also usually referred to as "uncertainty bands." Note that the observed values always lie within the confidence intervals predicted by the model. This indicates that the uncertainty characterization for the model predictions provided by GP is consistent.

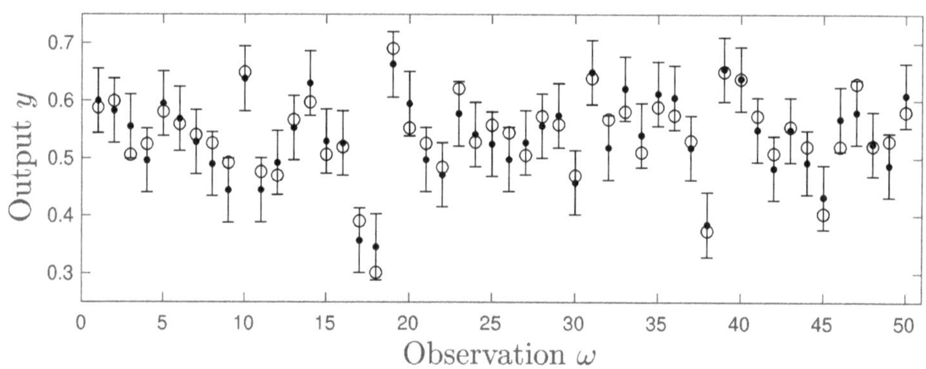

Figure 6.36 Observed outputs (white dots), mean outputs (black dots), and 95% confidence intervals for GP model predictions of Gibbs reactor.

Script to reproduce results: `ch6_gibbs_kriging` □

Example 6.25 Experimental Design Using GP Models

In this example, we illustrate how the uncertainty quantification capabilities of GP can be used to guide the collection of experimental data. Imagine that you have a system in which you can manipulate an input x (e.g., temperature) to optimize the performance of an output y (e.g., cost). You start with three experimental data points, shown in Figure 6.37(a), that you use to train your GP. The dashed line is the mean model prediction $\mu(x)$ of the resulting GP and the gray areas are the 95% confidence bounds for the predictions, obtained with the prediction uncertainty $\sigma^2(x)$. Note how the model prediction has no uncertainty at the location of the experimental points, but it has significant uncertainty in the unexplored regions of the input.

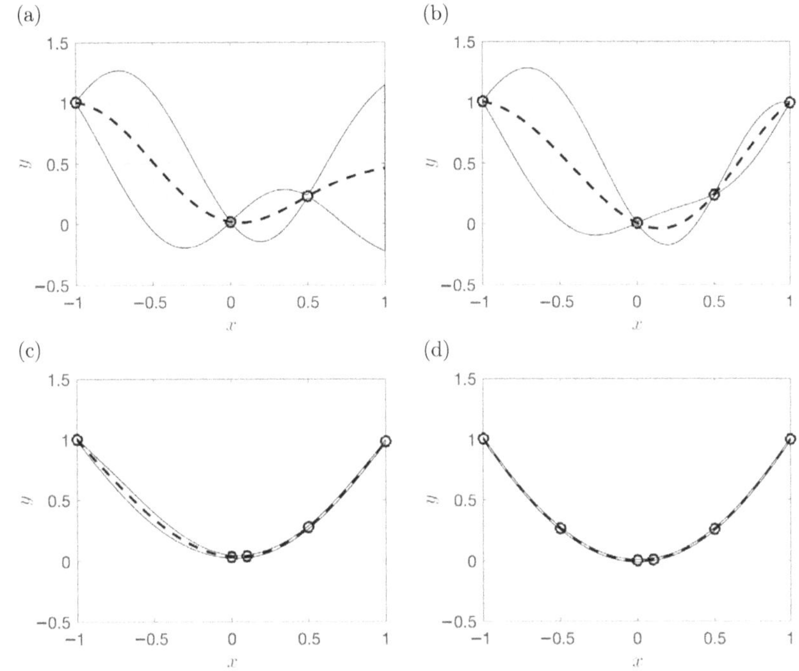

Figure 6.37 Evolution of model prediction uncertainty for GP with three data points (a), four data points (b), five data points (c), and six data points (d).

Because you notice that the model has high uncertainty on the right boundary of the input, you decide to explore this region and collect a new experimental point at $x_\omega = 1$. After retraining the GP, you find the prediction shown in Figure 6.37(b). Note how this new data point significantly reduces the uncertainty of the model in the right boundary. Now that you have a better model, imagine that you want to exploit/use your model to identify an experimental condition that minimizes the output; according to your mean model prediction $\mu(x)$, the output should be minimized at approximately $x_\omega = 0.1$. After collecting this new data point, you retrain the GP (see Figure 6.37(c)) and now you notice that your model has narrow uncertainty bands throughout the entire domain; as such, you feel confident that you have identified the true location of your optimum. To gain further confidence that your model is correct, you propose a final experiment in the left boundary of the domain, which is sparsely populated. Specifically, you select $x_\omega = -0.5$; after obtaining the new point and retraining the GP, you notice that your uncertainty bands are nearly gone, and your model appears to define a quadratic function. Indeed, the true system cost is a quadratic model.

This example illustrates the basic principle behind a closed-loop experimental technique known as Bayesian optimization. We will explore this technique in more detail in Chapter 7. This technique leverages the predictions of the GP model and their uncertainty to recommend experiments. This is done by alternating the exploration of regions in which there is a significant model uncertainty and the exploitation of the GP model to identify optimal experimental conditions. It is also important to note that the GP framework recommends experiments that mitigate prediction uncertainty (as opposed to parameter uncertainty, as done in parametric models).

Script to reproduce results: `ch6_kriging_uq` □

6.7 Neural Networks

We now explore a powerful set of predictive models known as neural networks (neural nets or NNs for short). The development of these models is motivated by the following questions:

- How do humans *learn* from observations?
- How do humans establish *connections* (relationships) between variables in their brains to make predictions and decisions?

NNs are the most general and powerful form of *parametric* models that currently exist. These models aim to mimic the way the human brain (a complex network of neurons) learns from observations (data collected with our senses). It can be shown that NNs are universal function approximations; in other words, they are so flexible that they can capture virtually any interrelationship between variables (e.g., nonlinear and multivariable).

The flexibility of NNs helps (to some extent) explain why humans are remarkably good at processing and learning from vast amounts of data. On the other hand, we will see that this flexibility can come at the expense of requiring an enormous number of observations and parameters; as a result, one has to carefully verify that NN models generalize well (predict outside the domain of the observations).

The study of NNs also gives rise to interesting questions on how humans combine prior knowledge and observations to make predictions. For example, a 4-year-old kid can learn to ride a bicycle by using only observations (without any knowledge of any physical laws), but they might require a lot of practice (lots of observations). However, if the kid has a prior understanding of physics (e.g., speed helps maintain equilibrium), then they might need less practice. This prior knowledge can be conveyed to the kid, for instance, by their parents or school teachers. Physical laws are also derived by humans as a result of observations that are synthesized in a compact form (e.g., mathematical equations), and we routinely exploit such laws as a form of prior knowledge (e.g., to engineer systems). As such, the human brain has flexible ways to gather, summarize, and leverage/use knowledge.

There are many variants of NN models and implementations that currently exist. The most basic form is fully connected NNs, which aim to make predictions from input data that is represented as vectors. Convolutional NNs are a more sophisticated model variant that can take input data that is represented as matrices and tensors (e.g., images and videos). Graph NNs (GNNs) are model variants that take input data that is represented as networks (e.g., molecules, supply chain networks, and social networks). Autoencoder NNs are a variant that aims to reduce input data into a latent/information space, which is analogous to PCA. Our discussion here will focus on fully connected NNs and CNNs, as this will allow us to discuss general concepts that are applicable to other types of NNs. In addition, we will aim to provide some insights (based on the concepts that we have learned in previous chapters) into the benefits and drawbacks of NNs.

6.7.1 Fully Connected Neural Nets

Fully connected NNs are parametric models of the form:

$$Y = m(X,\theta) + \epsilon. \tag{6.127}$$

What is different about NNs is that the model $m(X,\theta)$ is automatically constructed by using a set of basis functions that are mixed in a *hierarchical* manner, see Figure 6.38(a). The hierarchical model mimics how the brain works; in general, the deeper the hierarchy, the more complexity that can be captured. Neural nets provide a flexible approach to capture virtually any type of relationship between X and Y. A key advantage of this is that we do not need to postulate a model (e.g., nonlinear, linear, logistic, and mechanistic).

The goal is to train the NN model (estimate its parameters θ) to fit the observations:

$$y_\omega = m(x_\omega,\theta) + \epsilon_\omega,\ \omega \in \mathcal{S}. \tag{6.128}$$

The determination of θ from data mimics the *learning* process of a human. Specifically, NNs seek to mimic how the brain responds when exposed to data and how these responses accumulate to make predictions.

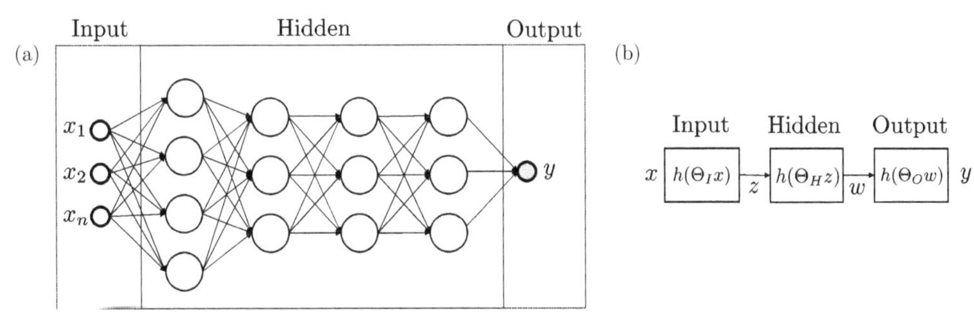

Figure 6.38 Sketch of the hierarchical structure of a neural network (a). Simplified visualization showing sequential transformations of data (b).

The brain is a hierarchical network that has neurons as basic processing units that interact with one another through signals. The mathematical description of a neuron is called a *perceptron.* A perceptron generates a signal if the evidence:

$$\theta^T x = \sum_{i=1}^{n} \theta_i x_i + \theta_0, \tag{6.129}$$

is strong enough. Here, x_i are the inputs with associated weights θ_i, and the parameter θ_0 is the *bias*. The weights capture how much emphasis we put on a given component of the evidence, while the bias is a parameter that captures situations in which we make predictions based on our inherent biases (ignoring evidence). Note how the notion of bias and evidence is the same as that discussed in logistic regression.

The response of a perceptron can be modeled using the following binary logic:

$$a = \begin{cases} 1 & \theta^T x \geq \tau \\ 0 & \theta^T x < \tau \end{cases}, \tag{6.130}$$

where τ is a threshold value. We thus have that a perceptron is a prediction unit that decides to "fire" or not when exposed to evidence. How to respond to different pieces of evidence is captured by the parameters θ. The binary behavior of the perceptron is similar to that used in logistic regression; in fact, perceptrons are often modeled using logistic (sigmoidal) functions:

$$a = \frac{1}{1 + e^{-\theta^T x}}. \tag{6.131}$$

Other functions typically used to capture perceptron logic include the hyperbolic tangent function and the rectified linear unit (ReLU) function. These *activation functions* play a critical role in capturing nonlinear behavior .

The NN mixes perceptron signals to construct *complex logic*; here, logic is constructed in a hierarchical manner (outputs of some perceptrons become inputs of other perceptrons). The components of the NN hierarchy (also known as the NN architecture) are:

- The input layer contains perceptrons that take the input data x_ω, $\omega \in \mathcal{S}$ to generate signals. This layer captures basic logic (i.e., I predicted this because of that).
- The hidden layers contain perceptrons that take signals from the input layer to generate additional signals. This layer captures abstract logic (i.e., I predicted this because of this, this, and that). As one goes deeper in the hierarchy, the hidden layers capture logic that is difficult for humans to rationalize (i.e., why did I predict that?).
- The output layer contains perceptrons that take signals from the hidden layer to generate a final output signal y_ω, $\omega \in \mathcal{S}$. The output signal can be continuous or discrete; consequently, NNs can be used for regression (e.g., predict actual temperature) or classification (e.g., predict hot or cold).

The design of the NN architecture is an important aspect of building these models. For instance, it is important to determine the number of hidden layers, how many perceptrons to include in each layer, and what activation functions to use. We now proceed to show the inner workings of NNs; for simplicity, we will assume an NN architecture that contains one hidden layer, see Figure 6.38(b).

To simplify the notation, define $x \in \mathbb{R}^n$ as the input observation x_ω. The input x is fed to the *input layer*, which is composed of $j = 1,\ldots,n_I$ perceptrons. The evidence in perceptron j is a mixture of the inputs:

$$a_j = \sum_{i=1}^{n} \theta^I_{j,i} x_i + \theta^I_{j,0},\ j = 1,\ldots,n_I. \tag{6.132}$$

Here, $\theta^I_{j,i}$ are the parameters (weights) of the perceptron ($\theta^I_{j,0}$ is the bias). Given evidence, the perceptron j generates the signal:

$$z_j = h(a_j),\ j = 1,\ldots,n_I, \tag{6.133}$$

where $h(a_j)$ is the activation function of the perceptron. One can write the mixture and activation of this layer in the following compact form:

$$\begin{aligned} a &= \Theta_I x \\ z &= h(a), \end{aligned} \tag{6.134}$$

where $a \in \mathbb{R}^{n_I}$ is a vector that contains all the perceptron evidences a_j, $z \in \mathbb{R}^{n_I}$ is a vector that contains all signals, and $\Theta_I \in \mathbb{R}^{n_I \times n}$ is the matrix that contains all the parameters of the layer. Note that this is a matrix that maps the inputs to the evidence; this parameter matrix is also known as the weighting matrix. We can also express the input layer operations in nested form as $z = h(\Theta_I x)$.

The *hidden layer* takes the output signals z of the input layer; the hidden layer is composed of $k = 1,\ldots,n_H$ perceptrons. The evidence in perceptron k is:

$$e_k = \sum_{j=1}^{n_I} \theta^H_{k,j} z_j + \theta^H_{k,0},\ k = 1,\ldots n_H. \tag{6.135}$$

Given this evidence, the signal generated is:

$$w_k = h(e_k),\ k = 1,\ldots,n_H. \tag{6.136}$$

We can express the operations of the hidden layer in compact form as $w = h(\Theta_H z)$, where $w \in \mathbb{R}^{n_H}$ contains the output signals of the model and $\Theta_H \in \mathbb{R}^{n_H \times n_I}$ is the weighting matrix of the layer that maps the input signals z to the output signals w.

The *output layer* takes the signals of the hidden layer and can have one or multiple perceptrons (depending on how many output variables y we have). If there is only one output, the evidence is:

$$a = \sum_{k=1}^{n_H} \theta^O_k w_k + \theta^O_0. \tag{6.137}$$

The final output (model prediction) is $y = h(a)$, where $h(a)$ is a sigmoidal function if the output is binary (classification) or $h(a) = a$ if the output y is continuous (regression). We can express the operation of this layer in a compact form as $\hat{y} = h(\Theta_O w)$, where $\Theta_O \in \mathbb{R}^{n_H}$ is a weighting vector (matrix if we have multiple output variables) that maps the output signals of the hidden layer to the final output prediction.

The propagation of the input x through the layers to obtain the output can be expressed as a nested, *forward propagation* mapping:

$$\begin{aligned} \hat{y} &= h\left(\Theta_O h\left(\Theta_H h(\Theta_I x)\right)\right) \\ &= m(x, \theta), \end{aligned} \tag{6.138}$$

where $\theta = (\Theta_I, \Theta_H, \Theta_O) \in \mathbb{R}^p$ is a vector that contains all the parameters of the NN. The nested mapping highlights how the NN is a hierarchical structure that conducts sequential transformations of data. It is important to note that the same types of transformations are applied in each layer and that such transformations are quite simple (linear mixtures and activation); however, we will see shortly that such simple transformations can capture rich nonlinear behavior.

Input data points $x_\omega, \omega \in \mathcal{S}$ are propagated through the NN to obtain the predictions:

$$\hat{y}_\omega = m_\omega(\theta) = m(x_\omega, \theta), \; \omega \in \mathcal{S}. \tag{6.139}$$

We can use an MLE framework to estimate θ; for regression problems, we solve the problem:

$$\min_\theta \; L(\theta) = \frac{1}{2} \sum_{\omega \in \mathcal{S}} (y_\omega - m_\omega(\theta))^2 + \kappa \rho(\theta). \tag{6.140}$$

In ML, the objective function $L(\theta)$ is known as the *loss function*. For classification problems, we find θ that maximizes the log loss (as discussed in Section 6.4). It is also possible to use a Bayesian framework to incorporate prior knowledge; for instance, in the form of a regularizer $\rho(\theta)$.

To solve the training problem, it is necessary to compute derivatives of the loss function with respect to the parameters $\nabla_\theta L(\theta) \in \mathbb{R}^p$. The derivatives can be computed by using a technique called *backpropagation*; this technique marches in the backward direction across the layers of the network (from the output to the input). The ability to compute derivatives in a sequential manner (layer by layer) is key in building NNs that contain thousands to millions of hidden layers. The ability to scale NNs to have this many layers gave rise to the area of *deep learning*, and this was considered a major breakthrough in the field of ML.

The derivative information is used to progressively refine the parameter estimates using a family of algorithms known as *gradient descent*. In a gradient descent algorithm, the parameters are updated as:

$$\theta_{\ell+1} = \theta_\ell - \alpha \nabla_\theta L(\theta_\ell), \tag{6.141}$$

where ℓ is the iteration index and $\alpha \in [0, 1]$ is a step size. In the context of ML, the step size is known as the *learning rate*. Computing the gradient $\nabla_\theta L(\theta_\ell)$ when one has many observations S can be impossible (even with backpropagation). For instance, some advanced applications of NNs can contain millions to billions of observations. As such, there are many variants of gradient descent (e.g., stochastic gradient descent) that systematically select subsets of observations to compute approximations of the gradient $\nabla_\theta L(\theta_\ell)$.

We make the following important observations about the advantages and limitations of NNs:

- A key advantage of NNs is that they are completely data-driven (no mechanistic understanding of a system is needed). As such, these models are completely empirical (based on observations). This explains how humans are capable of making complex predictions and decisions without any explicit mechanistic understanding of a phenomenon. However, the fact that NNs ignore mechanistic knowledge can also be a disadvantage.
- The NN parameters have no physical meaning, and it is thus difficult to explain what they do and to convey prior knowledge about them (e.g., incorporate constraints).
- Deep NNs might need large amounts of data to estimate parameters (which can be on the order of thousands to millions).
- A typical issue with NNs is that of *overfitting*. Regularization terms (e.g., $\rho(\theta) = \lambda\theta^T\theta$) are often added to the loss function to reduce the number of parameters. Parameters of NNs are also often nonunique, but this can actually be an advantage, as it means that the NN is flexible (different values of the parameters give a good fit). However, conducting cross-validation is particularly critical in making sure that NNs generalize well to unseen data.
- It is extremely difficult (or impossible) to compute the Hessian matrix $\nabla_\theta^2 L(\theta_\ell)$ for an NN due to the large number of parameters. This complicates uncertainty quantification procedures (e.g., obtaining the covariance matrix of the parameters) and the use of Newton-based optimization algorithms (which are more efficient than gradient descent). For extremely large models, one can resort to using approximations of the Hessian matrix (e.g., Gauss-Newton) and in obtaining only its diagonal elements to approximate variances. Quantifying the uncertainty of parameters and predictions for large NNs is, in general, quite challenging.

We know that humans benefit from having mechanistic knowledge about the world. For instance, the predictions of an engineer designing a reactor become much better when they use a mixture of physical knowledge (thermodynamics and kinetics) and empirical knowledge. We also know that the human brain is capable of fusing physical knowledge with data-driven empirical knowledge. As we discussed for GP models, we can seek to merge this behavior by postulating *hybrid* models of the form:

$$Y = m(X,\theta) + NN(X,\theta), \tag{6.142}$$

where $m(X,\theta)$ is the physical model and $NN(X,\theta)$ is the neural network model. There are many other variants of hybrid model structures; for instance, we can replace some elements of a physical model with an NN. This type of model has the structure:

$$Y = m(X,\theta, NN(X,\theta)). \tag{6.143}$$

Note that this is a way to impose prior physical knowledge on the NN model. As a result, we expect that the NN model will require fewer parameters, and we expect the model to be more generalizable. There are many variants of NNs that aim to incorporate physical knowledge; one such variant is physics-informed NNs (PINNs). This family of NNs constructs the loss function in a way that it forces the NN to behave as a physical model (this can be seen as a form of regularization/penalization).

NNs can also be used as *surrogates* of complex physical models. This is beneficial when the simulation of a physical model (e.g., a process simulation) is time-consuming; by training an NN that mimics the physical model, we can drastically accelerate simulations. Other types of ML models such as GP models can also be used as surrogates.

Example 6.26 Neural Network Model for Gibbs Reactor

We now compare the performance of NN models against GP models. The NN model used here is simple, and it contains one layer with a couple of perceptrons (modeled as logistic functions). Figure 6.39 shows that the NN model fits the data extremely well, indicating that it is capable of capturing the high nonlinearity of the model. The fact that this can be done with one layer and a few perceptrons is quite remarkable.

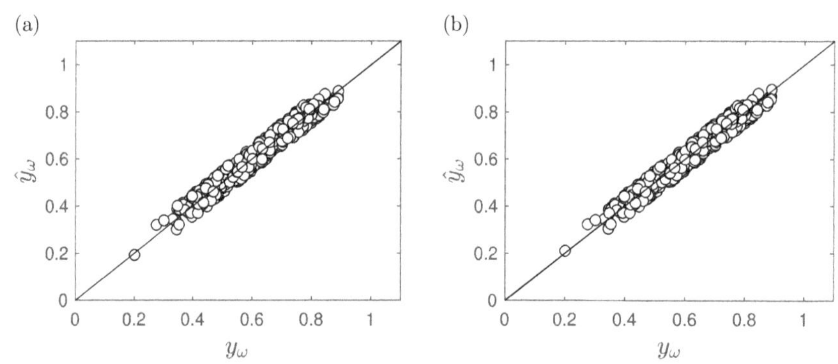

Figure 6.39 Fit of GP (a) and NN (b) models for Gibbs model.

Script to reproduce results: `ch6_gibbs_neuralnet` □

Example 6.27 NNs as Basis Function Models

The previous example indicates that a simple NN model can capture nonlinear behavior. So why are NNs so effective at capturing such behavior? An important observation is that NNs are sophisticated *basis function* models, as those discussed in Section 6.5; specifically, NNs use activation functions as basis functions.

To see this, we construct a simple NN model with one layer and N perceptrons; the perceptrons use sigmoidal activation functions of the form:

$$h(a_j) = \frac{1}{1 + e^{-a_j}}, \; j = 1, \ldots, n_I,$$

where $a_j = \theta^I_{j,1} \cdot x + \theta^I_{j,0}$ is the evidence of perceptron j. The NN output is a weighted mixture of the perceptron activations:

$$\hat{y} = \sum_{j=1}^{n_I} \theta^O_j h(a_j).$$

The NN model has a total of 18 parameters (corresponding to the weights of the evidence and of the output layer).

The fit of this NN model to observations generated from the nonlinear function $y = \sin(4x) + \sin(8x) + \exp(-x)$ is shown in Figure 6.40. We can see that the NN model can capture the nonlinearity fairly well. In Figure 6.41, we present the sigmoidal

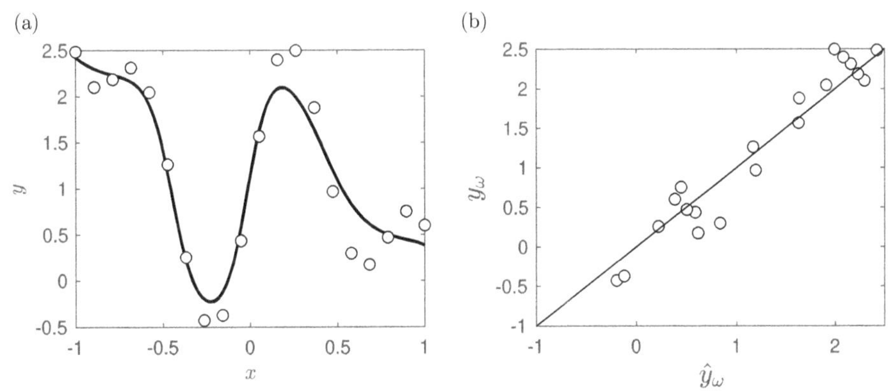

Figure 6.40 Fit of NN model to data (a) and parity plot (b).

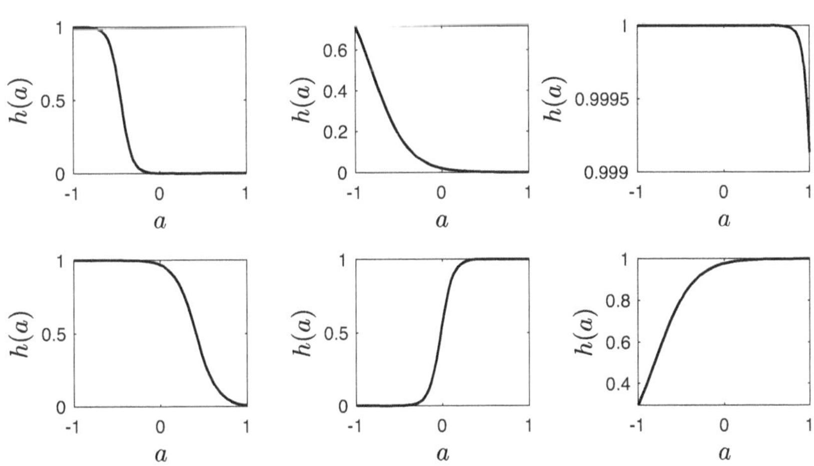

Figure 6.41 Sigmoidal functions of NN for example model.

functions (with estimated parameters) of the NN model. From this, we can see that the basis functions capture a wide range of behavior. We can thus see that the NN model captures complex behavior by mixing basis functions of different characteristics. The key difference with standard basis models (as those covered in Section 6.5) is that the shape of the basis functions is determined by the NN training process. In standard basis models, on the other hand, the basis functions are fixed, and we only estimate the mixing weights; as such NN models are more flexible/adaptive.

When we inspect the Hessian of the NN model $\mathcal{H}(\hat{\theta})$, we find that this has five eigenvalues that are close to zero. This indicates that the surface of the loss function $L(\hat{\theta})$ is flat at the estimated parameters $\hat{\theta}$ (the parameters of the NN model are not unique). This also indicates that the parameters will be exhibiting high uncertainty. This highlights that NN models are highly flexible but also tend to be overparameterized and lead to overfitting.

Script to reproduce results: `ch6_neuralnet_basis` □

6.7.2 Convolutional Neural Networks

Convolutional neural nets are NNs that are specialized to take *structured data* as inputs in the form of vectors $\mathbf{x}_\omega$ (for 1D) or matrices $\mathbf{X}_\omega$ (for 2D) to make output predictions y_ω (e.g., classify or predict a value). Input data vectors can be a sequence of numbers that encode a certain pattern, and matrices can be images or other objects containing spatial patterns. For example, we can train a CNN to predict the temperature of a furnace from a picture captured by an infrared (IR) camera, to predict a property of a molecule by looking at the sequence of its atoms, or to predict the failure of a system from a time series. For convenience, we will discuss the 2D case (since the 1D case is a special type of matrix). We will also see that CNNs can be generalized to process data that includes color images (by processing different channels associated with images).

Convolutional NNs are designed to extract patterns from the data using convolution operations, and this is what makes them different from regular NNs. For instance, we could be tempted to take an image $\mathbf{X}_\omega$, convert it into a vector $\text{vec}(\mathbf{X}_\omega)$, and feed this into an NN to make a prediction. However, this approach can hide spatial patterns present in the data. Convolutional NNs will extract patterns from the data by using different types of convolution filters, and the extracted features will then be summarized/reduced and used as evidence in perceptrons that generate activation signals. This process seeks to mimic how our brain makes predictions based on perceived patterns (e.g., if I perceive a rounded orange object, I can predict that it is a basketball).

In a CNN, filter matrices and the parameters of the activation functions are learned from data. In other words, it is not necessary for us to specify filters, as the CNN will automatically determine/learn the filters (and associated patterns) that best predict the label associated with the input image. This is important

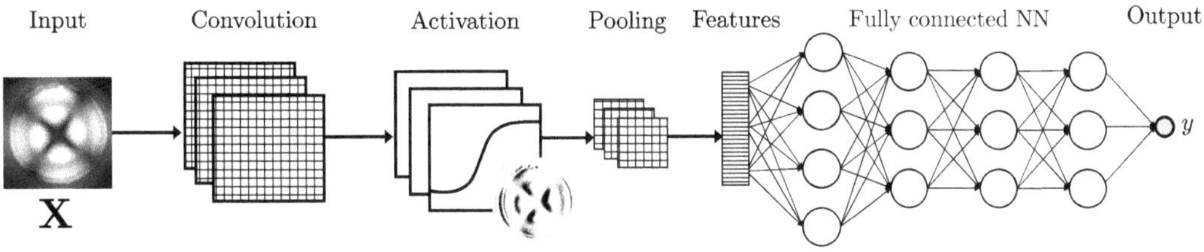

Figure 6.42 Sketch of a convolutional neural network architecture.

because it is not always obvious what features of an image or object best explain its nature.

The components of a CNN architecture are: input layer, hidden layers, and fully connected layer (see Figure 6.42). Input and hidden layers include convolution, activation, and pooling sublayers, while the fully connected layer contains a standard NN (with its own inputs, hidden, and output layers). The input and hidden layers seek to extract features from the input matrices (raw data), and these features are passed to the fully connected layer to make a prediction.

To simplify notation, we use $\mathbf{X}$ as the input of observation $\mathbf{X}_\omega$; you can think of this input as a grayscale image. The input is fed into an *input layer* composed of $j = 1,\ldots,n_I$ perceptrons (units). Each unit performs convolution, activation, and pooling operations. The evidence in the unit j is:

$$\mathbf{A}_j = \mathbf{X} * \Theta_j^I + \theta_j^I,\ j = 1,..,n_I. \tag{6.144}$$

Here, Θ_j^I is the convolution filter (a matrix) of unit j and θ_j^I is the bias of the unit. The filter matrices are usually small matrices (3×3, 5×5, or 7×7). In general, large filters tend to capture global features of a matrix (its scanning window is larger), while smaller filters capture local features. As part of the architecture, we can also decide how the scanning window of the filter is shifted by setting an option known as the *stride*. A stride value of 1 shifts the moving window by one entry at a time, while a value of 2 shifts by two entries and so on. One can think about the stride as the level of detail with which we scan a matrix. Increasing the stride value allows us to scan matrices faster to accelerate computations, but this might come at the expense of missing some features.

The output of the convolution in a given perceptron j is the evidence $\mathbf{A}_j$ (a matrix). This evidence is activated to produce an output signal:

$$\mathbf{Z}_j = h(\mathbf{A}_j),\ j = 1,..,n_I, \tag{6.145}$$

where $h(\mathbf{A}_j)$ is an activation function that is applied component-wise to the matrix $\mathbf{A}_j$ (to each entry) to obtain the activation matrix $\mathbf{Z}_j$. The activation function is typically a sigmoidal, ReLU, or linear function.

To accelerate computations in subsequent layers (and reduce computer memory), the activation matrices $\mathbf{Z}_j$ are reduced in size by using an operation called *pooling*. Pooling looks at small windows of the matrix (typically 2×2) and reduces the values in the window to a single value by taking the maximum value (called max-pooling) or the average value (called average pooling). We denote this reduction procedure as:

$$\bar{\mathbf{Z}}_j = p(\mathbf{Z}_j),\ j = 1,..,n_I, \tag{6.146}$$

where $p(\mathbf{Z}_j)$ is the pooling function and $\bar{\mathbf{Z}}_j$ is the reduced matrix.

The *hidden layer* is composed of $k = 1,\ldots,n_H$ perceptrons (here we describe an architecture with a single hidden layer, but multiple layers are possible). The evidence in perceptron k is:

$$\mathbf{E}_k = \sum_{j=1}^{n_I} \Theta^H_{k,j} * \bar{\mathbf{Z}}_j + \theta^H_k,\ k = 1,\ldots,n_H. \tag{6.147}$$

Given the evidence, the signal generated is:

$$\mathbf{W}_k = h(\mathbf{E}_k),\ k = 1,\ldots,n_H, \tag{6.148}$$

and this is reduced using pooling as $\bar{\mathbf{W}}_k = p(\mathbf{W}_k)$. The matrices $\bar{\mathbf{W}}_k$ are flattened (converted into vectors) as $\mathbf{w}_k = \text{vec}(\bar{\mathbf{W}}_k)$ and stacked into a *feature vector* $\mathbf{w} = (\mathbf{w}_1,\ldots,\mathbf{w}_{n_H})$ (this process is known as *flattening*). We can summarize the application of convolutions, activation, and pooling in the input and hidden layers as:

$$\mathbf{w} = f(\mathbf{X},\Theta,\theta), \tag{6.149}$$

where Θ contains all convolution filters and θ contains all biases. The goal of the function f is to extract feature information $\mathbf{w}$ from the raw data $\mathbf{X}$.

The *fully connected layer* is a standard NN that takes the feature vector $\mathbf{w}$ from the last hidden unit and maps this to a final prediction output. The NN contains its own inlet, hidden, and output layers. The mapping of the NN from the feature vector to the output is summarized as:

$$\hat{y} = NN(\mathbf{w},\gamma), \tag{6.150}$$

where γ are the parameters of the NN layers. This layer can generate an output that is discrete (classification) or continuous (regression).

Given all the parameters $\beta = (\Theta,\theta,\gamma)$, the CNN propagates an input $\mathbf{X}_\omega$ through all layers to generate an output:

$$\hat{y}_\omega = m_\omega(\beta) = m(\mathbf{X}_\omega,\beta). \tag{6.151}$$

We use an MLE framework to estimate β. For regression problems, we solve:

$$\min_\beta\ L(\beta) = \frac{1}{2}\sum_{\omega\in\mathcal{S}}(y_\omega - m_\omega(\beta))^2 + \kappa\rho(\beta). \tag{6.152}$$

As with standard NNs, we compute derivatives $\nabla_\beta L(\beta)$ with respect to all parameters using backpropagation. A gradient descent algorithm uses this information to learn the parameters that minimize the loss function.

We highlight the following important aspects of CNNs:

- Convolutional NNs are conceptually similar to NNs; the key difference is that CNNs preprocess the input data to extract features prior to making a prediction. As such, CNNs are significantly more difficult to train than standard NNs (convolutions involve repeated matrix operations).
- Convolutional NNs often have a large number of parameters; this is particularly evident from the fact that the parameters are matrices (the filter matrices). For example, a single 3×3 filter has nine parameters. As such, CNNs are prone to *overfitting*, and conducting cross-validation is also critical to make sure that the model generalizes well (to inputs not used for training.
- Convolutional NNs can be generalized to take input objects that have multiple channels. For example, a color image is the combination of three matrices corresponding to red, blue, and green (RGB) channels. That is, a color image is an object $\mathbf{X}_\omega$ that contains three matrices $\mathbf{X}_\omega^R, \mathbf{X}_\omega^B$, and $\mathbf{X}_\omega^G$. This indicates that an input object $\mathbf{X}_\omega$ is actually a *tensor*. Think of a tensor as a cube in which each slice is a matrix; in the case of a color image, the tensor has three slices. Convolutional NNs process tensor objects by applying filters to individual channels and then merging the results in the feature vector. This allows CNNs to extract information from color (which is an important feature that the human vision uses to make predictions).
- Convolutional NNs that use matrices as input data and filters are called 2D CNNs. If the data and filters are vectors, they are called 1D CNNs (which are used to process sequences, such as time series). It is possible to generalize CNNs to higher dimensions; for instance, in a 3D CNN, the filters are not matrices but tensors (which are used to process volumes, such as medical images and video). In 3D CNNs, one scans for features over volumes (while in 2D CNNs, we search for patterns over flat surfaces). The principles behind high-dimensional CNNs are similar to those of 2D CNNs; what is important is to recognize how the input data object should be provided to the CNN.
- After training the CNN, one can analyze the filter matrices (Θ) to determine what patterns they are searching for in the images. Some knowledge of how typical filters (e.g., Gaussian, Sobel, and Laplacian) work can be useful in explaining what information the CNN is looking for; however, it can be difficult to make sense of the filter matrices themselves. As such, one often uses advanced techniques such as saliency maps that aim to highlight aspects of the input data that the filters are activating.
- The repetitive application of convolutions in CNNs makes them computationally intensive. Techniques such as Fourier transforms are used to accelerate computations. In many cases, it is also necessary to use parallel computing architectures to accelerate the training process.

Example 6.28 Classification of Flow Cytometry Data

Endotoxins generated from bacteria can be detected by using an emulsion that contains small droplets that are filled with an LC [7]. The endotoxin molecules bind to the surface of the droplets and generate an optical change in the configuration of the LC inside the droplets. The change in configuration of the droplets can be measured using a flow cytometer; here, thousands of droplets are passed through the cytometer, and this generates a light-scatter field (see Example 3.3). The *shape* (spatial pattern) of the scatter field changes with the concentration of the endotoxin (which changes the number of droplets that have a new configuration) and also with the type of endotoxin (different bacterial species generate different endotoxins, and these induce different configurational changes).

We are interested in training a CNN that automatically analyzes light-scatter data to detect the presence of an endotoxin in a solution and to determine the bacterial species that the endotoxin is associated with. These problems can be cast as classification problems. We begin by training a CNN that aims to classify the presence or absence of an endotoxin in a solution; to do so, we collect 98 observations (49 with no endotoxin present and 49 with endotoxin present). In Figure 6.43, we present nine examples of scatter fields obtained for solutions that contain (labeled as radial) and do not contain (labeled as bipolar) an endotoxin. We can see that the presence of the endotoxin creates a perceptible change in the shape of the scatter field. Our goal is to train a CNN that automatically detects differences in the shape of the field. Here, we train a CNN with a single hidden layer that contains a convolutional layer with a single 3×3 filter (this is a simple CNN). We use a ReLU function to activate the convolved images. We use 30 data samples to train the CNN and 68 data samples to validate that the CNN predicts well. In Figure 6.44, we show the confusion matrix for the training and validation set; here, we can see that the CNN perfectly classifies the scatter fields (it achieves 100% accuracy for training and validation). This indicates that the spatial pattern of the scatter fields contains significant information to tell them apart.

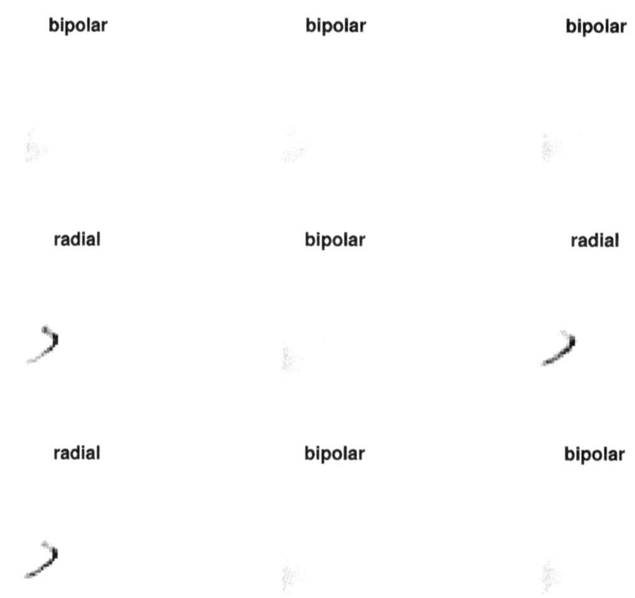

Figure 6.43 Examples of scatter fields for contaminated/clean solutions.

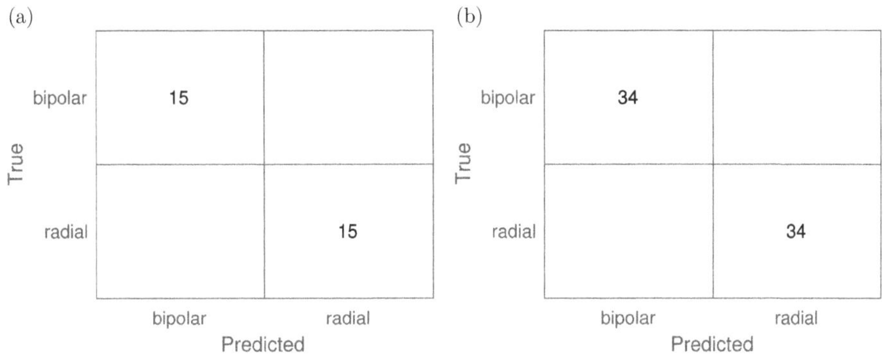

Figure 6.44 Confusion matrix for the training set (a) and validation set (b).

When we inspect the filter matrices, we can see that these do not have a structure that we are familiar with (e.g., Sobel filter or Gaussian filter). To determine what patterns the filter matrices are looking for, we visualize the activated images that result from convolution. The activated images for a given filter are shown in Figure 6.45 (these activations correspond to the fields shown in Figure 6.43. We can see that the CNN filter highlights the shape of the scatter field. Specifically, fields with a bipolar shape (no endotoxin present) occupy a larger volume.

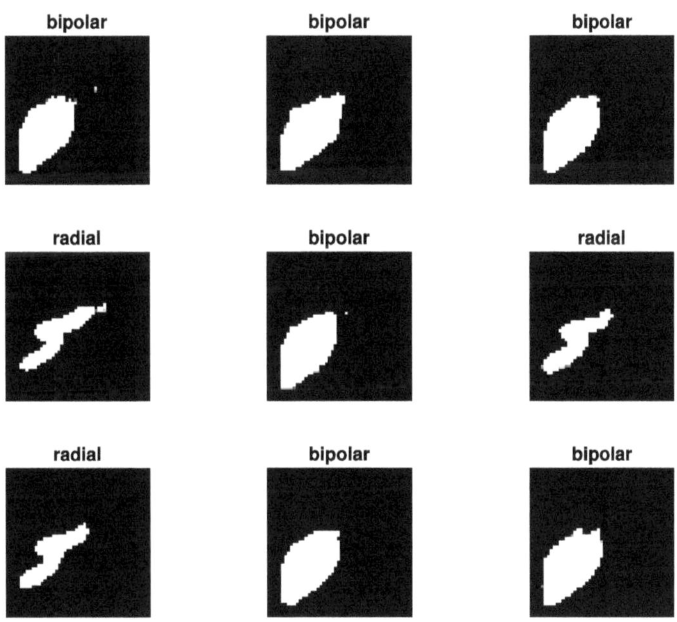

Figure 6.45 Activations of images shown in Figure 6.43 using a CNN filter.

Scripts: `ch6_cnn_endotoxins_caseI` □

Example 6.29 Detection of Water Contaminants

Sodium dodecyl sulfate (SDS) is an anionic surfactant widely used in industrial and consumer products, such as detergents, shampoos, and cosmetics [19]. Unlike certain industrial chemicals, SDS is not considered highly persistent in the environment due to its capacity to biodegrade. Typically, SDS decomposes within days to weeks under aerobic conditions through microbial processes. Despite its biodegradability, the extensive use of SDS raises environmental and health concerns, especially in regions with high usage or discharge where it can temporarily accumulate. Exposure to SDS can lead to health issues, such as skin and eye irritation, respiratory problems, and potential adverse effects on aquatic life. Consequently, monitoring SDS is vital for managing health risks and implementing effective remediation measures. Ongoing research aims to enhance SDS detection methods, which have traditionally relied on sophisticated laboratory techniques.

An innovative method for detecting SDS in water solutions involves using micrographs of LC droplets. Recent studies have shown that LC droplets undergo optical changes when exposed to SDS; these responses encode rich information in the form of spatial, topological, and color patterns. This unique property provides a basis for developing low-cost LC-based sensing methods for SDS.

Here, we are interested in training CNNs that automatically analyze LC droplet data to detect the presence of SDS. We input the micrographs in the form of color images; such color images are represented as a three-channel tensor. We begin by training a CNN that aims to classify the presence or absence of SDS in water (a binary classifier); to do so, we collect 100 observations (50 with no SDS present and 50 with 1 mM SDS present). In Figure 6.46, we present nine example micrographs of LC droplets.

We can see that the presence of the SDS creates a perceptible change in the shape of the droplet. Our goal is to train a CNN that automatically detects these differences. Here, we train a CNN with a single hidden layer that contains a convolutional layer with a single 3×3 filter (this is a simple CNN). We use a ReLU function to activate the convolved images. We use 80 data samples to train the CNN and use 20 data samples to validate that the CNN predicts well.

In Figure 6.47, we show the confusion matrix for the training and validation set; we can see that the CNN perfectly classifies the LC droplets (it achieves 100% validation accuracy). This indicates that the spatial pattern of the LC droplets encodes significant information to tell them apart.

To identify what patterns the CNNs are looking for, we visualize the activated images that result from convolution with a selected filter (see Figure 6.48). These activations correspond to the LC droplets shown in Figure 6.46. We can see that the filter selected highlights the shape (topological defect) of the droplet.

We now train a CNN to distinguish a much lower concentration of SDS (0.05 nM); this is done in order to explore the detection limit of the contaminant. Here, we use the previous 50 observations (LC micrographs) without SDS present and collect 50 new observations of LC micrographs treated with 0.05 nM SDS. We again use 80 observations for training and 20 for validation. In Figure 6.49, we present nine micrographs of LC droplets; we can see that it is quite difficult to distinguish them with our eyes.

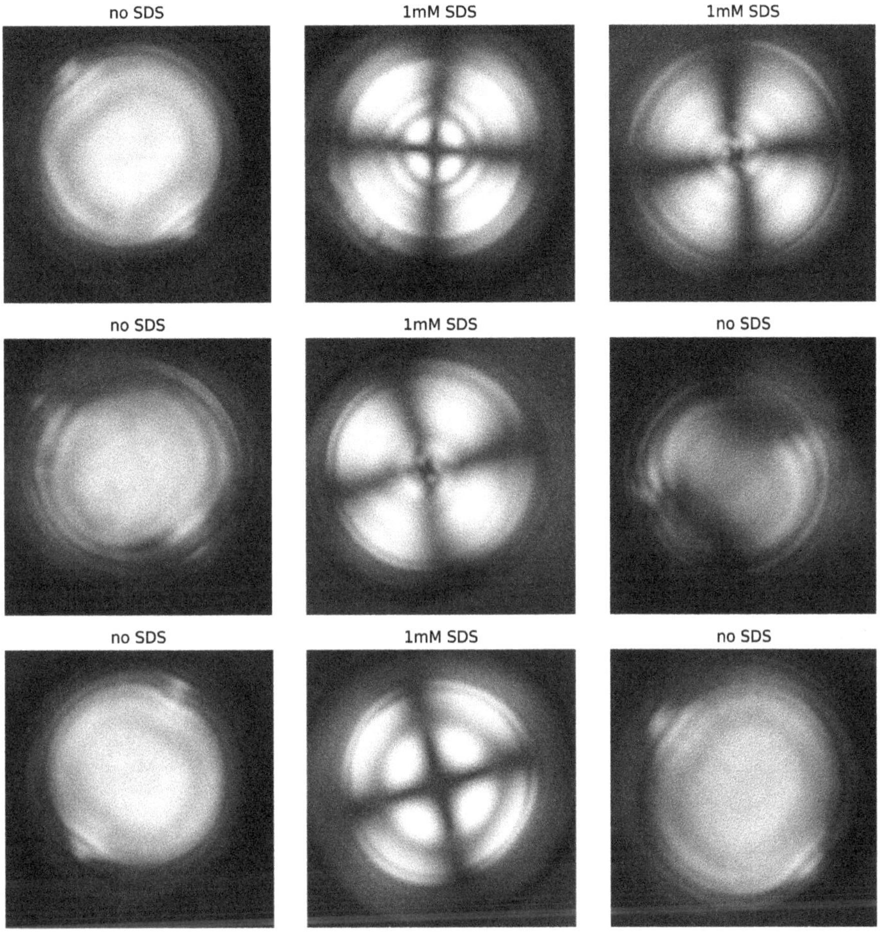

Figure 6.46 Example micrographs of liquid crystal droplets when exposed or not to a contaminant.

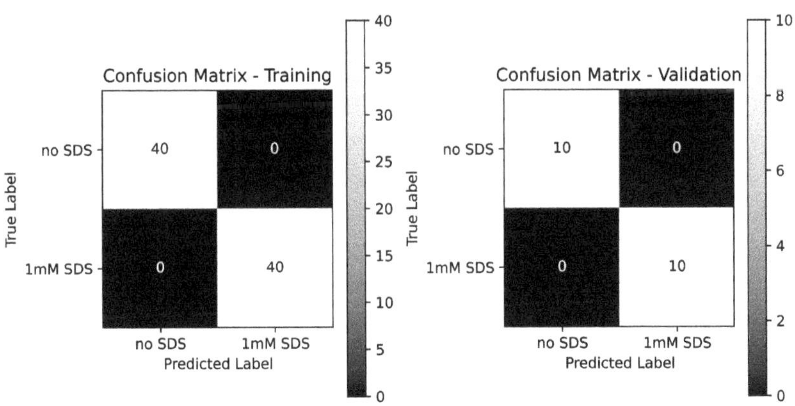

Figure 6.47 Confusion matrix for the training and validation sets for CNN.

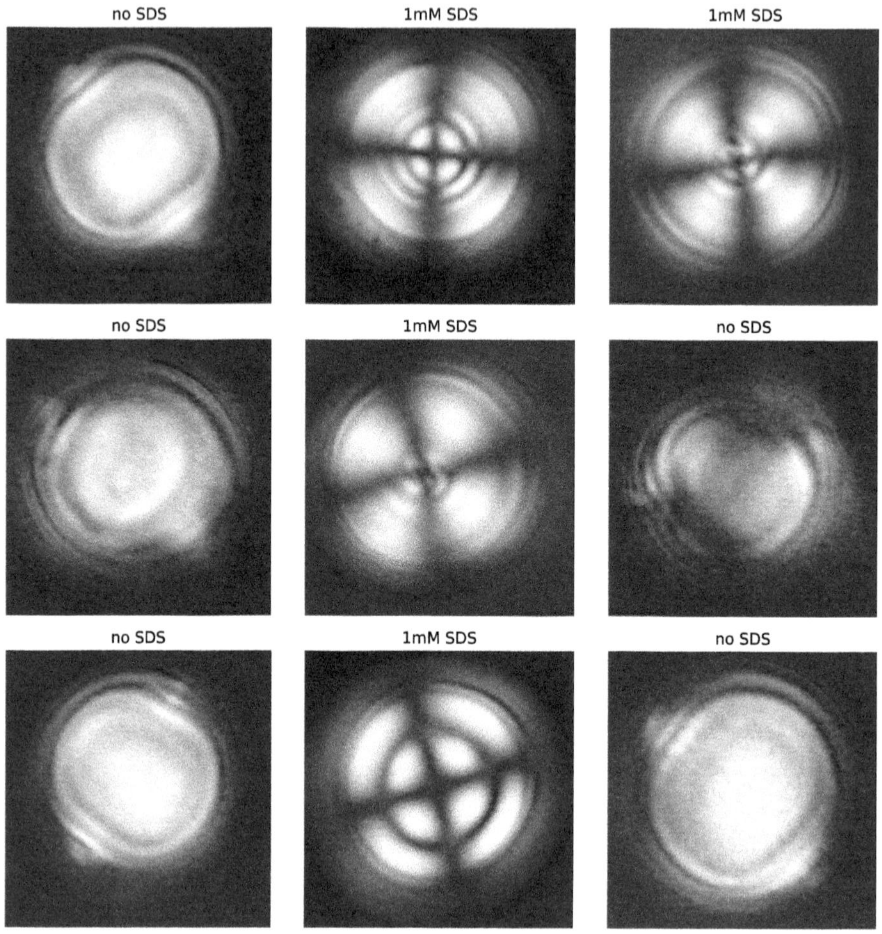

Figure 6.48 Activated images with a selected filter showing patterns that the CNN is focusing on.

Specifically, it would be quite difficult for a human to predict the LC droplet treated with 0.05 nM SDS by looking at the image itself.

To automatically classify the images, we train a CNN with a single hidden layer and two 3×3 filters, and we use a ReLU function for activation. In Figure 6.50, we show the confusion matrix for training and validation; the CNN achieves a prediction accuracy of 80% for validation. Although the predictions of the CNN are not perfect, it is remarkable that it can achieve such high accuracies. This indicates that the CNN is detecting nonobvious patterns in the LC micrographs. The reduction in accuracy is a clear indication that detecting contaminants at lower concentrations is significantly more challenging.

Scripts for reproducing the results: `ch6_cnn_sds_caseI` and `ch6_cnn_sds_caseII` □

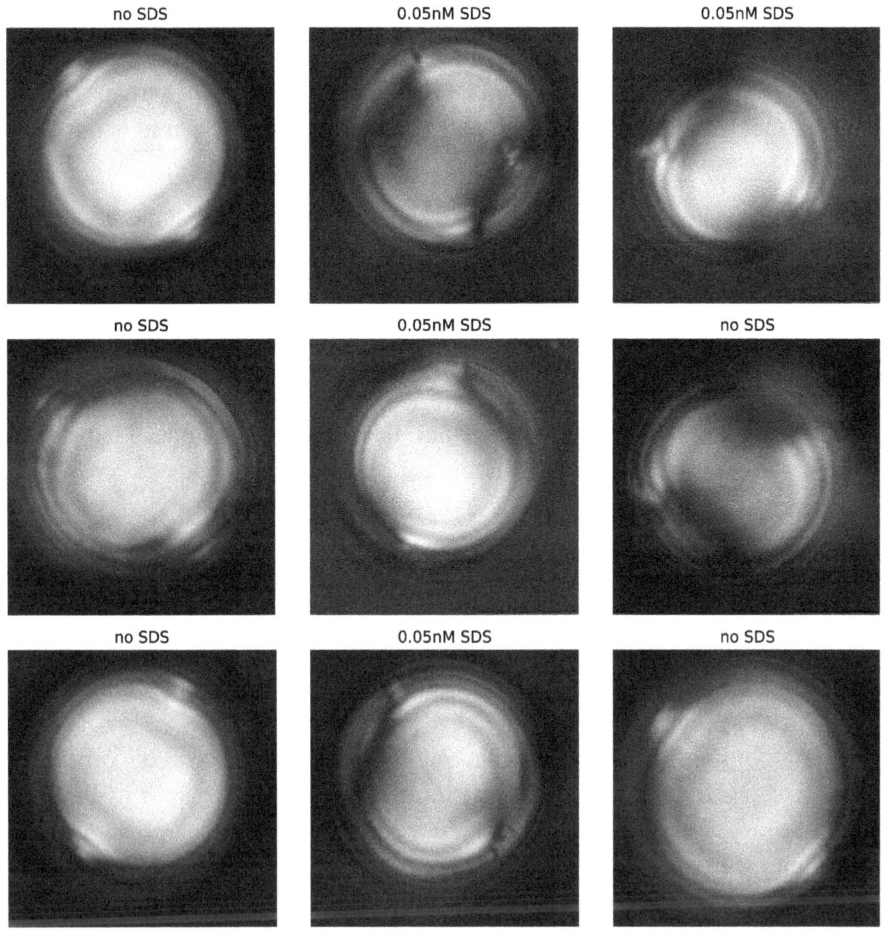

Figure 6.49 Example micrographs of liquid crystal droplets when exposed (or not) to a contaminant at low concentration.

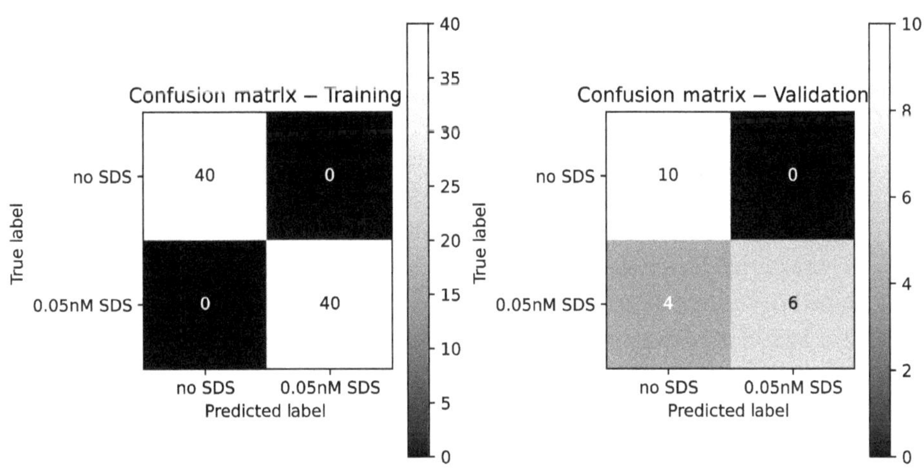

Figure 6.50 Confusion matrix for the training and validation sets for CNN for contaminant at low concentration.

6.8 Summary

In this chapter, we have discussed how to apply principles of statistics to advanced techniques of data science and ML. The following are important take-home messages:

- One can reduce the dimension of large datasets using techniques such as PCA and SVD. These dimensionality reduction techniques aim to transform our data from a physical space into a smaller latent/information space that facilitates data visualization and processing. In PCA, information is quantified in the form of variance.
- Principal component analysis and SVD decompose our input data matrix using eigenvalues and singular values; these values (and associated eigenvectors) encode complementary information of our input data and are ordered based on information content. We can use these values to determine what information we want to preserve (or not) when reducing our data.
- We have seen that PCA and SVD can be used to preprocess the data that is fed into regression models. This is very useful in identifying data redundancies, which can originate from linear dependencies and strong correlations between variables. The preprocessing of data can also be used to identify dominant variables of our regression models.
- Convolution is a powerful technique that can be used to transform our data in order to extract valuable information (features or patterns) or to eliminate noise (filtering). The transformation that we perform is influenced by our choice of a filter; filters can be designed to extract gradients of light from images or to conduct smoothing.
- It is also possible to transform our data to a frequency space using a technique known as the Fourier transform. This technique is useful for analyzing images and time-dependent data. We have also seen that operating in the frequency domain can facilitate convolution operations; specifically, we have seen that the convolution theorem allows us to conduct image convolution in the frequency domain by using simple products.
- Machine learning provides powerful techniques for building predictive models. We have seen that we can generate highly flexible nonlinear models that do not require parameters (known as non-parametric models) by using kernel functions. These kernel functions can be seen as generators of basis functions that are combined to capture relationships between input and output variables. GP modeling is a kernel modeling technique that generates models by treating the model as a random function. We have seen that GP models can capture strong nonlinearities and facilitate quantification of prediction uncertainty, which can be used to guide data collection.
- Neural network models are highly flexible parametric models that can capture any type of relationship between input and output variables. This is enabled by using a hierarchical network that conducts sequential data transformations (via

variable mixing and activation). We have seen that these simple operations are sufficient to capture highly complex nonlinear behavior. One of the reasons for this is that activation functions are flexible functions that act as basis functions.

- Convolutional neural networks are a variant of NNs that can process input data that appears in structured forms (e.g., images and tensors). This enables the processing, for instance, of image and video data. CNNs aim to mimic how human vision extracts information from our senses to make predictions. Specifically, CNNs aim to extract features/patterns from data to make predictions.

The discussion of this chapter only scratches the surface of the fields of data science and ML. At the time this is written, these fields continue to grow at an exponential rate (as predicted by one of my favorite movies, Terminator 2). There are some excellent introductory textbooks devoted to these topics, including [57, 58, 59]. There is a broad range of important techniques that are not discussed here, such as autoencoder networks for dimensionality reduction [60], graph NNs and geometric deep learning [61], reinforcement learning [62], and random forest models [63]. However, many of the fundamentals covered in the book can help understand how these techniques work and their advantages/limitations; for instance, autoencoder networks are generalizations of PCA.

6.9 Exercises

Exercise 6.1. Plastic Spectra Clustering Using PCA.
Plastic waste is a significant and persistent environmental issue. It has been estimated that per-capita generation of plastic waste in the United States is 200 lbs, and less than 10% of plastic waste is currently recycled. One of the primary challenges in plastic recycling is the difficulty in characterizing plastic effectively, particularly when present in waste mixtures.

Infrared (IR) spectroscopy is a technique that can be used to characterize mixed plastic waste [20]; however, the spectra signals obtained from the instrument are noisy and have significant overlap, making classification difficult (see Figure 6.51). Use the data provided in `plastic_ir.mat` and PCA to determine if there is enough separation between the spectra. Based on this, comment on whether IR spectra can be used to reliably identify plastic types. Comment also on what plastic types could be more difficult to distinguish.

Exercise 6.2. Multi-class Classification Using Support Vector Machine.
We would like to build a classification model known as an SVM to classify different types of plastics based on the IR spectra data provided in `plastic_ir.mat`. Address the following:

(a) *Data splitting.* Shuffle your dataset randomly and divide it into training and testing sets. The training set refines the parameters of the model, while the testing set evaluates its generalization. Use 60% of your data points for training and 40% for testing.

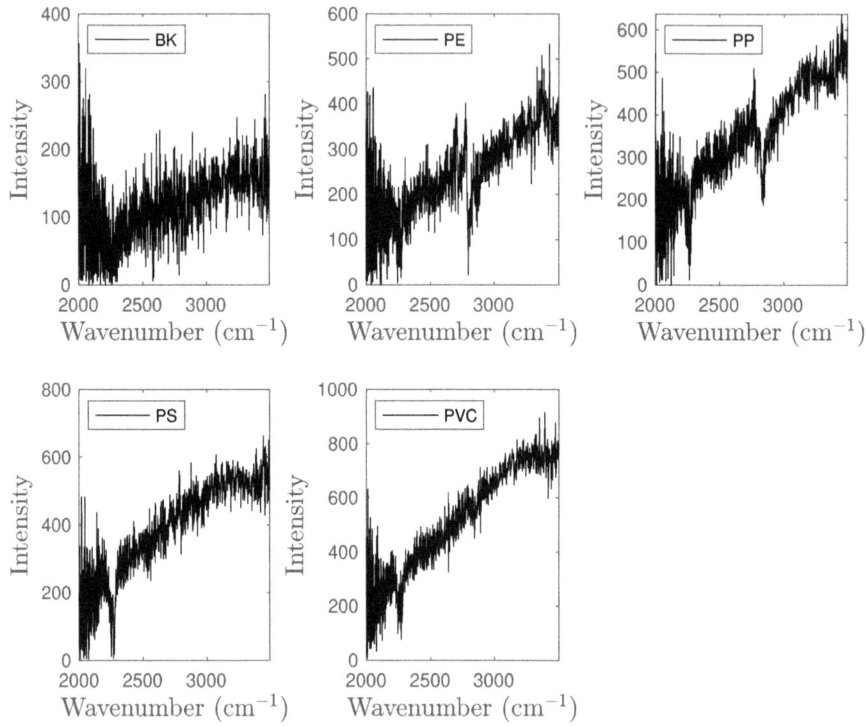

Figure 6.51 Examples of infrared spectra of different plastic types, including polystyrene (PE), polypropylene (PP), polyethylene (PS), and polyvinyl chloride (PVC). BK is the IR spectrum of the background.

(b) *Data scaling.* Before building the model, data scaling, such as standardization, is required. Without scaling, inputs/features with different scales/magnitudes may dominate the model. Scale your data to ensure that all feature variables have a zero mean and a standard deviation of one.

(c) *Model training and testing.* Build a *multi-class* classification model for different types of plastics. Compute the prediction accuracy and F_1 score for the test data. Explain the difference between accuracy and the F_1 score.

(d) *Result visualization.* Obtain a confusion matrix to visualize the SVM performance (based on the test data). Comment on how effective the SVM model is in predicting the plastic-type class from IR data.

Exercise 6.3. Multi-class Classification Using CNN.

Convolutional neural networks are useful for classifying data; in this problem, we aim to use 1D CNNs for classifying plastic waste based on the IR spectra provided in `plastic_ir.mat`. Address the following:

(a) *Data splitting and scaling.* Use a random split of your data into a training and testing set and perform scaling. Follow the same procedure used for building the SVM model of the previous exercise. Use 20% of your training data for validation.

(b) *Training accuracy and loss.* Design a 1D CNN that contains two convolution layers, that uses ReLU activation functions, and that uses average pooling. Plot the evolution of training accuracy and loss for the CNN with respect to the number of iterations. Comment on how many iterations the training process takes to converge.

(c) *Result visualization.* Obtain a confusion matrix to visualize the CNN performance (based on test data). Compare its performance with the SVM model of the previous exercise.

Exercise 6.4. Water Electrolysis Modeling.
Hydrogen is used in many chemical manufacturing processes and is currently being explored as an energy carrier. A prominent method for producing hydrogen is through water electrolysis, where electrical power is used to split water into hydrogen and oxygen. The productivity of electrolyzers depends on temperature and power [64]. You have data for hydrogen production at different temperatures and power availability over time, as shown in the figure below.

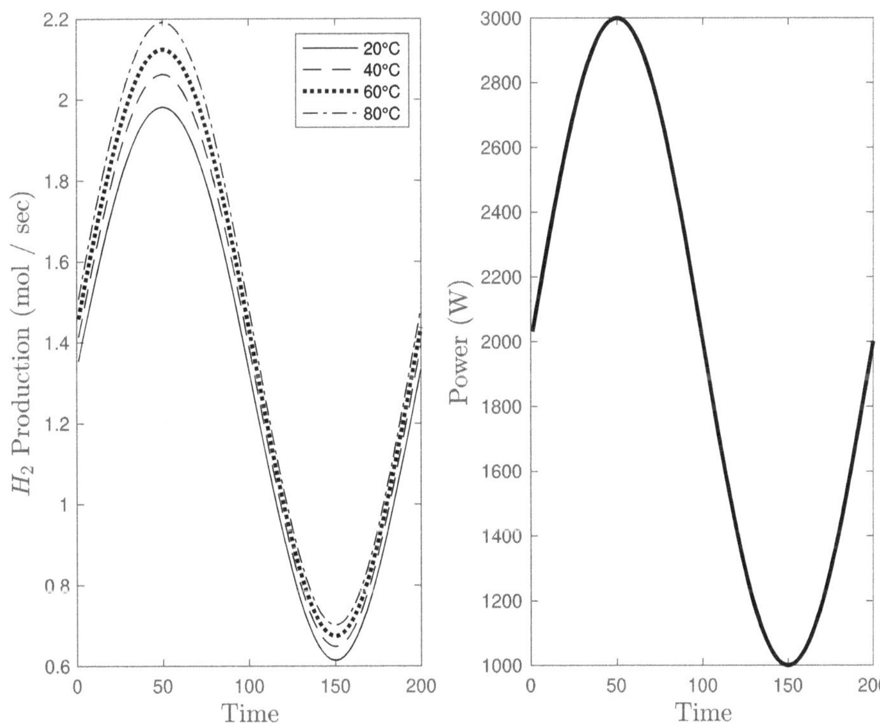

Load the data in `h2_data.csv`; the first eight columns correspond to the hydrogen production (in mol/sec) at temperatures of 20–90°C in 10-degree increments (i.e., column 1 is at 20°C, column 2 at 30°C, and so on). The last column corresponds to the available power (in Watts – W). Train an NN with one hidden layer (and 10 neurons in this layer) to predict hydrogen production as a function of power and temperature. Keep out the 50°C dataset for testing. Is your NN able to predict the productivity of the electrolyzer? Report the mean squared error for the 50°C dataset.

Exercise 6.5. Optimal Power Asset Placement.
As owners of a power asset company, you are responsible for investing in various types of assets (a combination of generators that produce power and devices that consume power). These assets can be used to sell and buy electricity in electricity markets at strategic times and locations. The profitability of your investment decision depends on electricity prices, which are random and depend on location. You want to make investments at five different locations, in a way that maximizes your total profit while minimizing the risk associated with price volatility/variability. The electricity price at each location is an RV; the prices across different times are assumed to be uncorrelated, but the prices across different locations are correlated. In other words, prices are not correlated in time but are correlated in space.

In this exercise, we will learn how covariance information and PCA can be used in investment decisions. Specifically, imagine that we want to find how much power should be injected or withdrawn at the five different locations; this is modeled using the allocation/loading vector $w \in \mathbb{R}^5$. Each entry of this allocation vector can be positive (power selling) or negative (power purchasing). The total profit is defined as $P = w^T X$, where X is the vector of electricity prices at the five locations. Because the prices are random, we have that the total profit is random.

One can show that the variance of the total profit (the risk) for a given allocation w is $\mathbb{V}[P] = w^T \Sigma w$, where $\Sigma = \text{Cov}[X]$ is the covariance of the prices. One can also show that the expected (mean) profit of an allocation w is $\mathbb{E}[P] = w^T \mathbb{E}[X]$.

Load the data in `asset_price_data.dat`; this is a matrix that contains hourly electricity prices at five locations and over 720 hr. This data was obtained from the California's power grid operator (CAISO) [21]. Use this data to address the following:

(a) Compute the mean price at each location and the covariance matrix. Comment on any correlations observed.
(b) The covariance matrix contains information about the spatial correlation of prices, and thus, we can exploit it to invest in different locations so as to minimize the profit risk. One way to do so is to find the eigenvectors and eigenvalues of the covariance matrix. Each eigenvector w gives an allocation vector and associated risk, given by the eigenvalue $\lambda = w^T \Sigma w$. Find all eigenvalues and eigenvectors of the covariance matrix and compute the risk and the mean profit associated with each allocation.
(c) How do risk and expected profit change with the different allocations? What is the best way to invest if one wants to maximize expected profit (should you invest in generators, loads, or both)? What is the best way to invest if one wants to minimize risk (should you invest in generators, loads, or both)?

Exercise 6.6. Process Drift.
You are given a data history of 400 time points for a process that is monitored by using five operating variables. Operators of the process seem to believe that these variables face occasional drifts from their normal operation and thus the process

exhibits multiple operating modes. These modes cannot be easily identified from direct inspection of time series data.

Use the historical dataset contained in `datadrift.dat` to address the following:

(a) Visualize the time series data of the five operating variables. Can you observe any operating modes?
(b) Implement PCA (do *not* use built-in functions in software packages) to project the data into an information space. Report the eigenvector and eigenvalue matrix.
(c) Define the total information as the sum of all eigenvalues. Determine % of the total information contained in every principal component.
(d) Use the first two dominant principal components to determine the number of operating modes of the process.
(e) Use sparse PCA (you can use any available package for this) to determine the two variables that dominate the first principal component and the two variables that dominate the second principal component. These variables are the ones driving a change in the modes (the ones drifting).

Exercise 6.7. Developing a Surrogate Model for a Chemical Process.
An engineer has developed a detailed chemical process simulator to predict how temperature and pressure conditions of a reactor affect overall process profit. Unfortunately, the model is complex and requires significant computational resources to run. The engineer would like to develop a surrogate model that is faster to simulate, and they are particularly interested in developing a GP model. The detailed process model has been run to obtain the dataset in `reactor_data.mat`; column 1 is temperature (K), column 2 is pressure (bar), and column 3 is profit (USD). Use this data to address the following:

(a) Use the data to generate a contour plot to visualize the effect of temperature and pressure on profit. Comment on how well the data is covering the domain of the temperature (673–823 K) and of the pressure (200–300 bar).
(b) Train a GP model using the available data; discretize the temperature and pressure domains using 20 equidistant points and predict profit at all these points. Generate a contour plot using the model predictions and determine the operating conditions that maximize profit.
(c) Generate a contour plot in which you visualize the uncertainty of the GP predictions. Comment on the overall magnitude of the uncertainty relative to the profit; can you trust the GP model? Comment also on how uncertainty is affected by the sparsity of data used for training the GP model; are there specific domains where uncertainty is lower/higher? Is the point of maximum profit in a region of high or low uncertainty? In what areas should the engineer collect more data using the detailed process model?

Exercise 6.8. Sulfide Removal.
Hydrogen sulfide is a potent chemical that can cause corrosion to industrial equipment and that has a highly unpleasant odor. A team of scientists has collected

experimental data to evaluate the potential of removing hydrogen sulfide from wastewater using a fluidized bed bioreactor [22].

The team has identified that the removal efficiency (% removed) is strongly affected by the operating conditions of the reactor, which include the upflow velocity (m/hr), hydraulic retention time (min), reactor operation time (hr), and pH.

The team would like to develop a data-driven model to predict removal efficiency as a function of operating conditions; the team is familiar with linear regression models but would also like to try more powerful NN models.

Use the data provided in `sulfide_data.csv` (the columns are upflow velocity, hydraulic retention time, reactor operation time, pH, and % removed) to address the following:

(a) Construct a linear regression model and determine the RMSE that the model can obtain. For the purpose of this study, it is OK to use all the data for training the model.
(b) Construct an NN model and play around with the design of the architecture (number of layers, the number of neurons per layer, and the type of activation functions). Determine a design that minimizes the RMSE of the NN. Compare how this RMSE value compares to the value obtained with the linear regression model. For the purpose of this study, it is OK to use all the data for training the model.
(c) Comment on the ability of the NN model to fit the data (compared to the linear regression model). Compare the parity plots of the linear regression and NN models. Is it worth the effort to develop a more sophisticated NN model or is a linear regression model sufficient? What are issues that can arise with the use of an NN model?

Exercise 6.9. Removing Data Redundancies.
Product flows in a supply chain network (see the figure below) are carefully monitored in order to ensure that the system is operating reliably. Data has been collected over 100 months for eight different product flows. Due to the tight connectivity of the product flows in the system, it is suspected that there might be redundancies in the data (these are also known as data collinearities); as such, it is suspected that not all flows need to be monitored (this can reduce unnecessary data storage and cost).

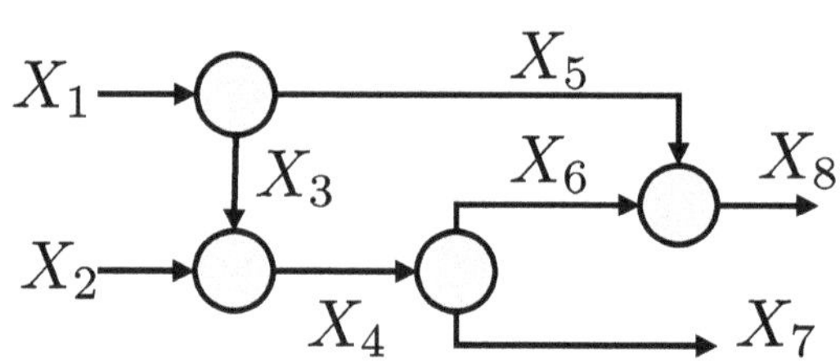

Load the data in `flow_data.mat` (this is a data matrix that contains a time series for the eight product flows). Address the following:

(a) Use SVD to determine if there are any redundancies in the data and if all the flows need to be monitored.
(b) If there are redundancies in the data, use SVD to determine which flows can be safely eliminated. Confirm that removing such flows leads to a dataset that contains no redundancies (use the rank of the data matrix to determine this).

Exercise 6.10. Optimizing Footwear Materials.
The use of natural rubber in the footwear industry has gained increasing attention in recent years [23]. Unlike its synthetic analog, natural rubber is generated from plants and thus provides a more sustainable solution to companies that aim to reduce their carbon footprint.

Determining natural rubber blend formulations with target properties is currently done via experimental trial and error, which can be inefficient and costly. We aim to develop a data-driven model to optimize the blend formulation for a footwear material. The target property is hardness, which is an important design parameter measured by a durometer. The hardness of a material depends on the amount of crosslinker, plasticizer, and the quantity of voids. More specifically, we vary the sulfur content (influences crosslink density), the paraffin oil content (serves as a plasticizer that changes viscosity), and the void percentage (generated by the addition of bicarbonate to create the porous inner structure). These three input variables can be used to predict the hardness of the material.

Load the data in `footwear.mat` and address the following:

(a) Explore the dataset and describe its general features. Identify the input and output variables. How many samples are in the dataset? Provide a table with the summarizing statistics of all variables (mean, standard deviation, median, min, and max).
(b) Data visualization is an important step in the initial understanding and analysis of the dataset. Using a 3D scatter plot, visualize the entire dataset (the axes are paraffin content, sulfur content, and percent void) and color-code each data point by the level of hardness. Describe how hardness changes with the sulfur, paraffin, and percent void contents.
(c) Standardize the input variables to obtain mean values of 0 and a standard deviation of 1. Randomly split the dataset into 80% training and 20% testing. Fit a linear regression model and report the RMSE for both training and testing. Also, report the values of the model coefficients/parameters. Comment on the magnitude of the parameters; are they consistent with your previous observations on how different inputs affect the output?
(d) Using the *fitrnet* routine of MATLAB (or an equivalent routine in other software packages), fit an NN model (one hidden layer with eight neurons) on the same standardized training data. Use the testing data for validation and use the default settings. Look into the documentation; what is the loss function and the stopping criteria of the default setting? Show a convergence plot (loss vs. iteration) for both training and validation. Did the loss converge? Are there any differences between the training and validation loss?

(e) Report the training and test RMSE. Compare with the results from linear regression.

(f) Now that we have a trained NN model for durometer reading prediction, we proceed to search for the optimal blend composition. The table `footwear_grid` (contained in `footwear.mat`) includes blend compositions that are experimentally feasible. Use the trained NN model to predict the durometer readings for these feasible compositions. If our target durometer value is 6, what is the optimal blend composition? **Note:** Do not forget to standardize `footwear_grid` using the same mean and standard deviation used to standardize the data used for developing your model.

Exercise 6.11. Applying Convolutional Filters.

Convolutional filters are used extensively in signal and image processing and form the basis of powerful ML tools such as CNNs. There are various types of filters available, each having distinct characteristics and uses. To understand how these work, we will apply different filters to 1D and 2D signals consisting of a time series and an image.

(a) *1D noise reduction*: At a chemical process, a sensor is used to monitor the concentration of a component in a fluid stream (see Figure 6.52(a)). Unfortunately, the raw data collected by the sensor contains a significant amount of noise. Your objective is to apply a Gaussian filter to smooth out the noise with 1D convolution. The data are contained in the data file `concentration_convolution.mat` and represent the component concentration in parts per million (ppm).

(b) *2D noise reduction*: In Figure 6.52(b), we present the response of an LC sensor in the presence of a gas contaminant. This image is noisy, and this might hide its overall features. Use a 2D Gaussian filter to smooth out the noise of the image `liquid_crystal_convolution.png` and comment on the effect that the filter has on the image.

(c) *2D edge detection*: Detecting edges in images is often needed to identify important features, such as phase change or object boundaries. One common method for edge detection is using Sobel filters along the horizontal and vertical directions. Another method is using the Laplacian filter. Apply Sobel and Laplace filters to the image, compare the convolved images with these filters, and comment on the effect that these filters have.

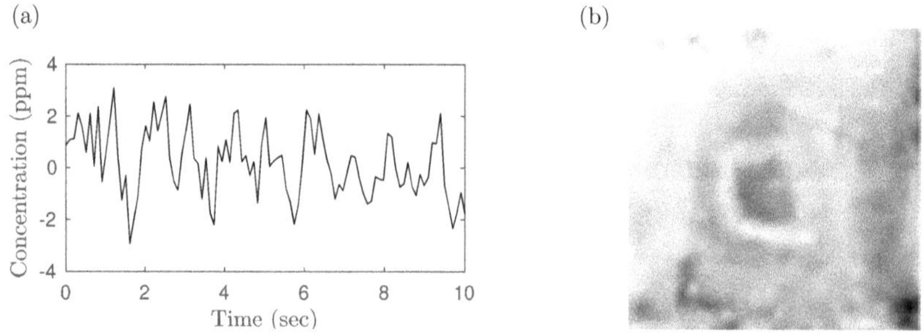

Figure 6.52 (a) Time evolution of concentration. (b) Sensor image.

7 Decision-Making Under Uncertainty

We have discussed techniques that use data to model random behavior and to develop predictive models; the combination of these techniques allows us to model the known (predictable) and unknown (uncertain) aspects of a system. We now would like to use these capabilities to *make decisions*. We will see that anticipating and mitigating uncertainty is essential when making decisions if we want our decisions to remain robust when confronted with unexpected situations.

We will also see that making decisions under uncertainty boils down to the following fundamental questions:

- How can we *compare* random variables (RVs)? How can we say that an RV is better than another?
- How can we *shape/control* the probability distribution of an RV? What aspects of the distribution do we focus on?

These questions lead to a couple of important paradigms for making decisions under uncertainty (see Figure 7.1).

To address these questions, we will make extensive use of *summarizing statistics*, and we will see how these quantities (also known as *risk measures*) can be used to model different attitudes that decision-makers (DMs) take when they face risk/uncertainty. We will also use risk measures to formulate optimization problems that allow us to find the best decisions to anticipate and mitigate/exploit risk.

Our discussion will also aim to address the following specific questions:

- Why is it necessary to account for uncertainty when making decisions?
- What can happen if we ignore uncertainty?
- How do we measure risk/uncertainty?
- How can we obtain a decision that anticipates and withstands risk?

Concepts and techniques that answer these questions are studied in the area of *stochastic optimization* (a.k.a. stochastic programming). This is a broad area of

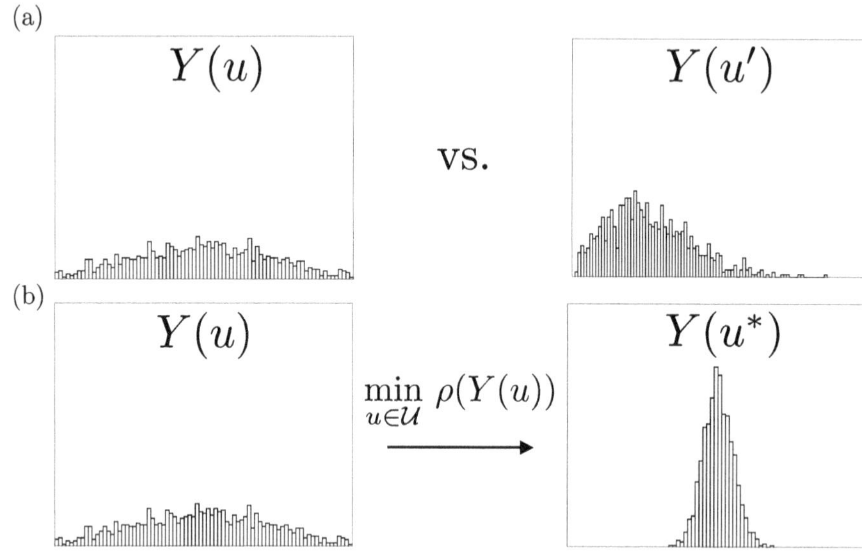

Figure 7.1 Paradigms for decision-making under uncertainty. Comparison of decisions u,u' and associated output random variables $Y(u),Y(u')$ (a). Find the best decision u^* that shapes $Y(u^*)$ in a desirable way (b).

optimization that has many applications in engineering and finance; our goal here will be to cover the fundamental elements of this area.

7.1 Deterministic Decision-Making

In order to begin our discussion, we revisit the system discussed in Chapter 1:

$$Y(u) = \varphi(X,u), \tag{7.1}$$

where $X = (X_1,X_2,\ldots,X_n)$ is a multivariate input RV with pdf f_X and Y is a univariate output RV with pdf f_Y. The RV X is our input uncertainty and captures aspects that we cannot control (e.g., electricity prices, weather, equipment failures, fluctuations of an input flow, and sensor noise). The system model is given by the function φ; u is a decision variable that we (the DM) have control over and that we can use to manipulate/shape the pdf of Y (our output uncertainty). The system model allows us to predict the impact of X and u on Y. As an example, our decisions u can be the operating conditions of a chemical process that are used to manipulate the total cost of the process Y and with this counteract the effect of the input uncertainty X. For convenience, we will highlight the dependence of Y on u by using the notation $Y(u)$.

It is convenient to think about how realizations/scenarios of X propagate through the system to generate realizations of $Y(u)$:

$$y_\omega(u) = \varphi(x_\omega,u),\ \omega \in \mathcal{S}. \tag{7.2}$$

This highlights how variability/uncertainty in the input X propagates through the system to generate variability in the output $Y(u)$. For instance, if the input represents the prices of a raw material, such prices will affect the cost of the process. Our goal is thus to find a decision u that can help us protect the system against the variability/uncertainty of the random inputs; this will be done by selecting u in such a way that we control the shape of the pdf of $Y(u)$; for instance, we may want to choose u so as to minimize the variance of $Y(u)$, but we will see that many other options are possible.

Decision-makers often make decisions in real life by *ignoring uncertainty* altogether. This so-called *deterministic* decision-making approach uses the following logic; we assume that the input X takes a *single* representative value such as the empirical mean:

$$x = \hat{\mathbb{E}}_X = \frac{1}{S}\sum_{\omega \in \mathcal{S}} x_\omega. \tag{7.3}$$

This implicitly assumes that the input is a deterministic variable (with a known value) and thus the output $Y(u)$ is deterministic with value:

$$y(u) = \varphi(x,u). \tag{7.4}$$

Another representative value commonly selected for the RV X is the worst-case value:

$$x = \max_{\omega \in \mathcal{S}} x_\omega. \tag{7.5}$$

For example, instead of using all the historical observations/data for raw material prices, a DM might simply select the historical mean or the historical worst-case value.

Consider now that we would like to compare a couple of decisions u and u' with associated output values $y(u) = \varphi(x,u)$ and $y(u') = \varphi(x,u')$. If we want to select the decision that minimizes the output $y(u)$, clearly, we will select u over u' if $y(u) \leq y(u')$ or we will select u' over u if $y(u') \leq y(u)$. Comparing decisions is *easy* in a deterministic setting because we are comparing *numbers*.

A fundamental issue that appears when facing uncertainty is that it is nontrivial to compare decisions. To see this, consider that the outputs for a couple of decisions are not numbers but are actually RVs $Y(u)$ and $Y(u')$; in such a case, how can we say that $Y(u)$ is better than $Y(u')$? This is not directly possible because the output RVs can take different values (different realizations). In fact, the notation $Y(u) \leq Y(u')$ and $Y(u') \leq Y(u)$ does not make mathematical sense; our traditional approach to comparing variables does not work anymore.

Another useful way to think about issues encountered when comparing RVs is to try to compare their realizations. Imagine that we have a set of observations/data for $Y(u)$ (given by $y_\omega(u) = \varphi(x_\omega,u)$, $\omega \in \mathcal{S}$) and for $Y(u')$ (given by $y_\omega(u') = \varphi(x_\omega,u')$, $\omega \in \mathcal{S}$). In what sense can we say that the observations of $Y(u)$ are better than the observations of $Y(u')$? One possibility could be to take the sample average of both observations and declare that $Y(u)$ is better than $Y(u')$ if:

$$\hat{\mathbb{E}}_{Y(u)} \leq \hat{\mathbb{E}}_{Y(u')}. \tag{7.6}$$

In other words, we select the decision that yields the smallest average/mean cost. Effectively, what we have done here is to summarize the outputs into a scalar value in order to make a comparison. However, summarizing the RVs using the average does not account for variability and possible extreme values. As such, one could also envision selecting the decision that yields the lowest sample variance (declare that $Y(u)$ is better than $Y(u')$ if $\hat{\mathbb{V}}_{Y(u)} \leq \hat{\mathbb{V}}_{Y(u')}$). The emergence of multiple ways of comparing RVs introduces *ambiguity* in the decision-making process; by using a deterministic approach, a DM does not have to deal with such ambiguity.

We now go back to the deterministic setting and assume that we would like to find a decision $u \in \mathcal{U}$ that minimizes the cost $y(u) = \varphi(x,u)$, where $\mathcal{U}$ is a set of possible decisions ($\mathcal{U}$ is the domain of u). This problem can be tackled by solving the optimization problem:

$$\min_{u \in \mathcal{U}} \varphi(x,u), \tag{7.7}$$

which gives an optimal decision u^*_D. Because we have *assumed* that the input and output variables are deterministic, the objective function is a scalar function $y(u) = \varphi(x,u)$ that we can optimize to find the optimal decision u^*_D. When the output is an RV $Y(u) = \varphi(X,u)$, the objective function is no longer a scalar function (as it can take possible values under different realizations of X); as such, the notation:

$$\min_{u \in \mathcal{U}} \varphi(X,u), \tag{7.8}$$

does not make mathematical sense. We will see later that, to properly pose the problem, we need to summarize the RV $\varphi(X,u)$ by using a summarizing statistic (e.g., the average or variance). For instance, we can aim to solve the problem:

$$\min_{u \in \mathcal{U}} \mathbb{E}[\varphi(X,u)]. \tag{7.9}$$

However, here again, we have ambiguity, as it is not clear what summarizing statistic we should use to formulate the optimization problem.

It appears that the deterministic way of thinking simplifies decision-making, as it is much easier to compare decisions and formulate optimization problems. However, a fundamental limitation of deterministic decision-making is that it ignores the inherent variability of X that is seen in real life. For instance, the "optimal" decision u^*_D obtained with Eq. (7.7) is only optimal for the assumed value of the input x (e.g., the mean raw material prices); as such, this decision might be vulnerable when it confronts other possible outcomes of X seen in real life (fluctuations/spikes in prices). In particular, it is important to emphasize that, even if we assume that X takes value x when making a decision, X can take other possible values (ignoring uncertainty/variability does not mean that this does not exist). For instance, if we design a process to satisfy average demand, this does not mean that demand cannot fluctuate over time and therefore my process might not be able to satisfy demand when this is higher than the average. As such, it is important to *validate* how the deterministic decision u_D^* would behave when exposed to other possible outcomes of X; this can be done via simulations:

$$y_\omega(u_D^*) = \varphi(x_\omega, u_D^*),\ \omega \in \mathcal{S}. \tag{7.10}$$

By doing this, it is possible to evaluate the performance statistics (e.g., mean, variance, and worst-case) or the output $Y(u_D^*)$ when it is exposed to different outcomes of X. This validation analysis would provide a more complete picture of how *vulnerable/robust* the decision u_D^* is when it confronts uncertainty in X; this can help us determine whether assuming a representative value of X was adequate and whether we should select other representative values (e.g., use the worst-case value of X instead of the average).

Example 7.1 Deterministic Decision-Making

Consider a setting in which a manufacturing facility is exploring the possibility of purchasing an on-site power generator in order to mitigate the rising costs of electricity of the local power grid. The power demand of the facility is random and is given by L; historical data indicates that the load behaves as a Weibull RV (see Figure 7.2).

The installed capacity of the generator is the decision u. The residual demand to be purchased from the power grid is a function of the capacity and is given by $R(u) = \max(L - u, 0)$; here, we use the max function because $R(u)$ cannot be negative (i.e., excess power cannot be used). The cost of the residual demand obtained from the power grid is a quadratic function $\alpha_R \cdot R(u)^2$, and the investment cost for the generator is also a quadratic function $\alpha_u \cdot u^2$; as such, the total cost of the generator is given by:

$$Y(u) = \alpha_R \cdot R(u)^2 + \alpha_u \cdot u^2. \tag{7.11}$$

Consider now that a DM would like to simplify the calculations and adopt a deterministic view of the problem; to be safe in the design process, the DM takes a representative value of the random load that equals the worst (largest) demand that they have seen historically, which is given by $l = \max_{\omega \in \mathcal{S}} l_\omega = 27.5$. As such, the cost is deterministic and given by $y(u) = \alpha_R \cdot r(u)^2 + \alpha_u \cdot u^2$, with $r(u) = \max(l - u, 0)$. The DM focuses on the highest historical demand because they are interested in decreasing the highest cost (this is known as "peak shaving").

The DM has the possibility of selecting generators of different (discrete) capacities, and the choices are given by the set $\mathcal{U} = \{0, 1, 2, \ldots, 10\}$. Here, the case for $u = 0$ denotes the "do-nothing" case and $u = 10$ is the generator of maximum capacity. The generator design problem can be cast as the optimization problem:

$$\min_{u \in \mathcal{U}} \ \alpha_R \cdot r(u)^2 + \alpha_u \cdot u^2. \tag{7.12}$$

In Figure 7.2, we can see that the cost is minimized with a generator of capacity $u_D^* = 6$.

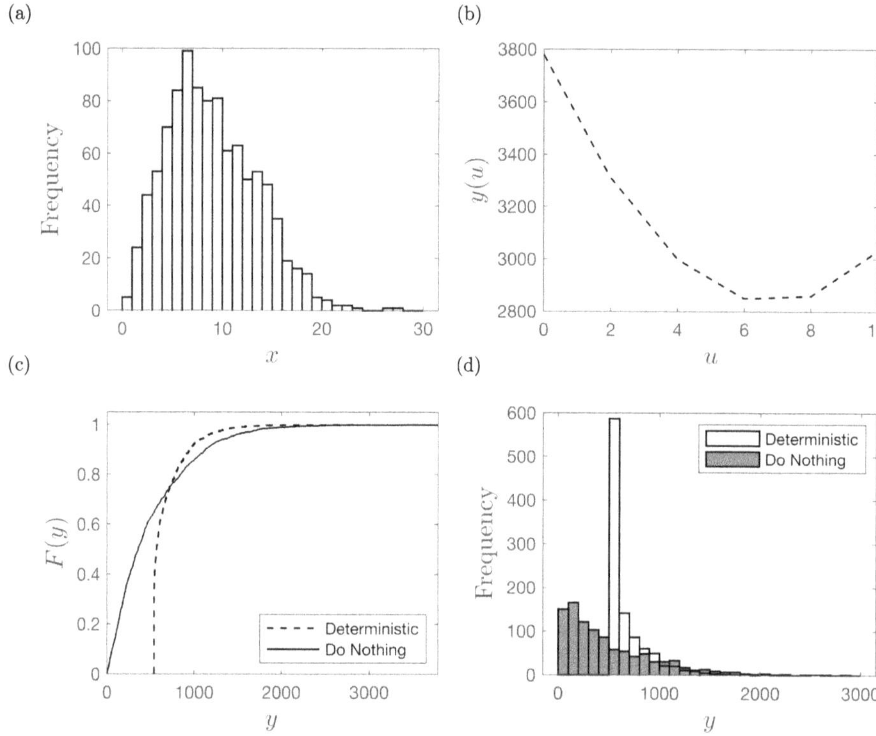

Figure 7.2 Random load of facility follows a Weibull RV (a); deterministic cost as a function of capacity (b); comparison of costs for the deterministic and do-nothing case using the cdf (c) and the pdf (d).

The DM then proceeds to do Monte Carlo simulations to validate how this generator would perform when exposed to the different possible outcomes of the load; to do so, the DM computes the costs $y_\omega(u) = \alpha_R \cdot r_\omega(u_D^*)^2 + \alpha_u \cdot (u_D^*)^2$ for $\omega \in \mathcal{S}$. To establish a benchmark, the DM also compares the costs for the do-nothing scenario (with $u = 0$), which corresponds to buying all power from the grid. The outcomes are summarized in the empirical pdf/cdf shown in Figure 7.2. By comparing the cdfs and pdfs, we can see that the existing system (do-nothing) has costs below a threshold value of 500, while the new system (deterministic) does not. However, we can see that the new system has lower cost values more frequently than the do-nothing case for costs that are higher than a threshold value of 500; for instance, the probability that the new system has a cost less than 1,500 is close to 1, while the probability that the existing system has a cost less of 1,500 is around 0.9. This indicates that the new system indeed has the effect of mitigating high/peak costs (as desired by the DM). On average, the costs for the new system are 674.1, while for the existing system are 501.3; as such, overall, the operation of the new system is significantly more expensive. The ability to shave peak costs can be beneficial for budgetary reasons; for example, if the facility has a tight budget at specific times, it might be beneficial to have lower peak costs than lower average costs. In addition, shaving peak demands of electricity can help the power grid.

Scripts to reproduce results: `ch7_generator_det` □

Assuming deterministic behavior can simplify the decision-making process, but it can also lead to a decision u_D^* that does not properly protect us against uncertainty (or that is expensive). A common approach used in practice to account for uncertainty in decision-making is to explore a range of decisions obtained for different realizations of uncertainty. For instance, consider the solution to the optimization problem:

$$\min_{u \in \mathcal{U}} \varphi(x_\omega, u), \tag{7.13}$$

which gives the solution u_ω^* for a realization x_ω of X. In other words, the approach computes a decision u_ω^* that is optimal for the realization x_ω; this approach is known as *sensitivity analysis* or Monte Carlo analysis and provides insight into how our decision changes under possible scenarios of X. However, an important limitation of this approach is that it does not deliver a single decision u that is optimal for all outcomes of X. As such, this approach does not deliver a decision that can be implemented; in other words, the sensitivity approach provides a range of possible solutions u_ω^*, $\omega \in \mathcal{S}$ that the DM needs to select from to make a final decision. In the discussion that follows, we will explore how to automatically select a decision u that properly takes into account possible outcomes of X.

7.2 Attitudes toward Risk

Quantifying/modeling risk is at the core of decision-making under uncertainty, as this will allow us to compare decisions and formulate optimization problems. At this point, we have all the fundamentals of statistics needed to define and quantify risk. In our discussion, we will consider a multivariate input RV X with pdf f_X and cdf F_X and a scalar *cost* $\varphi(X,u)$ that depends on the input RV X and a decision u. Because X is an RV, the cost is also an RV, which we represent as univariate RV $Y(u) = \varphi(X,u)$ with pdf $f_{Y(u)}$ and cdf $F_{Y(u)}$. As such, the pdf and cdf of the cost $Y(u)$ depend on the decision u; consequently, u can be used to shape/control the pdf and cdf. In our discussion, we will assume that outcomes of $Y(u)$ with low values are preferred (we want to minimize the cost); as such, we can also think of $Y(u)$ as a random *loss* (we want to minimize losses). If $Y(u)$ was associated with profit, we would prefer outcomes with high values; in such a case, we could simply consider $-Y(u)$ as the cost/loss.

In *popular culture*, risk is associated with a decision or situation (e.g., "this is a risky investment," "the reactor is operating at risky conditions," or "we are facing too much risk"). Risk is also commonly associated with *extreme events* or *large/catastrophic losses* (e.g., "I am going bankrupt" or "the system is about to collapse"). When making decisions, we want *precise* mathematical definitions that capture this logic; such definitions will help us determine what is a risky decision and how risky it is.

Unfortunately, there is no *unique* mathematical definition of risk because humans tend to think about risk in different ways (e.g., probability of failure vs. magnitude of failure). Moreover, humans tend to have different levels of risk tolerance (e.g., what seems risky to me might not seem risky to you), and such tolerance depends on the nature of the decision-making setting (e.g., I am more conservative when spending my own money than when spending the money of other people). In fact, we will see that some DMs value risk, as they might see extreme outcomes as an opportunity (even if unlikely) to get a reward (e.g., gambling and extreme sports). The broad spectrum of perceptions of risk is analogous to defining happiness and fairness, as humans will naturally disagree on what it means to be happy or on what is a fair decision.

Because there is no unique definition of risk, what we want to explore here is not what risk is, but rather what definitions of risk exist and what makes a definition of risk *proper* or consistent. Ultimately, establishing a definition of risk for a particular decision at hand is often based on *consensus* between multiple DMs. As such, it is often important to consider a range of possibilities that allow DMs to reach consensus.

Example 7.2 Decision-Maker Disagreement

We consider a simple example to illustrate how disagreements can naturally arise when comparing RVs. Assume that you need to decide between the competing options u (Option I) and u' (Option II) with associated random costs $Y(u)$ and $Y(u')$. The distributions for the costs are shown in Figure 7.3; here, we see that option u yields a lower mean cost but higher variance/uncertainty, while option u' yields a higher mean cost but lower variance/uncertainty.

Which option is better? A first observation that we make is that there is no unique answer to this question and that the notion of better depends on how the DM attributes value to different aspects of the cost distributions. If the DM values average cost, the decision u is clearly better than u', but this decision fails to capture the fact that u has a higher uncertainty and has a higher probability of incurring a large cost. In other words, it seems that option u is more "risky." On the other hand, if the DM wants more certainty (less "risk") and seeks to protect themselves against outcomes with high cost, then they would clearly prefer u', but then this decision also leads to a higher average cost. We thus have that the mean and variance lead to contradictory decisions; ideally, we would like a quantity that captures multiple aspects (features) of the cost distributions simultaneously. We will see shortly some options for doing this.

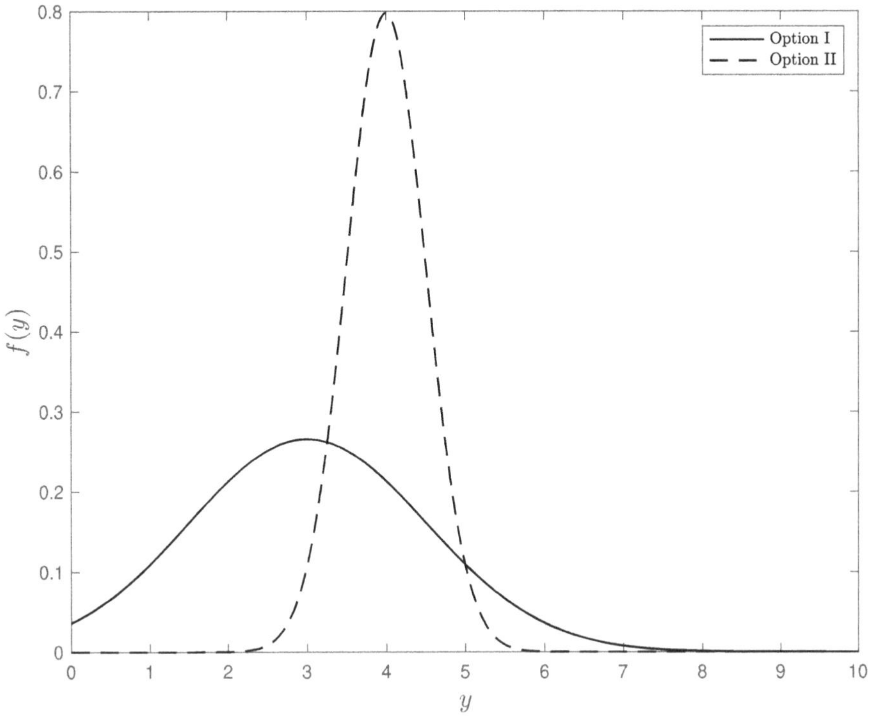

Figure 7.3 Probability density functions for the cost of competing decisions.

Scripts to reproduce results: `ch7_mean_var_comparison` □

To motivate the definitions of risk that we might consider, we explore different attitudes toward risk that DMs might take:

- *Risk-neutral*: The DM prefers $Y(u)$ over $Y(u')$ if:

$$\mathbb{E}[Y(u)] \leq \mathbb{E}[Y(u')].$$

  This DM worries about the average costs of the decisions and is not concerned about the variability/uncertainty of the decisions; for instance, the DM does not worry about the fact that $Y(u)$ might have outcomes with large cost compared to those of $Y(u')$. Here, we can say that the DM is *indifferent* to risk.
- *Risk-conscious*: The DM prefers $Y(u)$ over $Y(u')$ if:

$$\mathbb{P}(Y(u) \leq \bar{y}) \geq \mathbb{P}(Y(u') \leq \bar{y}),$$

  for some threshold value $\bar{y}$. This DM wants a decision that is *more likely* to achieve low outcomes (relative to a threshold value). This type of DM might also prefer decisions with lower variance ($\mathbb{V}_{Y(u)} \leq \mathbb{V}_{Y(u')}$) because u has lower uncertainty than u'.
- *Risk-averse*: The DM prefers $Y(u)$ over $Y(u')$ if:

$$\max Y(u) \leq \max Y(u').$$

Here, $\max Y(u)$ is the worst possible outcome of $Y(u)$; as such, this DM focuses on the worst possible outcomes of the decisions u and u' and prefers the decision that has the smallest worst outcome. In other words, this DM has a *pessimistic view* of uncertainty (uncertainty is bad because it can lead to extremely high-cost values); as such, this DM aims to *mitigate/minimize uncertainty*.

- *Risk-taker*: The DM prefers $Y(u)$ over $Y(u')$ if:

$$\min Y(u) \leq \min Y(u').$$

Here, $\min Y(u)$ is the best possible outcome of $Y(u)$; as such, this DM focuses on the best possible outcomes of u and u' and prefers the decision that leads to the smallest possible outcome (even if this comes at the expense of high uncertainty). In other words, this DM has an *optimistic view* of uncertainty (uncertainty is good because it can lead to extremely low values for the output); as such, this DM aims to *exploit/maximize uncertainty*.

It is important to highlight that DMs can take different attitudes toward risk under different situations; for instance, some DMs might be conservative when investing but might enjoy practicing extreme sports. We also highlight that sometimes there is no strict categorization of attitudes, but rather there is a continuous spectrum of attitudes that can be explored. Finally, from a mathematical standpoint, we highlight that different DMs focus on different aspects of the distribution of the cost $Y(u)$.

Example 7.3 Attitudes toward Risk

We revisit Example 7.2 in which we need to select between a couple of options. In Figure 7.4, we present the pdfs of the costs associated with such options. Here, we also present the cdfs because this will help visualize important aspects associated with risk.

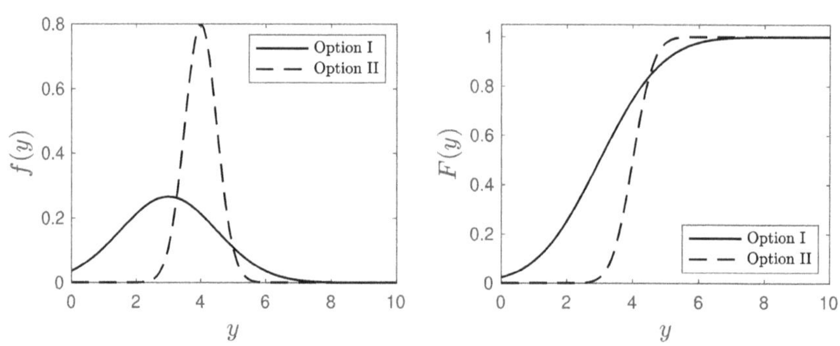

Figure 7.4 Probability and cumulative density functions for the cost of competing decisions.

As we have seen, the option to be selected depends on how the DM attributes value to different aspects of the cost distribution. More precisely, the choice is based on the attitude that the DM takes toward risk. Consequently, a more proper question than “Which option is better?” is “Which option is better if you are a risk-neutral, risk-conscious, risk-averse, or risk-taker DM?”

- If you are a *risk-neutral* DM, you attribute no value to uncertainty. By inspecting the pdf, you would select option u because this has the lower expected cost.
- If you are a *risk-conscious* DM, you select a threshold for the cost that you are willing to tolerate (e.g., say 5) and then you select the option with the highest probability of having a lower cost than the threshold. By inspecting the cdf, you would select option u' because this has a higher probability of achieving a cost that is below a threshold value of 5 MUSD. Note that if we change the threshold value to 4, then you would select option u instead if u'.
- If you are a *risk-averse* DM, you focus on outcomes of high cost. By inspecting the pdf, you would select u' because this has no outcomes with high cost (e.g., above 6). This type of DM focuses on the right tail of the pdf and is similar to a risk-conscious DM that assigns a high threshold value.
- If you are a *risk-taking* DM, you focus on outcomes of low cost. By inspecting the pdf, you would select u because this has outcomes of very low cost (e.g., below 1). In fact, decision u yields outcomes with a negative cost (generates revenue), and thus, the DM sees an opportunity in exploiting risk. This type of DM focuses on the left tail of the pdf and is similar to a risk-conscious DM that assigns a low threshold value.

Scripts to reproduce results: `ch7_mean_var_comparison` □

Example 7.4 Comparing Investment Options

Assume that we have a couple of investment opportunities:

- Initial capital $c_I = 25{,}000$ USD with random annual rate of return $R_I \sim \mathcal{N}(0.5, 0.1)$ (as a fraction of c_I).
- Initial capital $c_{II} = 50{,}000$ USD with random annual rate of return $R_{II} \sim \mathcal{N}(0.4, 0.2)$ (as a faction of c_{II}).

We compare the net present value (NPV) for both options (10 years, interest rate i of 5%):

$$NPV_I = c_I - R_I \cdot c_I \cdot \frac{(1-(1+i)^{-10})}{i}$$

$$NPV_{II} = c_{II} - R_{II} \cdot c_{II} \cdot \frac{(1-(1+i)^{-10})}{i}.$$

We want the NPV to be as *negative* as possible (as small as possible); specifically, a negative NPV indicates that the investment makes money, while a positive NPV indicates that the investment loses money.

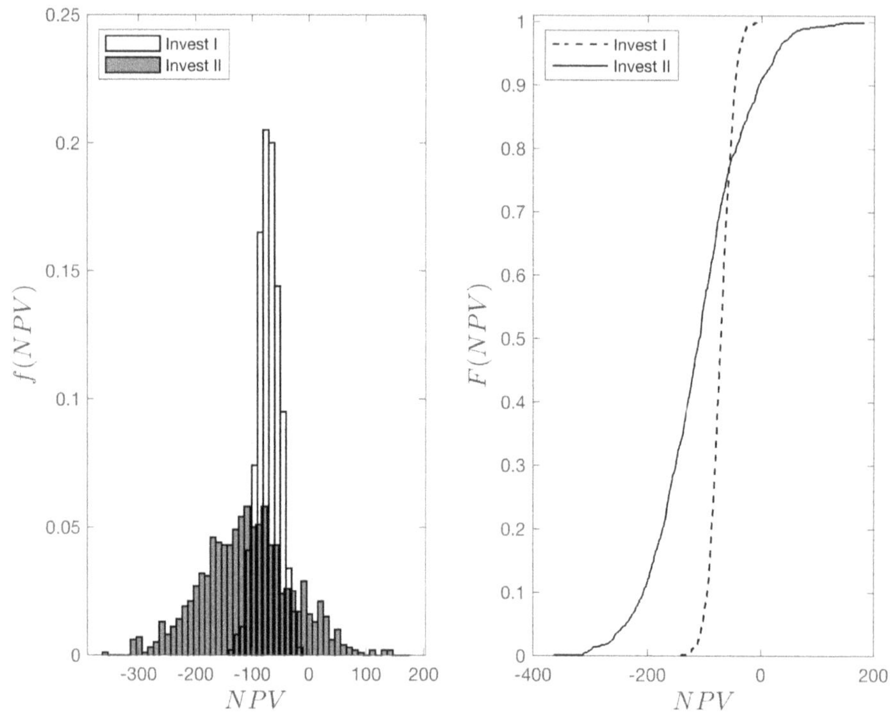

Figure 7.5 Probability and cumulative density functions for net present values (NPVs) for competing investment options.

In Figure 7.5, we compare the pdfs/cdfs for the NPVs. Here, we have scaled the NPV values by a factor of 1,000 to facilitate visualization. We can see that the first investment has a higher mean, but it is always negative (the investment always makes money, and it is thus a safe investment). Moreover, we can see that the first investment has a lower variance (has low uncertainty); this can be seen as a good thing.

In comparison, the second investment exhibits high variance and can lose money about 10% of the time, and in some instances, it can lose more than 150,000 USD. However, with this investment, we can also make money 90% of the time, and it is possible to make a lot of money (at levels that the first investment cannot reach). Specifically, note that there is a 50% probability that the second investment makes more than 170,000 USD.

Assuming that we are conscious about risk (and specifically we do not want to lose money), the first investment seems like a better option. On the other hand, if we are risk-takers, we might be willing to take a chance of potentially losing money 10% of the time because this investment can make lots of money in some outcomes.

Scripts to reproduce results: `ch7_npv_comparison` □

To make a decision, it is necessary to quantify risk; this is done using functions that are known as *risk measures*. A risk measure is a scalar function $\rho(Y(u))$ that maps the univariate RV $Y(u)$ to a scalar quantity (a number). A risk measure is nothing but a *summarizing statistic*; the word *summarizing* is important because it highlights the fact that a risk measure aims to extract key aspects/features of the RV (which is a complicated object that is characterized by its pdf and cdf).

Common risk measures used in practice are:

- Expected value:

$$\mathbb{E}[Y(u)]. \tag{7.14}$$

- Variance:

$$\mathbb{V}[Y(u)] = \mathbb{E}[(Y(u) - \mathbb{E}[Y(u)])^2]. \tag{7.15}$$

- Mean-variance:

$$\mathbb{MV}_\kappa[Y(u)] = \mathbb{E}[Y(u)] + \kappa \cdot \mathbb{V}[Y(u)], \tag{7.16}$$

  for some trade-off weight $\kappa \in \mathbb{R}_+$.
- Probability of loss:

$$\mathbb{P}(Y(u) > \bar{y}), \tag{7.17}$$

  for some threshold value $\bar{y} \in \mathbb{R}$. Depending on context, the probability of loss is also often expressed as $\mathbb{P}(Y(u) \leq \bar{y})$.
- Value-at-risk:

$$\mathbb{Q}_{Y(u)}(\alpha), \tag{7.18}$$

  for some probability level $\alpha \in [0, 1]$. This is also known as the α-quantile and is also often denoted as $\text{VaR}_{Y(u)}(\alpha)$.
- Conditional value-at-risk (CVaR):

$$\mathbb{E}[Y(u)|Y(u) \geq \text{VaR}_{Y(u)}(\alpha)], \tag{7.19}$$

  for some probability level $\alpha \in [0, 1]$. This is also known as the expected shortfall; this risk measure quantifies large values about the quantile $\text{VaR}_{Y(u)}(\alpha)$.
- Mean absolute deviation:

$$\mathbb{E}[|Y(u) - \mathbb{E}[Y(u)]|]. \tag{7.20}$$

- Worst-case and best-case value:

$$\max Y(u) \quad \text{and} \quad \min Y(u). \tag{7.21}$$

Different measures are used to model different risk attitudes taken by DMs and are essential in comparing decisions and making optimal decisions. The probability of loss is perhaps the measure that more closely resembles our colloquial definition of risk; this has a natural interpretation that DMs find useful in practice (i.e., I want a small probability that my cost $Y(u)$ goes above a certain value $\bar{y}$). The probability of loss, however, has important caveats; for instance, it fails to capture certain important aspects of $Y(u)$ (e.g., the magnitude of the potential losses).

There is a broad selection of risk measures that one can select and these have different properties (which we will discuss shortly); in fact, any summarizing statistic can in principle be used as a risk measure (e.g., mode, quantiles, and moments). Also note that some risk measures depend on parameters (e.g., the mean-variance depends on κ and CVaR depends on α); such parameters can be tuned to capture a spectrum of risk attitudes. It is also important to highlight that the selection of a risk measure should be based on the nature of the RV to be compared; for instance, comparing Gaussian RVs (these are symmetric RVs) is not the same as comparing exponential RVs (these are asymmetric RVs).

Example 7.5 Measuring Risk for Investment Optimization

We go back to Example 7.4 and compute different risk measures that facilitate the comparisons. In doing so, it is important to note how these quantities relate to the pdfs and cdfs shown in Figure 7.5 and how this can inform the selection of different types of DMs. Here, we denote the NPV of investment I as $Y(u)$ and the NPV of investment II as $Y(u')$ and recall that our goal is to make the NPV as negative as possible.

- Expected value:
  $\mathbb{E}[Y(u)] = -$ 70,891 USD, $\mathbb{E}[Y(u')] = -$ 109,048 USD.
- Standard deviation:
  $\mathbb{SD}[Y(u)] =$ 19,284 USD, $\mathbb{V}[Y(u')] =$ 79,288 USD.
- Probability of loss:
  $\mathbb{P}(Y(u) > 0) = 0\%$ and $\mathbb{P}(Y(u') > 0) = 9\%$.
- Probability of gain:
  $\mathbb{P}(Y(u) \leq \bar{y}) = 6.4\%$ and $\mathbb{P}(Y(u') \leq \bar{y}) = 55.90\%$ for $\bar{y} = -$ 100,000 USD.
- Value-at-risk:
  $\mathbb{Q}_{Y(u)}(\alpha) = -$ 46,699 USD and $\mathbb{Q}_{Y(u')}(\alpha) = -$ 4,935 USD for $\alpha = 0.9$.
- Conditional value-at-risk:
  $\text{CVaR}_\alpha[Y(u)] = -$ 37,530 USD and $\text{CVaR}_\alpha[Y(u')] =$ 34,810 USD for $\alpha = 0.9$.
- Worst-case:
  $\max Y(u) = -$ 9,129 USD and $\max Y(u') =$ 181,413 USD.
- Best-case:
  $\min Y(u) = -$ 140,609 USD and $\min Y(u') = -$ 363,708 USD.

A risk-neutral DM will prefer u' because it has a lower mean cost; a risk-conscious DM will prefer u because it has low variance (standard deviation), a lower value of CVaR, and a lower probability of loss; a risk-conscious DM would prefer u because it has a lower worst-case value; a risk-taking DM would prefer u' because this has a lower best-case value and a higher probability of gain (if the goal is to have large gains above 100,000 USD).

Scripts to reproduce results: `ch7_npv_comparison` □

7.3 Stochastic Dominance

Given that there is a wide range of possible risk measures that we can use, we can ask the following questions:

- What are the desirable features of a risk measure?
- What is a proper risk measure?

To answer these questions, it is necessary to discuss what makes an RV better than another RV (so as to understand how can we properly compare RVs). An important concept that arises here is that of *stochastic dominance* (SD); we will see that this concept provides a useful perspective on how to select risk measures that enable proper comparisons.

We say that Y *stochastically dominates* Y' (written mathematically as $Y \preceq Y'$) if:

$$\mathbb{P}(Y \leq \bar{y}) \geq \mathbb{P}(Y' \leq \bar{y}) \text{ for all } \bar{y} \in \mathcal{D}_Y, \tag{7.22}$$

where $\mathcal{D}_Y \subseteq \mathbb{R}$ is the common domain of Y and Y'.

We now highlight some important observations regarding the definition of SD:

- The mathematical notation $Y \preceq Y'$ is a generalization of the more familiar notation $y \leq y'$ (used for comparing deterministic variables) that we will use for stating that the RV Y is better than the RV Y'. The sense of better defined by SD is *probabilistic* and dictated by Eq. (7.22).
- Stochastic dominance says that Y is better than Y' if the probability of Y being below a threshold value $\bar{y}$ is higher than the probability of Y' being below the same threshold for all the possible threshold values $\bar{y}$. Note that this is analogous to comparing RVs using the probability of loss, but the RV needs to be better than the other consistently for all possible threshold values.
- We can also express SD using the cdf as:

$$F_Y(\bar{y}) \geq F_{Y'}(\bar{y}) \text{ for all } \bar{y} \in \mathcal{D}_Y. \tag{7.23}$$

In simple terms, we have that $Y \preceq Y'$ indicates that the cdf curve of Y is above that of Y' for any threshold value $\bar{y}$.

It is important to note how SD is *different* from more traditional dominance concepts that we are familiar with. For instance, we often think that a decision dominates another one if it is *always* better (in all possible situations). In statistical terms, this would imply that outcomes of Y are always lower than or equal to those of Y':

$$y_\omega \leq y'_\omega, \quad \omega \in \Omega. \tag{7.24}$$

Note that this requirement is quite strict and is unlikely to occur in practice. Moreover, this requirement does not take into consideration the probability of different outcomes. SD is a more *statistically consistent* requirement because it indicates that the outcomes of Y should be lower than the threshold $\bar{y}$ more frequently than the outcomes of Y.

We recall that a risk-conscious DM would require that $\mathbb{P}(Y \leq \bar{y}) \geq \mathbb{P}(Y' \leq \bar{y})$ holds *for a single* threshold value $\bar{y}$; this notion, however, requires that the DM selects a threshold. SD is a much stricter requirement because it requires $\mathbb{P}(Y \leq \bar{y}) \geq \mathbb{P}(Y' \leq \bar{y})$ to hold *for any* threshold $\bar{y}$; SD thus eliminates the need to select a threshold, but the requirement can be quite strict and might be impossible to satisfy. As such, there are *weakened (relaxed)* versions of SD that are also often used in practice; for example, we might require that $\mathbb{P}(Y \leq \bar{y}) \geq \mathbb{P}(Y' \leq \bar{y})$ holds *for some* values $\bar{y}$. Alternatively, we can say that the cdf of Y is above that of Y more often than not.

To illustrate these concepts, compare the cost cdfs of Figure 7.4. Note that we cannot say that option u dominates u' (i.e., $Y(u) \preceq Y(u')$) because the cdf of $Y(u)$ is not above that of $Y(u')$ for all threshold values $\bar{y}$. However, we can say that u dominates u' at several threshold values. In fact, we can say that u dominates u' over a wider range of threshold values (compared to the range of values over which u' dominates u).

The definition of SD makes it clear that, in order to properly compare RVs, it is necessary to compare their *probability density functions*; moreover, we note that the comparison is based on the cdfs (and not the pdfs). These observations are important, as DMs are often inclined to compare RVs based on pdfs. Comparing cdfs is more intuitive, as it simultaneously captures the likelihood and magnitude of different events.

Our next goal will be to define risk measures that are consistent with the definition of SD. Also, motivated by SD, we will also explore how to *measure distances* between the cdfs of different RVs, which will allow to compare RVs and shape cdfs in a desirable form.

7.4 Coherent Risk Measures

There are a number of fundamental properties that a proper (a.k.a. coherent) risk measure should satisfy. These properties have been proposed based on their usefulness in *actual practical applications* and based on mathematical consistency. Interestingly, and as we will see shortly, these properties are analogous to properties of vector norms (which are used to quantify the magnitude of vectors). One can think of the following properties as a set of rules (a.k.a. mathematical axioms) that give us guidance into how to select a suitable risk measure. Similar types of axioms exist for defining proper measures of fairness.

We say that the risk measure ρ is proper if it satisfies the properties:

- Monotonicity:

$$\text{If } Y \preceq Y', \text{ then } \rho(Y) \leq \rho(Y').$$

This property requires that if Y stochastically dominates Y', then its risk should also be lower. This indicates that a proper risk measure should be consistent with SD; this makes sense because if an RV dominates another, then the risk measure should indicate this.

- Translation invariance:

$$\rho(Y + c) = \rho(Y) + c \quad \text{for } c \in \mathbb{R}.$$

This property requires that adding a constant (deterministic) value to an RV should result in adding the same constant to the risk measure. In other words, this indicates that adding a constant does not alter the inherent distribution properties of the RV. For instance, adding a constant quantity to a random cost only displaces the distribution (but does not alter the shape of its distribution); as such, a proper risk measure should be consistent with this fact.

- Positive homogeneity:

$$\rho(c \cdot Y) = c \cdot \rho(Y) \quad \text{for } c \in \mathbb{R}_+.$$

This property is analogous to translation invariance and requires that scaling an RV by a constant (deterministic) factor should result in a scaled version of the risk measure. In other words, scaling an RV does not affect the inherent distribution properties of the RV; a proper risk measure should be consistent with this fact. For instance, the risk of a random cost expressed in USD or in euros should be the same up to a factor.

- Subadditivity:

$$\rho(Y + Y') \leq \rho(Y) + \rho(Y').$$

This property requires that the risk measure is such that the risk of a combined pair of RVs should not exceed the sum of their individual risks. For instance, adding a couple of random costs (e.g., for a couple of investments) should result in a risk that is no greater than the sum of the individual risks of the investments. In other words, a proper risk measure does not arbitrarily magnify risk.

7.5 Properties of Risk Measures

Now that we understand how DMs think about risk and what are the properties that a risk measure should satisfy, we can proceed to more formally analyze the benefits and limitations of different risk measures.

7.5.1 Expected Value

- The expected value $\mathbb{E}[Y]$ captures the magnitude of Y; i.e., $\mathbb{E}[Y] \leq \mathbb{E}[Y']$ indicates that Y is, on average, smaller than Y'. The expected value has a natural and intuitive interpretation, and it is thus a common measure used in practice. The expected value is associated with the location of the distribution of Y.

- The expected value is a proper risk measure (it satisfies all properties); as such, it is consistent with SD.
- The expected value ignores outcomes with extreme values. For instance, even if $\mathbb{E}[Y] \leq \mathbb{E}[Y']$ holds, Y can have outcomes of larger magnitude that Y' does not have. As a result, making a decision based solely on the expected value can put us in a vulnerable position (e.g., it is possible that an outcome will have an extreme value).

7.5.2 Variance and Mean-Variance

- The variance $\mathbb{V}[Y] = \mathbb{E}[(Y - \mathbb{E}[Y])^2]$ is a measure of variability; i.e., $\mathbb{V}[Y] \leq \mathbb{V}[Y']$ indicates that Y has less variability than Y'. The variance has a natural and intuitive interpretation; specifically, variability is naturally associated with the degree of uncertainty of an RV, and thus, variance can be interpreted as a measure that tells us how uncertain is a given decision. The variance is also associated with the spread of the distribution of Y.
- An important limitation of the variance is that it does not capture the magnitude of the RV (as the expected value does). As a result, making decisions solely based on variance can be *misleading*. To remedy this issue, the variance is often used in conjunction with the expected value by using the mean-variance $\mathbb{E}[Y] + \kappa\mathbb{V}[Y]$, where $\kappa \in \mathbb{R}_+$ is a weight that trades off magnitude and uncertainty of a decision. The trade-off weight helps span a range of attitudes toward risk, from risk-neutral ($\kappa = 0$) to risk-conscious ($\kappa > 0$) and to risk-averse ($\kappa \gg 0$).
- The mean-variance measure was proposed by Harry Markowitz in the 1950s and became a standard tool for many years to optimize portfolio investments (Markowitz earned the Nobel prize in economics for this work). Its main appeal is the ability to capture a wide range of DM behavior.
- Reducing variance results in reducing large undesired outcomes (in the upper tail of the pdf); however, the connection between variance and the tail is not direct, and this results in several important inefficiencies. Specifically, recall that the variance is centered around the expected value and penalizes deviations from it symmetrically (on both sides of the pdf). Symmetric penalization is often undesirable in applications because we are often only interested in accounting for outcomes of large magnitude (and not with those of low magnitude). This is particularly an issue in applications where the RV does not have a symmetric pdf (e.g., exponential or Weibull RV), and therefore, the variance can fail to properly capture tail effects. In such applications, we want a risk measure that helps us control only one side of the pdf; we will see that the expected shortfall and the probability of loss can achieve this.
- Importantly, the variance is *not a proper risk measure*. Specifically, it does not satisfy monotonicity (it is not consistent with SD). From a practitioner standpoint, this does not make sense; specifically, the fact that the variance is not compatible with SD tells us that, even if Y dominates Y', the variance of Y is not

necessarily lower than the variance of Y'. This incompatibility has made practitioners question the use of variance as a measure of risk.

- The mean-variance is a proper risk measure for sufficiently small values of κ; this makes sense because the expected value is a proper measure of risk.
- The variance and expected value do not have the same units. Consequently, it is often preferred to use the standard deviation $\mathbb{SD}[Y] = \sqrt{\mathbb{V}[Y]}$ when trading off with the expected value. There is an interesting risk measure that simultaneously captures expected value and standard deviation, known as the *coefficient of variation*:

$$\frac{\mathbb{SD}[Y]}{\mathbb{E}[Y]}. \tag{7.25}$$

This can be seen as a trade-off between variability and expected value.

7.5.3 Conditional Value-at-Risk

The CVaR (or expected shortfall) overcomes many deficiencies of the variance and has recently become a widely used risk measure [65]. This measure has many important properties and connections with other risk measures that should be highlighted.

Recall that the expected shortfall of Y at a given probability α is:

$$\mathbb{E}[Y|Y \geq \mathbb{Q}_Y(\alpha)], \tag{7.26}$$

where $\mathbb{Q}(\alpha)$ is the α-quantile of Y; this is also often referred to as the value-at-risk $\text{VaR}_{Y(u)}(\alpha)$. Recall also that the α-quantile of Y is the threshold value $\bar{y}$ satisfying

$$\mathbb{P}(Y \leq \bar{y}) = \alpha. \tag{7.27}$$

The expected shortfall is the expected value of costs above $\bar{y} = \mathbb{Q}_Y(\alpha)$; the location of the threshold is dictated by the choice of α.

- Conditional value-at-risk captures the magnitude of outcomes of high value in the right tail while ignoring those of small magnitude in the left tail (it is an asymmetric risk measure). This provides an important benefit over the variance, which is a symmetric risk measure.
- If we set $\alpha = 0$, the quantile $\mathbb{Q}_Y(\alpha)$ is the minimum value of Y, and therefore, the CVaR is the expected value $\mathbb{E}[Y]$. If we set $\alpha = 1$, the quantile $\mathbb{Q}_Y(\alpha)$ is the maximum value of Y and therefore the CVaR is the worst-case value $\max Y$. Consequently, CVaR captures a range of risk attitudes (from risk-neutral to risk-conscious and to risk-averse). In this respect, CVaR is a *unifying* risk measure, analogous to the mean-variance risk measure.
- Conditional value-at-risk is a proper risk measure; as such, it is consistent with SD.
- A caveat of CVaR is that it does not offer direct control on the probability of loss, which is a risk measure of interest to many DMs. In other words, reducing

CVaR does not necessarily imply reducing the probability of loss. Moreover, the DM needs to specify α, and this selection can lead to disagreement.

7.5.4 Probability of Loss

The probability of loss $\mathbb{P}(Y > \bar{y})$ (also known as the probability of failure) is one of the most widely used measures of risk and has the following properties.

- What constitutes a loss is defined by the threshold value $\bar{y}$ (defined by the DM). Consequently, it is more appropriate to call this measure the "probability of acceptable loss."
- The meaning of the probability of loss is intuitive in many applications; the probability of loss is the most widely used notion of risk.
- Some important caveats of the probability of loss is that the DM needs to select a threshold value; moreover, this measure does not take into consideration the actual magnitude of the losses. For example, even if $\mathbb{P}(Y > \bar{y}) \leq \mathbb{P}(Y' > \bar{y})$, one might still prefer Y' because losses incurred are not that large compared to those of Y. In other words, Y' is less "catastrophic" than Y. Recall that CVaR captures the magnitude of losses, and it is thus preferred when the magnitude of the losses is important in making a decision.
- The probability of loss is *not a coherent risk measure*; specifically, it does not satisfy translation invariance. This is rather surprising, given how widely used this risk measure is. However, note that the probability of loss is directly compatible and related to SD (as it compares probabilities).
- The DM needs to specify a threshold value to express what constitutes an acceptable loss, and this selection often leads to disagreement of DMs. In other words, what is an acceptable loss limit for one DM might not be acceptable for another DM.
- Note that the probability of loss is an event probability (probability of the event $Y > \bar{y}$). It is possible to define more general event probabilities and use these as risk measures; for instance, we can define the probability of the event $\underline{y} \leq Y \leq \bar{y}$ (probability that cost is in a specific range). As we discussed in Chapter 3, we can use event probabilities to model complex decision-making logic that DMs might follow.

7.5.5 Risk Measures as Vector Norms

Coherency properties of risk measures resemble coherency properties of *vector norms* (seen in linear algebra); viewing coherency properties from the vector perspective can be useful in gaining insight into how risk measures work.

Norm properties give us guidance into how we should measure and compare vectors (which are collections of values); to highlight this, we ask ourselves the following question:

How can we tell if the vector $\mathbf{y} = (y_1, y_2, \ldots, y_S)$ is better than the vector $\mathbf{y}' = (y_1', y_2', \ldots, y_S')$?

For simplicity, here we will assume that all the entries of the vectors are non-negative and define the set $\mathcal{S} = \{1, \ldots, S\}$. The previous question reveals similar difficulties that arise when comparing RVs; specifically, we can consider the following alternatives.

We can say that $\mathbf{y}$ is better than $\mathbf{y}'$ if the total magnitude of its entries is lower:

$$\sum_{\omega \in \mathcal{S}} y_\omega \leq \sum_{\omega \in \mathcal{S}} y_\omega'. \tag{7.28}$$

This can be written as $\frac{1}{S} \sum_{\omega \in \mathcal{S}} y_\omega \leq \frac{1}{S} \sum_{\omega \in \mathcal{S}} y_\omega'$ and is analogous to the expected value risk measure.

We can also say that $\mathbf{y}$ is better than $\mathbf{y}'$ if:

$$\max_{\omega \in \mathcal{S}} y_\omega \leq \max_{\omega \in \mathcal{S}} y_\omega'. \tag{7.29}$$

This is analogous to the worst-case risk measure.

Another possibility is to say that $\mathbf{y}$ is better than $\mathbf{y}'$ if:

$$y_\omega \leq y_\omega' \quad \text{for all } \omega \in \mathcal{S}. \tag{7.30}$$

This is analogous to saying that the RV Y is *always better* than the RV Y'. This notion is not consistent because the entries of $\mathbf{y}$ and $\mathbf{y}'$ may not be arranged in order (by magnitude); as such, maybe an entry of $\mathbf{y}$ can be larger than any of the entries of $\mathbf{y}'$. This inconsistency is analogous to that found when comparing RVs. A more consistent approach would be to first arrange entries in decreasing order and then compare the rearranged entries. This is *precisely analogous to SD*; we establish a threshold on the magnitude of the entries and count how many entries for the vectors are below that threshold value. We can then say that $\mathbf{y}$ is better than $\mathbf{y}'$ if its arranged entries are more frequently above the arranged entries of the competing vector (for any threshold value).

Another possibility is to arrange the entries in increasing order and sum the largest k entries; we can then declare that a vector is better if it has a smaller sum for the largest k entries. The sum of the largest k entries is in fact a vector norm that is often called the *k-max norm*; this norm is *directly analogous to CVaR* (which takes the average of all values above the quantile). In the CVaR case, the k entries are selected by setting the probability level α.

Vector norms are often used to compare vectors. We recall that the norm $\rho(\mathbf{y})$ of a vector $\mathbf{y}$ is a measure of its size. The most used vector norm is the ℓ_p norm:

$$\|\mathbf{y}\|_p = \left(\sum_{\omega \in \mathcal{S}} y_\omega^p \right)^{1/p}, \tag{7.31}$$

for some value $p = 1, 2, \ldots, \infty$. This norm has the following special cases:

- For $p = 1$, $\|\mathbf{y}\|_1 = \sum_{\omega \in \mathcal{S}} y_\omega$ and note that this is S times the average magnitude of the entries (analogous to the expected value).
- For $p = \infty$, $\|\mathbf{y}\|_\infty = \max_{\omega \in \mathcal{S}} y_\omega$ and note this is the largest value (analogous to the worst-case).

Once we have defined a norm, we can use them to compare vectors. For instance, we can say that a vector $\mathbf{y}$ is better than another vector $\mathbf{y}'$ if $\|\mathbf{y}\|_p \leq \|\mathbf{y}'\|_p$.

As in the case of risk measures, one can define many types of norms to measure and compare vectors in different ways. For instance, the k-max norm was proposed in the machine learning community to capture outliers and is defined as the sum of the k largest entries of $\mathbf{y}$. Recently, researchers in the stochastic optimization community have proposed the CVaR norm of a vector $\mathbf{y}$, which is the k-max norm divided by k (the average of the largest k entries) [66]. For the CVaR norm, the α level is given by $\alpha = 1 - k/S$. The CVaR norm has an interesting averaging behavior that is analogous to the p-norm: When $k = S$, we take the average of all the entries (as in the $p = 1$ norm and equivalent to setting $\alpha = 0$); moreover, when $k = 1$, we have that the CVaR norm becomes the worst value of the vector (as in the $p = \infty$ norm and equivalent to set $\alpha = 1$).

So which of all the previous options for comparing vectors is consistent/proper? As we can imagine, there are many possibilities because there are many types of vector norms; therefore, we need a set of rules/axioms that help us come up with proper norms.

A *proper* norm of a vector must satisfy the following properties:

- Homogeneity:
$$\rho(c \cdot \mathbf{y}) = c \cdot \rho(\mathbf{y}) \text{ for } c \in \mathbb{R}_+.$$
- Subadditivity:
$$\rho(\mathbf{y} + \mathbf{y}') \leq \rho(\mathbf{y}) + \rho(\mathbf{y}').$$

The ℓ_p norm and the CVaR norm satisfy these properties and are thus proper. The homogeneity property is analogous to that of a proper risk measure. In the context of vector norms, the subadditivity property is known as the *triangle inequality*, and we note that this is directly analogous to the subadditivity property of risk measures. Although monotonicity is not a required property of vector norms, we should also expect this to hold in practical settings. Specifically, if the entries of a vector are frequently below the entries of another vector, we should expect that a proper norm captures this fact. Similarly, translation invariance is not a required property of vector norms, but we should expect this to hold in practical situations (as adding a constant quantity should not affect comparisons). There are other properties that are specific to norms and that must be satisfied by a proper norm; we do not mention those here to keep the discussion simple. Our discussion here only aims to highlight the interpretation of risk measures as vector norms.

Example 7.6 Risk Measures as Vector Norms

Consider the vectors:

$$\mathbf{y} = (100, 70, 50, 20, 10)$$
$$\mathbf{y}' = (110, 80, 40, 10, 0).$$

The mean and worst-case values for such vectors are:

- Mean is:
$(1/5) \cdot (100 + 70 + 50 + 20 + 10) = 50$ and $(1/5) \cdot (110 + 80 + 40 + 10 + 0) = 48$.
- Worst-Case is: 100 and 110.

From these values, we can compare the vectors and determine which one is "better." Using the mean, we have that $\mathbf{y}$ is better (because the mean is lower); using the worst-case, we also have that $\mathbf{y}$ is better (because the largest value is smaller).

The CVaR values for such vectors is:

- If $\alpha = 1 - 5/5$, then CVaR is:
$(1/5) \cdot (100 + 70 + 50 + 20 + 10) = 50$ and $(1/5) \cdot (110 + 80 + 40 + 10 + 0) = 48$.
- If $\alpha = 1 - 4/5$, then CVaR is:
$(1/4) \cdot (100 + 70 + 50 + 20) = 60$ and $(1/4) \cdot (110 + 80 + 40 + 10) = 60$.
- If $\alpha = 1 - 3/5$, then CVaR is:
$(1/3) \cdot (100 + 70 + 50) = 73$ and $(1/3) \cdot (110 + 80 + 40) = 76$.
- If $\alpha = 1 - 1/5$, then CVaR is:
$(1/1) \cdot (100) = 100$ and $(1/1) \cdot (110) = 110$.

The CVaR norm extracts the largest elements of the vector and has the mean and worst-case values as extremes (ℓ_1 and ℓ_∞ norms). Note also that different values of α lead to different conclusions on which vector is better. □

7.6 Stochastic Optimization

The previous tools allow us to systematically compare decisions when we are facing uncertainty. Consider now the situation in which we do not want to compare decisions (and associated random costs), but rather we would like to find the *best decision possible* that helps us mitigate (or exploit) uncertainty. To do so, we will aim to formulate optimization problems that explicitly take uncertainty into account. This will require us to use risk measures, which will aim to shape the distribution of the cost in some desired form. The discussion will also introduce interesting and important notions of *conflict resolution* (trading-off different risk measures and constraints) and *recourse* (corrective) decisions.

7.6.1 Basic Formulation

Our ultimate goal is to find a decision u that optimizes the random cost $\varphi(X, u)$. In order to formulate the problem properly, we aim to optimize a risk measure;

we recall that a risk measure summarizes the random cost into a scalar quantity. The selection of the risk measure gives rise to a family of possible optimization problems; some examples include:

- Mean-variance:

$$\min_{u \in \mathcal{U}} \; \mathbb{E}[\varphi(X,u)] + \kappa \cdot \mathbb{V}[\varphi(X,u)]. \tag{7.32}$$

- Probability of loss:

$$\min_{u \in \mathcal{U}} \; \mathbb{P}(\varphi(X,u) > y). \tag{7.33}$$

- Expected shortfall (CVaR):

$$\min_{u \in \mathcal{U}} \; \mathbb{E}[\varphi(X,u) \mid \varphi(X,u) \geq \mathbb{Q}(\alpha)]. \tag{7.34}$$

These formulations (known as stochastic optimization formulations) are different from deterministic formulations discussed previously because they *explicitly* take the uncertainty of X into account (as opposed to assuming that X takes a specific value x). As such, stochastic formulations tend to obtain decisions u^* that protect us against uncertainty (or that exploit uncertainty to achieve gain). What type of protection/exploitation we are achieving depends on the risk measure that we use.

It is also important to highlight that the *ultimate goal* of a stochastic optimization formulation is to manipulate the shape of the pdf/cdf of the cost $Y(u)$; this observation is key because, in general, it is quite difficult (if not impossible) to have full control over the shape of the pdf/cdf by optimizing a risk measure. For instance, by selecting to minimize a specific aspect of the RV (e.g., its average), we might need to sacrifice some other aspects (e.g., probability of loss or variance). This tends to introduce conflicting trade-offs that we need to resolve.

Another approach to try to control the shape of the pdf of $Y(u)$ is to directly go back to the definition of SD and find a decision that dominates or that approaches a given benchmark (given by a reference RV Y^*). In other words, we seek u such that $Y(u) \preceq Y^*$; an important issue with this approach is that the problem might have no solution; in other words, it might be impossible to find a decision such that $\mathbb{P}(Y(u) \leq \bar{y}) \geq \mathbb{P}(Y^* \leq \bar{y})$ for all $\bar{y} \in \mathcal{D}_y$. An alternative would be to *relax* our requirement of SD and ensure that $Y(u)$ is close to Y'; we need to define a notion of closeness/distance between RVs (we will discuss how to do this shortly).

Example 7.7 Deterministic vs. Stochastic Decision-Making

Consider the generator sizing setting of Example 7.1. Assume now that the DM would like to select the best size of the system in a way that it captures the uncertainty of the facility demand. To do so, the DM formulates an optimization problem that minimizes the CVaR with a probability value of $\alpha = 0.9$ (which penalizes the 10% highest values of the cost). With this, the DM is trying to control the high costs. The problem is given by:

$$\min_{u \in \mathcal{U}} \text{CVaR}_{\alpha}[Y(u)]. \tag{7.35}$$

The solution to the stochastic problem selects a generator with a capacity of $u^* = 4$, which is smaller than that of the deterministic solution $u_D^* = 6$.

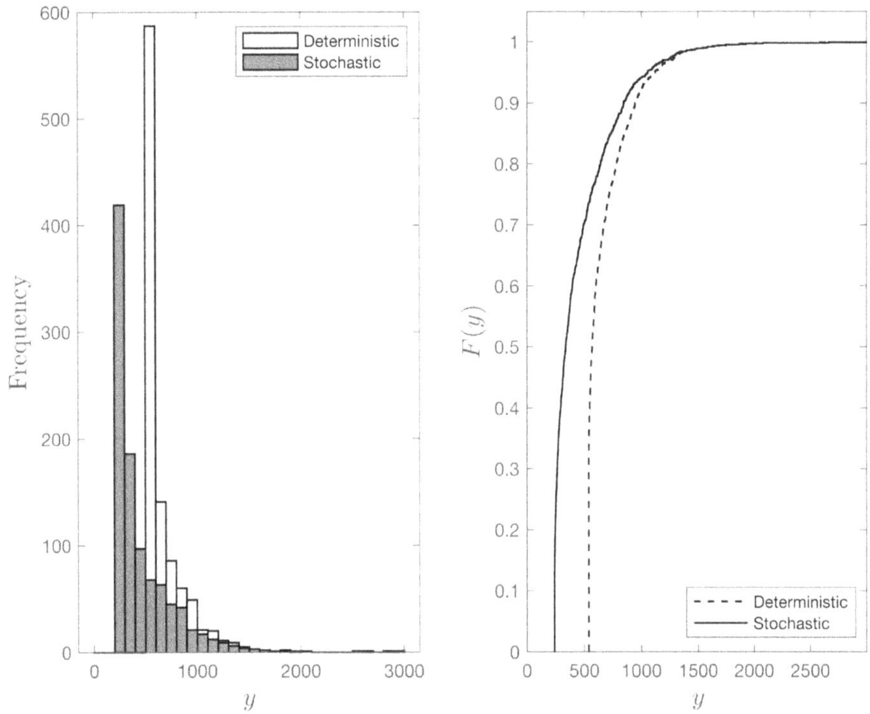

Figure 7.6 Pdf and cdf for generator sizing problem using deterministic and stochastic decision-making.

In Figure 7.6, we compare the pdf/cdf for the deterministic and stochastic design; from the pdf, it is clear that accounting for uncertainty has the effect of reducing overall costs; specifically, we highlight that the average cost achieved with the stochastic setting is 460.5, and we recall that the average cost for the deterministic setting is 674.1. In addition, it is interesting that the stochastic setting offers the same level of protection to extreme costs as the deterministic setting discussed in Example 7.1; specifically, from the cdfs, we see that the stochastic solution dominates the deterministic solution and has the same probability of achieving costs lower than a value of 1,500. This indicates that the deterministic solution is *unnecessarily conservative*.

The ability to *systematically* control the shape of the pdf/cdf of the cost is a key benefit of using stochastic optimization techniques (compared to deterministic techniques). This is particularly true for CVaR because we can tune the probability level α to explore different levels of protection to high costs.

Scripts to reproduce results: `ch7_generator_stoch` □

7.6.2 Conflict Resolution

The different advantages and disadvantages encountered with risk measures highlight the difficulty in controlling and comparing RVs. In particular, different DMs might use different risk measures, and such measures are often *conflicting* (e.g., reducing one measure often results in increasing another measure). To deal with this situation, we might be interested in minimizing a given risk measure while constraining the value of another risk measure. This can be done by formulating *constrained* stochastic optimization problems.

Consider the situation in which we have a random cost $Y(u)$ for which we want to minimize the risk measure $\rho_o(Y(u))$ while keeping the value of another risk measure $\rho_c(Y(u))$ below a certain value $\bar{\rho} \in \mathbb{R}$. This can be formulated as the optimization problem:

$$\min_{u \in \mathcal{U}} \ \rho_o(Y(u)) \tag{7.36a}$$

$$\text{s.t.} \ \rho_c(Y(u)) \leq \bar{\rho}. \tag{7.36b}$$

The solution to this problem for different values of $\bar{\rho}$ gives a family of optimal solutions that is known as the *trade-off (Pareto) frontier*. From the trade-off frontier, a DM can select a decision that gives the best possible value of $\rho_o(Y(u))$ while obtaining an acceptable value $\rho_c(Y(u))$. This solution is often called a *compromise solution*. Note that this also imposes a priority/preference/ranking over the risk measures.

In most practical situations, the expected value is used as a risk measure in the objective, and risk measures that mitigate extreme events (e.g., variance or CVaR) are used as a constraint. In this context, the parameter $\bar{\rho}$ can be interpreted as a *risk budget* or *risk tolerance*. One typically observes that setting $\bar{\rho} \to 0$ (less risk tolerance) offers better protection to extreme events but also increases the expected cost; this highlights the fact that, typically, obtaining more robustness comes at a cost. For example, ensuring that a chemical process will satisfy demand with high probability might require that we build large inventory tanks. Similarly, ensuring that the power grid satisfies the demands of customers all the time might require that we build a network with many power lines and that we add spare generators.

As we discussed in the context of estimation in Chapter 5, one can also construct a trade-off frontier by minimizing a weighted sum of the risk measures:

$$\min_{u \in \mathcal{U}} \ \rho_o(Y(u)) + \kappa \cdot \rho_c(Y(u)), \tag{7.37}$$

where $\kappa \in \mathbb{R}_+$ is a weighting factor. Note that this formulation is analogous to the mean-variance formulation; in fact, when we select $\rho_o(Y(u)) = \mathbb{E}[Y(u)]$ and $\rho_c(Y(u)) = \mathbb{V}[Y(u)]$, we obtain the mean-variance formulation. The constrained and weighted formulation allows us to navigate the trade-offs between different risk measures; however, the constrained formulation is often more intuitive, as we can have direct control over the acceptable level of risk that we are willing to tolerate. It is also worth highlighting that some risk measures naturally capture trade-offs between different aspects of $Y(u)$; for instance, minimizing the coefficient of variation $\mathbb{SD}[Y(u)]/\mathbb{E}[Y(u)]$ has the effect of minimizing variability while also maximizing the expected value.

In many applications, a DM often encounters the situation in which they want to minimize the risk of cost $Y(u)$ while constraining the risk of another RV $Y'(u)$ (such as environmental impact or safety). This can be done by formulating the problem:

$$\min_{u\in\mathcal{U}} \ \rho_o(Y(u)) \tag{7.38a}$$

$$\text{s.t.} \ \rho_c(Y'(u)) \leq \bar{\rho}. \tag{7.38b}$$

As before, one can construct trade-off frontiers to determine a suitable solution. *Risk constraints* of the form $\rho_c(Y(u)) \leq \bar{\rho}$ can take diverse forms (based on the selection of the risk measure). Viewing risk measures through the lens of constraints provides additional perspectives on how to interpret their behavior. Some common risk constraints include the following:

- Expected value constraint:

$$\mathbb{E}[Y(u)] \leq \bar{\rho}. \tag{7.39}$$

  This constraint controls the overall magnitude of the random cost but might lead to severe violations for outcomes that are much larger than the expected value.
- Probability of loss constraint:

$$\mathbb{P}(Y(u) > \bar{y}) \leq \bar{\rho}. \tag{7.40}$$

  This constraint offers direct control over the probability of surpassing a value above a given threshold. This type of constraint is also known as a *probabilistic or chance constraint.*

  When we set $\bar{\rho} = 0$, we have $\mathbb{P}(Y(u) > \bar{y}) = 0$; in other words, we are completely forbidding that the cost $Y(u)$ takes any values above $\bar{y}$. This is a strict constraint that is also known as an *almost surely* constraint. In simple terms, this ensures that $Y(u) \leq \bar{y}$ for all possible outcomes of $Y(u)$; because of this, this is equivalent to enforcing that $\mathbb{P}(Y(u) \leq \bar{y}) = 1$. In this case, it is said that the constraint $Y(u) \leq \bar{y}$ holds with probability one (w.p.1. for short).
- CVaR constraint:

$$\mathbb{E}[Y(u)|Y(u) \geq \mathbb{Q}(\alpha)] \leq \bar{\rho}. \tag{7.41}$$

  This constraint aims to keep CVaR below a certain threshold. Because of the properties of the expected shortfall, this constraint can capture a range of behaviors that span from the expected value constraint $\mathbb{E}[Y(u)] \leq \bar{\rho}$ to the worst-case constraint $\max Y(u) \leq \bar{\rho}$; note that the latter ensures that $\mathbb{P}(Y(u) \leq \bar{\rho}) = 1$ (equivalent to the almost surely constraint when $\bar{\rho} = \bar{y}$). As such, CVaR constraints offer different levels of protection.

Example 7.8 Conflict Resolution in Generator Sizing

We revisit the generator sizing setting of Example 7.1. The DM is now interested in exploring the trade-off between expected cost and cost variability (standard deviation). The standard deviation is selected as a way to penalize extreme/peak costs. The DM formulates the stochastic optimization problem with a mean-variance objective:

$$\min_{u \in \mathcal{U}} \mathbb{E}[Y(u)] + \kappa \cdot \mathbb{SD}[Y(u)]. \tag{7.42}$$

The DM explores the trade-off between the objectives by trying different values of the parameter $\kappa \in [0, 20]$; when $\kappa = 0$, the generator is sized so as to minimize the mean cost (while ignoring large costs); when κ is large, the generator is sized to try to reduce large costs. To simplify the solution to the problem, the generator capacity is assumed to be continuous and given by the domain $\mathcal{U} = [0, 10]$.

The DM is aware that $\mathrm{CVaR}_\alpha[Y(u)]$ is a risk measure that can capture similar behavior; specifically, when $\alpha = 0$, CVaR becomes the mean cost $\mathbb{E}[Y(u)]$ and, when $\alpha = 1$, CVaR becomes the worst-case cost $\max Y(u)$. The DM thus solves the problem:

$$\min_{u \in \mathcal{U}} \mathrm{CVaR}_\alpha[Y(u)], \tag{7.43}$$

for different values of $\alpha = [0, 1]$ and explores the behavior of $\mathbb{E}[Y(u)]$ and $\mathbb{SD}[Y(u)]$ as α is varied.

The results are presented in Figure 7.7; in Figure 7.7(a), we have the trade-off curve of mean cost and standard deviation for different values of κ. We can see that there is indeed a conflict between these risk measures; specifically, if we aim to reduce the mean cost, we must increase the standard deviation (and thus increase peak costs). In Figure 7.7(b), we see how the optimal generator capacity changes as we put more emphasis on the standard deviation until we reach a minimum value of around $\mathbb{SD}[Y(u^*)] = 112$; here, we can see that the capacity must increase for us to reduce the variability of the costs. To completely minimize the variability, we need to have a generator of the maximum available capacity ($u^* = 10$). As such, the standard deviation cannot be reduced further if we keep increasing κ.

In Figure 7.7(c and d), we have the trade-off curve of the mean cost and standard deviation as we span α for CVaR. Note how this has a similar effect as κ; specifically, the standard deviation is decreased and the optimal generator capacity increases as we increase α. This highlights how CVaR is a risk measure that can be naturally used to explore trade-offs. However, when comparing the trade-off curve with that of the mean-variance risk measure, we can observe a discrepancy; when we reach the maximum value of $\alpha = 1$, the standard deviation achieved ($\mathbb{SD}[Y(u^*)] = 200$) is higher than that of the limiting case obtained with the mean-variance objective ($\mathbb{SD}[Y(u^*)] = 112$). Moreover, when $\alpha = 1$, we have that the largest generator selected has a capacity of approximately $u^* = 7$ (much smaller than that obtained with a mean-variance risk measure).

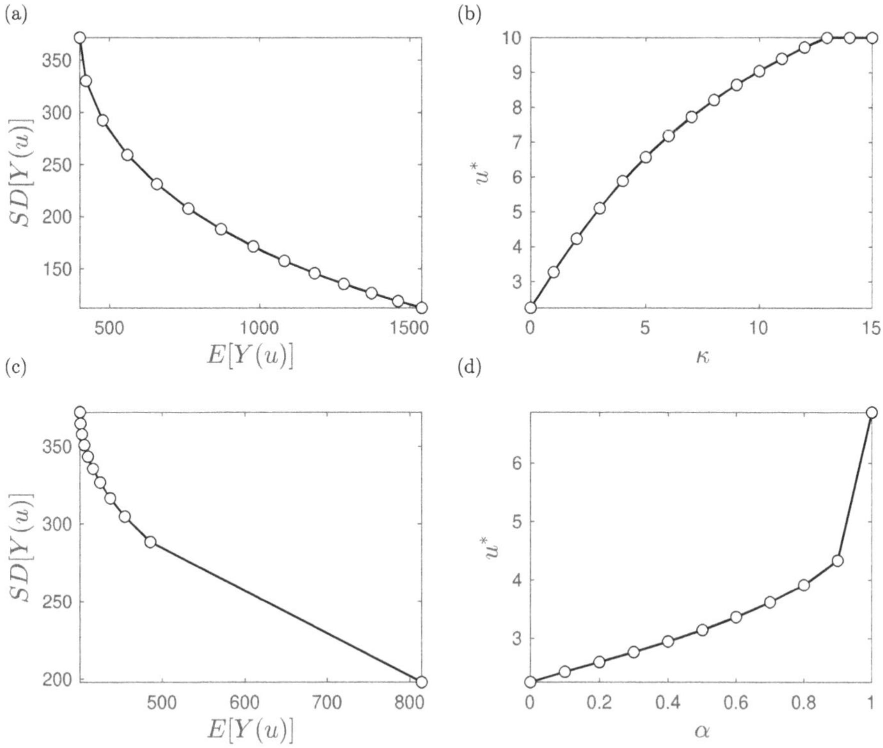

Figure 7.7 Trade-off curve obtained as we span κ in mean-variance risk measure (a and b). Trade-off curve obtained as we span α in CVaR risk measure (c and d).

So what is going on here? The differences observed are due to the asymmetric penalization nature of CVaR and the symmetric penalization nature of the standard deviation (variance). Specifically, the standard deviation penalizes the extreme costs on both tails of the distribution at the same time, while CVaR only penalizes the extreme costs on the right tail of the distribution. This difference is extremely important and highlights the danger of using the standard deviation as a measure of risk. In particular, minimizing the standard deviation is an *indirect* approach to mitigate large costs (does not offer direct control to mitigate large costs). This can result in inconsistent and/or extremely conservative solutions (in this case, this is manifested in having a large generator). CVaR, on the other hand, offers direct control over the large costs (for $\alpha = 1$, it becomes the worst-case cost), and this leads to more consistent control of extreme costs.

Scripts to reproduce results: `ch7_generator_conflict` □

7.6.3 Formulations with Recourse

The goal of a stochastic optimization formulation is to find a decision u that protects us against uncertainty. This decision is taken in *anticipation* of uncertainty; however, in many practical situations, we have the ability to protect ourselves against uncertainty by using *corrective actions* that are taken once we observe a particular outcome of our uncertainty. Such corrective actions are also known as *recourse or feedback*.

Consider, for instance, the situation in which we design a chemical process based on a set of possible market conditions. The design process sets the size of equipment (this is our decision made in anticipation of uncertainty); however, once the process is operational and we observe the actual market conditions, we might be able to still correct how we run the chemical process (these are our corrective decisions/actions). As an another example, consider the situation in which you select a given driving route based on expected traffic conditions; once you observe the actual traffic on that route, you might decide that such route is not the best one and decide to change routes. This type of *sequential* decision-making is inherent in how we, as humans, anticipate and respond to uncertainty.

Stochastic optimization techniques can capture sequential decision-making; in order to explain this concept, we consider a system model of the form:

$$Y = \varphi(X, u, W(X)). \tag{7.44}$$

Here, u is a decision that is made in anticipation of uncertainty and $W(X)$ is a decision that is made once uncertainty is realized (once we observe uncertainty). Note that $W(X)$ is an RV because it is a function of the RV X; this highlights the fact that this decision is taken depending on our observations of X. This differs from the decision u, which we are not allowed to correct once the decision has been made.

If we consider outcomes of X, the system can also be written as:

$$y_\omega = \varphi(x_\omega, u, w_\omega),\ \omega \in \mathcal{S}. \tag{7.45}$$

This reveals that there is a corrective decision w_ω for every observed realization of X (given by x_ω). This set of corrective decisions is known as a decision *policy*.

One can formulate a stochastic optimization problem that gives us the best pre-emptive decision u and the best corrective decision policy $W(X)$; for instance, we consider:

$$\min_{u \in \mathcal{U}, W(X) \in \mathcal{W}(u)} \rho_o(\varphi(X, u, W(X))), \tag{7.46}$$

where $\mathcal{W}(u)$ is the set of possible corrective decisions and note that this depends on the decision u. Analogous formulations can be considered that capture risk objectives and risk constraints.

The optimization formulation (7.46) is also known as a two-stage formulation because it takes decisions in a couple of stages (before uncertainty is observed and after uncertainty is observed). This formulation highlights the fact that, when facing uncertainty, it is important to take our first step (u) carefully to ensure that we are in a good enough position to correct errors. This is important because if we do not take a careful first step, we might be put in a position in which we cannot correct our actions (this is highlighted by the dependence of $\mathcal{W}(u)$ on u) or it might be too costly to correct (captured in the objective). For instance, when selecting a driving route, we want to select a route that is not too far from other routes (or it will be difficult to make a transition). Similarly, when designing a chemical process, we need to ensure that the equipment is flexible enough so that we can respond to dynamic market conditions (if the equipment is too rigid, we might not be able to adjust, and we might lose money).

It is possible to generalize a two-stage decision-making setting to a multistage setting in which decisions are progressively taken as uncertainty reveals itself over time. This more advanced setting is not discussed here, but there are a plenty of useful resources that cover how to formulate and solve such problems [67].

Example 7.9 Two-Stage Formulation for Generator Sizing

We consider a variant of the generator sizing problem; assume that we would like to size a generator (by selecting capacity u) to help control the costs for a couple of random loads $X = (X_1, X_2)$ (both of them are Weibull RVs), see Figure 7.8. In addition, assume that the generator can be operated flexibly; specifically, the power produced by the generator (denoted as $W(X)$) satisfies $W(X) \leq u$.

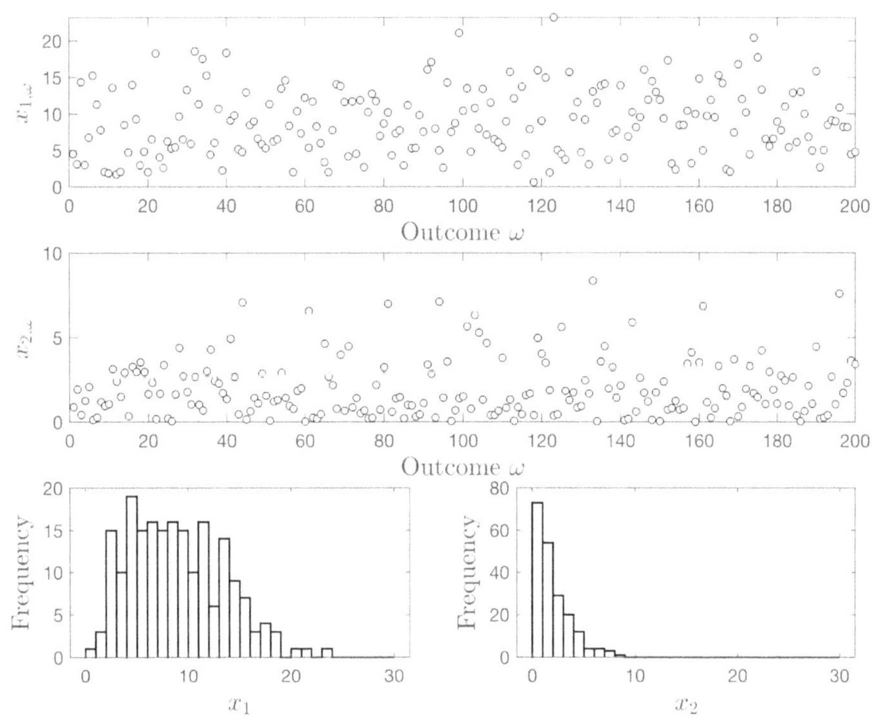

Figure 7.8 Data for random loads X_1 and X_2 and empirical pdfs.

The power of the generator can be adjusted depending on the load conditions X observed; however, the capacity of the generator cannot be adjusted once this is selected (since this is the designed capacity of the system). The residual load requested for the power grid is:

$$R = X_1 + X_2 - W(X). \tag{7.47}$$

The cost function of the system is:

$$\varphi(X,u,W(X)) = \alpha_R \cdot R^2 + \alpha_u \cdot u^2. \tag{7.48}$$

We assume that we can only select between a couple of generator sizes $\mathcal{U} = \{1, 10\}$. In addition, in order to prevent power waste, we assume that the power generator output is bounded as $0 \le W(X) \le X_1 + X_2$. As such, the constraint set for the generator output is:

$$\mathcal{W}(u) = \{0 \le W(X) \le X_1 + X_2 \;\&\; W(X) \le u\}. \tag{7.49}$$

We note that this depends on the decision u; specifically, note that the decision u dictates how flexible the system can be to reduce costs (a smaller generator is less flexible). As such, the decision u has to be made carefully; in order to do this, we formulate the stochastic optimization problem:

$$\min_{u \in \mathcal{U}, W(X) \in \mathcal{W}(u)} \quad \mathrm{CVaR}_\alpha[\varphi(X,u,W(X))]. \tag{7.50}$$

Here, we aim to protect the system against large costs, and thus, we set $\alpha = 0.9$. The solution to this problem yields an optimal generator size of $u^* = 1$ and an optimal CVaR value of 1,684.

In order to establish a comparison, we consider the solution to the problem using a deterministic setting; here, we consider that both loads take their worst-case values $x_1 = \max_{\omega \in \mathcal{S}} x_{1,\omega}$ and $x_2 = \max_{\omega \in \mathcal{S}} x_{2,\omega}$, and we find the optimal size for the system by solving the deterministic optimization problem:

$$\min_{u \in \mathcal{U}} \; \alpha_R \cdot r^2 + \alpha_u \cdot u^2 \tag{7.51a}$$

$$\text{s.t. } r = x_1 + x_2 - w \tag{7.51b}$$

$$0 \le w \le x_1 + x_2 \tag{7.51c}$$

$$w \le u. \tag{7.51d}$$

We note that this problem is much simpler; however, we note that this cannot anticipate the inherent variability of the loads. Specifically, it tries to anticipate this by selecting their worst-case values. This is a highly conservative assumption and directly manifests in the optimal solution; specifically, the optimal sizing that results from the problem is $u_D^* = 10$. When we evaluate the performance of this generator in the realizations of the random loads, we find that the CVaR value obtained is 1,937. As can be seen, there is a significant penalty that has to be paid for the failure to properly anticipate the uncertainty using the deterministic approximations x_1, x_2 of the RVs X_1, X_2.

These types of inconsistencies can be encountered when neglecting uncertainty, particularly when many RVs are present; specifically, it is not always obvious how to properly select combinations of representative values of the RVs.

Scripts to reproduce results: `ch7_generator_twostage` □

7.6.4 Shaping Formulations

Recall that the ultimate goal of a stochastic optimization formulation is to *shape* the distribution of the random cost $Y(u)$. We also recall that, ideally, we want a solution u^* such that its random cost $Y(u^*)$ stochastically dominates the random cost $Y(u)$ of any other available decision $u \in \mathcal{U}$.

Consider that the decision set $\mathcal{U}$ is discrete and thus contains a finite number of decisions $u_1, u_2, \ldots, u_K$ with associated random costs

$$Y(u_1), Y(u_2), \ldots, Y(u_K). \tag{7.52}$$

We want to find a decision $u \in \mathcal{U}$ such that:

$$Y(u) \preceq Y(u_k), \quad k = 1, 2, \ldots, K. \tag{7.53}$$

As we have seen, it might be impossible for such an ideal decision to exist (because SD is a strict requirement); as such, we need to relax our requirements. One possibility is to consider a *target* random cost Y^* that captures some *ideal behavior* that we want and consider minimizing the distance between our cost $Y(u)$ to such ideal behavior. The distance can be measured as:

$$\mathbb{D}(Y(u), Y^*) = \int_{y \in \mathcal{D}_Y} |F_{Y(u)}(y) - F_{Y^*}(y)| dy. \tag{7.54}$$

Here, $\mathcal{D}_Y$ is the domain for the cost $Y(u)$ and the target Y^*. This distance measure is known as the *Wasserstein distance*; this is a powerful tool that can be used to quantify how similar/dissimilar a couple of RVs are. Note that the degree of similarity is given by the integral distance between their cdfs, highlighting the critical role that the cdfs play in characterizing the behavior of RVs. Note also that the Wasserstein distance is different from the distance that we used in the Kolmogorov–Smirnov (KS) test of Chapter 5; specifically, the KS test compares RVs by measuring the maximum distance between their cdfs. This highlights how one can use different distance measures to determine how similar/dissimilar are RVs.

Having defined a sense of distance to the ideal target, we can formulate a stochastic optimization problem that finds a decision that has minimum distance:

$$\min_{u \in \mathcal{U}} \mathbb{D}(Y(u), Y^*). \tag{7.55}$$

This formulation aims to bring the cdf of $Y(u)$ as close as possible to the reference/target cdf of Y^*. This formulation is analogous to how optimal dynamic trajectories are obtained (e.g., for rockets); specifically, one often aims to compute a trajectory that is as close as possible to some target trajectory. This formulation is intuitive and highlights the inherent nature of stochastic optimization, which ultimately aims to shape the distribution of the cost. We note that our ability to shape the distribution is dictated by the decisions that we have available and by the nature of the underlying RVs (e.g., aiming to shape an exponential RV as a Gaussian RV will be difficult). As such, properly selecting a target cdf is important; this is analogous to the selection of a dynamic trajectory for a rocket (a rocket trajectory is constrained by physics and might not be capable to hit certain types of targets).

7.6.5 Quantifying Flexibility and Robustness

In decision-making, we might be interested in quantifying how *flexible/robust* a given system is. We now explore how to use concepts of stochastic optimization to assess this; this discussion will open new perspectives on how uncertainty impacts system behavior.

Consider a system that is modeled using the random performance function $\varphi(X,u)$ and assume that system operation is deemed *feasible* whenever the event $\varphi(X,u) \leq \bar{\varphi}$ holds; conversely, system operation is deemed *infeasible* if the event $\varphi(X,u) > \bar{\varphi}$ holds. Note how, in this setting, the constraint $\varphi(X,u) \leq \bar{\varphi}$ is interpreted as an *event*; this is analogous to our discussions on event probabilities of Chapter 3.

Now consider that we want to determine the probability that the system will have feasible operation under decision u; this is given by:

$$\delta(u) = \mathbb{P}(\varphi(X,u) \leq \bar{\varphi}). \tag{7.56}$$

Here, the quantity $\delta(u) \in [0,1]$ can be interpreted as a *measure of flexibility/robustness* of the system; the higher this value is, the more flexible the system (the more ability it has to withstand/absorb uncertainty). This flexibility index was proposed by Straub and Grossmann [68]. We can also determine the probability that the system will have feasible operation under an alternative decision u', which is given by $\delta(u') = \mathbb{P}(\varphi(X,u') \leq \bar{\varphi})$. We thus say that u is a more robust/flexible decision than u' if it has a higher probability of feasible operation:

$$\delta(u) \geq \delta(u'). \tag{7.57}$$

Consider now that the system operation is described by a *set of performance functions* $\varphi_k(X,u)$, $k \in \mathcal{K}$, which can capture diverse factors such as cost, carbon emissions, energy use, water use, and safety. We say that the system exhibits feasible operation if the event $\varphi_k(X,u) \leq \bar{\varphi}_k$ holds for all $k \in \mathcal{K}$. In other words, the system is feasible if all our factors are below their corresponding threshold values. We can measure the flexibility of this system by using the event probability:

$$\delta(u) = \mathbb{P}(\varphi_k(X,u) \leq \bar{\varphi}_k,\ k \in \mathcal{K}). \tag{7.58}$$

We can thus see that event probabilities are a useful tool to quantify and optimize the flexibility of systems. In this context, for instance, we might aim to determine a decision u that maximizes flexibility $\delta(u)$:

$$\max_{u \in \mathcal{U}} \; \delta(u). \tag{7.59}$$

This will provide a decision u^* that can tolerate as much uncertainty as possible (e.g., increase the size of our generator so as to guarantee high probability of feasible operation). Moreover, if we find that $\delta(u^*) = 1$, we have that the optimal system fully tolerates uncertainty, while, if $\delta(u^*) < 1$, we have that the optimal system cannot fully withstand uncertainty. It is also possible to derive a formulation that finds a decision u that maximizes flexibility under a given cost budget (as increasing flexibility/robustness will often involve higher costs) and to build a Pareto frontier that allows us to explore the trade-off between flexibility and cost.

7.6.6 Numerical Solution

Determining decisions in the face of uncertainty is computationally challenging because one often encounters multidimensional integrals when computing risk measures (when RVs are continuous). In addition, computing some risk measures involves complex operations (e.g., CVaR and quantiles), and one might need to evaluate system performance over a wide range of scenarios. Here, we discuss some strategies to deal with these issues.

To motivate our discussion, consider the following stochastic optimization problem:

$$\min_{u \in \mathcal{U}} \; \mathbb{E}[\varphi(X,u)], \tag{7.60}$$

with solution u^*. Here, we assume that $X = (X_1, X_2, \dots, X_n)$ is a continuous and multivariate RV; as such, the expectation of the cost is given by a multidimensional integral of the form:

$$\mathbb{E}(u) = \int_{X_n} \cdots \int_{X_2} \int_{X_1} \varphi(X_1, X_2, \dots, X_n, u) dX_1 dX_2 \cdots dX_n. \tag{7.61}$$

Here, we use the simplified notation $\mathbb{E}(u) = \mathbb{E}[\varphi(X,u)]$. As we discussed in Chapter 4, the above integral is impossible to compute explicitly in most applications; as such, the optimal decision u^* is often *impossible to compute*. However, as we saw in our discussion on Monte Carlo methods, we can compute empirical approximations for this integral, using sampling, and we can assess the quality of such approximations.

The idea is to generate a set of random samples x_ω, $\omega \in \mathcal{S}$ (or use available data) to compute the sample (empirical) mean:

$$\hat{\mathbb{E}}_{\mathcal{S}}(u) = \frac{1}{S} \sum_{\omega \in \mathcal{S}} \varphi(x_\omega, u). \tag{7.62}$$

We use this to compute the decision $\hat{u}_S$ from:

$$\min_{u \in \mathcal{U}} \hat{\mathbb{E}}_S(u). \tag{7.63}$$

The decision $\hat{u}_S$ is an approximation of the optimal decision u^*; the quality of this approximation depends on the number of samples S used. From the law of large numbers, we have that $\hat{\mathbb{E}}_S(u) \to \mathbb{E}(u)$ as $S \to \infty$; as such, we have that $\hat{u}_S \to u^*$ as $S \to \infty$.

We have seen that we can assess the quality of the sample estimate $\hat{\mathbb{E}}_S(u)$ (with respect to $\mathbb{E}(u)$) by using the central limit theorem (CLT); the CLT allows us to estimate the variance of the sample estimate. However, in an optimization setting, we are not interested in the quality of the cost $\hat{\mathbb{E}}_S(u)$ per se, but we are more interested in the quality of the decision $\hat{u}_S$.

To assess the quality of the candidate solution $u = \hat{u}_S$ (obtained from the solution of Eq. (7.63)), we can use a *bounding procedure*. We obtain multiple, independent sample sets $\mathcal{S}_j$ for $j = 1, \ldots, B$; each sample set $\mathcal{S}_j$ has S samples. For a given sample set $\mathcal{S}_j$, we evaluate the sample cost:

$$e_S^j(u) = \frac{1}{S} \sum_{\omega \in \mathcal{S}_j} \varphi(x_\omega, u), \tag{7.64}$$

and repeat the process for all sample sets $j = 1, \ldots, B$. This approach evaluates the performance of the candidate solution u in independent sets of samples. From this, we can compute the mean and variance of $e_S^j(u),\ j = 1, \ldots, B$; these quantities are given by:

$$\bar{e}_S = \frac{1}{B} \sum_{j=1}^{B} e_S^j(u) \tag{7.65a}$$

$$\bar{v}_S^2 = \frac{1}{B-1} \sum_{j=1}^{B} (e_S^j(u) - \bar{e}_S)^2. \tag{7.65b}$$

The idea behind this procedure is that, if the number of samples S used for computing the candidate decision u is large, we expect that $\bar{v}_S^2$ will be small and that $\bar{e}_S$ will be close to $U = \mathbb{E}(u)$, which is an upper bound of the actual optimal cost $\mathbb{E}(u^*)$ (recall that it is not possible to compute $\mathbb{E}(u)$ or $\mathbb{E}(u^*)$). The variance $\bar{v}_S^2$ is expected to be small if computing u originally took into consideration a large number of possible outcomes, and thus, when exposed to new situations, the decision will lead to similar system behavior. In other words, we can think of this procedure as a *validation/verification procedure*, in which we test a decision in alternative scenarios. As such, this validation procedure can give interesting insights into the robustness of our decision; this is analogous to the validation procedures we used to test how robust/generalizable are model predictions obtained with parameter estimates that are inferred from data (see Chapter 5).

The previous verification approach does not address the variability of the decision itself as we change scenarios. In order to evaluate this, we use the sample sets $\mathcal{S}_j,\ j = 1, \ldots, B$ again to solve the following family of optimization problems:

$$\min_{u \in \mathcal{U}} \hat{\mathbb{E}}_S^j(u), \tag{7.66}$$

with solutions $\hat{u}_S^j$, $j = 1, \ldots, S$ and where:

$$\hat{\mathbb{E}}_S^j(u) = \frac{1}{S} \sum_{\omega \in \mathcal{S}_j} \varphi(x_\omega, u). \tag{7.67}$$

The optimal objective value of Eq. (7.66) is denoted as $e_S^{j,*}$, and we use it to compute the mean and variance:

$$\underline{e}_S = \frac{1}{B} \sum_{j=1}^{B} e_S^{j,*} \tag{7.68a}$$

$$\underline{v}_S^2 = \frac{1}{B-1} \sum_{j=1}^{B} (e_S^{j,*} - \underline{e}_S)^2. \tag{7.68b}$$

Here, the variance $\underline{v}_S^2$ captures the variability of the optimal decision (of its optimal objective value) as we change the samples used. The idea is that, if the number of samples S is large, the optimal decision will not change much, and thus, the variance $\underline{v}_S$ is small and $\underline{e}_S$ is close to $L = \mathbb{E}[\min_{u \in \mathcal{U}} \varphi(X, u)]$, which is a lower bound of the optimal cost $\mathbb{E}(u^*)$.

The above procedure provides quantities Eq. (7.65a) that can be used to construct a statistical upper bound for the optimal cost U, while the quantities Eq. (7.68) can be used to construct a statistical lower bound for the optimal cost L. Specifically, the $1 - \alpha$ confidence interval for the upper bound is:

$$U \in \left[\overline{e}_S \pm z_{\alpha/2} \cdot \overline{v}_S / \sqrt{B}\right], \tag{7.69}$$

and the $1 - \alpha$ confidence interval for the lower bound is:

$$L \in \left[\underline{e}_S \pm z_{\alpha/2} \cdot \underline{v}_S / \sqrt{B}\right]. \tag{7.70}$$

Here, we recall that $z_{\alpha/2}$ is the quantile $\mathbb{Q}(1 - \alpha/2)$ of $\mathcal{N}(0, 1)$.

Having a lower and upper bound is useful, as this allows us to know how far is our candidate solution u from the actual solution u^*. If the lower and upper bounds are not close enough, this is an indication that the candidate solution u is not of high quality and we might need to refine it by increasing the number of scenarios S. It is also often useful to visualize the region defined by the extremes of the lower and upper confidence intervals:

$$\left[\underline{e}_S - z_{\alpha/2} \cdot \underline{v}_S / \sqrt{B}, \quad \overline{e}_S + z_{\alpha/2} \cdot \overline{v}_S / \sqrt{B}\right]. \tag{7.71}$$

This is the region where we expect our optimal objective value $\mathbb{E}[\varphi(X, u^*)]$ to lie on. One also often uses the bounds to define the solution *gap*, which is given by $\epsilon_S = |\overline{e}_S - \underline{e}_S|$. We also highlight that, in some cases, it is possible to construct bounds for the solution u^* itself; this can be done using the samples obtained during the computation of the lower bound $\hat{u}_S^j$, $j = 1, \ldots, B$. Empirical evidence reported in the work of Linderoth, Shapiro, and Wright [69] indicates that solutions of high quality can be obtained with few scenarios. One of the reasons for this is that stochastic optimization problems typically have constraints, and this limits the variability of their solution.

Assessing the quality of a constrained and two-stage stochastic optimization problems is more complicated, but the basic ideas are similar. Constraints play an essential role because they restrict the space over which the solution and the cost can be; as such, problems that are more constrained tend to converge faster to the solution as we increase the sample size S. However, constraints can also complicate the analysis because in some cases the solution might be infeasible for a given set of samples; as such, specialized procedures are needed to penalize constraint violations. The key insight to remember is that we need to test how the candidate solution behaves in alternative scenarios that were not used when computing this solution. These procedures are analogous to how we test regression and machine learning models using data that were not used for training them; specifically, we typically obtain large confidence bounds when we have few data, and such bounds tend to shrink as we add more data into the problem.

Example 7.10 Assessing the Quality of a Stochastic Solution

We explore how to assess the quality of the solution to stochastic optimization problems. Specifically, we study the simple problem:

$$\min_{u} \mathbb{E}[(u - X)^2], \tag{7.72}$$

with $X \sim \mathcal{N}(1, 10)$. The solution to this problem can be obtained analytically and is given by $u^* = 1$ and $\mathbb{E}(u^*) = 100$.

We would like to assess the quality of solutions obtained from the sample approximation:

$$\min_{u} \hat{\mathbb{E}}_S(u) = \frac{1}{S} \sum_{\omega \in \mathcal{S}} (u - x_\omega)^2, \tag{7.73}$$

as we increase the number of samples S. To do so, we explore the solution at values that span the range $[1, 10{,}000]$; specifically, we compute the candidate solution $u = \hat{u}_S$ and its associated cost estimate $\hat{\mathbb{E}}_S(u)$. We also compute the 95% confidence bounds (7.71) where we expect the optimal cost to be $\mathbb{E}(u^*)$ and the confidence bounds where we expect the optimal solution u^* to be. The results are shown in Figure 7.9; here, we can see large variance of the cost and candidate solution for small sample sizes, but there is quick convergence of the candidate solution and associated cost to the optimal counterparts as we increase S.

Overall, we can see that the solution settles at a value of $S = 1{,}000$. We can also see that the confidence bounds are quite wide for small sample sizes S and quickly shrink as we increase the sample size; in addition, we observe that these provide good guidance of where the actual solution might be (although in some cases, we see that the estimate might fall outside the intervals). We also computed the gap ϵ_S and found that

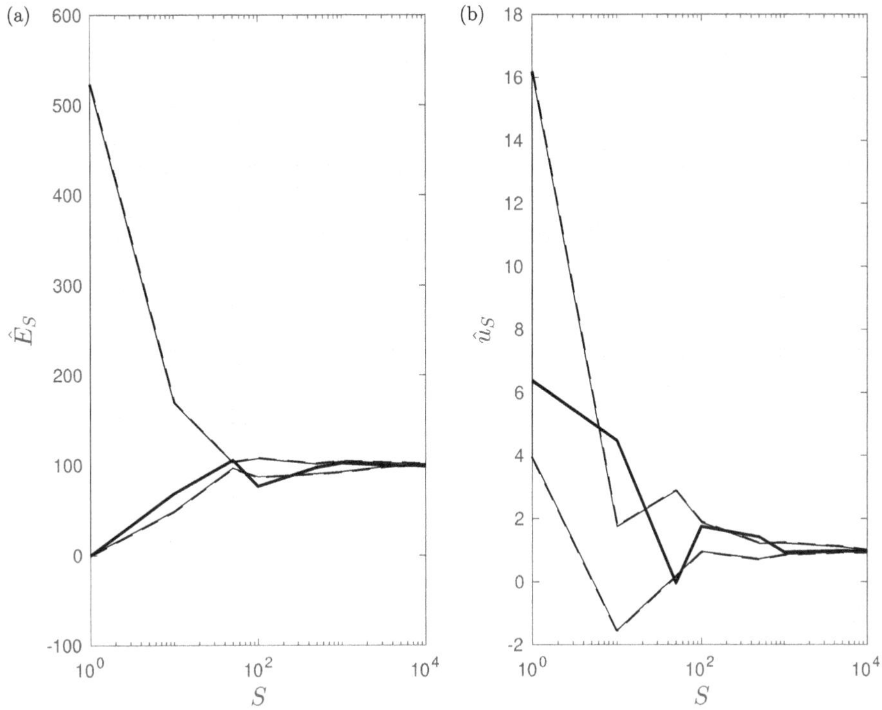

Figure 7.9 Solution estimate $\hat{\mathbb{E}}_S(u)$ and confidence bounds for optimal cost $\mathbb{E}[u^*]$ (a). Solution $\hat{u}_S$ and confidence bounds for solution u^* (b).

this value is 318 for $S = 1$, 60 for $S = 10$, 1.5 for $S = 100$, 3.6 for $S = 1{,}000$, and 0.54 for $S = 1{,}000$; this further confirms the quick convergence of the solution.

Scripts to reproduce results: `ch7_example_SAA` □

As we saw in Chapter 5, Monte Carlo can also be used to approximate diverse risk measures and thus provides a general framework to numerically solve stochastic optimization problems and to compute flexibility measures. For instance, the probability of loss can be expressed as:

$$\mathbb{P}(u) = \mathbb{P}(\varphi(X,u) \leq \bar{y}) = \mathbb{E}[\mathbf{1}[\varphi(X,u) \leq \bar{y}]], \tag{7.74}$$

where we recall that $\mathbf{1}$ is the indicator function; as such, this risk measure can approximated as:

$$\hat{\mathbb{P}}_S(u) = \frac{1}{S} \sum_{\omega \in \mathcal{S}} \mathbf{1}[\varphi(x_\omega,u) \leq \bar{y}]. \tag{7.75}$$

We can also compute an empirical approximation for the flexibility measure (7.58) as:

$$\hat{\delta}_S(u) = \frac{1}{S} \sum_{\omega \in \mathcal{S}} \mathbf{1}\left[\varphi_k(x_\omega,u) \leq \bar{\varphi}_k,\ k \in \mathcal{K}\right]. \tag{7.76}$$

It is important to highlight, however, that the indicator function is a discontinuous function that requires specialized numerical techniques to be handled. For instance, optimization solvers that rely on Newton's method or on gradient methods cannot handle these types of functions. In such cases, one can *smooth out* the indicator function using a sigmoidal function (as done in logistic regression).

The algorithm that we discuss here for solving stochastic optimization problems follows an *external* sampling approach. Here, we assume that the expectation is approximated using Monte Carlo samples, and we solve this approximation to obtain a candidate solution. If we are not satisfied with the quality of the solution, we increase the number of Monte Carlo samples and solve the new approximation. There is another family of algorithms that use *internal* sampling; here, the samples are increased adaptively as the search for the optimum progresses; this class of algorithms includes the so-called stochastic gradient method, as discussed in Chapter 6 in the context of neural networks (NNs). Unfortunately, this class of methods cannot handle constraints and complex risk measures, but they provide an interesting approach to solving basic stochastic optimization problems.

7.7 Bayesian Optimization

We now discuss Bayesian optimization (BO), which is a special class of a stochastic optimization algorithm. This algorithm is a powerful "closed-loop" paradigm that integrates elements of modeling, uncertainty quantification, experimental design, and decision-making. We discussed some of the basics of this approach in Chapter 5; here, we analyze this algorithm from the perspective of decision-making under uncertainty and stochastic optimization.

To explain the BO paradigm, we consider an input–output system of the following form:

$$Y(u) = \varphi(u) + \epsilon. \tag{7.77}$$

Here, $\varphi(u)$ is a function that represents the system performance, ϵ is random noise, u is an input decision, and $Y(u)$ is the observed output of the system. For now, we will assume that the input decision represents the conditions under which we run the system (e.g., the experimental or operating conditions of our system); we have full control over this decision, and this is thus a deterministic variable. Random aspects that we cannot control are captured by the noise ϵ.

The ultimate goal of BO is to find decisions u that maximize the system performance:

$$\max_{u \in \mathcal{U}} \varphi(u), \tag{7.78}$$

where $\mathcal{U}$ is our decision or design space (e.g., our set of possible conditions). The challenge here is that we *do not know* the system function $\varphi(u)$; in other words, the system is a *black box*. As such, we need to build a *data-driven* model of the performance function using input–output data (u_ω, y_ω), $\omega \in \mathcal{S}$. One often calls this model a *surrogate* model $m(u)$.

In principle, we could use any of the models discussed to build a surrogate model such as linear regression, neural nets, and basis function models. The key requirements for the modeling approach used are that it needs to capture potentially complex behavior of the system (e.g., nonlinearity) and that we need to be able to quantify the uncertainty of the model predictions. Uncertainty information is desired because this will be used to guide the selection of new operating conditions, and this in turn can help us refine our model (e.g., help us design new experiments). As we have seen, a modeling approach that satisfies these requirements is *kriging* (Gaussian Process [GP] models) and will thus be our model of choice.

Imagine that we have already built a GP model using a set of preliminary data (u_ω, y_ω), $\omega \in \mathcal{S}$. Recall that GP provides a mean prediction model $\mu_\ell(x)$ and an uncertainty model $\sigma_\ell(u)$. We define our initial data set as $\mathcal{D}_\ell$ and our initial model obtained with such data as $(\mu_\ell(u), \sigma_\ell(u))$; here, ℓ is an iteration counter.

We can use the available model for a couple of tasks:

- Compute a decision $u_{\ell+1}$ that maximizes the expected performance:

$$\max_{u \in \mathcal{U}} \ \mu_\ell(u). \tag{7.79}$$

- Compute a decision $u_{\ell+1}$ that maximizes information gain:

$$\max_{u \in \mathcal{U}} \ \sigma_\ell(u). \tag{7.80}$$

The first task is known in the machine learning community as *exploitation*, as it exploits the knowledge of the system (embedded in the prediction model $\mu_\ell(u)$) to maximize performance. The second task is known as *exploration*, as it seeks to explore the design space so as to collect information (maximize variance $\sigma_\ell(u)$); here, the reasoning is that regions of the space $\mathcal{U}$ that have not been explored will have high uncertainty, and we thus want to visit such regions. This is important because, if the model $\mu_\ell(u)$ is not particularly accurate (it cannot predict the actual performance $\varphi(u)$ well), then the decision made in the exploitation task will not be particularly useful. On the other hand, if we trust the model $\mu_\ell(u)$ (it has low uncertainty), then we should use it to optimize the system.

We can select a decision $u_{\ell+1}$ that balances exploration and exploitation by solving the optimization problem:

$$\max_{u \in \mathcal{U}} \ \mu_\ell(u) + \kappa \cdot \sigma_\ell(u). \tag{7.81}$$

The objective here is known as the *acquisition function* $AF_\ell(u) = \mu_\ell(u) + \kappa \cdot \sigma_\ell(x)$; $\kappa \in \mathbb{R}_+$ is a parameter that balances performance and information. Note that the acquisition function is the mean-variance risk measure.

Having a new decision $u_{\ell+1}$ obtained from Eq. (7.81), we can run this experiment in the system to obtain a new data point (u'_ω, y'_ω), and we augment our dataset $\mathcal{D}_{\ell+1} \leftarrow \mathcal{D}_\ell \cup (u_\omega, y_\omega)$.

We use this new dataset to retrain our GP model (recompute its parameters), and we use the new model $\mu_{\ell+1}(u), \sigma_{\ell+1}(u)$ to obtain a new decision that maximizes the acquisition function $AF_{\ell+1}(u)$. This loop is repeated over multiple iterations

$\ell, \ell + 1, \ldots$ and this will aim to converge to a point that maximizes the actual performance of the system $\varphi(u)$.

The BO algorithm is useful because it automatically selects experiments/decisions to progressively build and improve the model and to ultimately drive the system to conditions that maximize performance. This approach is different from traditional experimental design in which one designs experiments to only maximize information (e.g., minimize the model uncertainty). The problem with such an approach is that it can actually overestimate the amount of experiments (the amount of data) that is actually needed. In practice, it is observed that few experiments are actually needed to have a useful model that identifies conditions that maximize performance.

BO is closely related to a design of experiments technique known as the response-surface methodology (proposed by George Box) (see Exercise 5.11). In such an approach, a set of experiments is designed to build a model of the system, and then, one uses the model to identify optimal operating conditions. This approach, however, does not refine the experiments if the model is not accurate.

The BO framework explained here uses a nonparametric GP model to build the system surrogate. However, one can envision many possible variants of this approach that use parametric models such as linear regression, basis/kernel models, and NNs; the framework will be similar, but the main difference will be in how we estimate the model uncertainty that guides exploitation (e.g., how we determine $\sigma(u)$). It is also possible to combine a physics model with GP; as discussed in Chapter 6, the use of a physics model can help guide the BO search [70].

Note also that the BO approach discussed maximizes the mean-variance, but we can also use other risk measures when building the acquisition function. For instance, there is a variant of BO that maximizes the probability of improvement (which is an event probability), but we can also envision optimizing CVaR.

The BO paradigm can also be used to optimize the performance of algorithms; this is an area in computer science that is known as *hyperparameter tuning*. For instance, imagine that you want to *tune* the regularization parameter of an estimation problem, tune the number of layers in an NN, tune the settings of a thermostat, or tune the settings of a microscope; BO can be used to automatically find the best settings.

The BO paradigm is powerful because it combines predictive modeling, experimental design, uncertainty quantification, and stochastic optimization. In particular, note that the only fundamental difference between experimental design and stochastic optimization is the goal that they follow; the former aims to explore and collect information, while the latter aims to exploit available information to optimize performance. Understanding the synergies between experimental design, estimation, and optimization is important because modern decision-making tools developed in fields such as artificial intelligence and machine learning leverage fundamentals of such tools. Many of such tools ultimately aim to more closely capture how humans navigate the world by collecting data via their senses, extract knowledge from such data (learn), and leverage such knowledge to make decisions.

Example 7.11 Bayesian Optimization for Closed-Loop Experimental Design

We revisit the experimental design problem of Exercise 5.11. A team of scientists wants to determine optimal conditions to extract pectin from sunflower heads using oxalic acid as a solvent [17]. They need to identify how pectin yield (denoted as Y) is influenced by three input variables u_1 (extraction temperature), u_2 (extraction time), and u_3 (oxalic acid concentration). Unfortunately, there is no physical understanding of these dependencies, and they need to optimize performance directly from data. The design space for the input variables has been scaled to lie in domain $\mathcal{U} = [-1, 1]^3$. We apply BO to automatically identify experimental conditions that maximize the yield.

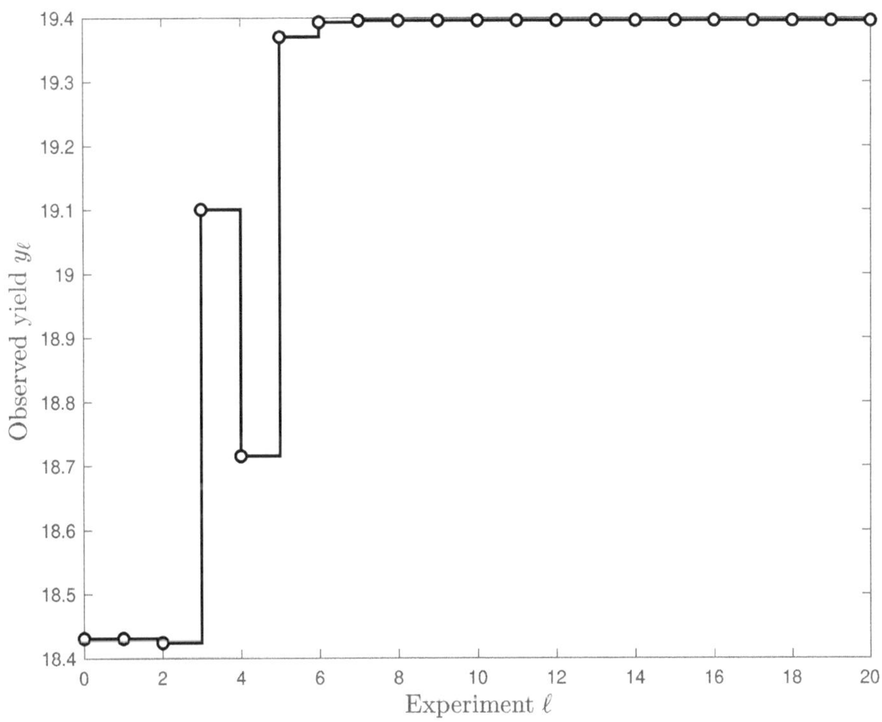

Figure 7.10 Evolution of pectin yield using experiments proposed by BO.

We construct an initial GP model using only three experiments $(0, 0, 0)$, $(1, 1, 1)$, and $(-1, 1, 1)$. A subset of the experiments proposed by BO is shown in Table 7.1, and the evolution of the yield is shown in Figure 7.10; here, we can see some interesting trends and behavior. Note that BO first runs the three inputs at their highest allowable values, and it does this even if that experiment was already included in the initial dataset. This indicates that, initially, BO still observes high uncertainty in this region that it aims to resolve before moving to a new region. BO then progressively lowers the values of the inputs but eventually realizes that lowering temperature (u_1) is not a good strategy; as such, it decides to increase temperature again while continuing to lower extraction

Table 7.1 Experiment history suggested by BO.

1.0000	1.0000	1.0000
1.0000	1.0000	1.0000
0.9946	0.9946	0.9692
0.8355	0.5647	0.5133
0.8697	0.7806	0.3191
1.0000	0.6770	0.2967
⋮	⋮	⋮
1.0000	0.4505	0.2219
1.0000	0.4426	0.2283
1.0000	0.4356	0.2338
1.0000	0.4300	0.2380
1.0000	0.4259	0.2411
1.0000	0.4259	0.2411

time (u_2) and oxalic acid concentration (u_3). We can see that performance is improved quickly, and it eventually saturates, converging to the conditions $u_1 = 1$, $u_2 = 0.426$, and $u_3 = 0.24$. BO identifies such conditions in less than 30 experiments (accounting for the initial set of experiments).

Scripts to reproduce results: `ch7_bayesopt_yield` □

7.8 Summary

In this chapter, we discussed techniques for making decisions under uncertainty. The following are important take-home messages:

- We can leverage predictive models to make decisions, but such predictions will invariably contain some degree of uncertainty. Uncertainty can originate from different sources such as uncertainty in the parameter estimates of our model or uncertainty in factors that are not in our control (e.g., weather and market prices).
- It is important to properly account for uncertainty when making a decision. Specifically, not accounting for uncertainty (e.g., following a deterministic view of the world) can lead to decisions that are vulnerable to outcomes that we cannot predict. Many engineered systems fail simply because the design did not properly account for uncertainty.
- Techniques for making decisions under uncertainty rely on our ability to compare RVs and to shape the probability densities of RVs in desirable ways.
- Risk measures are an important tool that allows us to quantify uncertainty/risk of decisions and with this find optimal decisions. Risk measures are summarizing statistics (e.g., mean, variance, quantiles, and event probabilities) that help model attitudes that decision-makers take toward risk. We have seen that DMs can be

risk-neutral (indifferent to risk), risk-averse (aim to mitigate risk), and risk-takers (aim to exploit risk).

- There are well-defined rules/axioms that a proper risk measure must follow. Understanding such rules is important to understand the advantages and limitations of risk measures.
- Stochastic dominance is an important concept in decision-making under uncertainty; this is a concept that is used to formally compare RVs. At the core, this concept focuses on comparing the cdfs of RVs.
- Stochastic optimization is an area of decision-making that provides tools to make decisions under uncertainty by using optimization techniques. The key idea is to find decisions that shape the distribution of a desired target RV (e.g., cost) by minimizing a risk measure and by imposing constraints. Stochastic optimization also provides tools to quantify system flexibility/robustness and to model decisions that are made in a sequential manner.
- Bayesian optimization is an important stochastic optimization algorithm that combines elements of predictive modeling, experimental design, uncertainty quantification, and decision-making under uncertainty. BO is widely used as an algorithm for closed-loop experimental design and for solving optimization problems in which the objective function needs to be learned progressively from data. BO captures the data→models→decisions pipeline.

Our discussion in this chapter scratches the surface of the area of stochastic optimization. Some excellent resources for gaining more in-depth knowledge of this subject are the books of Birge and Louveaux [67], Grossmann [71], and Garnett [50]. The area of stochastic optimization continues to grow as the communities of operations research, optimization, and machine learning continue to fuse.

7.9 Exercises

Exercise 7.1. Data Center in Electricity Markets.
The power grid operates a day-ahead market (DAM) and a real-time market (RTM) for electricity. A company is seeking to determine in which market it is most convenient to plug a data center to. The data center has a large electricity demand of 10 MWh, which is approximately the electricity demand of 10,000 homes in the United States.

Use the historical price data (in USD/MWh) provided in `DAM_datacenter.csv` and `RTM_datacenter.csv` to address the following:

(a) Determine the total hourly cost of the data center and compare their cumulative densities.
(b) Determine the probability that the hourly cost is higher than 200 USD/hr in both markets.
(c) Determine the mean hourly cost in each market.
(d) Which market stochastically dominates in terms of hourly cost?

Exercise 7.2. Optimizing Yield under Uncertainty.
Revisit the experimental design problem of Example 7.11. A team of scientists is aiming to determine optimal operating conditions to extract pectin from sunflower heads using oxalic acid. Pectin is a complex polysaccharide that is contained in plant primary cell walls and is widely used in industry (e.g., food, pharma, and cosmetics).

The scientists are seeking to identify the influence of three variables (also known as factors in the design of experiments literature): X_1 (extraction temperature), X_2 (extraction time), and X_3 (oxalic acid concentration) on the extraction yield (denoted as Y). They propose a 2nd-order polynomial model:

$$Y = \theta_0 + \theta_1 X_1 + \theta_2 X_2 + \theta_3 X_3 + \theta_4 X_1^2 + \theta_5 X_2^2 + \theta_6 X_3^2 \\ + \theta_7 X_1 X_2 + \theta_8 X_1 X_3 + \theta_9 X_2 X_3.$$

The experimental data is available in `box_behnken.dat`. The scaled values (-1, 0, and 1) of factor X_1 (extraction temperature) represent 70, 85, and 100°C. The scaled values ($-1, 0,$ and 1) of factor X_2 (extraction time) represent 1.0, 1.5, and 2.0 hr. The scaled values (-1, 0, and 1) of factor X_3 (oxalic acid concentration) represent 0.1, 0.2, and 0.3%. In the table, you can also find the yield recovery outcomes obtained for these experiments (in %).

Use the available data to address the following:

(a) Use maximum likelihood estimation to determine estimates for the model parameters. Determine the experimental conditions that maximize the predicted yield by the model. Note that this approach is deterministic (it ignores uncertainty of the model parameters obtained from MLE).
(b) To account for the uncertainty of the parameters, determine the experimental conditions that maximize the mean yield. Generate 1,000 realizations of the parameters from their probability density to obtain a Monte Carlo approximation of the mean yield.
(c) Determine the experimental conditions that minimize the standard deviation of the yield (minimize yield uncertainty). Use the same realizations used before to estimate the standard deviation.
(d) Determine the experimental conditions that maximize the standard deviation of the yield (maximize yield uncertainty). Use the same realizations used before to estimate the standard deviation. By comparing with the previous case, what does this tell you about explored/unexplored regions in previous experiments?
(e) You are concerned that running the new experiment that maximizes expected yield is risky because the model has uncertainty. As such, you want to be a bit conservative and select an experiment that maximizes mean yield and also minimizes uncertainty of the yield. What is an experiment that achieves this?

Exercise 7.3. Monte Carlo vs. Analytical Expressions.
In the previous exercise, you used realizations and Monte Carlo to estimate the mean and standard deviation of the yield. This approach can become

computationally expensive as we increase the number of realizations; as such, we should try to derive analytical expressions whenever possible.

(a) Derivate analytical expressions for the mean yield and for its standard deviation. Why is it possible to determine analytical expressions for this problem?
(b) Show that Monte Carlo converges to the analytical expressions as you increase the number of realizations.
(c) Use the analytical expressions to determine the experimental conditions that maximize and minimize uncertainty.
(d) Use Monte Carlo to determine the experimental conditions that maximize and minimize uncertainty. How many realizations are needed to obtain solutions that are close to the analytical expressions?

Exercise 7.4. Flare Optimization.
Flares are key pieces of industrial equipment that help combust excess material generated at manufacturing facilities [26]; combustion helps mitigate environmental hazards associated with the direct release of such materials. Flares are also used as safety devices, as controlled combustion of gases can be used to control equipment pressure; this technique is widely used for controlling pressure in natural gas transport networks.

The design of a flare is a complex process that involves three key variables: height of the flare, diameter of the flare, and the input flow of combustible material. The height and diameter of the flare need to be carefully selected so as to minimize investment cost; in addition, these variables need to be selected to ensure that the radiation that emanates from combustion remains below a safe level (to prevent impacting nearby personnel) and so that the Mach number remains below a safe level (to prevent vibrations and damaging of equipment). Specifically, it has been determined that the maximum allowable radiation is 3,000 BTU/hr – ft^2, and the maximum Mach number allowable is 0.7.

Unfortunately, the design process is complicated by inherent randomness of the input flow of combustible material, which varies substantially throughout the year.

Use the model `ch7_flare_sharing.m` and input flow data in `flow_flare.dat` to address the following:

(a) Use Monte Carlo to determine the impact of the height and diameter of the flare on the radiation, Mach number, and cost. To do so, test flare heights of 20 ft and 50 ft (for a fixed diameter of 1 ft) and test a flare diameter of 5 ft (for a fixed height of 20 ft). For all the designs, report the probabilities that the radiation and Mach numbers stay below the allowable limits. Comment on any trends that you observe.
(b) Formulate an optimization problem that determines the flare height and diameter that minimizes design cost and that ensures that the radiation and Mach number stay below the allowable limits for all possible values of the input flow. Note that the flare height should be in the range of 20–150 ft and the flare diameter should be in the range 1–5 ft.

(c) Upon solving the previous optimization problem, confirm that the design obtained indeed satisfies the constraints for all possible values of the input flow. Do this by visualizing the pdf or cdf of the radiation and Mach numbers.

Exercise 7.5. Flexibility Analysis.
Flexibility can be defined as the ability of a system to withstand disturbances/disruptions. From a statistical standpoint, flexibility can be defined as the probability that a system maintains acceptable operation in the face of random factors. In the context of the flare system of the previous exercise, we can define flexibility as the probability that the system satisfies the safety constraints on radiation and Mach numbers in the face of variations of the input flow.

(a) An engineer is interested in evaluating the flexibility of a couple of flare designs; the first design has a diameter of 3 ft and a height of 100 ft, and the second design has a diameter of 3 ft and a height of 20 ft. The flare cannot exceed a radiation of 3,000 BTU/hr-ft^2 and a Mach number of 0.7. Use the historical data and flare model provided to in Exercise 7.4 determine the probability of maintaining feasible operation for both designs. Comment on which design is more flexible.
(b) You are becoming increasingly concerned about the effect of radiation on the operating personnel of the flare facility. As such, you would like to tighten the maximum allowable radiation to 1,500 BTU/hr-ft^2. Determine the impact that this tightening of constraints has on the flexibility of the flare designs.
(c) Find a suitable limit for radiation that gives a probability of feasible operation of 0.9 for the first design. Determine the critical value of flow in which the system transitions from feasible to infeasible operation.

Exercise 7.6. Flexible and Inflexible Operation of an Electrolyzer.
A hydrogen production station has a constant hydrogen demand of 200 kg/hr. The station has traditionally produced hydrogen from natural gas (via steam methane reforming) at an estimated production cost of 1.2 USD per kilogram of hydrogen. The station is considering adding the option of producing hydrogen via water electrolysis in order to exploit favorable electricity market prices and in order to reduce its carbon footprint.

To do so, they are considering purchasing an electrolyzer, which has a capital cost given by the quadratic cost $\alpha \cdot x^2$ ($\alpha = 2,500$ USD per MW, x is the capacity of the electrolyzer in MW). It is estimated that each kilogram of hydrogen produced requires an electricity consumption of 0.05 MWh.

Due to limitations of the local power grid, the hydrogen station cannot consume more than 20 MW to run the electrolyzer. You are also given historical data for the hourly electricity prices at this location; see `RTM.csv` (prices are in USD per MWh).

(a) Formulate an optimization problem to determine the optimal design capacity of the electrolyzer that satisfies the hydrogen demand of 200 kg/hr using a combination of hydrogen produced from natural gas and hydrogen produced from electrolysis. For simplicity, assume that the electrolyzer always operates at its installed capacity (the electrolysis system is inflexible). The installed design capacity should minimize the capital cost plus the expected value of the total annual operating cost (associated with operating the methane reformer and purchasing electricity). Assume that the electrolyzer operates 8,760 hr in a year. Report the total cost and comment on whether it makes economic sense to combine the electrolyzer with the methane reformer.

(b) Consider now the more complex operational situation in which the electrolysis system can flexibly adjust its power demand (below its installed capacity) so as to respond to electricity prices and with this cut down operational costs. Formulate an optimization problem to determine the capacity of the electrolyzer to minimize capital cost plus the expected total cost. Comment on how the ability to operate flexibly cuts down (or not) installation and operational costs.

Exercise 7.7. Chance Constraints.
In decision-making under uncertainty, it is common to formulate optimization problems of the following form:

$$\min_{u} f(u)$$
$$\text{s.t. } \mathbb{P}(\varphi(u,X) \leq \overline{\varphi}) \geq \alpha.$$

where φ is a scalar function, X is a multivariate RV, $\overline{\varphi} \in \mathbb{R}$ is a threshold value, and $\alpha \in [0,1]$ is a desired probability level. The constraint of the optimization problem is known as a chance constraint (or probabilistic constraint) that aims to keep the scalar function below the threshold with high probability.

One can show that the chance constraint can be expressed as a constraint on the quantile of φ as:

$$\mathbb{Q}_{\varphi(u,X)}(\alpha) \leq \overline{\varphi}.$$

Show that, when $X \sim \mathcal{N}(\mu,\Sigma)$ and $\varphi(u,X) = a^T u + b^T X$, the quantile constraint can be expressed as:

$$\mathbb{E}[\varphi(u,X)] + z_{\alpha/2}\mathbb{SD}[\varphi(u,X)] \leq \overline{\varphi}.$$

Exercise 7.8. Investment Risk.
You are evaluating a couple of investments with quite different risk profiles. Specifically, the competing options have a return on investments (ROIs) given by the Weibull RVs: Weibull(0.3, 3) (scale of 0.3 and shape of 3) and Weibull(0.2, 1) (scale of 0.2 and shape of 1). The ROI is the fraction of your initial investment that will be recovered in a year; as such, investments with high ROIs are preferable.

(a) From the perspective of a risk-neutral DM, you want the option that has the highest mean ROI. Which option would you select?

(b) From the perspective of a risk-averse DM, you prefer the option that will yield an ROI of low uncertainty. Which option would you select?

(c) From the perspective of a risk-averse DM, you prefer the option that will yield ROIs that are at least 0.1 with high probability. Which option would you select?
(d) From the perspective of a risk-taking DM, you prefer the option that has a higher chance of achieving ROIs that are higher than 0.5. Which one would you select?
(e) Comment on whether one investment clearly dominates the other one. Comment on whether one option dominates the other at specific ranges of ROIs.

Exercise 7.9. Statistics, Fairness, and Social Justice.
The state government is launching an initiative to remove algae from its 100 most important lakes because such algae are affecting recreational activities, tourism, and property values. The increasing algae production has been the result of excessive fertilizer use in agricultural land surrounding the lakes. The current budget to fund this environmental remediation effort is 1 billion USD (1,000 MUSD).

Distributing the available budget in a fair manner is not straightforward, as the size and level of pollution of the lakes vary substantially (by orders of magnitude). As such, a local government representative of each lake has been requested to deliver a target budget; the associated empirical pdf and cdf for the requested budgets is shown below. The total requested budgets can be found in `budgets.dat`.

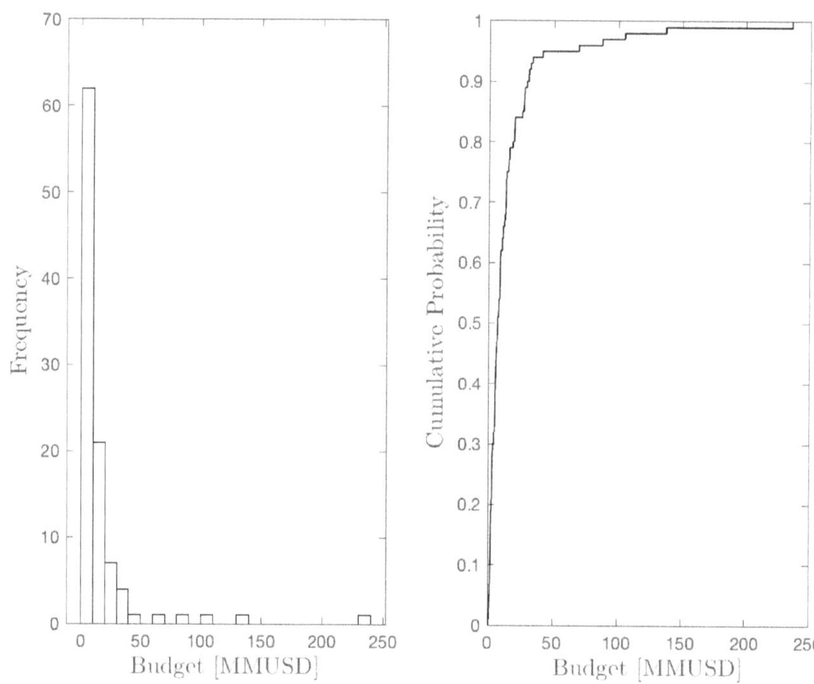

The total requested budget is above the available budget; as such, the government needs to determine a strategy to prioritize efforts in a way that it maximizes fairness.

Fairness is an ambiguous notion (analogous to risk) because what is perceived as fair depends on the preferences of DMs. For instance, our instinct might tell us that protecting large lakes is a priority, but this would leave small lakes vulnerable (this might affect small communities that rely on such lakes). As such, the government will assess different strategies to determine the best allocation.

(a) Denote x_s as the budget delivered to lake $s \in \mathcal{S}$, denote $\bar{x}_s$ as the budget requested by lake s, and denote $\bar{x}_s - x_s$ as the served budget. Formulate an optimization problem that identifies the delivered budgets x_s, $s \in \mathcal{S}$ that maximize the average served budget:

$$\frac{1}{S}\sum_{s \in \mathcal{S}}(\bar{x}_s - x_s). \tag{7.82}$$

Compare the cdf of the optimal delivered budgets to the cdf of the requested budgets. Comment on how the budget is allocated to small requests and large requests; is the allocation favoring small or large requests? Comment also on how many lakes get their full budget requested.

(b) Fairness can be measured using the geometric mean:

$$\left(\prod_{s \in \mathcal{S}} \frac{x_s}{\bar{x}_s}\right)^{1/S},$$

where $x_s/\bar{x}_s$ is the fraction of the requested budget allocated for lake s. Note that this measure better scales the budget requests (which is useful in dealing with small and large requests more evenly) and note that the geometric mean is one when all budgets are perfectly allocated. This measure is also known as the Nash allocation measure and has desirable fairness properties. Formulate an optimization problem that optimizes the geometric mean measure. Compare the cdf of the delivered budgets to the cdf of the requested budgets (use a log transformation of the x-axis to facilitate visualization). Comment on how the budget is allocated to small requests and large requests and comment on how many lakes get their full budget requested. Compare any discrepancies observed with the allocation that maximizes the average served budget.

Exercise 7.10. Bayesian Optimization of Pyrolysis Reactor.
Pyrolysis is a promising pathway to thermally decompose plastic waste and turn the resulting products into valuable chemicals [25]. Specifically, pyrolysis decomposes plastic (polyolefins) into a mixture composed of liquids, waxes, and gases. Pyrolysis involves a highly complex kinetic mechanism that can be difficult to model (see below); as such, it is difficult to optimize the operating conditions of the reactor. Moreover, conducting pyrolysis experiments involves significant effort, and it is thus not always possible to find optimal conditions by trial-and-error and to obtain data to estimate kinetic parameters.

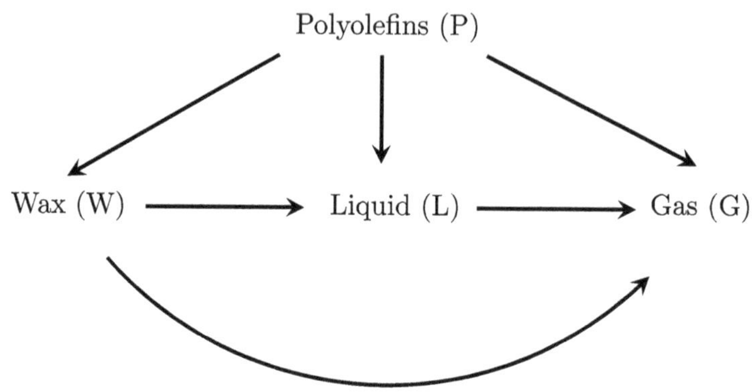

Develop a BO algorithm to find the optimal operating conditions for a pyrolysis reactor. Specifically, a couple of critical conditions to optimize are the heating rate of the reacting mixture and the residence time of the reactor. These conditions can be optimized in order to maximize the composition of the product mixture. For the purposes of this exercise, you may assume that the model provided (see `ch7_plastic_pyrolysis.m`) is the ground truth; use this model as a black-box to simulate the performance of the experimental reactor.

(a) Your goal is for your BO algorithm to find the heating rate and residence time for the reactor that maximizes the final concentration of the liquid products. Due to constraints in the experimental effort, you cannot run the experiment more than 20 times. Report the best concentration of liquid and the associated experimental conditions. Use the model to visualize the time evolution of all the products. Comment also on how many experiments BO requires to identify the optimal conditions (when BO settles).

(b) Your team has identified a new technology to obtain valuable chemicals from the wax products. Find the conditions to maximize the final concentration of liquid and wax. Due to constraints in experimental budgets, you cannot run the experiment more than 20 times. Report the concentration found (for wax and liquid) and the associated experimental conditions. Comment on whether the yield of liquids is reduced relative to the previous case; specifically, is the production of wax competing with the production of liquid? Use the model to visualize the time evolution of all the products.

References

[1] W. D. Seider, D. R. Lewin, J. Seader, and S. Widagdo, *Product and Process Design Principles: Synthesis, Analysis, and Evaluation*. John Wiley and Sons, 2009.

[2] Mettler Toledo Inc., "Sieve analysis," https://www.mt.com/us/en/home/applications/Laboratory_weighing/sieve-analysis.html, 2023 accessed: May 27, 2024.

[3] D. C. Harris, *Materials for Infrared Windows and Domes: Properties and Performance*, vol. 158. SPIE Press, 1999.

[4] E. Salamin, "The Weibull distribution in the strength of glass," https://wp.optics.arizona.edu/optomech/wp-content/uploads/sites/53/2016/10/SalaminTutorial1.pdf, 2016, accessed: June 25, 2024.

[5] H. S. Fogler, *Elements of Chemical Reaction Engineering*, 4th ed., Pearson, 2009.

[6] E. J. Harbron and P. F. Barbara, "The poisson distribution and single-molecule spectroscopy. An undergraduate analytical laboratory experiment," *Journal of Chemical Education*, vol. 79, no. 2, pp. 211–13, 2002.

[7] S. Jiang, J. Noh, C. Park, A. D. Smith, N. L. Abbott, and V. M. Zavala, "Using machine learning and liquid crystal droplets to identify and quantify endotoxins from different bacterial species," *Analyst*, vol. 146, no. 4, pp. 1224–33, 2021.

[8] B. A. Ogunnaike, *Random Phenomena: Fundamentals of Probability and Statistics for Engineers*. CRC Press, 2011.

[9] Electric Reliability Council of Texas, "Ercot market prices," https://www.ercot.com/mktinfo/prices, 2024, accessed: May 27, 2024.

[10] H. Richardson, G. Lisandrelli, A. Riobueno-Naylor, and R. Saxe, "Mri data of 3-12 year old children and adults during viewing of a short animated film," https://openneuro.org/datasets/ds000228/versions/1.1.1, 2024, accessed: May 27, 2024.

[11] Reliawiki, "Arrhenius relationship," https://reliawiki.com/index.php/Arrhenius_Relationship, 2024, accessed: May 27, 2024.

[12] N. L. Carr, "Kinetics of catalytic isomerization of n-pentane," *Industrial & Engineering Chemistry*, vol. 52, no. 5, pp. 391–96, 1960.

[13] Wisconsin Department of Natural Resources, "Wisconsin lakes," https://apps.dnr.wi.gov/lakes/waterquality, 2024, accessed: June 25, 2024.

[14] R. A. Granberg and Å. C. Rasmuson, "Solubility of paracetamol in pure solvents," *Journal of Chemical & Engineering Data*, vol. 44, no. 6, pp. 1391–95, 1999.

[15] S. Qin, T. Jin, R. C. Van Lehn, and V. M. Zavala, "Predicting critical micelle concentrations for surfactants using graph convolutional neural networks," *The Journal of Physical Chemistry B*, vol. 125, no. 37, pp. 10 610–20, 2021.

[16] X. Lan and R. Keeling, "Trends in atmospheric carbon dioxide," https://gml.noaa.gov/ccgg/trends/data.html, 2024, accessed: June 25, 2024.

[17] X. Peng, G. Yang, Y. Shi, Y. Zhou, M. Zhang, and S. Li, "Box–Behnken design based statistical modeling for the extraction and physicochemical properties of pectin from sunflower heads and the comparison with commercial low-methoxyl pectin," *Scientific Reports*, vol. 10, no. 1, p. 3595, 2020.

[18] R. G. Dastidar, M. S. Kim, P. Zhou, Z. Luo, C. Shi, K. J. Barnett, D. J.

McClelland, E. Y.-X. Chen, R. C. Van Lehn, and G. W. Huber, "Catalytic production of tetrahydropyran (THP): A biomass-derived, economically competitive solvent with demonstrated use in plastic dissolution," *Green Chemistry*, vol. 24, no. 23, pp. 9101–13, 2022.

[19] F. Wang, S. Qin, C. Acevedo-Vélez, R. C. Van Lehn, V. M. Zavala, and D. M. Lynn, "Decoding optical responses of contact-printed arrays of thermotropic liquid crystals using machine learning: Detection and reporting of aqueous amphiphiles with enhanced sensitivity and selectivity," *ACS Applied Materials & Interfaces*, vol. 15, no. 43, pp. 50532–45, 2023.

[20] F. Long, S. Jiang, A. G. Adekunle, V. M Zavala, and E. Bar-Ziv, "Online characterization of mixed plastic waste using machine learning and mid-infrared spectroscopy," *ACS Sustainable Chemistry and Engineering*, vol. 10, no. 48, pp. 16064–69, 2022.

[21] C. ISO, "Open access same-time information system (oasis)," http://oasis.caiso.com/mrioasis/logon.do, 2024, accessed: June 25, 2024.

[22] V. Midha, M. K. Jha, and A. Dey, "Neural network prediction of fluidized bed bioreactor performance for sulfide oxidation," *Korean Journal of Chemical Engineering*, vol. 30, pp. 385–91, 2013.

[23] A. J. Román, S. Qin, J. C. Rodríguez, L. D. González, V. M. Zavala, and T. A. Osswald, "Natural rubber blend optimization via data-driven modeling: The implementation for reverse engineering," *Polymers*, vol. 14, no. 11, p. 2262, 2022.

[24] S. Jiang, N. Bao, A. D. Smith, S. Byndoor, R. C. Van Lehn, M. Mavrikakis, N. L. Abbott, and V. M. Zavala, "Scalable extraction of information from spatiotemporal patterns of chemoresponsive liquid crystals using topological descriptors," *The Journal of Physical Chemistry C*, vol. 127, no. 32, pp. 16 081–98, 2023.

[25] G. Jiang, R. Fenwick, J. Seville, H. B. Mahood, R. B. Thorpe, S. Bhattacharya, D. S. Monsalve, and G. A. Leeke, "Lumped kinetic modelling of polyolefin pyrolysis: A non-isothermal method to estimate rate constants," *Journal of Analytical and Applied Pyrolysis*, vol. 164, p. 105530, 2022.

[26] J. Tovar-Facio, Y. Cao, J. M. Ponce-Ortega, and V. M. Zavala, "Scalable solution strategies for chance-constrained nonlinear programs," *Industrial & Engineering Chemistry Research*, vol. 57, no. 23, pp. 7987–98, 2018.

[27] M. Baldea, T. F. Edgar, B. L. Stanley, and A. A. Kiss, "Modular manufacturing processes: Status, challenges, and opportunities," *AIChE Journal*, vol. 63, no. 10, pp. 4262–72, 2017.

[28] C. Carrillo, A. O. Montaño, J. Cidrás, and E. Díaz-Dorado, "Review of power curve modelling for wind turbines," *Renewable and Sustainable Energy Reviews*, vol. 21, pp. 572–81, 2013.

[29] C. G. Hill and T. W. Root, *Introduction to Chemical Engineering Kinetics and Reactor Design*. John Wiley & Sons, 2014.

[30] M. D. Graham and J. B. Rawlings, *Modeling and Analysis Principles for Chemical and Biological Engineers*. Nob Hill Publishing, 2022.

[31] V. Venkatasubramanian, Y. Luo, and J. Sethuraman, "How much inequality in income is fair? A microeconomic game theoretic perspective," *Physica A: Statistical Mechanics and Its Applications*, vol. 435, no. 1, pp. 120–38, 2015.

[32] S. J. Blott, D. J. Croft, K. Pye, S. E. Saye, and H. E. Wilson, "Particle size analysis by laser diffraction," *Geological Society, London, Special Publications*, vol. 232, no. 1, pp. 63–73, 2004.

[33] J. Sun, F. Wang, Y. Sui, Z. She, W. Zhai, C. Wang, and Y. Deng, "Effect of particle size on solubility, dissolution rate, and oral bioavailability: Evaluation using coenzyme q10 as naked nanocrystals," *International*

Journal of Nanomedicine, vol. 7, pp. 5733–44, 2012.

[34] Y. Kudo, M. Yasuda, and S. Matsusaka, "Effect of particle size distribution on flowability of granulated lactose," *Advanced Powder Technology*, vol. 31, no. 1, pp. 121–27, 2020.

[35] L. C. Wolstenholme, *Reliability Modelling: A Statistical Approach*. Routledge, 2018.

[36] T. A. Azlarov, N. Volodin, and I. Olkin, *Characterization Problems Associated with the Exponential Distribution*. Springer, 1986.

[37] W. Weibull, *Fatigue Testing and Analysis of Results*. Elsevier, 2013.

[38] K. Conradsen, L. Nielsen, and L. Prahm, "Review of Weibull statistics for estimation of wind speed distributions," *Journal of Climate and Applied Meteorology*, vol. 23, no. 8, pp. 1173–83, 1984.

[39] F. M. Dekking, *A Modern Introduction to Probability and Statistics: Understanding Why and How*. Springer Science & Business Media, 2005.

[40] E. G. Schilling and D. V. Neubauer, *Acceptance Sampling in Quality Control*. Chapman and Hall/CRC, 2009.

[41] L. McKay, T. Bondelid, T. Dewald, C. Johnston, R. Moore, and A. Rea, "Nhdplus version 2: User guide," https://www.epa.gov/system/files/documents/2023-04/NHDPlusV2_User_Guide.pdf, accessed: May 29, 2024.

[42] L. A. Escobar and W. Q. Meeker, "A review of accelerated test models," *Statistical Science*, vol. 21, no. 4, pp. 552–77, 2006.

[43] L. T. Biegler, *Nonlinear Programming: Concepts, Algorithms, and Applications to Chemical Processes*. SIAM, 2010.

[44] J. Nocedal and S. J. Wright, *Numerical Optimization*. Springer, 1999.

[45] O. A. Hougen and K. M. Watson, "Solid catalysts and reaction rates," *Industrial and Engineering Chemistry*, vol. 35, no. 5, pp. 529–41, 1943.

[46] M. S. Grewal and A. P. Andrews, *Kalman Filtering: Theory and Practice with MATLAB*. John Wiley & Sons, 2014.

[47] E. L. Haseltine and J. B. Rawlings, "Critical evaluation of extended kalman filtering and moving-horizon estimation," *Industrial & Engineering Chemistry Research*, vol. 44, no. 8, pp. 2451–60, 2005.

[48] K. Chaloner and I. Verdinelli, "Bayesian experimental design: A review," *Statistical Science*, vol. 10, no. 3, pp. 273–304, 1995.

[49] J. Wang and A. W. Dowling, "Pyomo. doe: An open-source package for model-based design of experiments in python," *AIChE Journal*, vol. 68, no. 12, p. e17813, 2022.

[50] R. Garnett, *Bayesian Optimization*. Cambridge University Press, 2023.

[51] I. T. Jolliffe and J. Cadima, "Principal component analysis: A review and recent developments," *Philosophical Transactions of the Royal Society A: Mathematical, Physical and Engineering Sciences*, vol. 374, no. 2065, p. 20150202, 2016.

[52] I. T. Jolliffe, "A note on the use of principal components in regression," *Journal of the Royal Statistical Society Series C: Applied Statistics*, vol. 31, no. 3, pp. 300–303, 1982.

[53] P. Geladi and B. R. Kowalski, "Partial least-squares regression: A tutorial," *Analytica Chimica Acta*, vol. 185, pp. 1–17, 1986.

[54] S. De Jong and H. A. Kiers, "Principal covariates regression: Part I. Theory," *Chemometrics and Intelligent Laboratory Systems*, vol. 14, no. 1–3, pp. 155–64, 1992.

[55] H. Zou, T. Hastie, and R. Tibshirani, "Sparse principal component analysis," *Journal of Computational and Graphical Statistics*, vol. 15, no. 2, pp. 265–86, 2006.

[56] M. Owen, *Practical signal processing*. Cambridge University Press, 2007.

[57] C. M. Bishop, "Pattern recognition and machine learning," *Springer*, vol. 2, pp. 645–78, 2006.

[58] K. P. Murphy, *Probabilistic Machine Learning: An Introduction*. MIT Press, 2022.

[59] G. James, D. Witten, T. Hastie, R. Tibshirani, *An introduction to statistical learning*. Springer, vol. 112, 2013.

[60] M. A. Kramer, "Autoassociative neural networks," *Computers & Chemical Engineering*, vol. 16, no. 4, pp. 313–28, 1992.

[61] M. M. Bronstein, J. Bruna, Y. LeCun, A. Szlam, and P. Vandergheynst, "Geometric deep learning: Going beyond euclidean data," *IEEE Signal Processing Magazine*, vol. 34, no. 4, pp. 18–42, 2017.

[62] R. S. Sutton and A. G. Barto, *Reinforcement Learning: An Introduction*. MIT Press, 2018.

[63] T. Hastie, R. Tibshirani, J. H. Friedman, and J. H. Friedman, *The elements of statistical learning: Data mining, inference, and prediction*. Springer, vol. 2, 2009.

[64] M. Rizwan, V. Alstad, and J. Jäschke, "Design considerations for industrial water electrolyzer plants," *International Journal of Hydrogen Energy*, vol. 46, no. 75, pp. 37120–36, 2021.

[65] R. T. Rockafellar and S. Uryasev, "Conditional value-at-risk for general loss distributions," *Journal of Banking & Finance*, vol. 26, no. 7, pp. 1443–71, 2002.

[66] K. Pavlikov and S. Uryasev, "CVaR norm and applications in optimization," *Optimization Letters*, vol. 8, no. 7, pp. 1999–2020, 2014.

[67] J. R. Birge and F. Louveaux, *Introduction to Stochastic Programming*. Springer Science & Business Media, 2011.

[68] D. A. Straub and I. E. Grossmann, "Design optimization of stochastic flexibility," *Computers & Chemical Engineering*, vol. 17, no. 4, pp. 339–54, 1993.

[69] J. Linderoth, A. Shapiro, and S. Wright, "The empirical behavior of sampling methods for stochastic programming," *Annals of Operations Research*, vol. 142, no. 1, pp. 215–41, 2006.

[70] Q. Lu, L. D. González, R. Kumar, and V. M. Zavala, "Bayesian optimization with reference models: A case study in MPC for HVAC central plants," *Computers & Chemical Engineering*, vol. 154, p. 107491, 2021.

[71] I. E. Grossmann, *Advanced Optimization for Process Systems Engineering*. Cambridge University Press, 2021.

Index

www.ingramcontent.com/pod-product-compliance
Ingram Content Group UK Ltd.
Pitfield, Milton Keynes, MK11 3LW, UK
UKHW052355240426
470365UK00009B/335
* 9 7 8 1 0 0 9 5 4 1 8 9 3 *